Sediment Transport

Sediment Transport

Editors

Gordon Gilja
Manousos Valyrakis
Panagiotis Michalis
Thomas Pahtz
Oral Yagci

MDPI

Basel • Beijing • Wuhan • Barcelona • Belgrade • Novi Sad • Cluj • Manchester

Editors

Gordon Gilja
Department of Hydroscience and Engineering, Faculty of Civil Engineering
University of Zagreb
Zagreb
Croatia

Manousos Valyrakis
Department of Civil Engineering
Aristotle University of Thessaloniki
Thessaloniki
Greece

Panagiotis Michalis
Institute of Communication and Computer Systems
National Technical University of Athens
Athens
Greece

Thomas Pahtz
Ocean College
Zhejiang University
Zhoushan
China

Oral Yagci
Division of Hydraulics,
Faculty of Civil Engineering
Istanbul Technical University
Istanbul
Turkey

Editorial Office
MDPI
St. Alban-Anlage 66
4052 Basel, Switzerland

This is a reprint of articles from the Special Issue published online in the open access journal *Applied Sciences* (ISSN 2076-3417) (available at: https://www.mdpi.com/journal/applsci/special_issues/Sediment).

For citation purposes, cite each article independently as indicated on the article page online and as indicated below:

Lastname, A.A.; Lastname, B.B. Article Title. *Journal Name* **Year**, *Volume Number*, Page Range.

ISBN 978-3-03928-602-7 (Hbk)
ISBN 978-3-03928-601-0 (PDF)
doi.org/10.3390/books978-3-03928-601-0

Contents

Preface

Understanding the morphodynamic behavior of riverbeds is one of the most fundamental problems in fluvial geomorphology, especially in rivers suffering from extensive hydromorphological pressures. Sediment transport patterns present a continuous research challenge because of their impact on hydraulic structures, infrastructure, waterways, confluences, or naturally morphologically variable zones like confluences and deltas. The significance of the physical alterations of rivers—natural or anthropogenic—is correlated with the increasing frequency, intensity, and duration of natural hazards driven by climate change. The sustainable management of rivers and risk management related to them are based on the predictive tools emerging from state-of-the-art research conducted under relevant environment-protection scenarios. Continuous advancements in the development of laboratory instruments, remote sensing techniques, numerical modeling of turbulent flows, and data processing methods improve the understanding of the fluid–structure interaction underlying the sediment transport process. Such processes can vary across a wide range of scales, from the particle to the landscape, which can directly impact both the form (geomorphology) and function (ecology and biology) of natural systems and the built infrastructure surrounding them. This Special Issue presents a collection of recent advancements in sediment transport research and hydraulic engineering that contribute to filling the knowledge gap in morphodynamic processes at different scales, with an outlook towards innovative river engineering approaches, monitoring systems, available technologies, and management practices tackling rapidly changing environmental conditions.

Gordon Gilja, Manousos Valyrakis, Panagiotis Michalis, Thomas Pahtz, and Oral Yagci
Editors

applied sciences

MDPI

Article

Advances in Monitoring and Understanding the Dynamics of Suspended-Sediment Transport in the River Drava, Slovenia: An Analysis More than a Decade-Long

Janja Kramer Stajnko, Renata Jecl and Matjaž Nekrep Perc *

Faculty of Civil Engineering, Transportation Engineering and Architecture, University of Maribor, Smetanova 17, 2000 Maribor, Slovenia; janja.kramer@um.si (J.K.S.); renata.jecl@um.si (R.J.)
* Correspondence: matjaz.nekrep@um.si

Abstract: Managing sediment transport in streams is crucial to the surface water resource development strategy and has several implications for flood risk and water management, hydropower use, and balancing river morphology. This paper summarises the movement and behaviour of suspended sediment within the Slovenian portion of the River Drava, covering a span of thirteen years from 2005 to 2018. An analysis of relevant data collected during this period is also presented. Suspended-sediment dynamics strongly depend on flow velocity, seasonal variations in sediment sources, and human interventions in the riverbed. The transportation of material in the River Drava results in the accumulation of sediments in reservoirs and riverbeds, consequently impeding the natural hydrological cycle by reducing the outflow into aquifers. The 2018 high-water event is analysed in terms of the dependence of concentration of suspended sediments on discharge, where counterclockwise hysteresis was observed, providing an essential clue to the origin of sediment. Sediments from the River Drava in Slovenia are managed with some conventional processes and are mainly deposited or reintegrated into rivers and aquatic ecosystems. Some additional sediment management strategies with long-term solutions for efficient and comprehensive water management, hydropower, and ecological problems are proposed.

Keywords: suspended-sediment dynamics; River Drava; sediment management

Citation: Kramer Stajnko, J.; Jecl, R.; Nekrep Perc, M. Advances in Monitoring and Understanding the Dynamics of Suspended-Sediment Transport in the River Drava, Slovenia: An Analysis More than a Decade-Long. *Appl. Sci.* **2023**, *13*, 9036. https://doi.org/10.3390/app13159036

Academic Editor: Kelin Hu

Received: 21 June 2023
Revised: 27 July 2023
Accepted: 3 August 2023
Published: 7 August 2023

1. Introduction

Sediment transport in surface waters is a complex dynamic process and has multiple implications, particularly in the areas of environmental protection, hydropower exploitation of rivers, flood risk management, and water resources management. Recently, suspended-sediment dynamics have been investigated intensively, as suspended sediment represents the majority of material that is in motion within a stream. The deposition of sediment in water bodies has significant impacts on the accumulation of silt in river channels, flood-prone areas, and surface waters such as lakes and coastal areas, as well as environmental impacts due to sediment-bound pollutants. Sediments found in rivers are a result of the weathering processes occurring on the upper layers of the Earth's crust, which interact with the forces of erosion that cause the movement of material and its transportation along river channels. Suspended-sediment concentration depends on the current flowrate, climate changes, glacial melting, and intense erosional processes. Human intervention in riverbeds, such as the construction of hydropower plants and other hydraulic structures, can have a significant impact on the transport dynamics and quantity of sediment in a watercourse. As a consequence of material transport, reservoirs and the riverbed become filled, which ultimately affects the natural hydrological cycle by reducing the discharge into aquifers [1–6].

Previous studies investigating suspended sediment in Slovenian rivers [7–13] have indicated that the majority of the overall material is transported during periods of high water flow. However, not every high-water event increases the amount of suspended sediment, as the dynamics of sediment depend on the sum of hydrologic events in the river, the previous hydrologic condition, the intensity of precipitation in the hinterland, and many other factors. The source of suspended sediment can be predicted from the relationship between water discharge and the concentration of suspended sediments, which usually takes the form of a hysteresis loop during a high-water event. Depending on various factors during the event, e.g., runoff, erosion processes, and human activities, different forms of hysteresis loops are possible, e.g., a single-value (straight or curved) relationship, clockwise loop (positive), counterclockwise loop (negative), single-value plus a loop, and figure of eight [14–18].

The dynamics of suspended-sediment transport in streams has been intensively studied recently, as they are closely related to runoff formation and flooding. Erosion processes are particularly intense in agricultural catchments, where suspended-sediment transport dominates during flood events [19]. A comprehensive study of suspended-sediment transport using a log-time dataset to obtain more information about the factors that determine suspended-sediment transport in headwater watersheds was presented in [20]. In [21], the effect of urbanization on suspended-sediment flux in a selected watershed was investigated, and it became clear that sediment dynamics vary spatially and temporally. The high spatiotemporal variability in suspended-sediment transport allows the determination of sediment sources and a general understanding of erosion in the catchment, which is why the statistical analysis of hydrometeorological data is crucial, as presented in [6,22]. In [23], different hysteresis relationships between suspended-sediment concentration and discharge due to catchment characteristics in a multiple catchment of the middle Yellow River were reported. A comprehensive understanding of the spatial and temporal dynamics of suspended-sediment concentrations requires properly calibrated and tested numerical models, as described in [24].

Measuring the amount and dynamics of suspended sediment in streams remains a complex problem, despite the existence of several established methods. Sediment transport is a highly dynamic process that cannot be fully captured by any measurement technique due to constantly changing conditions in rivers. Therefore, when selecting an appropriate measurement model, it is crucial to consider the temporal and spatial variability in sediment transport in both longitudinal and transversal cross-sections. If an accurate determination of the fractional volume of sediment is required, a combination of direct and indirect measurement techniques is necessary [1,2]. Establishing a comprehensive monitoring system and modelling suspended-sediment transport are critical to finding a systematic, long-term solution to the problem of sedimentation in streams and are the starting point for developing the management strategy.

The article provides an overview of suspended-sediment measurements on the River Drava in Slovenia between the years 2005 and 2018 and an analysis of suspended-sediment dynamics at different temporal scales (between different years, seasonal, and during a high-water event). Furthermore, an analysis of suspended-sediment dynamics during a high-water event is presented and a prediction of the origin of sediments in the river is made. In addition, an overview of the current management of suspended sediment in the River Drava is given and some future topics are proposed.

2. Study Area

The River Drava, which is the fourth-largest tributary on the right bank of the Danube, is extensively used for hydropower generation, with almost the entirety of its water potential energy being harnessed. Originating in South Tyrol, Italy, it flows through Austria, Slovenia, and Croatia, and ultimately merges with the Danube at Osijek [25]. The River Drava stretches for a length of 117.7 km in Slovenia and has a catchment area of 4662 km^2. The river has an average discharge of 292 m^3/s and is joined by several tributaries, including the Meža and Dravinja, with respective discharge rates of 12 m^3/s and 11 m^3/s. Additionally, the river has numerous torrential tributaries along its entire length. The Drava is home to a total of 22 hydropower plants located in Austria, Slovenia, and Croatia, with eight plants situated in Slovenia. The Zlatoličje and Formin areas in Slovenia feature two artificially constructed side channels (Figure 1).

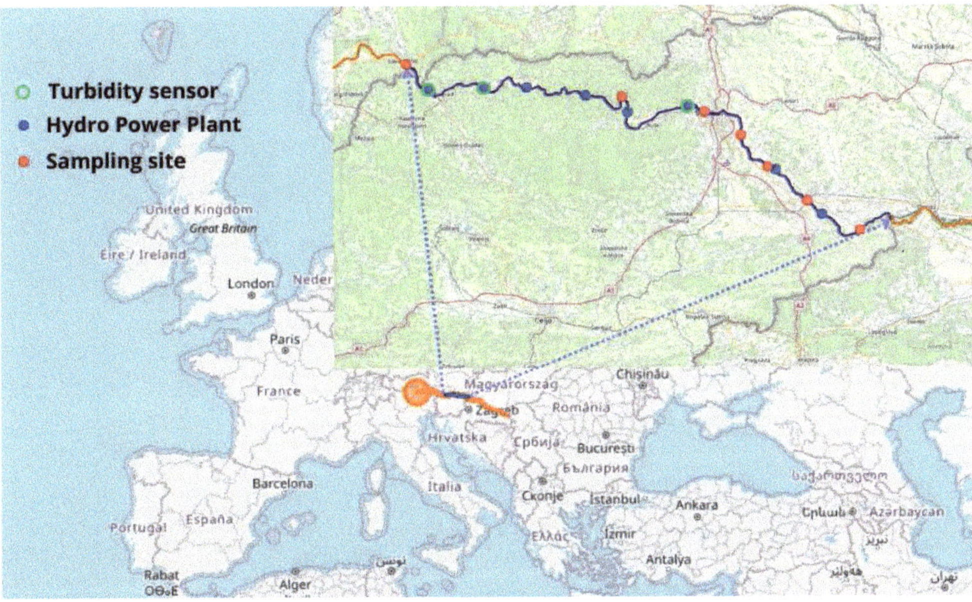

Figure 1. The River Drava in Slovenia; hydropower plants on the River Drava in Slovenia (blue dots); turbidity sensors (green circles); and monitoring sites presented in the current study (red dots).

The River Drava is significantly regulated along its entire path through Slovenia, which also strongly impacts suspended-sediment dynamics. The historical political structure of the countries through which the Drava flows affected the management and research activities on this river, which were neglected until the political upheavals of 1989–1990 and the end of the Serbian–Croatian war in 1995. However, there are few published studies that incorporate parameters of the river environment that have been published in the past [26]. Recently, some studies have been carried out on the River Drava on the possible reuse of sediments in the construction sector [13,27,28].

The national hydrological monitoring of surface waters required by the European Union under the European Water Framework Directive (Water Framework Directive 2000/60/EC) includes the measurement of water level, water velocity, flow, geometry of cross-sections, water temperature, as well as the suspended-sediment concentration in the water. The Slovenian Environment Agency regularly monitors suspended sediment in only the primary watercourses in Slovenia. However, the River Drava is not included in this monitoring program.

3. Materials and Methods

3.1. Monitoring of the Transport of Suspended Sediment in the River Drava in Slovenia

The first investigation of suspended sediment on the Slovenian part of the River Drava was undertaken at the monitoring station in Ptuj in the period from 1956 to 1975. Sampling was not performed randomly. At the Ptuj water monitoring station, suspended-sediment samples were taken between 1965 and 1975. The maximum value of concentration of suspended sediments was recorded during the flood on 20 August 1966, at 2802 g/m^3. A detailed analysis of the dependence of concentration of suspended sediments on other hydrologic parameters was not performed. The total amount of suspended sediment was not determined due to a lack of samples [29]. Thereafter, no further measurements were made as part of the national hydrological monitoring.

Today, the monitoring of suspended sediment is carried out by Dravske elektrarne (DEM)—the operator of the hydropower plants on the River Drava—at four sites using turbidity sensors. Turbidity measurements [30] at the four sites in the hydropower plants reservoirs started gradually in 2011 in the reservoir of the Dravograd hydropower plant, in 2012 at two other sites in Vuzenica and Mariborski otok, and in 2013 at the fourth site in Markovci (Figure 2) [10,29].

Figure 2. (**a**) Point-integrated sampler; (**b**) nozzle enabling isokinetic sampling; (**c**) crane with winch and steel cable; (**d**) lowering the sampler to take a sample; (**e**) measuring with ADCP; (**f**) simultaneous sampling and measurement with an ADCP.

Within the Slovenian–Austrian Cross-border Cooperation Program 2008–2013 (SI-AT), entitled "Drava-Mura Crossborder Initiative—DRA-MUR-CI", within one of the work packages, the transport of suspended sediment (measurements and modelling) was carried out at selected monitoring sites of the River Drava, namely Šturmov potok, Zavrč, Ptuj, and Studenška brv Maribor (Figure 1). Furthermore, in the year 2018, measurements of suspended sediment were conducted at Zgornji Duplek and Zlatoličje locations (Figure 1) persistently for one year, including the high-water occurrence in October 2018.

3.1.1. Suspended-Sediment Samplers

Suspended-sediment samplers are used to collect representative water samples in the stream. These samples are then analysed in a laboratory setting to determine the suspended-sediment concentration levels. To ensure the most reliable sampling, some conditions should be met, such as the following: water should enter the nozzle isokinetically (where the velocity and direction of water entering the sampler are the same as in the environment); stream flow patterns should be disturbed as little as possible; standard bottle sizes should be used (e.g., 1 pint or 1 L); and the equipment should be easy to operate and maintain [4,31,32].

There are two different types of samplers for suspended-sediment collection, namely depth-integrated and point-integrated samplers. The depth-integrated sampler is designed to obtain water samples from a vertical column of water by lowering it to the bottom of a stream and, afterwards, raising it back to the surface. The point-integrated sampler, on the other hand, collects the water at a specific point in the stream, which is made possible by the use of a valve that can be activated electrically. Figure 2 shows a point-integrated sampler US P-72 with appropriate equipment, which can be used for the isokinetic sampling of water at any point in the stream and was used for sampling at some of the monitoring sites investigated in this study.

The sampler is made of aluminium, has a length of 0.7 m, and weighs 18 kg. The hydrodynamic shape of the sampler allows it to rotate in the direction of the flow and be immersed at any depth. The diameter of the nozzle is 5 mm. Inside the housing there is a space for a collection bottle with a capacity of 0.95 L. On the side, there is a hole through which air can escape from the bottle while it fills with water. The sampler can be used in streams with a velocity between 0.5 and 2.0 m/s and a depth of 20 m.

3.1.2. Acoustic Method

Acoustic methods in hydrology were developed primarily for measuring the velocity and discharge of watercourses [33,34]. However, we can use the backscatter and attenuation of acoustic signals in water to monitor suspended sediment in water bodies. A calibration within situ measurements is necessary to determine the concentration of suspended sediment from the backscatter. Acoustic backscatter depends on the concentration and size of the scatterers and is also impacted by attenuation due to suspended matter along the water column [31,35,36].

The most commonly used device for measuring water velocity and indirectly further parameters is the Acoustic Doppler Current Profiler (Figure 2). All results presented in this article were collected with a 1200 kHz Teledyne RDI WorkHorse Rio Grande. In this ADCP model, five different frequencies (75–1200 kHz) are possible, which differ in sound absorption (sound absorption increases with frequency). The measuring device working with this frequency can be used in rivers with a minimum depth of 0.4 m to 15 m and velocities up to 10 m/s, which are also the conditions at the selected measuring points on the River Drava. By crossing the stream, the water depth and velocity can be measured. With a computer connected to the ADCP via a radio link, all data were recorded and collected in real time. The parameters such as discharge, cross-sectional area, and average velocity are known immediately after the crossing.

Based on the backscattered signal, the information about the concentration of suspended sediment can be calculated. The conversion of the backscattered signal into a concentration of suspended sediments is possible with another piece of conversion software, e.g., Aqua Vision, VISEA-PDT 4.0 (PDT—Plume Detection Toolbox), which allows a direct conversion of the measured data into a concentration of suspended sediments. In addition, the software allows a calibration of the measured backscatter with actual values of concentration of suspended sediments obtained with the direct methods. The theoretical background of the backscatter measured with ADCP is represented by the acoustic equation which links the discrepancy between the received and transmitted sound energy and the lost energy during the propagation of the sound wave [37]:

$$S_v = C + \log 10\,[T_T\,R^2/L\,P_T] + 2\,\alpha\,R + K_C\,(E - E_r) \tag{1}$$

The equation for acoustic backscatter in dB (S_v) is determined by several factors, including a constant value (C) in dB, the temperature (T_T) of the Acoustic Doppler Current Profiler (ADCP) transducer in °C, the slant range (R) along the beam to the scatterers in m, the transmit pulse length (L) in m, the transmit power (P_T) in W, the attenuation coefficient (α) in dB/m, a scale factor (K_C) in dB/count, the relative backscatter (E) equal to echo intensity in the count, and the received noise (E_r) in the count. The values of C, TT, R, and PT are either recorded with the ADCP or provided by the factory. The value of E is derived from the Received Signal Strength Indicator of the receivers, and Er is the noise value when no signal is present. The typical value of Er is 40 counts, and the values of factor KC range from 0.35 to 0.55 dB/count. The attenuation coefficient α is the sum of the water absorption α_w and particle attenuation α_s, which are obtained based on empirical formulas that can be found in the literature [37].

The procedure of conversion of backscatter into a sediment concentration can be summarized in three steps: the conversion of backscatter into dB, the standardization of the instrument, the standardization of the area.

The most important part of the measurement is the calibration with real values of sediment concentration. This can be performed with data from the optical backscatter sensors or with water samples where the sediment concentration is determined in the laboratory and then manually entered into the programme.

4. Results and Discussion

The paper presents the results of suspended-sediment measurements that were collected over a period of 13 years (2005–2018) from four monitoring sites: Šturmov potok, Ptuj, Studenška brv Maribor, and Zgornji Duplek during a high-water event. (Figure 1) During the measurements with different measurement techniques, some experience was gained in order to propose the optimal technique for the monitoring and management of suspended sediments on the River Drava. The hydrological data were obtained from Dravske elektrarne (DEM)—the operator of the hydroelectric power plants on the River Drava—who provide data on discharge; at the Ptuj measuring point, the discharge and water level were measured by the Slovenian Environment Agency site. Different measurement methods were used to determine the suspended-sediment concentration, depending on the available equipment and the conditions at the measurement sites. The suspended-sediment concentration was determined by analysing the collected water samples. A variety of water samplers were used, from the simple water samplers used under severe conditions, such as floods, to the depth-integrated samplers described and shown in the previous chapter. However, to obtain the most representative results of suspended-sediment dynamics in a cross-section, the basic measuring technique was supplemented with the acoustic method. The results presented are divided into three parts: First, the dynamics of suspended sediment at selected sites at mean discharge of the River Drava are presented. In addition, a case study is presented where a combination of direct and indirect methods was applied. Finally, the results of a high-water event in October 2018 are presented.

4.1. Suspended-Sediment Dynamics at Average Discharge

Analysis of the correlation between the concentration of suspended sediments (c in g/m^3) and flow rate (Q in m^3/s) in the River Drava showed that the dependence of these quantities can be adequately described by quadratic functions that vary depending on the measurement location. The concentration of suspended sediment can vary both longitudinally and transversely along the river. In the example of the River Drava, the amount of suspended sediment can be compared at three different distances from the bank

(Profile1, Profile2, and Profile3; Table 1) and at two depths (2 m and 4 m) on a river cross-section at the Šturmov potok monitoring site (Figure 3), where the results from October 2009 and May 2005 are shown.

Table 1. Locations and dates of water sampling from the River Drava, maximum and minimum water flows, and concentrations of suspended material.

Sample No.	Location	Depth (m)	Date	Flow Q (m^3/s)	Suspended-Sediment Concentration c (g/m^3)
			DRAVA Šturm		
1	Profile 1	2	14 October 2009	441	12.0
2	Profile 1	4	14 October 2009	441	12.0
3	Profile 2	2	14 October 2009	441	14.0
4	Profile 2	4	14 October 2009	441	12.0
5	Profile 3	2	14 October 2009	441	11.0
6	Profile 3	4	14 October 2009	441	14.0
7	Profile 1	2	14 January 2010	314	4.5
8	Profile 1	4	14 January 2010	314	3.0
9	Profile 2	2	14 January 2010	314	4.0
10	Profile 2	4	14 January 2010	314	4.0
11	Profile 3	2	14 January 2010	314	4.0
12	Profile 3	4	14 January 2010	314	5.0
13	Profile 1	2	20 April 2010	239	4.0
14	Profile 1	4	20 April 2010	239	3.0
15	Profile 2	2	20 April 2010	239	4.0
16	Profile 2	4	20 April 2010	239	4.0
17	Profile 3	2	20 April 2010	239	3.0
18	Profile3	4	20 April 2010	239	4.0
19	Profile 1	0	17 June 2010	556	54.0
20	Profile 1	1	17 June 2010	556	49.0
21	Profile 2	0	17 June 2010	556	53.0
22	Profile 2	1	17 June 2010	556	54.0
23	Profile 3	0	17 June 2010	556	50.0
24	Profile 3	1	17 June 2010	556	49.0
25	Profile 1	2	16 September 2010	429	17.0
26	Profile 1	4	16 September 2010	429	13.0
27	Profile 2	2	16 September 2010	429	16.0
28	Profile 2	4	16 September 2010	429	15.0
29	Profile 3	2	16 September 2010	429	15.0
30	Profile 3	4	16 September 2010	429	15.0
31	Profile 1	0	31 January 2011	312	5.0
32	Profile 1	1	31 January 2011	312	1.0
33	Profile 2	0	31 January 2011	312	3.0
34	Profile 2	1	31 January 2011	312	1.0
35	Profile 3	0	31 January 2011	312	1.0
36	Profile 3	1	31 January 2011	312	3.0
37	Profile 1	0	22 March 2011	169	6.0
38	Profile 1	1	22 March 2011	169	5.0
39	Profile 2	0	22 March 2011	169	14.0
40	Profile 2	1	22 March 2011	169	14.0
41	Profile 3	0	22 March 2011	169	20.0
42	Profile 3	1	22 March 2011	169	20.0
43	Profile 1	0	15 May 2012	301	19.0
44	Profile 1	2	15 May 2012	301	19.0
45	Profile 2	0	15 May 2012	301	14.0
46	Profile 2	2	15 May 2012	301	14.0
47	Profile 3	0	15 May 2012	301	20.0
48	Profile 3	2	15 May 2012	301	20.0

Table 1. *Cont.*

Sample No.	Location	Depth (m)	Date	Flow Q (m^3/s)	Suspended-Sediment Concentration c (g/m^3)
			DRAVA—Ptuj		
61	left	0	5 May 2011	391	8.0
62	left	2	5 May 2011	391	6.0
63	middle	0	5 May 2011	391	8.0
64	middle	2	5 May 2011	391	9.0
65	right	0	5 May 2011	391	7.0
66	right	2	5 May 2011	391	9.0
67	left	0	9 June 2011	536	67.0
68	middle	0	9 June 2011	536	53.0
69	right	0	9 June 2011	536	50.0
70	left	0	20 June 2011	631	185.0
71	middle	0	20 June 2011	631	234.0
72	right	0	20 June 2011	631	246.0
73	left	0	18 August 2011	345	15.0
74	left	2	18 August 2011	345	13.0
75	middle	0	18 August 2011	345	31.0
76	middle	2	18 August 2011	345	34.0
77	right	0	18 August 2011	345	40.0
78	right	2	18 August 2011	345	40.0
79	left	0	22 July 2011	988	97.0
80	middle	0	22 July 2011	988	110.0
81	right	0	22 July 2011	988	110.0
82	left	0	13 September 2012	796	95.0
83	middle	0	13 September 2012	796	95.0
84	right	0	13 September 2012	796	64.0

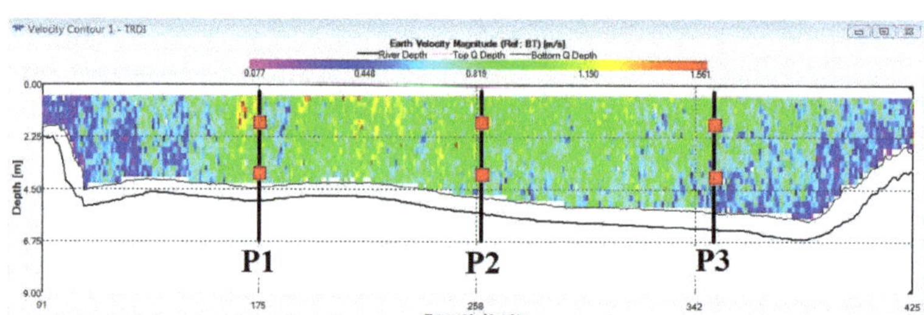

Figure 3. Measured velocity profile, profiles and points where the water samples were taken (red squares).

The results presented in Figure 4 show that the amount of suspended sediment is almost the same at both depths of the river and that there is no significant difference between the selected profiles. Evaluations of the relationship between the discharge and concentration of suspended sediments and of the annual suspended-sediment amounts were performed.

The grain size distribution curve of a composite sample showing the weight fractions of the individual material fractions smaller than the specified grain sizes at the Drava—Šturm measuring point is shown in Figure 5. The examination of the grain size shows that most of the suspended material in the Drava belongs to silt (0.002–0.060 mm), less than 10% is clay fraction (<0.002 mm), and about 8% is fine sand (>0.06 mm).

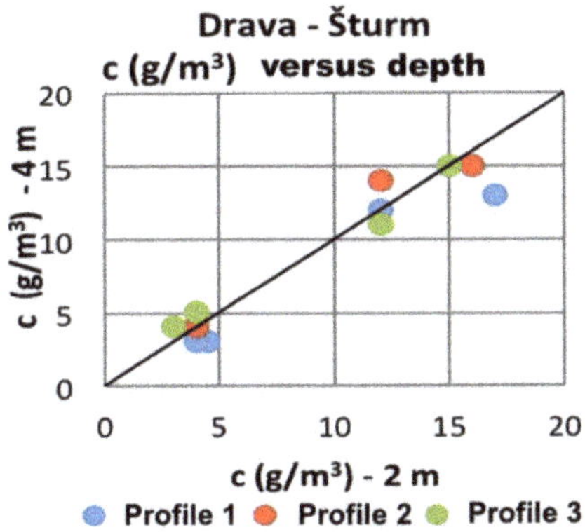

Figure 4. The ratio between the concentrations of suspended material at depths of 2 m and 4 m, measured in three profiles of the River Drava (Profile 1, Profile 2, and Profile 3) at the monitoring site Šturmov potok.

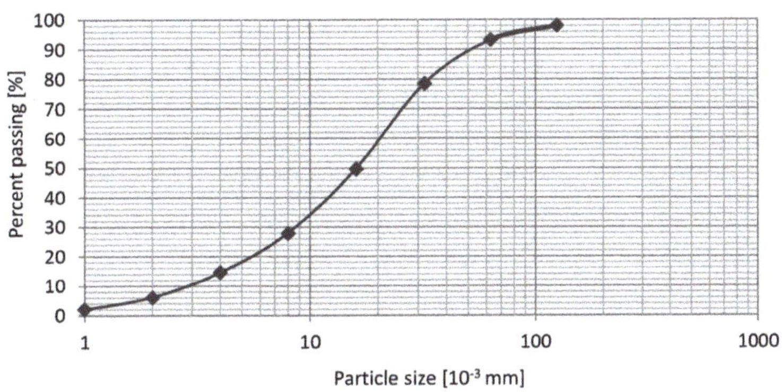

Figure 5. Grain-size distribution curve of suspended material of a composite sample at the Drava Šturm measuring point.

4.2. Prediction of Suspended-Sediment Transport

4.2.1. Q/c Curve

Based on long-term measurement results, we can predict the transport of suspended sediments in a selected cross-sectional profile of a watercourse. If we know the water flow (Q) in a geometrically defined cross-section and with measured velocity vectors, as well as simultaneously determined concentrations (c) of suspended particles at selected points of the profile, we can predict the concentration over the entire profile and simulate the Q/c diagram. With the known Q/c function and a time-dependent flow Q(t), we can determine the annual balance (B) of transport of suspended particles and use the annual trends of flow Q to statistically predict the transport of suspended material through the selected profile. The described interdependencies are shown in Figure 6 for the Drava—Ptuj and Drava—Šturm locations, where the graphs also include the functional dependence, with R^2 (coefficient of determination) representing the fit estimate of the function to the measured values.

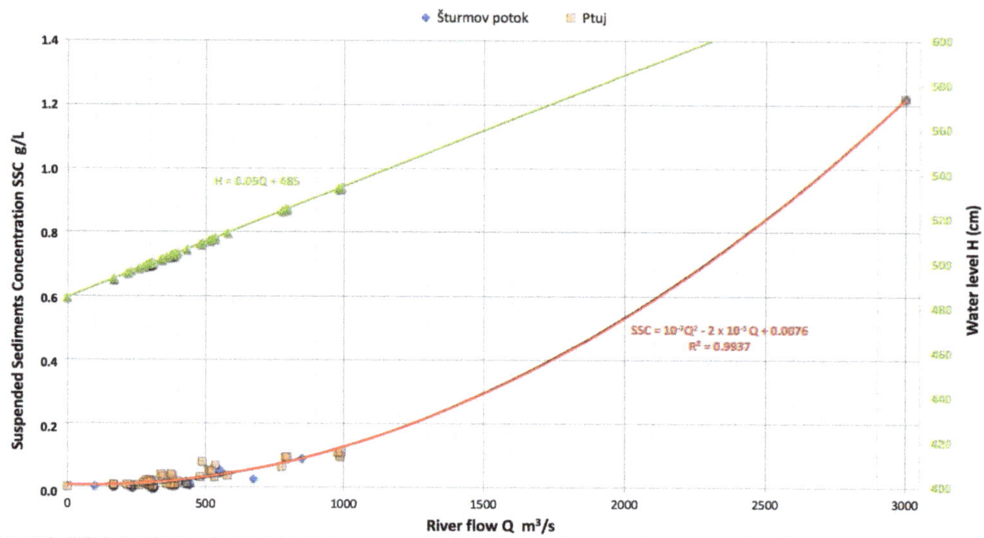

Figure 6. Q/c and Q/H (water level) relationships for Šturmov potok and Ptuj measuring sites.

4.2.2. Annual Quantities of Suspended Sediments

The cumulative amounts of suspended-sediment transport can be determined using a suspended-sediment transport model and measured and historical runoff data. Figure 7 shows the hourly measurements of the flow rate of the River Drava in the years 2005 and 2010 in m^3/s (data from ARSO) in the first blue diagram. From the discharge data, the total of suspended sediment c in the water body in kg was calculated for each hour. The middle diagram shows the quantities. The bottom diagram summarizes these quantities, and the final value gives the aggregate annual total of suspended sediment in the river for 2005 and 2010 in the Šturm profile.

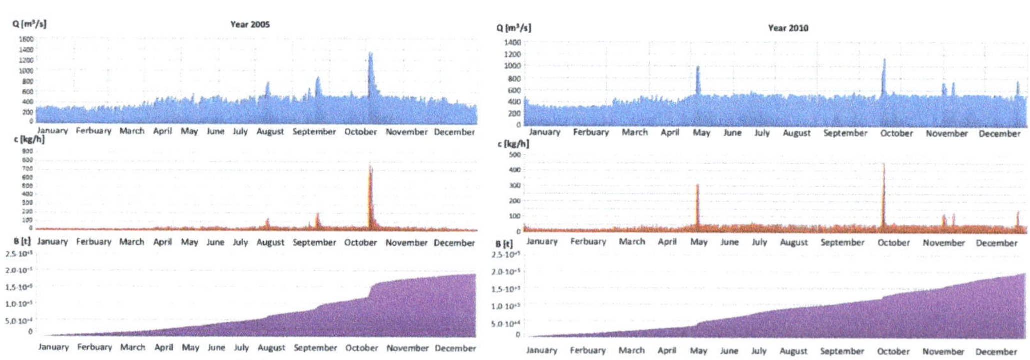

Figure 7. Flow rate, suspended-sediment concentration, and suspended-sediment annual balance B(t) at the Šturm monitoring station in 2005 and 2010.

4.2.3. Studenška Brv, Maribor Measuring Site

A case study of the suspended-sediment concentration in a selected River Drava profile, using indirect acoustic measurements (with an ADCP) and point sampling (direct method), is presented. Combining these methods is necessary to obtain a representative picture (temporal and spatial distribution) of sediment transport through the river profile.

The measurement was performed on the profile of the River Drava in the area of Studenška brv in Maribor. The ADCP was operated from a bridge (Figure 2) with data collected and monitored in real time on a connected laptop. A wireless connection was established between the instrument and the laptop. Simultaneously, water samples were collected using a spot sampler (Figure 2) and later analysed in a laboratory.

The measured velocity profile and the profiles where the water samples were collected are shown in Figure 8. The actual flow rate at the time of measurement was 287 m^3/s. The streambed was characterized by uneven depths, with the greatest depth of 7.5 m reached near the left bank. There were two prominent points in the profile where the depth was shallowest (5.5 m), namely in the area of the bridge piers. These piers visibly influenced the dynamics of the water flow and caused significant sediment deposition at this location.

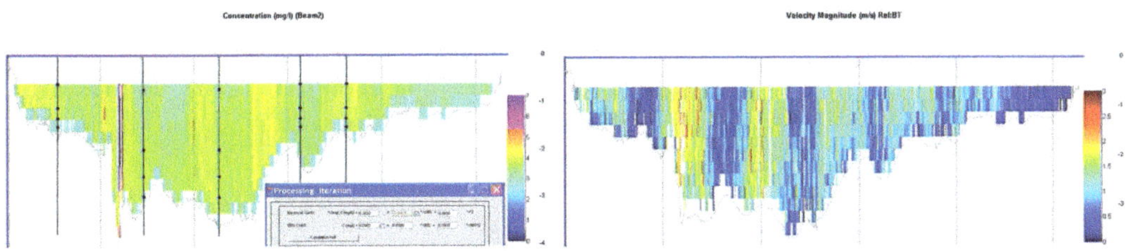

Figure 8. Sediment concentration in mg/L showing sample collection sites and iteration with calculation of coefficients A and B in Equation (1) and velocity profile at Studenška brv measuring site.

The concentrations of suspended sediments in the collected samples were measured and are presented in the Table 2. The concentration was determined using a method in accordance with the Slovenian standard SIST EN 872:2005, which is further described in [11].

Table 2. Suspended-sediment concentration *c* in collected samples according to SIST EN 872:2005.

Sample	Depth [m]	Concentracion [mg/L]
V1-1	0.3	0.0090
V1-2	0.9	0.0066
V1-3	1.2	0.0116
V1-4	1.4	0.0098
V2-1	0.4	0.0052
V2-2	1.2	0.0038
V2-3	1.6	0.0038
V2-4	1.9	0.0030
V3-1	0.7	0.0022
V3-2	2.1	0.0014
V3-3	2.8	0.0020
V3-4	3.3	0.0048
V4-1	0.7	0.0020
V4-2	2.1	0.0028
V4-3	2.8	0.0034
V4-4	3.3	0.0026
V5-1	0.4	0.0104
V5-2	1.2	0.0060
V5-3	1.6	0.0066
V5-4	1.9	0.0026

The backscatter signal and the concentration of suspended particles both depend on the attenuation of the particles, and conversely, the attenuation of the particles depends on the concentration of the suspended particles. Therefore, the concentration values must be optimized in an iterative process, which was performed automatically by the program VISEA-PDT [37].

The relationship between the absolute value of the backscatter signal and the concentration of suspended sediments is given by the expression:

$$10 \log (C) = A\, I + B, \tag{2}$$

C is the concentration of suspended sediments in mg/L, I represents the measured absolute backscatter in dB, and A and B are constants resulting from the measured values of the sediment concentration of the samples.

The actual value of the concentration of suspended sediments was determined in an iterative substep, since both the strength of the sound signal reflection and the concentration of suspended sediments depend on the attenuating particles. The loss of sound energy due to absorption by particles and scattering in the water was taken into account.

The use of an Acoustic Doppler Current Profiler to determine the concentration and distribution of suspended sediment over a river profile has proven to be a good alternative to other methods when used in conjunction with appropriate software that allows the conversion of acoustic signals to the concentration of suspended sediments. This approach allows the presentation of a comprehensive picture of sediment transport through a selected cross-section of a water body.

4.3. High-Water Event Dynamics of Suspended-Sediment Concentration

The aim of this case study was to investigate the concentration of suspended sediment in the River Drava during a flood event. The detailed description of the measurement procedure and the results have already been published in [38], but in order to give a comprehensive picture of sediment transport in the Drava, some main conclusions are summarised below.

During high-water events characterized by large flow rates, measuring sediment transport in rivers poses significant challenges due to hazardous conditions, large woody debris, and other floating objects on the river surface. It is a known fact that more than 50% of suspended sediment is transported during high-water events. Since such situations usually occur in very short periods of time, predicting them and collecting samples are very difficult if the system is not automated. However, when conditions are optimal, classical sampling methods are uneconomical, and errors in sampling and calculating the final concentration of sediment increase. Traditional measurement techniques such as Acoustic Doppler Current Profilers, hydrometric wings, and sediment samplers face considerable difficulties or are even impractical. Therefore, the introduction of invasive and automated methods for determining suspended-sediment concentration is crucial [2,39–43]. As alternative measurement approaches or modifications to existing techniques, some novel technologies, such as remote sensing methods or automated sampling systems, may provide potential solutions by enabling measurements from a safe distance or offering enhanced instrument resilience to debris interference.

The investigation of a high-water event occurred at the Duplek Bridge site, located between the cities of Maribor and Ptuj. At this site, the outflow channel is separated from the riverbed at the Melje Dam to supply the Zlatoličje Hydropower Plant. The maximum capacity of the derivation channel is 500 m^3/s, and when this capacity is exceeded, the extra water flows into the natural riverbed. This area's required ecological river discharge is 10 m^3/s in winter and 20 m^3/s in summer.

In the autumn of 2018, a high-water event commenced on 29 October when the discharge of the River Drava surpassed 500 m^3/s. The discharge within the natural riverbed began to escalate from 10 m^3/s and reached a peak discharge of 1391 m^3/s, lasting for a duration of three days. Discharge data were obtained from the Melje Dam's automated measuring station at the Dravske elektrarne site and were accessible online in real time.

In order to determine the concentration of suspended sediments relative to discharge during the high-water event, water samples were collected manually using 1000 mL plastic bottles from the bridge. The sampling period continued until 4 November 2018, with only one sample per day collected during the last four days when the discharge was below

500 m^3/s and decreased significantly. The collected samples were filtered through glass microfiber filters, and the concentration of suspended sediments was determined according to the standard SIST EN 872:2005 [44]. Filter trays were dried in an oven and weighed using an analytical balance (RADWAG AS 220.R2).

The results of the suspended-sediment measurements during the high-water event in October 2018 are shown in Figure 9. The minimum discharge value was 92 m^3/s on 5 November 2018, at 15:13, while the maximum was 1391 m^3/s on 30 October 2018, at 18:33. The minimum concentration value was 105 mg/L on 5 November 2018, at 15:13, while the maximum was 2450 mg/L on 31 October 2018, at 00:28. The suspended-sediment concentrations were found to be different at rising water levels from those measured at the same water level at falling water levels. The suspended-sediment dynamics in watercourses rely on various factors, including energy conditions determined by discharge levels, sediment availability, precipitation intensity, land use, and vegetation cover. These factors influence the typical hysteresis effects that are observable when studying the relationship between runoff and sediment concentration. Figure 9 displays the correlation between the concentration of suspended sediments and discharge for a specific stretch of the River Drava during a high-water event in October 2018. A counter-clockwise hysteresis loop was observed, indicating that sediment sources were located at the area's far end, contributing to surface runoff, which suggests that sediments reached the stream during the falling limb [38]. However, the hysteresis loop may also have occurred due to material eroded in a catchment that did not reach the stream during the previous flood event and settled on the slope. The surface runoff then transported these sediments during the next high-water event. Reports of high values of suspended-sediment concentration during a falling limb are associated with bank collapse [45].

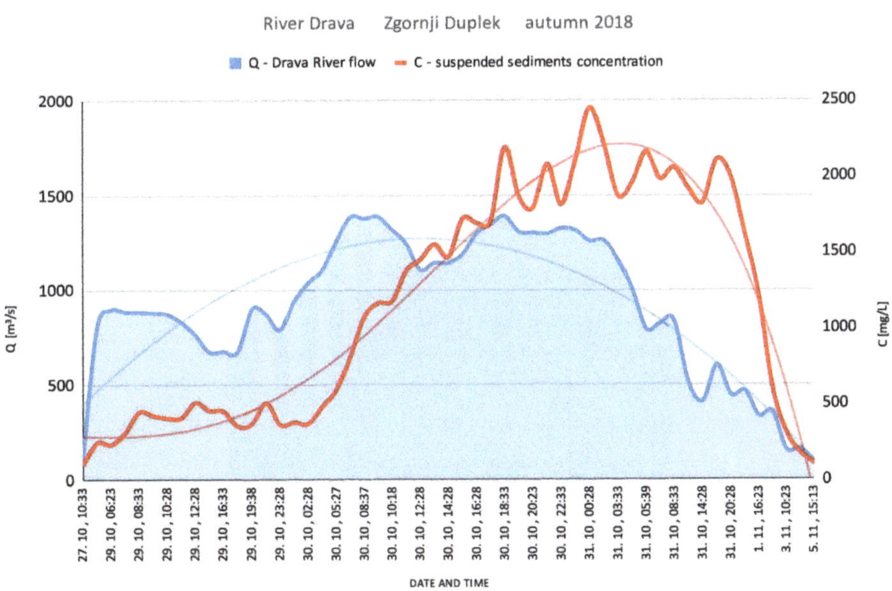

Figure 9. The relationship between the concentration of suspended sediments (C) and water discharge values (Q) was investigated. Time evolution of the water discharge Q and concentration of suspended sediments C for the high-water event in October 2018 (based on [38]).

To determine the main factors contributing to the increase in concentration of suspended sediments during the falling limb, additional land use and precipitation intensity data must be analysed.

From the results, it is clear that on this part of the River Drava, most of the material is transported during high-water events, which can lead to various problems and damages in the flooded areas, e.g., the pollution of agricultural land and reduced performance of returning waves. Special attention and additional measures to protect these areas are required.

4.4. Possible Measures for the Management of Suspended Sediment in the River Drava

Sedimentation in the reservoirs of the Slovenian River Drava is an important issue for hydropower plant operators as it affects the efficiency of the plants. It is estimated that between 30,000 and 40,000 m^3 of sediment accumulate in the reservoirs annually [27]. It was determined from the observations that the volume of reservoirs is decreasing. It was noted that the total volume of the reservoirs decreased by 32% from the commissioning of the hydropower plants to 2014. In addition, the usable volume of the reservoirs has reduced by about 33,000 m^3 annually [13].

Historically, river sediments have been treated as waste and managed in a variety of ways, including landfilling or reintegration into rivers and aquatic systems [13,28,46]. However, some alternative solutions for the treatment and reuse of river sediments from the River Drava have been developed [27,28]. An investigation into whether the sediments from the River Drava are appropriate as a filling material in the construction of levees and its further potential was given in reference [27].

In reference [13], a solution to the problem of sedimentation in the reservoirs of the River Drava in Slovenia was proposed. Sedimentation in reservoirs due to changed hydraulic conditions strongly affects the efficiency of hydropower plants. Based on the particle size distribution of the suspended sediment, most of the material is fine-grained, so flooding the sediment into the core of the stream at higher flow velocities and further flushing would be the most appropriate action. However, the proposed solution does not comprehensively solve the sedimentation problem, as the problem is only shifted downstream.

An effective sediment management measure in streams is sediment fences [47], which are temporary sediment barriers made of permeable fabric that trap and retain small amounts of sediment generated by construction activities or unsecured areas. These fences cause the water to slow down, allowing sediment to settle in a sediment basin in front of the fence. However, they are unsuitable for high waters and require regular maintenance, inspection, and sediment removal.

Sediment traps [48,49] are used to retain sediment from construction sites, but they are unsuitable for functioning water courses.

To minimize the sediment transport of fine-grained material, sediment curtains of impermeable materials [50–52] are an efficient alternative. They can extend to the bottom and prevent sediment passage over the entire height of the curtain or only to a certain depth to prevent the transport of suspended sediment in the upper part of the water column.

5. Conclusions

The suspended-sediment dynamics in the River Drava were studied using different measurement methods and under different conditions (mean discharge and flood events). The problem of sedimentation in the River Drava is of particular importance for hydropower utilisation, as it leads to the clogging of reservoirs and reduces the efficiency of hydropower plants.

The results at selected gauging stations show that the dependence of discharge and the concentration of suspended sediments can be described with the aid of rating curves, which can be used to determine the cumulative amounts of suspended-sediment transport. The concentration of suspended sediments in the River Drava is subject to temporal and spatial fluctuations, which can only be recorded with a suitable combination of indirect and direct measurement methods. An example of a combination of the acoustic method and point sampling in a selected cross-section was shown, giving a representative picture of suspended-sediment transport through the cross-section.

The findings from the analysis of the high-water event in October 2018 suggest that the correlation between the concentration of suspended sediments and runoff follows a clockwise hysteresis loop. This phenomenon can be attributed to several factors, including the sediment sources being located far away from the area where surface runoff is generated, or the sediment being deposited on the slope without being washed away. Nevertheless, additional research is necessary to identify the source of the suspended sediment in the stream during the high-water event.

In order to establish an efficient model of suspended-sediment dynamics in the River Drava, continuous monitoring should be established. In addition to the existing turbidity sensors at four hydropower stations, physical samples and other methods should be used to obtain a representative picture of the selected cross-section that could be used for further predictions of sediment transport. Based on an accurate model of sediment transport, sediment management can be determined in terms of flooding sediments into the core of the stream or constructing sediment fences, sediment traps, or sediment curtains. However, all measures should be sustainable and not adversely affect other users and life in the vicinity of the river.

Author Contributions: Conceptualization, J.K.S., R.J. and M.N.P.; Methodology, J.K.S. and M.N.P.; Investigation, J.K.S., R.J. and M.N.P.; Writing—original draft, J.K.S. and M.N.P.; Writing—review & editing, J.K.S., R.J. and M.N.P.; Visualization, J.K.S. and M.N.P.; Supervision, R.J. All authors have read and agreed to the published version of the manuscript.

Funding: This research received no external funding.

Institutional Review Board Statement: Not applicable.

Informed Consent Statement: Not applicable.

Data Availability Statement: Not applicable.

Conflicts of Interest: The authors declare no conflict of interest.

References

1. DHV Consultants; Delft Hydraulics. *Sediment Transport Measurements*; Delft: Delft, The Netherlands, 2003; Volume 5.
2. Habersack, H. *Schwebstoffe im Fliessgwaesser*; Bundesministerium für Land und Forstwirtschaft, Umwelt und Wasserwirtschaft: Wien, Austria, 2008.
3. Lick, W. *Sediment and Contaminant Transport in Surface Waters*; Willey: Hoboken, NJ, USA, 2009.
4. Mikoš, M. Metode Terenskih Meritev Suspendiranih Sedimentov v Rekah. *Gradb. Vestn.* **2012**, *61*, 151–158.
5. Khan, A.A.; Wu, W. *Sediment Transport: Monitoring, Modeling and Management*; Nova Science Publishers: Hauppauge, NY, USA, 2013; ISBN 9781626186835.
6. Vercruysse, K.; Grabowski, R.C.; Rickson, R.J. Suspended Sediment Transport Dynamics in Rivers: Multi-Scale Drivers of Temporal Variation. *Earth-Sci. Rev.* **2017**, *166*, 38–52. [CrossRef]
7. Ulaga, F. Monitoring Suspendiranega Materiala v Slovenskih Rekah. *Acta Hydrotech.* **2005**, *23/39*, 117–128.
8. Ulaga, F. Vsebnost in Premeščanje Suspendiranega Materiala v Slovenskih Rekah. *Agencija Repub. Slov. Okolje* **2005**, 1–7.
9. Ulaga, F. Transport Suspendiranega Materiala v Slovenskih Rekah/Suspended Sediment Transportation in Slovene Rivers. *Ujma* **2006**, *20*, 144–150.
10. Gregorc, B. Spremljanje Vsebnosti Suspendiranega Materiala v Rečni Vodi Drave s Pomočjo On-Line Meritev. *Ekolist* **2010**, *10*, 8–10.
11. Dolinar, B. Suspendirani Sedimenti v Reki Dravi. *Gradb. Vestn.* **2014**, *63*, 94–100.
12. Bezak, N.; Šraj, M.; Mikoš, M. Pregled Meritev Vsebnosti Suspendiranega Materiala v Sloveniji in Primer Analize Podatkov. *Gradb. Vestn.* **2013**, *62*, 274–280.
13. Knapič, M.; Ulaga, F.; Preglau, A. Upravljanje S Sedimenti V Akumulacijskih Bazenih Hidroelektrarn Na Reki Dravi: Predstavitev Pilotnega Projekta Plavljenja Sedimentov V Matico. *Mišičev Vodarski Dan* **2019**, 171–176.
14. Williams, G.P. Sediment Concentration Versus Water Discharge During Single Event. *J. Hydrol.* **1989**, *111*, 89–106. [CrossRef]
15. Asselman, N.E.M. Suspended Sediment Dynamics in a Large Drainage Basin: The River Rhine. *Hydrol. Process.* **1999**, *13*, 1437–1450. [CrossRef]
16. Bogen, J. The Hysteresis Effect of Sediment Transport Systems. *Nor. Geogr. Tidsskr. Nor. J. Geogr.* **1980**, *34*, 45–54. [CrossRef]
17. Bača, P. Hysteresis Effect in Suspended Sediment Concentration in the Rybárik Basin, Slovakia/Effet d'hystérèse Dans La Concentration Des Sédiments En Suspension Dans Le Bassin Versant de Rybárik (Slovaquie). *Hydrol. Sci. J.* **2008**, *53*, 224–235. [CrossRef]

18. Eder, A.; Strauss, P.; Krueger, T.; Quinton, J.N. Comparative Calculation of Suspended Sediment Loads with Respect to Hysteresis Effects (in the Petzenkirchen Catchment, Austria). *J. Hydrol.* **2010**, *389*, 168–176. [CrossRef]
19. Sherriff, S.C.; Rowan, J.S.; Melland, A.R.; Jordan, P.; Fenton, O.; Huallacháin, D.O. Investigating Suspended Sediment Dynamics in Contrasting Agricultural Catchments Using Ex Situ Turbidity-Based Suspended Sediment Monitoring. *Hydrol. Earth Syst. Sci.* **2015**, *19*, 3349–3363. [CrossRef]
20. Rodríguez-Blanco, M.L.; Taboada-Castro, M.M.; Taboada-Castro, M.T. An Overview of Patterns and Dynamics of Suspended Sediment Transport in an Agroforest Headwater System in Humid Climate: Results from a Long-Term Monitoring. *Sci. Total Environ.* **2019**, *648*, 33–43. [CrossRef]
21. Ferreira, C.S.S.; Walsh, R.P.D.; Kalantari, Z.; Ferreira, A.J.D. Impact of Land-Use Changes on Spatiotemporal Suspended Sediment Dynamics within a Peri-Urban Catchment. *Water* **2020**, *12*, 665. [CrossRef]
22. Vercruysse, K.; Grabowski, R.C. Temporal Variation in Suspended Sediment Transport: Linking Sediment Sources and Hydro-Meteorological Drivers. *Earth Surf. Process. Landf.* **2019**, *44*, 2587–2599. [CrossRef]
23. Wang, B.; Wang, C.; Jia, B.; Fu, X. Spatial Variation of Event-Based Suspended Sediment Dynamics in the Middle Yellow River Basin, China. *Geomorphology* **2022**, *401*, 108115. [CrossRef]
24. Alpaos, A.D.; Tognin, D.; Tommasini, L.; Alpaos, L.D.; Rinaldo, A.; Carniello, L.; Alpaos, A.D.; Tognin, D.; Tommasini, L.; Alpaos, L.D.; et al. Statistical Characterization of Erosion and Sediment Transport Mechanics in Shallow Tidal Environments. Part 1: Erosion Dynamics. *Authorea* **2023**. *preprint*.
25. Lóczy, D. *The Drava River*; Springer International Publishing: Cham, Switzerland, 2019; ISBN 978-3-319-92815-9.
26. Novák, T.J.; Lóczy, D. (Eds.) The Drava River: Environmental Problems and Solutions. In *Hungarian Geographical Bulletin*; Springer: Cham, Switzerland, 2019; pp. 99–101. ISBN 9783319928159.
27. Ducman, V.; Bizjak, K.F.; Likar, B.; Kolar, M.; Robba, A.; Imperl, J.; Božič, M.; Gregorc, B. Evaluation of Sediments from the River Drava and Their Potential for Further Use in the Building Sector. *Materials* **2022**, *15*, 4303. [CrossRef]
28. Mikoš, M. Rečni Sedimenti in Mineralni Agregati v Gradbeništvu. *Gradb. Vestn.* **2017**, *64*, 28.
29. Horvat, U.; Konečnik Kotnik, E. *Geografije Podravja*; Univerza v Mariboru, Filozofska Fakulteta: Maribor, Slovenia, 2017; ISBN 9789612860745.
30. Vas, L.; Tamás, E.A. Surrogate Method for Suspended Sediment Concentration Monitoring on the Alluvial Reach of the River Danube. *Appl. Sci.* **2023**, *13*, 5826. [CrossRef]
31. Simpson, M.R. *Discharge Measurements Using a Broad-Band Acoustic Doppler Current Profiler*; United States Geological Survey Open-File Report 01-1; U.S. Geological Survey: Sacramento, CA, USA, 2001; 134p.
32. Dipas, P.; Kuhnle, R.; Graj, J.; Glysson, D.; Edwards, T. Sediment Transport Measurements. *USGS* **1999**, 305–352.
33. Trček, R. Meritve Pretoka z Akustičnim Dopplerjevim Merilnikom (ADMP). *MVD* **2004**, 211–217.
34. Trček, R. Uporaba Horizontalnega Merilnika Hitrosti (H-ADCP) Za Določitev Pretoka Rek. *Mišičev Vodarski Dan* **2005**, 1–8.
35. Kim, Y.H.; Gutierrez, B.; Nelson, T.; Dumars, A.; Maza, M.; Perales, H.; Voulgaris, G. *Using the Acoustic Doppler Current Profiler (ADCP) to Estimate Suspended Sediment Concentration*; University of South Carolina: Columbia, SC, USA, 2004.
36. Dinehart, R.L.; Burau, J.R. Repeated Surveys by Acoustic Doppler Current Profiler for Flow and Sediment Dynamics in a Tidal River. *J. Hydrol.* **2005**, *314*, 1–21. [CrossRef]
37. Aardoom, J.H. Quantification of Sediment Concentrations and Fluxes from ADCP Measurements. *IXemes Journées Natl. Génie Civ. Génie Côtier* **2006**, 501–510. [CrossRef]
38. Kramer Stajnko, J.; Jecl, R.; Nekrep Perc, M. Measuremet of Suspended Sediment Concentration in the Drava River during High-Water Event. In *Nanos u Vodnim Sustavima*; Oskoruš, D., Rubinić, J., Eds.; Hrvatsko Hidrološko Društvo: Zagreb, Croatia, 2020.
39. Lalk, P.; Haimann, M.; Habersack, H. Sediment Monitoring: Application of a New Monitoring Strategy and Analysis Concept of Suspended Sediments in Austrian Rivers. In *Sediment Matters*; Springer: Cham, Switzerland, 2015.
40. Gray, J.R.; Gartner, J.W. *Surrogate Technologies for Continuous Suspended-Sediment Monitoring in the United States*; Tsinghua University Press: Beijing, China, 2004; pp. 2–515.
41. Gray, J.R.; Gartner, J.W. Technological Advances in Suspended-Sediment Surrogate Monitoring. *Water Resour. Res.* **2009**, *45*, W00D29. [CrossRef]
42. Landers, M.N. Review of Methods To Estimate Fluvial Suspended Sediment Characteristics From Acoustic Surrogate Metrics. In Proceedings of the 2nd Joint Federal Interagency Conference, Las Vegas, NV, USA, 27 June–1 July 2010; pp. 1–2.
43. Habersack, H.; Hauer, C. Sedimentforschung Und -Management. *Osterr. Wasser Abfallwirtsch.* **2019**, *71*, 108–110. [CrossRef]
44. *SIST EN 872*; Water Quality—Determination of Suspended Solids—Method by Filtration through Glass Fibre Filters. SIST: Ljubljana, Slovenia, 2005.
45. Russell, M.A.; Walling, D.E.; Hodgkinson, R.A. Suspended Sediment Sources in Two Small Lowland Agricultural Catchments in the UK. *J. Hydrol.* **2001**, *252*, 1–24. [CrossRef]
46. Nones, M. Dealing with Sediment Transport in Flood Risk Management. *Acta Geophys.* **2019**, *67*, 677–685. [CrossRef]
47. Øtrem, G.; Haakensen, N.; Olsen, H.C. Sediment Transport, Delta Growth and Sedimentation in Lake Nigardsvatn, Norway. *Geogr. Ann. Ser. A Phys. Geogr.* **2005**, *87*, 243–258. [CrossRef]

48. Hupp, C.R.; Kroes, D.E.; Noe, G.B.; Schenk, E.R.; Day, R.H. Sediment Trapping and Carbon Sequestration in Floodplains of the Lower Atchafalaya Basin, LA: Allochthonous Versus Autochthonous Carbon Sources. *J. Geophys. Res. Biogeosci.* **2019**, *124*, 663–677. [CrossRef]
49. Ralston, D.K.; Yellen, B.; Woodruff, J.D. Watershed Suspended Sediment Supply and Potential Impacts of Dam Removals for an Estuary. *Estuaries Coasts* **2021**, *44*, 1195–1215. [CrossRef]
50. Silt Curtains Assist in Contaminated Sediment Removal. *World Dredg. Min. Constr.* **1996**, *32*.
51. Li, Y.H.; Yu, G.L. Experimental Study on the Obliquity Angle of Suspended-Flexible-Curtain for Sediment Deposition. *Shanghai Jiaotong Daxue Xuebao/J. Shanghai Jiaotong Univ.* **2009**, *43*, 169–172 + 177.
52. Youn, S.; Jung, B.; Lee, S. Limited Installation Ranges of Silt Curtain in Ocean and River Hydrodynamic Environment. *J. Coast. Res.* **2021**, *114*, 106–110. [CrossRef]

applied sciences

MDPI

Article

A Novel Suspended-Sediment Sampling Method: Depth-Integrated Grab (DIG)

Joel T. Groten [1,*], Sara B. Levin [1], Erin N. Coenen [1], J. William Lund [1] and Gregory D. Johnson [2]

1 Upper Midwest Water Science Center, U.S. Geological Survey, St. Paul, MN 55108, USA
2 Minnesota Pollution Control Agency, St. Paul, MN 55155, USA
* Correspondence: jgroten@usgs.gov

Abstract: Measuring suspended sediment in fluvial systems is critical to understanding and managing water resources. Sampling suspended sediment has been the primary means of understanding fluvial suspended sediment. Specialized samplers, sampling methods, and laboratory methods developed by select U.S. Federal Agencies are more representative of river and stream conditions than commonly used grab sampling and total suspended solids (TSS) laboratory methods but are not widely used because they are expensive, time consuming, and not required as part of water quality standards in the United States. A new suspended-sediment sampling method called a depth-integrated grab (DIG) was developed by combining certain elements from both grab and depth-integrating sampling methods and suspended-sediment concentration (SSC) laboratory methods. The goal of the DIG method was to provide more accurate results than Grab-TSS while being easier and cheaper to sample than specialized samplers and methods. Approximately 50 paired comparison samples were collected at 9 sites in Minnesota from 2018 through 2019. Results showed no significant difference between the DIG and specialized sampling methods and a significant difference between both methods and the Grab-TSS method. The DIG-SSC provided an improved alternative to the Grab-TSS method, but additional research and testing is important to evaluate if this method is appropriate in different conditions than were observed in this study.

Keywords: sediment transport; suspended sediment; suspended-sediment concentration; isokinetic; depth-integrating; grab sampling; total suspended solids; water quality; Minnesota; Wilcoxon test

Citation: Groten, J.T.; Levin, S.B.; Coenen, E.N.; Lund, J.W.; Johnson, G.D. A Novel Suspended-Sediment Sampling Method: Depth-Integrated Grab (DIG). *Appl. Sci.* **2023**, *13*, 7844. https://doi.org/10.3390/app13137844

Academic Editors: Kelin Hu and Yu Tian

Received: 16 May 2023
Revised: 30 June 2023
Accepted: 2 July 2023
Published: 4 July 2023

1. Introduction

Scientists, engineers, and natural resource managers have a need for fluvial suspended sediment data because they are not always available at locations of interest. Using suspended-sediment samplers has been the primary method for collecting fluvial suspended sediment data, but not all sampling and laboratory methods are representative of a river's suspended-sediment characteristics. Sediment may not be well mixed horizontally across the river cross section, especially when a tributary enters with different sediment characteristics [1]. Fluvial suspended sediment and its grain size distributions are rarely distributed uniformly in a river's cross section and can vary with depth [2], especially when sand-size particles (0.0625 to 2 millimeters [mm]) are present [3]. The occurrence and concentration of sand is often greater near the streambed, and it can be transported as bedload and/or suspended in the water column. Sand is more likely than fines to be deposited in the channel or floodplain when the river's velocities decrease. Fine-sized particles, which consist of silt and clay (less than 0.0625 mm), are more homogenously mixed in the river's cross section and can stay in suspension longer than sand.

Grab sampling does not represent the horizontal and vertical distribution of sediment because it only incorporates the top of the water column (less than 0.5 meters [m]), and only one location in the cross section is sampled (Figure 1c). Additionally, the total suspended solids (TSS) laboratory method [4], which is often used to analyze grab samples, provides additional errors [5–9]. The TSS laboratory method involves subsampling the original water sample, so the amount of suspended sediment in the subsample can be filtered and measured [4]. The TSS laboratory method is not representative of the whole water sample because sand can settle during the subsample extraction given Stokes's law. Several studies have shown that Grab-TSS is not as accurate as using specialized samplers and laboratory methods developed by the Federal Interagency Sedimentation Project (FISP) and the U.S. Geological Survey (USGS) because Grab-TSS fails to measure sand [5–9]. Even though this knowledge exists, Grab-TSS is widely used by water quality agencies and organizations as a water quality standard [10] for rivers and streams, such as in Minnesota where this study took place. There needs to be a less expensive and feasible alternative that is more representative of river conditions than Grab-TSS yet easier and cheaper than specialized samplers and methods.

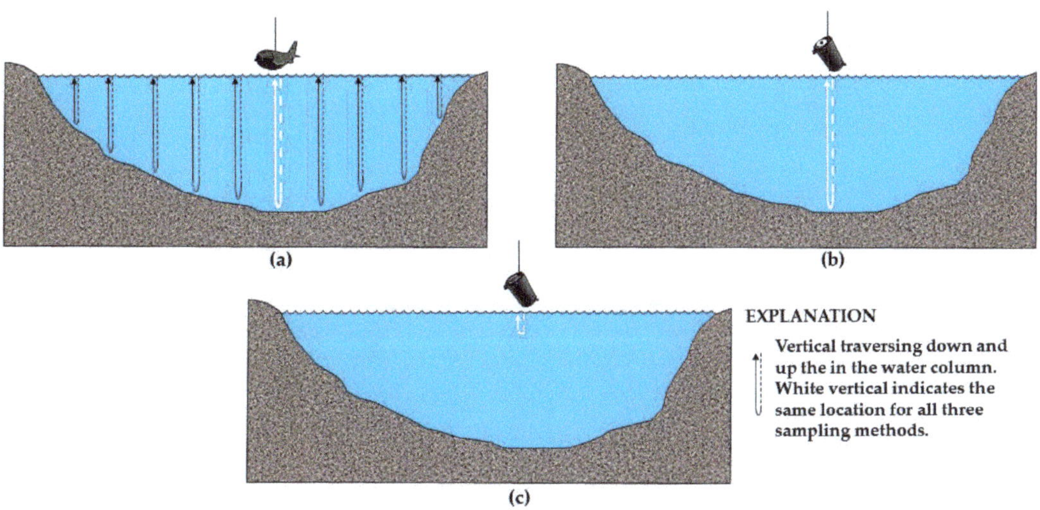

Figure 1. Three suspended-sediment sampling methods: (**a**) equal-width increment (EWI) suspended-sediment sampling method with 10 verticals and a D-74 suspended sediment sampler; (**b**) depth-integrated grab (DIG) suspended-sediment sampling method with 1 vertical and a US WBH-96 suspended sediment sampler with a hole in the bottle's cap; (**c**) Grab suspended-sediment sampling method with 1 vertical (near surface) and a US WBH-96 suspended-sediment sampler with an open-mouth bottle.

The US Federal agencies (USGS, U.S. Army Corps of Engineers, Bureau of Reclamation, and U.S. Department of Agriculture) that make up the FISP have specialized samplers, sampling methods, and laboratory methods to accurately represent fluvial suspended sediment in a river's cross section given that suspended sediment is not always uniformly distributed in fluvial systems due to the presence of sand. The primary suspended-sediment sampling and laboratory methods include equal-width increment (EWI; Figure 1a) or equal-discharge increment (EDI) sampling [11–13] and suspended-sediment concentration (SSC) laboratory methods [14,15]. These standard methods (EWDI-SSC) provide the best measures of suspended sediment because they depth-integrate nearly the entire water depth (except the bottom ~ 10 centimeters (cm) if the sampler touches the river's bed), measure the river's cross section at multiple verticals, and sample isokinetically. Isokinetic sampling is when water and sediment particles enter the sampler's nozzle at the same velocity as the stream velocity outside the sampler's nozzle. The SSC laboratory method measures the entire water-sediment mixture without a subsample extraction used with the TSS method. However, these samplers and methods are not widely used due to being more expensive, time consuming, and requiring more specialized equipment and training than grab sampling and TSS laboratory methods. For example, a weighted bottle sampler [16], commonly used for grab samples, is approximately 7 percent of the total cost of a US D-74 isokinetic sampler [17].

Technology and methods involving turbidity, acoustics, and laser diffraction (called surrogates) have shown great potential for accurately and cost-effectively estimating SSCs and particle sizes [18–20]. These methods and technology allow for increased spatial and/or temporal resolution in fluvial suspended sediment data and, once relations are developed, allow for rapid estimates of SSCs and particle sizes when collecting samples are not feasible or cost effective. However, those methods do not replace physical samples because physical samples are still used to calibrate the sensors.

There is a need for alternative samplers and methods that are more representative of river conditions than Grab-TSS methods but easier and cheaper to deploy than EWDI-SSC methods. The USGS Upper Midwest Water Science Center developed a suspended sediment sampling method that combined the elements from both grab and EWDI-SSC sampling methods. The goal of the new method was to improve the accuracy compared to Grab-TSS and reduce costs associated with EWDI-SSC sampling. The developed sampling method is called a depth-integrated grab (DIG; Figure 1b). It was originally hypothesized that the DIG would not be as accurate as an EWDI-SSC because it only consisted of one vertical measurement and was assumed to be non-isokinetic. However, the results suggest that there was not a significant difference between DIG-SSC and EWDI-SSC.

2. Materials and Methods

2.1. Study Area

The advancement and retreating of glaciers, predominantly the Wisconsinan glaciation occurring 85,000 to 11,000 years ago, has created diverse landforms and surface water conditions across Minnesota [21]. The resulting soil type and topographic relief created 5 sediment regions [22] within the state: southeast, southwest, middle, northeast, and northwest (Figure 2).

Glacial materials older than the Wisconsinan glaciation are found at the surface only in the southeastern and southwestern corners of the state [23]. The southeast region was the only region of the state not covered by the Wisconsinan glaciation, resulting in a lack of glacial till and shallower, poorer topsoil in the region [23]. Rivers in this region have created deep-cut valleys into underlying bedrock, resulting in more efficient drainage systems and more advanced erosion [23].

The southwest region is defined by the present-day Minnesota River valley and was formed from the drainage of glacial Lake Agassiz approximately 10,000 years ago. The tributaries to the Minnesota River flow through highly erodible knickpoints made up of fine-grained till, producing incised valleys throughout the region [24,25].

The middle region in central Minnesota contains a mixture of cultivated crops, pasture, and forests [25]. The middle region is in a transition zone between the agricultural land use of the south and the forested regions of the north [25]. The middle region contains the largest urban developed region in the state, the Minnesota-St. Paul metropolitan area. The northwest has a flat landscape, and the predominant land use is cultivated crops.

The predominantly forested northeast region has shallow bedrock, and steeper gradient rivers flow toward Lake Superior [26,27]. The northwest region contains predominantly cultivated crops, and the landscape is relatively flat compared to the rest of the regions. Overall, Minnesota's low-relief glaciated landscape contains differing sediment transport regimes that vary based on regional vulnerability to erosion, supplies, and controls [22]. The 9 sampling locations selected for this study represent these sediment regions present in Minnesota (Figure 2).

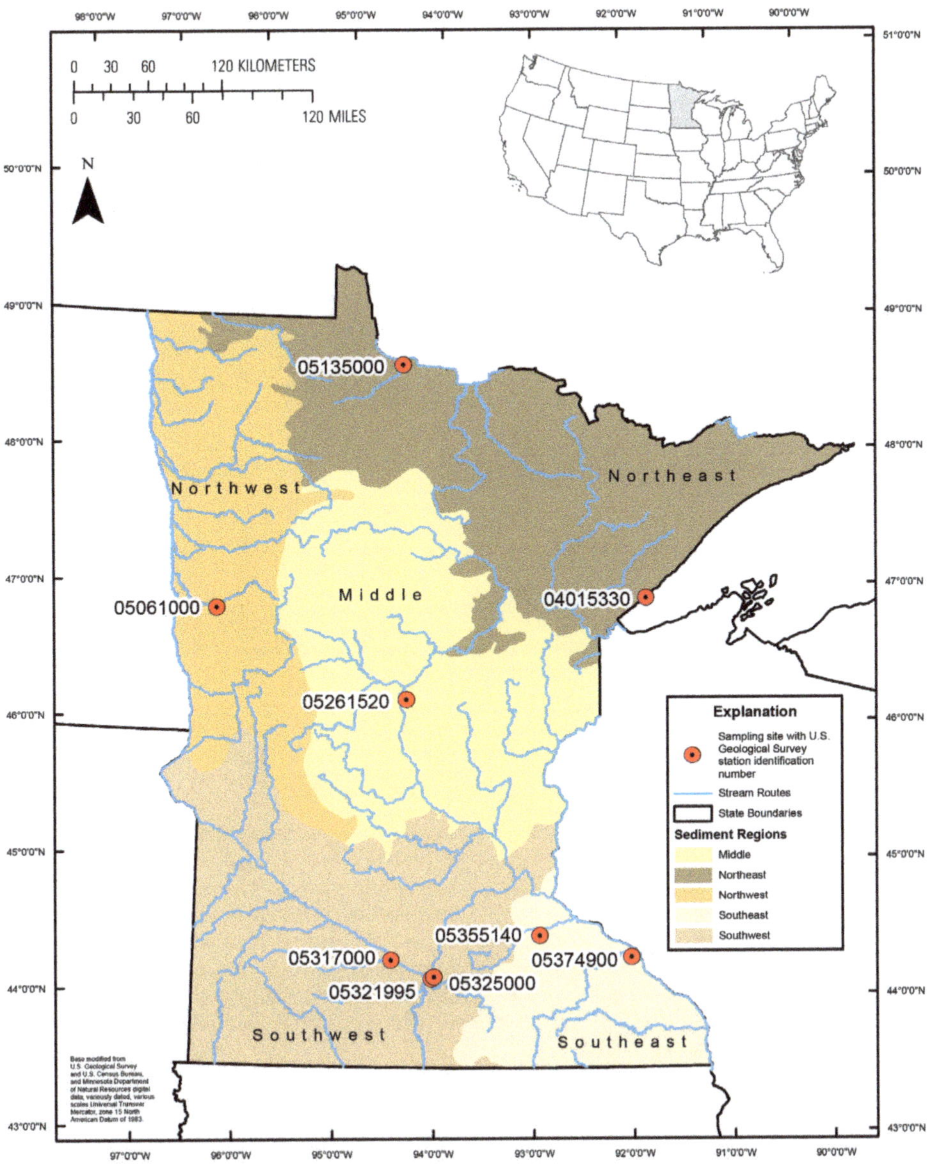

Figure 2. Study area with 9 suspended-sediment sampling sites in Minnesota.

2.2. Equal-Width Increment or Equal-Discharge Increment (EWDI) Sampling Method

Based on previous research, the EWDI-SSC was considered the representative measure of SSC and was used as a comparison to other sampling and laboratory methods [5–9]. Suspended-sediment samples were collected with isokinetic and depth-integrating samplers at EWIs or EDIs [12,13]. For the collection of samples, the stream was either divided into 10 EWIs (Figure 1a) or 5 EDIs. Each depth-integrated, isokinetic sample was collected at the centroid of each increment [12]. The EWI method had equal widths while the EDI method's width increments varied depending on the distribution of discharge during the time of sample collection. The EWI method was the primary sampling method used in this study.

Before an EWI sample could be collected, the sampler transit rate had to be determined. Different sampler transit rates (generally, 0.4 times the average velocity in the sampled vertical [12]) were tested by sampling the deepest and fastest part of the river, to ensure the sample container did not overfill during the final sample. The same transit rate was used for all the EWI sample verticals. Depending on the river depth and velocity, a US DH-48 bottle sampler (with a 0.5-L glass bottle), US D-74 bottle sampler (with a 0.5-L glass bottle or 1-L glass bottle) (Figure 3b), or US D-96 bag sampler (with a 3-L bag) [13]. The US D-74 sampler and 0.635 cm sampler nozzle were mostly used in this study.

Figure 3. The 3 suspended-sediment samplers: (**a**) depth-integrated grab (DIG) suspended-sediment sampler (US WBH-96 with a hole in the bottle's cap) attached to a D-74 suspended sediment sampler; (**b**) A D-74 suspended sediment sampler near the water surface; (**c**) A US WBH-96 grab suspended-sediment sampler with an open-mouth bottle.

The US DH-48 bottle sampler and US D-74 operate identically with the only difference being that the US D-74 can hold either a 0.5- or 1-L glass bottle while the US DH-48 can only hold a 0.5-L bottle. The US DH-48 and US D-74 have vents which are located on the side of the sampler and point downstream during sample collection. The vent allows air to escape as it is displaced by the sample being collected in the bottle. The US D-96 bag sampler can sample at deeper depths and higher velocities than the US DH-48 and US D-74. The US D-96 does not have a vent but, prior to sample collection, the person sampling collapses the bag with their hands which pushes the air out from the bag and nozzle prior to sample collection.

Contamination can occur when a sampler's nozzle digs into the streambed. A depth-sounding was taken before each sample was collected to determine the water depth. Each vertical was sampled approximately 0.15 to 0.3 m above the streambed to prevent bed contamination. The distance above the streambed varied depending on the estimated height of the river's sand dunes to avoid inadvertently sampling the sand dune. All samples collected from the centroids of the stream transect were composited into one sample and sent to the laboratory for analysis.

2.3. *Depth-Integrated Grab (DIG) Sampling Method*

A DIG consists of a single depth-integrating vertical. For this study, the location of the DIG sample was the centroid of the river's channel. The location of the DIG sample was always the same location as one of the EWDI verticals and the grab sampling location (white verticals in Figure 1a–c). A depth-sounding was taken before each sample was collected to determine the water depth. Each vertical was sampled approximately 0.15 to 0.3 m above the streambed to prevent bed contamination. The DIG method sampled the same depth of the water column as the one vertical of the EWDI sample.

The DIG deploys a weighted-bottle sampler (US WBH-96) with a 1-L high-density polyethylene plastic bottle secured to the inside of the sampler with a rubber band. For this study, the weighted bottle was attached directly to the isokinetic sampler to obtain a concurrent sample (Figure 3a) that was collected at the same time as 1 of the EWDI-SSC verticals. The one-liter bottle had a cap attached to the mouth of the bottle with a 1.27 cm hole in the center of the bottle cap (Figure 3a). The hole in the cap was created with a drill press. The purpose of the hole in the cap was to prevent the bottle from overfilling while depth integrating. The DIG sampler does not have a separate vent like the US DH-48 and US D-74 samplers, so air escapes from the hole in the cap during sample collection.

A transit rate for the DIG sample was determined at the same time as the EWDI transit rate for the isokinetic samplers (US DH-48, US D-74, or US D-96). It was important to determine a transit rate common to the DIG sampler and isokinetic sampler that prevented overfilling because the sample container's volume and nozzle opening sizes differed between samplers. The same transit rate was used with the DIG sample as the EWDI sample because the samplers were sent into the water column concurrently. After the sample was collected, the bottle was inspected to make sure the sample was not overfilled. This was done by making sure the collected sample was below the shoulder of the bottle. A cap without a hole in it was secured to the bottle and sent to the laboratory for analysis.

2.4. Grab Sampling Method (Grab)

A grab sample was collected using the same type of 1-L HDPE plastic bottle as the DIG but had no cap attached (open-mouth) to the bottle (Figure 3c). The plastic bottle was secured to the inside of a weighted-bottle sampler (US WBH-96) with a rubber band (Figure 3c). The grab sample was collected from the centroid of the river channel (same location as DIG and one EWDI vertical) at a depth ranging from directly below the water surface to less than approximately 0.5 m below the water surface, depending on water velocities (Figure 1). If water velocities were higher, the sampler was not able to go that far below the surface. The exact depth was not measured each time. A Grab-TSS sample was collected directly before (sequentially) and within minutes of when the DIG and EWDI samples were collected. The grab sample was not collected at the same time as the EWDI and DIG because only 1 weighted-bottle sampler was available during the sampling events. After the sample was collected, the bottle was secured with a cap and sent to the laboratory for analysis.

2.5. Suspended-Sediment Concentration Labratory Method (SSC, Fines, and Sands)

The SSC laboratory analyses consisted of two methods. Suspended-sediment samples collected with EWDI and DIG sampling methods were both analyzed for SSC following method D3977-97 [14,15] by the USGS Sediment Laboratory in Iowa City, Iowa. SSC is measured by measuring the dry weight of sediment from a known volume of a water-sediment mixture. SSC measures the entire water-sediment mixture. Percentages of fines (%Fines) were also determined from SSC samples at the same laboratory by wet sieving [14]. SSC and percent fines results are available in the USGS National Water Information System [28].

The suspended-sand concentration (Sands) was calculated from Equations (1) and (2) while the suspended-fines concentration (Fines) was calculated from Equation (1). First, %Fines was multiplied by the corresponding SSC value and dividing the product by 100 to obtain the Fines. Second, the calculated Fines (Equation (1)) was subtracted (minused) from the corresponding SSC to calculate Sands (Equation (2)).

$$\text{Fines} = (\%\text{Fines} \times \text{SSC})/100 \quad (1)$$

$$\text{Sands} = \text{SSC} - \text{Fines} \quad (2)$$

2.6. Total Suspended Solids Laboratory Method (TSS)

Suspended sediment samples collected with the grab sampling method were analyzed at the Minnesota Department of Health Laboratory in St. Paul, Minnesota for TSS following method 2540 D [4]. The results from the TSS laboratory analyses are available from the Water Quality Portal [29].

2.7. Data Analysis

Data analysis included summary statistics, visual inspection of the data, bootstrapped median and percentile values, and the Wilcoxon signed rank test [30]. The R statistical environment was used to produce the bootstrapped median and percentile values and perform the Wilcoxon signed rank tests [31].

3. Results

A total of 9 rivers were sampled in 2018 and 2019 (Table 1). The sampling campaign resulted in 48 pairs of EWDI-SSC and DIG-SSC with all but 2 of those pairs having Grab-TSS being collected at nearly the same time (sequential).

Table 1. Summary of sampling sites and number of samples collected.

Station Name	Station Number	Date from (yyyymmdd)	Date to (yyyymmdd)	Number of Samples
KNIFE RIVER NEAR TWO HARBORS, MN	04015330	20180531	20190415	3
BUFFALO RIVER NEAR HAWLEY, MN	05061000	20190401	20190716	6
NOKASIPPI RIVER NEAR FORT RIPLEY, MN	05261520	20190402	20190923	8
COTTONWOOD RIVER NEAR NEW ULM, MN	05317000	20190418	20190904	2
BLUE EARTH RIVER AT HWY 169 AT MANKATO, MN	05321995	20180608	20180921	2
MINNESOTA RIVER AT MANKATO, MN	05325000	20180421	20190529	9
LITTLE CANNON RIVER NEAR CANNON FALLS, MN	05355140	20190323	20190702	3
ZUMBRO RIVER AT KELLOGG, MN	05374900	20180422	20180921	4
EAST FORK RAPID RIVER NEAR CLEMENSTON, MN	05135000	20190424	20190710	11

Abbreviations: MN, Minnesota.

The rivers' sizes ranged from 12–126 m in width and 1–10 m in depth (Table S1; Supplemental Materials). The 9 rivers' streamflows ranged from 3–1823 cubic meters per second during sample collection (Table S1; Supplemental Materials). The range of EWDI-SSC measured was 4–1690 milligrams per liter (mg/L). The percentage of suspended sand in the EWDI-SSC samples ranged from 0–56 percent. The range of EWDI-Fines concentrations ranged from 2–1487 while the EWDI-Sands concentrations ranged from 0–278 mg/L (Table 2).

Table 2. Summary statistics for sampling and laboratory methods and river conditions.

	EWDI-SSC (mg/L)	DIG-SSC (mg/L)	EWDI-Sands (mg/L)	DIG-Sands (mg/L)	EWDI-Fines (mg/L)	DIG-Fines (mg/L)	Grab-TSS (mg/L)	Water Depth (m)	Stream Width (m)	Q (cms)
Min.	4	3	0	0	2	2	3	1	12	3
Max.	1690	1690	278	629	1487	1572	1600	10	126	1823
Mean	267	279	52	60	215	219	221	3	49	277
Med.	90	80	13	15	65	52	76	2	37	16
SD	362	400	71	113	307	320	317	3	39	517

Abbreviations: Min., minimum; Max., maximum; Med., median; SD, standard deviation; EWDI, equal-width or -discharge increment; SSC, suspended-sediment concentration; mg/L, milligrams per liter; DIG, depth-integrated grab; TSS, total suspended solids; m, meters; Q, streamflow; cms, cubic meters per second.

EWDI-SSC, EWDI-Sands, and EWDI-Fines were considered the representative measures of suspended sediment and were the primary samples to which the other methods were compared. Sediment sample distributions by sampling methods are shown in Figure 4. Sediment sample distributions and median concentrations of SSC, TSS, Fines, and Sands varied among rivers and the 3 sampling methods, EWDI, DIG, and grab (Figure 4). The highest measured EWDI-SSC, -Fines, and -Sands were in the southwestern and southeastern sediment regions of Minnesota at the Blue Earth River at Mankato (05321995), the Cottonwood River near New Ulm (0531700), and the Zumbro River at Kellogg (05374900). The northeastern sediment region site East Fork Rapid River near Clemenston (05135000) and middle sediment region site the Nokasippi River near Fort Ripley (05261520) had the lowest median SSC, Fines, and Sands values, regardless of sampling method. The Grab-TSS concentrations were lower than those obtained by EWDI and DIG methods at most of the sites (Figure 4).

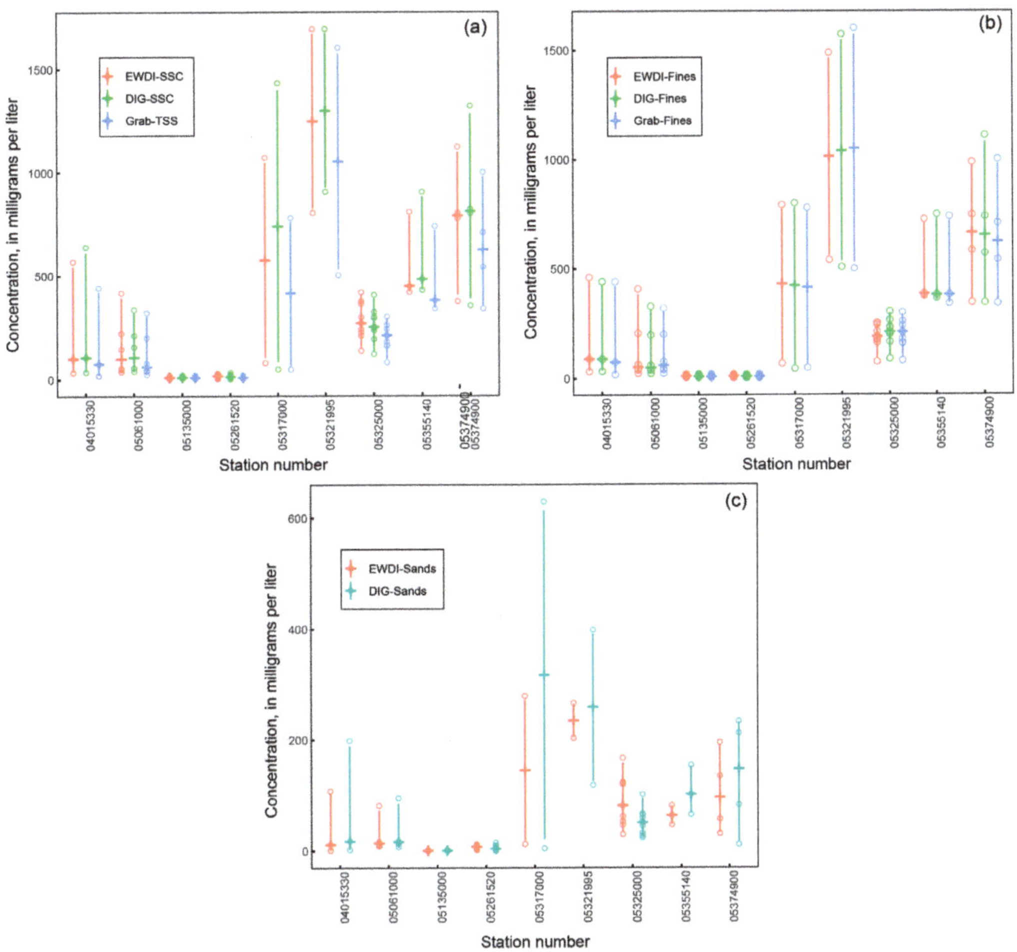

Figure 4. Bootstrapped median values (horizonal line) and the 2.5th and 97.5 percentiles (vertical line), and measured values (open circle) for sampling (equal-width or -discharge increment [EWDI], depth-integrated grab [DIG], and Grab) and laboratory methods at each of the 9 sampling sites: (**a**) Suspended-sediment concentration (SSC) and total-suspended solids (TSS) laboratory methods; (**b**) Suspended-fines concentration (Fines) laboratory method; (**c**) Suspended-sands concentration (Sands) laboratory method.

Visual inspection of the SSC data showed close agreement between EWDI and DIG sampling methods (Figure 5). Most of the data plotted near or on the 1:1 line. The paired data from site 0531700 which deviated the furthest from the 1:1 line was at 1070 mg/L on the x-axis (EWDI-SSC) and 1430 mg/L on the y-axis (DIG-SSC). This paired data had the greatest discrepancy due to there being much higher Sands in the DIG sample (629 mg/L) than the EWDI sample (278 mg/L) when compared to the whole dataset. These paired samples had the highest measured Sands observed during the study. This discrepancy in measured Sandse between the two methods could be due to only measuring one vertical with the DIG method and measuring it at a location with potentially higher sand transport. The location of the DIG samples was at the centroid of the channel which is generally the deepest and fastest part of the river because it was the same location where transit rates were tested and determined.

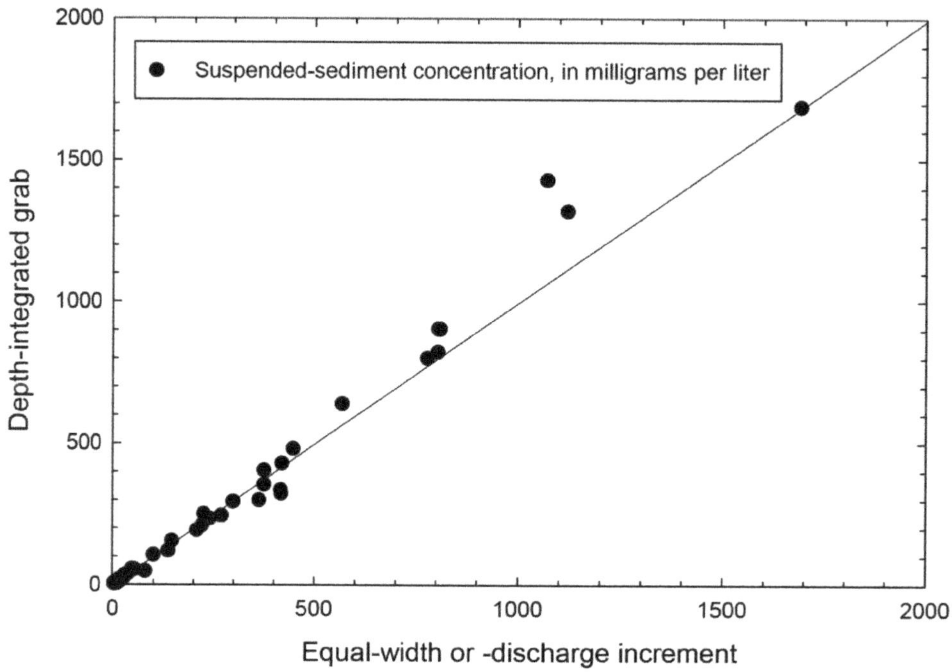

Figure 5. Relationship between depth-integrated grab (DIG) and equal-width or -discharge increment (EWDI) sampling methods using the suspended-sediment concentration (SSC) laboratory method. The black line is the 1:1 line.

When comparing the Fines and Sands between EWDI and DIG sampling methods, the range of Sands was generally less than the Fines. The maximum measured Sands was 629 mg/L (DIG) while the maximum measured Fines was 1572 mg/L (DIG). There was a closer agreement between the EWDI-Fines and DIG-Fines than the EWDI-Sands and DIG-Sands (Figure 6). Most of the Fines data plotted near or on the 1:1 line (Figure 6). There was more variability around the 1:1 line for Sands data. The same sample that deviated the furthest from the 1:1 line in Figure 5 also plotted the furthest above the 1:1 line, with the DIG sample having an additional 351 mg/L of Sands than the EWDI sample.

A paired Wilcoxon test, also known as the Wilcoxon signed-rank test, was used to compare samples collected by 2 different methods [30]. The paired test takes the sample differences from each collection method and tests the null hypothesis that the median difference is equal to zero. Distributions of paired samples are shown in Figure 7. The results from the paired Wilcoxon test are shown in Table 3.

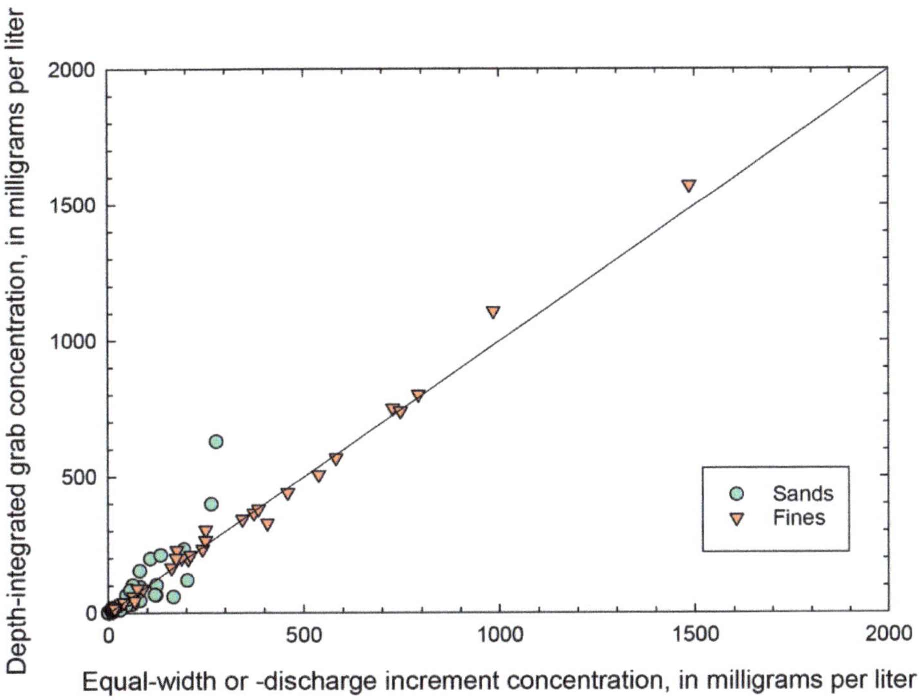

Figure 6. Relationship between Sands and Fines concentrations for equal-width or -discharge increment (EWDI) sampling method (x-axis) and depth-integrated grab (DIG) sampling method (y-axis). The black line is the 1:1 line.

Table 3. Results from the paired Wilcoxon test [30].

First Method	Second Method	Median	*p*-Value
EWDI-SSC	DIG-SSC	0	0.68
EWDI-SSC	Grab-TSS	28	<0.01
DIG-SSC	Grab-TSS	15	<0.01
EWDI-Sands	DIG-Sands	0.09	0.96
EWDI-Sands	Grab-TSS	−33	<0.01
DIG-Sands	Grab-TSS	−47	<0.01
EWDI-Fines	DIG-Fines	0.65	0.42
EWDI-Fines	Grab-TSS	1.72	0.06
DIG-Fines	Grab-TSS	1.67	<0.01

Abbreviations: *p*-value, probability value; <, less than; EWDI, equal-width or -discharge increment; SSC, suspended-sediment concentration; DIG, depth-integrated grab; TSS, total suspended solids.

Statistically significant (*p*-value less than 0.05) differences were not found between EWDI and DIG methods for any sediment type (SSC, Sands, and Fines; Table 1). There was not a statistically significant difference between EWDI-Fines and Grab-TSS (Table 3). Conversely, there were statistically significant differences between all the other comparisons (EWDI-SSC-Sands and DIG-SSC-Fines) and Grab-TSS. The Grab-TSS samples were generally lower than EWDI and DIG methods. The median difference between EWDI-SSC and Grab-TSS was 28 mg/L, and the median difference between DIG-SSC and Grab-TSS was 15 mg/L (Table 3). For Fines, the median differences were 1.72 and 1.67 between EWDI-Fines and Grab-TSS, and DIG-Fines and Grab-TSS, respectively (Table 3).

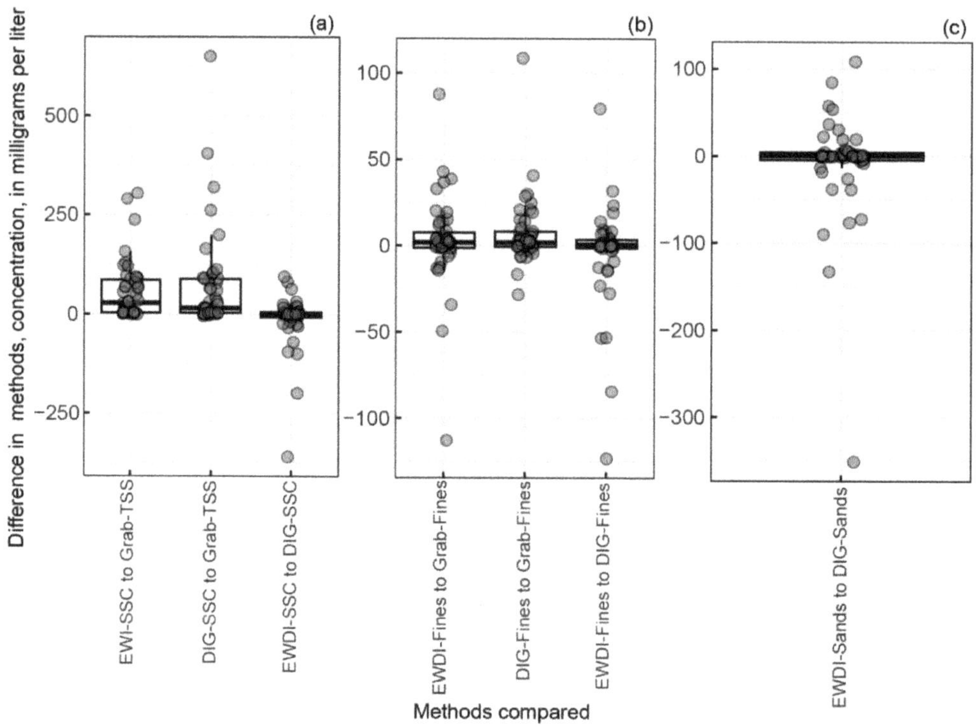

Figure 7. Box plots depict the minimum, first quartile, median, third quartile, and maximum, with measured values depicted as single points, and show the difference in the paired values between each sampling (equal-width or -discharge increment [EWDI], depth-integrated grab [DIG], and Grab) and laboratory method for a particular sample: (**a**) Suspended-sediment concentration (SSC) and total-suspended solids (TSS) laboratory methods; (**b**) Suspended-fines concentration (Fines) laboratory method; (**c**) Suspended-sands concentration (Sands) laboratory method.

4. Discussion

Since there was not a statistically significant difference between the EWDI-SSC and DIG-SSC sampling methods, the DIG-SSC shows promise as a more representative method than Grab-TSS. Furthermore, the DIG-SSC is easier and cheaper to use in the field than EWDI-SSC. However, the DIG-SSC was only tested at 9 sites within a specific range of river and suspended-sediment characteristics (Table 2). The DIG-SSC method would benefit from testing at other rivers with different streamflow and suspended-sediment characteristics that were not represented in this study. Additional results could help validate or challenge our study's results.

The close agreement between EWDI and DIG methods may be explained by the rivers in this study being relatively well mixed horizontally. The main limitation with the DIG method occurs when a river is poorly mixed horizontally. If deploying the DIG method and the river is poorly mixed horizontally, possible solutions would be to go further downstream where the river is fully mixed horizontally or modify the DIG method to sample at multiple verticals by using EWDI methods with the DIG sampler.

The DIG-SSC method would benefit from testing at sites with higher concentrations of sand than what was observed in this study. Sand caused greater variability when comparing EWDI and DIG sampling methods. The variability might be explained by the DIG method only sampling one vertical while the EWDI method sampled multiple verticals. Furthermore, the sample location of the DIG method was at the deepest and fastest part of the river, which probably had the highest concentrations of sand, and greater than the other EWDI sampling locations. This could be tested by comparing the same location and the same number of verticals with both the DIG and isokinetic samplers. This would make the sampling methods comparable to see if the variability seen in this study is from the different number of verticals used by DIG and EWDI methods.

It would also be beneficial to test if the DIG sampler is isokinetic or not. The measure of isokinetic sampling is defined as the ratio of the velocity through the nozzle entrance (Vn) to the ambient stream velocity (V):

$$\text{Isokinetic} = \text{Vn}/\text{V} \tag{3}$$

where Vn and V are averaged over the sample time and depth for each specific sample. This could be another possible explanation of the variability between EWDI and DIG methods. If additional sampling data were to be collected in the future, it would be beneficial to determine the DIG sampler's intake efficiency (IE) which is a measure of isokinetic sampling [32]. Equation (4) contains K which is indexed to the nozzle opening, volume (Vol), duration, and stream velocity (SV).

$$\text{IE} = \text{K} \times (\text{Vol}/\text{D})/\text{SV} \tag{4}$$

The average IE should fall within $0.75 < \text{IE} < 1.25$ [32]. Moreover, IE could be compared with temperature [33] and depth which would help determine the operational specifications of the DIG sampler. Furthermore, future research could entail testing the DIG sampler in a tow tank to see if it samples isokinetically and(or) by performing 3-dimensional (3D) modeling to see if the DIG samples isokinetically. Since there is no significant difference between the DIG-SSC and EWDI-SSC, the DIG-SSC could be isokinetic; however, the results in this study cannot confirm this without additional testing and evaluation.

The DIG-SSC might not be applicable everywhere but does provide a more representative estimate of suspended sediment than Grab-TSS due to its ability to capture and measure sand with sampling and laboratory methods. Even though the comparison of EWDI-Fines and Grab-TSS was only marginally above the threshold of significance, another study showed the similarity between EWDI-Fines and Grab-TSS [9]. Since there are nominal differences between Grab-TSS and measured Fines with EWDI methods, Grab-TSS might only be used as an estimate of Fines and may not be considered a complete measure of Sands and Fines.

When deploying the DIG method in future studies, there are some important considerations. Since the DIG method uses one vertical, it might only be used at rivers that are well mixed horizontally. If a river is poorly mixed horizontally, EWI and EWDI methods can be used with the DIG sampler. Before the final sample is collected, an independent transit rate should be determined, so the bottle is not overfilled. At higher water velocities, additional weight may be required to maintain position in the water column during the lowering and raising of the sampler (Figure 8a,b). A weight can be attached to the bottom of a weighted bottle sampler (Figure 8a,b). A reel and bridge board can assist in raising and lowering the sampler and weight to help maintain a consistent transit rate when operated correctly (Figure 8b,c).

The DIG method could replace grab sampling at lower flows and velocities when sampling conditions are non-isokinetic. FISP samplers are designed to sample at velocities equal to or greater than 0.457 meter/second [m/s; 13], and the DIG method allows for depth-integrating sample collection at velocities less than 0.457 m/s. Even though it is assumed there is no sand in transport at lower velocities, the DIG-SSC method would be able to capture any potential sand in suspension. At lower river velocities, a rope could be used to raise and lower the weighted-bottle sampler by hand (Figure 8d).

Figure 8. Suggested depth-integrated grab (DIG) suspended-sediment sampling setup for high (**a**,**b**) and low (**d**) stream velocities: (**a**) Depth-integrated grab (DIG) suspended-sediment sampler (US WBH-96 with a hole in the bottle's cap) attached to a sounding weight; (**b**) A bridge board, A-55 sounding reel, US WBH-96 with a hole in the bottle's cap attached to a sounding weight, and a person sampling from a bridge; (**c**) A bridge board, b-reel, and a person sampling from a bridge; (**d**) Depth-integrated grab (DIG) suspended-sediment sampler (US WBH-96 with a hole in the bottle's cap) attached to a rope.

5. Conclusions

Even with the advancement of technology, sampling suspended sediment still remains the primary means of understanding fluvial suspended sediment. The novel DIG-SSC sampling method provided more accurate results than Grab-TSS, and results showed there was no significant difference between the DIG and EWDI methods. The DIG method is easier and cheaper to sample than with isokinetic samplers and EWDI sampling methods. However, DIG-SSC was only tested at 9 sites (48 sample pairs) within a range of river and suspended-sediment characteristics, so this method would benefit from further testing at other rivers with different river and suspended-sediment characteristics not represented in this study.

Additional data collection could entail comparing the same number of verticals from EWDI and DIG. Collecting additional data to determine the IE, tow tank testing, and 3D modeling may help answer if the DIG sampler is isokinetic or not. Comparing the same number of verticals and isokinetic testing may help determine the cause of variability between EWDI and DIG methods observed in this study. Answering these questions could help potential users know if this method could be applied at rivers beyond this study area.

In addition to there being benefits of additional data collection, since this is a new sampling method, here are some important considerations when using the DIG method:

(1). The DIG method does not replace isokinetic samplers and EWDI methods but shows promise as an alternative if users are unable to use isokinetic samplers and EWDI sampling methods.
(2). The DIG method may only be used at sites that are well mixed horizontally. If a site is poorly mixed horizontally, EWDI methods with a DIG sampler can be used.
(3). Grab sampling often occurs at lower velocities as isokinetic samplers are not recommended. The DIG-SSC method could replace grab sampling at lower velocities.

If users are unable to use the DIG-SSC method, here is one important consideration when using the Grab-TSS method:

(1). Grab-TSS may only be used as an estimate of suspended fines.

Supplementary Materials: The following supporting information can be downloaded at: https://www.mdpi.com/article/10.3390/app13137844/s1, Table S1: All sampling-method comparison and river condition data for 9 river sites in Minnesota, 2018–2019.

Author Contributions: Conceptualization, J.T.G.; methodology, J.T.G., J.W.L. and G.D.J.; formal analysis, J.T.G. and S.B.L.; investigation, J.T.G. and J.W.L.; resources, J.T.G., G.D.J. and J.W.L.; data curation, E.N.C.; writing—original draft preparation, J.T.G., S.B.L. and E.N.C.; writing—review and editing, J.T.G., S.B.L., J.W.L. and G.D.J.; visualization, S.B.L., J.W.L. and E.N.C.; supervision, J.T.G.; project administration, J.T.G. and G.D.J.; funding acquisition, J.T.G. and G.D.J. All authors have read and agreed to the published version of the manuscript.

Funding: This research was funded by the Minnesota Clean Water Fund via the Minnesota Pollution Control Agency and U.S. Geological Survey Cooperator Matching Funds.

Informed Consent Statement: Not applicable.

Data Availability Statement: The results from SSC and percentage fines laboratory analysis are available in the USGS National Water Information System [28]. The results from the TSS laboratory analyses are available from the Water Quality Portal [29]. The data can also be found in the Supplementary Materials section.

Acknowledgments: Special thanks to Brett Savage and Christopher Ellison for originally conceiving of the idea of using a US WBH-96 with a hole in the bottle's cap and collecting samples during low flow, non-isokinetic suspended-sampling conditions in Minnesota. Gerald Storey, Brent Mason, and Josh Ayers are acknowledged for assisting with sample collection. Ben Adolphson is acknowledged for assisting with photos of the different samplers. We also thank the Journal's anonymous reviewers and Tim Straub, USGS, for their useful comments to improve the manuscript. This journal article has been peer reviewed and approved for publication consistent with USGS Fundamental Science Practices (https://pubs.usgs.gov/circ/1367/ (accessed on 30 June 2023)).

Conflicts of Interest: The authors declare no conflict of interest. The funders had no role in the design of the study; in the collection, analysis, or interpretation of data; in the writing of the manuscript, or in the decision to publish the results.

Disclaimer: Any use of trade, firm, or product names is for descriptive purposes only and does not imply endorsement by the U.S. Government.

References

1. Lewis, J.; Eads, R. *Implementation Guide for Turbidity Threshold Sampling: Principles, Procedures, and Analysis*; General Technical Report PSW-GTR-212; U.S. Department of Agriculture; Forest Service; Pacific Southwest Research Station: Albany, CA, USA, 2009; p. 87. [CrossRef]
2. Steegen, A.; Govers, G. Correction factors for estimating suspended sediment export from loess catchments. *Earth Surf. Process. Landf.* **2001**, *26*, 441–449. [CrossRef]

3. Guy, H.P. Fluvial Sediment Concepts. In *U.S. Geological Survey Techniques of Water-Resources Investigations*; Book 3, Chap. C1; U.S. Geological Survey: Reston, VA, USA, 1970; p. 55. Available online: https://pubs.usgs.gov/twri/twri3-c1/pdf/TWRI_3-C1.pdf (accessed on 6 February 2023).
4. Clesceri, L.S.; Greenberg, A.E.; Eaton, A.D. *Standard Methods for the Examination of Water and Wastewater*, 20th ed.; American Public Health Association; American Water Works Association; Water Environment Federation: Washington, D.C., USA, 1998; variously paged.
5. Gray, J.R.; Glysson, G.D.; Turcios, L.M.; Schwarz, G.E. *Comparability of Suspended-Sediment Concentration and Total Suspended Solids Data*; U.S. Geological Survey Water-Resources Investigations Report 00-4191; U.S. Geological Survey: Reston, VA, USA, 2000; p. 14. Available online: https://pubs.usgs.gov/wri/wri004191/ (accessed on 8 February 2023).
6. Selbig, W.; Bannerman, R. Ratios of Total Suspended Solids to Suspended Sediment Concentrations by Particle Size. *J. Environ. Eng.* **2011**, *137*, 1075–1081. [CrossRef]
7. Ellison, C.A.; Savage, B.E.; Johnson, G.D. *Suspended-Sediment Concentrations, Loads, Total Suspended Solids, Turbidity, and Particle-Size Fractions for Selected Rivers in Minnesota, 2007 through 2011*; U.S. Geological Survey Scientific Investigations Report 2013–5205; U.S. Geological Survey: Reston, VA, USA, 2014; p. 43. [CrossRef]
8. Groten, J.T.; Johnson, G.D. *Comparability of River Suspended-Sediment Sampling and Laboratory Analysis Methods*; U.S. Geological Survey Scientific Investigations Report 2018-5023; U.S. Geological Survey: Reston, VA, USA, 2018; p. 23. [CrossRef]
9. Groten, J.T.; Johnson, G.D. Comparability of different river suspended sediment sampling and laboratory analysis methods and the effect of sand. In Proceedings of the SEDHYD 2019 Conference on Sedimentation and Hydrologic Modeling, Reno Nevada, NV, USA, 24–28 June 2019; pp. 1–16. Available online: http://www.sedhyd.org/2019/proceedings/SEDHYD_Proceedings_2019_Volume3.pdf (accessed on 6 February 2023).
10. Minnesota Legislature. Available online: https://www.revisor.mn.gov/rules/7050.0222/ (accessed on 15 March 2023).
11. Ward, J.R.; Harr, C.A. *Methods for Collection and Processing of Surface-Water and Bed-Material Samples for Physical and Chemical Analyses*; U.S. Geological Survey Open-File Report 90-140; U.S. Geological Survey: Reston, VA, USA, 1990; p. 71. [CrossRef]
12. Edwards, T.K.; Glysson, G.D. Field Methods for Measurement of Fluvial Sediment. In *U.S. Geological Survey Techniques of Water-Resources Investigations*; Book 3, Chap. C2; U.S. Geological Survey: Reston, VA, USA, 1999; p. 89. Available online: https://pubs.usgs.gov/twri/twri3-c2/pdf/TWRI_3-C2.pdf (accessed on 7 February 2023).
13. Davis, B.E. *A Guide to the Proper Selection and Use of Federally Approved Sediment and Water-Quality Samplers*; U.S. Geological Survey Open File Report 2005-1087; U.S. Geological Survey: Reston, VA, USA, 2005; p. 20. [CrossRef]
14. Guy, H.P. Laboratory Theory and Methods for Sediment Analysis. In *U.S. Geological Survey Techniques of Water-Resources Investigations*; Book 3, Chap. C1; U.S. Geological Survey: Reston, VA, USA, 1969; p. 58. Available online: https://pubs.usgs.gov/twri/twri3-c1/pdf/TWRI_3-C1.pdf (accessed on 9 February 2023).
15. *ASTM D3977–97*; American Society for Testing and Materials (ASTM) International. Standard Test Methods for Determining Sediment Concentration in Water Samples. ASTM: West Conshohocken, PA, USA, 2000; Volume 11.02, Chap. Water (II). pp. 395–400.
16. Performance Results Plus, US WBH-96 Weighted Bottle Sampler. Available online: https://prph2o.com/us-wbh-96-weighted-bottle-sampler/ (accessed on 15 March 2023).
17. Performance Results Plus, D-74 Sediment Sampler. Available online: https://prph2o.com/d-74-sediment-sampler/ (accessed on 8 March 2023).
18. Rasmussen, P.P.; Gray, J.R.; Glysson, G.D.; Ziegler, A.C. Guidelines and Procedures for Computing Time-Series Suspended-Sediment Concentrations and Loads from in-Stream Turbidity-Sensor and Streamflow Data. In *U.S. Geological Survey Techniques and Methods*; Book 3, Chap. C4; U.S. Geological Survey: Reston, VA, USA, 2009; p. 53. Available online: https://pubs.usgs.gov/tm/tm3c4/ (accessed on 8 February 2023).
19. Wood, M.S.; Groten, J.T.; Straub, T.D.; Whealdon-Haught, D.R.; Griffiths, R.E.; Boldt, J.A.; Lucena, Z.N.; Brown, J.E.; Suttles, S.E.; Dickhudt, P.J. State of the Science and Decision Support for Measuring Suspended Sediment with Acoustic Instrumentation. In Proceedings of the SEDHYD 2023 Conference on Sedimentation and Hydrologic Modeling, St. Louis, MO, USA, 8–12 May 2023; pp. 1–16. Available online: https://www.sedhyd.org/2023Program/_program.html (accessed on 22 June 2023).
20. Muneer, M.; Czuba, J.A.; Curran, C.A. In-Stream Laser Diffraction for Measuring Suspended Sediment Concentration and Particle Size Distribution in Rivers: Insights from Field Campaigns. *J. Hydraul. Eng.* **2023**, *149*, 05022007. [CrossRef]
21. Minnesota Department of Natural Resources, Lands and Minerals. Available online: http://files.dnr.state.mn.us/lands_minerals/geologyhandbook.pdf (accessed on 3 March 2023).
22. Lund, J.W.; Groten, J.T.; Karwan, D.L.; Babcock, C. Using machine learning to improve predictions and provide insight into fluvial sediment transport. *Hydrol. Process.* **2022**, *36*, e14648. [CrossRef]
23. Gran, K.B.; Belmont, P.; Day, S.S.; Jennings, C.; Johnson, A.; Perg, L.; Wilcock, P.R. *Geomorphic Evolution of the Le Sueur River, Minnesota, USA, and Implications for Current Sediment Loading*; Geological Society of America Special Paper 451; Geological Society of America: Boulder, CO, USA, 2009; pp. 119–130. [CrossRef]
24. Gran, K.B.; Belmont, P.; Day, S.S.; Jennings, C.; Lauer, J.W.; Viparelli, E.; Wilcock, P.R.; Parker, G. *An Integrated Sediment Budget for the Le Sueur River Basin*; Minnesota Pollution Control Agency (MPCA) Report wq-iw7-29o; Minnesota Pollution Control Agency: St. Paul, MN, USA, 2011.

25. Ellison, C.A.; Groten, J.T.; Lorenz, D.L.; Koller, K.S. *Application of Dimensionless Sediment Rating Curves to Predict Suspended-Sediment Concentrations, Bedload, and Annual Sediment Loads for Rivers in Minnesota*; U.S. Geological Survey Scientific Investigations Report 2016–5146; U.S. Geological Survey: Reston, VA, USA, 2016; p. 68. [CrossRef]
26. Ojakangas, R.W.; Matsch, C.L. *Minnesota's Geology*, 1st ed.; University of Minnesota Press: Minneapolis, MN, USA, 1982; p. 255.
27. Sims, P.K.; Morey, G.G. *Geology of Minnesota: A Centennial Volume*; Minnesota Geological Survey: Minneapolis, MN, USA, 1972; p. 632. Available online: https://hdl.handle.net/11299/57318 (accessed on 6 February 2023).
28. U.S. Geological Survey. USGS Water Data for the Nation; U.S. Geological Survey National Water Information System Database. 2023. Available online: https://waterdata.usgs.gov/nwis (accessed on 7 March 2023).
29. National Water Quality Monitoring Council. Water Quality Portal. 2023. Available online: https://www.waterqualitydata.us/ (accessed on 15 March 2023).
30. Wilcoxon, F. Individual Comparisons by Ranking Methods. *Biom. Bull.* **1945**, *1*, 80–83. [CrossRef]
31. R Core Team. R: A Language and Environment for Statistical Computing. R Foundation for Statistical Computing. 2023. Available online: https://www.R-project.org/ (accessed on 8 February 2023).
32. Federal Interagency Sedimentation Project. Best practices for FISP Bag Sampler Intake Efficiency Tests and Operational Velocities; 2013. Federal Interagency Sedimentation Project Memorandum 2013.01. Available online: https://water.usgs.gov/fisp/docs/FISP_Tech_Memo_2013.01.pdf (accessed on 12 June 2023).
33. Manaster, A.E.; Landers, M.N.; Straub, T.D. *Intake Efficiency Field Results for Federal Interagency Sedimentation Project Bag Samplers*; U.S. Geological Survey Open-File Report 2022–1036; U.S. Geological Survey: Reston, VA, USA, 2022; p. 27. [CrossRef]

applied sciences

MDPI

Article

Surrogate Method for Suspended Sediment Concentration Monitoring on the Alluvial Reach of the River Danube (Baja, Hungary)

László Vas [1,*] and Enikő Anna Tamás [2]

1 Lower Danube Valley Water Directorate, H-6500 Baja, Hungary
2 Faculty of Water Sciences, University of Public Service, H-6500 Baja, Hungary; tamas.eniko.anna@uni-nke.hu
* Correspondence: vas.laszlo@aduvizig.hu

Featured Application: Riverine suspended sediment monitoring.

Abstract: Sediment balance is essential for understanding changes in river morphology and ecosystems and related services depending on them. However, the currently used methods to quantify riverine sediment processes are not adequate enough. We have examined the sediment regime of the Danube River, particularly the suspended sediment yield. This parameter can be calculated based on stage or discharge using a suspended sediment yield rating curve; however, the uncertainty of this method can reach even 150%. The suspended sediment yield of a section does not only depend on processes that take place in the riverbed; thus, it cannot be described by only one easily measurable parameter. An integrated surrogate method based on turbidity registration is tested in order to determine suspended sediment yield on the lower Hungarian (sand-bed alluvial) reach of the Danube River. The near-bank turbidity is converted into suspended sediment concentration and then into suspended sediment yield. The turbidity is measured with a built-in turbidity probe, while the suspended sediment yield is determined with traditional methods (discharge measurement, suspended sediment sampling, laboratory processing, and calculation). The traditional and integrated surrogate methods are compared based on the results of the measurements, and different aspect correlations are established between flow parameters, turbidity, and suspended load. The results achieved with the integrated method are promising, but more measurements are required in order to refine the relationships in a broader interval.

Keywords: Danube; suspended sediment; surrogate; turbidity; suspended sediment yield; monitoring

Citation: Vas, L.; Tamás, E.A. Surrogate Method for Suspended Sediment Concentration Monitoring on the Alluvial Reach of the River Danube (Baja, Hungary). *Appl. Sci.* **2023**, *13*, 5826. https://doi.org/10.3390/app13105826

Academic Editors: Manousos Valyrakis, Oleg S. Pokrovsky, Gordon Gilja, Thomas Pahtz, Panagiotis Michalis and Oral Yagci

Received: 7 April 2023
Revised: 30 April 2023
Accepted: 4 May 2023
Published: 9 May 2023

1. Introduction

Knowledge of river sediment transport conditions and sediment balance is essential for river management professionals. Sediment balance is very important for understanding changes in riverine ecosystems, channel morphology, and related habitat resources. That is why it is necessary to know if a certain reach of a river is in a state of sediment equilibrium, accumulation, or deficit [1]. In the case that there are gauging stations along the river where regular sediment monitoring takes place, the changes of the sediment balance over time or the influence of tributaries and floodplains can be observed, which makes it possible to investigate sediment transport processes, deposition, and erosion between the stations. However, in the European Union, operational sediment monitoring is unfortunately not yet well developed. As stipulated by the Common Implementation Strategy (CIS) for the Water Framework Directive (WFD), "*sediments provide important ecosystem services, such as balancing riverine and coastline morphology, contributing to the connection between surface water and groundwater, increasing soil fertility, contributing to natural water purification, mitigating the negative effects of extreme flow events, etc., and the importance of properly managing sediment to reach the environmental objectives of the WFD, but also of many other EU policies, has been now well recognised."* Turbidity as one of the surrogate measuerements for suspended sediment concentation determination is recommended [2].

The importance of sediment monitoring, especially along the second largest river of Europe, the Danube River, is also shown by the fact that in the last decade, several renowned European institutions (e.g., the University of Natural Resources and Life Sciences in Vienna, the Budapest University of Technology and Economics, and the Faculty of Water Sciences of the University of Public Service in Baja (UPS FWS)), all situated along the Danube River, have investigated the issue in the frames of different projects (e.g., SEDDON, DanubeSediment) [3–5].

Suspended sediment can be described by a couple of parameters, including dry matter content (weight [mg, g] of dry matter in an arbitrary volume of water sample), sediment concentration (mass of sediment in a unit volume of water [mg/L, kg/m^3]), and sediment yield (mass of sediment passing through the selected cross-section during a selected period of time [kg/s, t/year]), while turbidity can be defined as the reduced transparency of water caused by the particles in it by scattering and absorbing the passing-through light (in the form of, e.g., NTU—Nephelometric Turbidity Unit, FTU—Formazin Nephelometric Unit or FNU—Formazin Nephelometric Unit). Turbidity is now widely used as a surrogate measurement method to estimate suspended load [6–8]. This method is based on the quasi-continuous measurement of turbidity, where the turbidity sensor measurements are calibrated with (traditional) suspended sediment sampling during several flood events [9]. In situ turbidimeters are widely used because of the advantages (quasi-continuous measurement, automatic data transfer) of them. Their main disadvantage is biological fouling, which can be minimized by built-in methods (brush) or by regular maintenance [8].

The principles of sediment transport have long been known; however, quick and efficient measurement and accurate direct or indirect calculation methods require further development. In Hungarian practice, the sediment transport conditions of a river at a cross-section are described by the sediment yield rating curves, which give the sediment yield as the function of the discharge (G = f(Q)). The sediment yield rating curves can be created based on the results of the simultaneous suspended sediment sampling and discharge measurement. The dates of the official sediment measurements are regulated by the measurement plan of the General Directorate of Water Management and the water management directorates. The investigated reach belongs to the Lower Danube-Valley Water Directorate (LDVWD). In recent years, on average, five measurements are made annually at predetermined dates. Depending on the hydrological situation, additional measurements are carried out, so the number of measurements at a gauging station in a year is from 5 to 12.

As a result of the current practice, suspended sediment sampling is regular but rare in Hungary, and it is not well harmonized to flood events. Sampling is done with different pump samplers, and laboratory analyses are subcontracted to different firms. Based on this, the sediment datasets in Hungary are of rather questionable quality [10]. Because of the above, the sediment transport conditions can only be described in an approximate way with the sediment yield rating curve created based on the results.

To illustrate this, the difference between the sediment yield rating curve and the measured points, or even the difference between the measured points at nearly the same stage or discharge, can be significant. For example, at the Dunaújváros station (Danube), at a discharge of approximately 3000 m^3/s, 80 kg and 250 kg of suspended sediment can as well be measured.

Due to the above, the sediment yield calculated based on the sediment yield curve is fraught with uncertainties. The identification of the changes in the sediment regime of major rivers is heavily dependent on the availability of reliable data [8]. In the case of major rivers around the world (and on the Danube River as well), only a few sediment samples are collected each year, so the data are usually aggregated over a long time in order to create a suspended sediment yield rating curve for that period [11].

Therefore, it is necessary to have a method that provides more reliable data faster than the currently used one, with less resources (no direct sampling costs, i.e., boat and staff, no laboratory needs) as well as a better temporal resolution. There are many good practices in international literature [12], as well as examples from Hungary [13–15], but nothing like that along the alluvial sand-bed reach of the Danube.

Furthermore, in the regular sediment monitoring of the Water Authorities, the Baja gauging station (rkm 1479) was included in the past (1951–1965), but nowadays it is not. Thus, there is no regular sediment sampling, just occasional measurements in the frame of projects at this very important section, and one of the reasons why the current study focuses on this particular place is to re-establish sediment monitoring here. This gauging station is located on the left bank of the Danube at river kilometer 1479 (46°10′37″ N, 18°55′26″ E). The '0' point of the station is 80.99 m above the Baltic Sea's sea level. The water fluctuation range of the stage is 30–1000 cm, and discharge fluctuation is 900– m^3/s. The catchment of the station is 208,282 km^2.

Within the framework of the DanubeSediment Project, a number of scientific articles have been published, recommending good practices applicable to the entire reach of the Danube. An integrated method based on remotely registered turbidity measurement was recommended for the determination of the mass of sediment for the Danube River [4]. The essence of the method is that a sensor installed close to the riverbank continuously registers turbidity, which can be converted into a near-bank suspended sediment concentration using a calibration equation. The sediment concentration measured at a point close to the bank can be converted into a section mean concentration by using another calibration equation. By multiplying the section mean concentration with the discharge, the sediment yield of the entire section can be calculated, which, when integrated over time, results in the mass of the sediment [4,12]. There are some examples already regarding the successful application of the recommended method. The staff of the University of Natural Resources and Life Sciences (Vienna, Austria) have successfully established a multistage relationship between near-bank turbidity and the section mean concentration [12]. In Hungary, the staff of the Budapest University of Technology and Economics has also successfully applied the method on the gravel-bed Danube reach between Sződliget and Ráckeve (rkm 1674, rkm 1604), where a relationship was established between the suspended sediment concentration registered near-bank and the sediment yield of the cross-section [15]. The flowchart of the recommended method is shown in Figure 1.

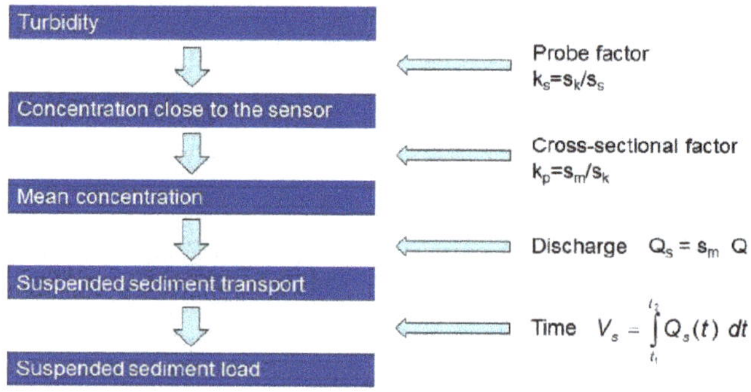

Figure 1. Determination of suspended sediment yield based on turbidity, based on [12]: k_s is the probe factor, which is the ratio of the concentration close to the probe (s_k) and the probe turbidity (s_s); k_p is the cross-sectional factor and can be calculated as the ratio of the mean suspended sediment concentration (s_m) and the concentration close to the probe (s_k).

2. Materials and Methods

2.1. Direct Sampling of Suspended Load

Until recent times, riverine suspended sediment data have been produced by gravimetric analyses performed on water sediment samples collected most of the time manually or, in the U.S., in some sites by automatic samplers. These methods tend to be expensive, difficult, labor intensive, and, under some conditions, e.g., during high floods, hazardous. Specialized equipment and considerable training are prerequisites for obtaining reliable samples and results [8]. However, in Hungary, direct sampling is still an integral part of the sediment monitoring standardization, and it stipulates sampling by a bottle (which is very inexact) or with a pump [16].

During pump sampling, the sample is brought to the surface (on the Danube River, aboard a measuring ship anchored in the vertical) with a pump. In international literature, pump sampling can as well be executed in a different way, when the sample at a monitoring station is taken automatically through a built-in pump. In Hungarian practice, the end of the pump suction tube is lowered into the sampling point using a suitable weight, and the sample of the required volume is sucked up. The disadvantage of the pumping method is that if the sampling velocity does not match the water velocity at the point, the sediment concentration of the sample will not, either. Depending on the relative sampling rate (sampling rate/water velocity), the difference in concentration can be −20–+60%. A deviation is also caused if the end of the intake is not parallel to the flow direction. The effects of divergence in various parameters were studied and published at the University of Iowa, U.S. [17].

Direct sampling methods can be further subdivided according to the number of sampling verticals, and the literature distinguishes between single- and multi-vertical methods. The location of the verticals in the case of the single-vertical method is determined based on different technical considerations (in the middle, at the maximum depth). In the selected vertical, the sampler is lowered and lifted up during the continuous sampling. Among the multi-vertical methods, the international literature recommends three: sampling in lamellae with the same water flow, lamellae of the same width, and lamellae of the same area. Sampling is carried out in a similar way for multi-vertical methods as for the single-vertical method (moving at the same speed); the difference is in the number of verticals. For the lamellae method with the same water flow, a vertical average sample should be taken in between 4 and 9 verticals, and for the method of lamellae of the same width, at least 10 verticals [18]. The Hungarian standard on suspended sediment sampling provides for a multi-vertical method where the lamellae are of equal width, but the average sample for a given vertical is to be prepared by taking 1 L of sample in the vertical at each of the 10 points of different depths, evenly distributed, and then pouring these samples together into one canister to form the vertical average sample with a volume of 10 L. In the case of the Danube, the standard prescribes for seven sampling verticals [16].

As stipulated by the Hungarian standard [16], "*during sampling it is very important to ensure that the sampling nozzle faces the flow, the pipe is not bent and to let enough time before taking samples to flush the pipe. Sampling needs to be carried out with care to adjust the revolutions per minute value (RPM) or the discharge of the pump for the velocity through the nozzle* V_{in} *should not differ much from the velocity of the flow* v *at the given point*

$$0.8\,v \leq V_{in} \leq 1.5\,v \tag{1}$$

In case the velocities are outside this range, the RPM of the pump should be accordingly adjusted, or a tap should be installed at the end of the pipe to ensure that intake velocities match. In order to determine intake velocity, the discharge of the pump (q_p) *has to be divided by the cross-section area of the nozzle* (f_n)

$$v_{in} = q_p/f_n \tag{2}$$

In practice, sampling is performed with a constant pumping discharge, assigning a fixed intake velocity to different velocity ranges of the flow, keeping the hydraulic coefficient between the values 0.8 and 2.0. This ensures a maximum 20% difference in concentrations, which is acceptable." [10]

For the purposes of the current study, direct sampling of the suspended sediment load was carried out in accordance with the relevant technical regulations. The method can be used to measure the suspended sediment yield of open watercourses occasionally, but not more than 1–2 times a day (from a boat or bridge). The method is applicable if the vertical mean velocity does not exceed 1.8 m/s and if the water depth at each sampling point is greater than 1 m [16].

Sampling was carried out simultaneously with discharge measurement based on the Doppler principle (using an ADCP instrument) with at least four crossings according to the relevant technical prescription [19]. After the discharge measurement, the same measuring group performed the suspended sediment sampling. As the samples belonging to one vertical are stored in one single container, during the laboratory analyses, the vertical average values of the sediment parameters are given. In some exceptional cases, when we also examined the differences in the sediment concentration/particle size distribution (PSD) within a vertical (e.g., for model calibration), we analyzed all the samples from each vertical individually, but later we averaged the concentration for each vertical in order to get similar results.

2.2. Laboratory Analyses

Before analysis, the samples were left to settle. During and after settling, the samples were stored in a place free of light and frost. After short storage, the turbidity of the samples was measured with a handheld turbidity meter. To measure turbidity, a Hach 2100QIS handheld turbidity meter was used, which determines turbidity based on the standard EPA 180.1. The instrument measures the 90° dispersion of the light transmitted through the sample in NTU. The turbidity meter can measure the turbidity of a 15 mL sample in the range from 0 to 1000 NTU. The correct results of the instrument can be checked regularly with a series of 6-part standard series of known turbidity and, if necessary, easily calibrated with the same series. During turbidity measurement, a sample is taken from the well-stirred vertical average sample with a smaller vessel, and then the instrument's cuvette is filled with it. After cleaning the cuvette and shaking it, it is inserted it into the instrument and the measurement is started. The turbidity is determined three times per vertical, taking new samples from the stirred vertical average each time. If there is a result significantly different from the average of the three measurements, a supplementary measurement is carried out. The turbidity of the vertical average sample will be the average of the three best measurement results.

After manual turbidity measurement, the sediment samples are settled, withdrawn to a volume of ~1 L, and sent to the laboratory, where their dry matter content and particle size distribution (PSD) curves are determined (PSDs are not discussed in the present study). The method of laboratory processing is regulated by the same [16] technical prescription as the sampling (it is prescribed that the dry matter contents of the samples must be determined after drying the samples at 105 °C for 24 h, and the PSDs of suspended sediments must be determined by a special settling device operating based on Stokes's law).

2.3. Sediment Yield Calculation

The basic equation for calculating the sediment yield is:

$$G = \sum_{i=1}^{n} q_i * c_i \quad (3)$$

where:

G—is the suspended sediment yield of the section [g/s]
n—is the number of sediment sampling verticals [pcs]
q_i—is the partial discharge of the sediment sampling lamella [m^3/s]

c_i—is the sediment concentration of the sediment sampling vertical [g/m^3]

The calculation shall be done for the lamellae of the sediment sampling verticals (bed width belonging to the sediment sampling vertical). A sediment measurement lamella consists of several discharge measurement lamellae; therefore, the discharge of the sediment sampling lamella is the sum of the partial discharges of the discharge measurement lamellae contained therein. The product of the partial discharge of the sediment measurement lamella and the sediment concentration of the sediment sampling vertical gives the sediment yield of the lamella. Summing up the sediment yields per lamellae, one can obtain the suspended sediment yield of the cross section (Figure 2, based on [20]).

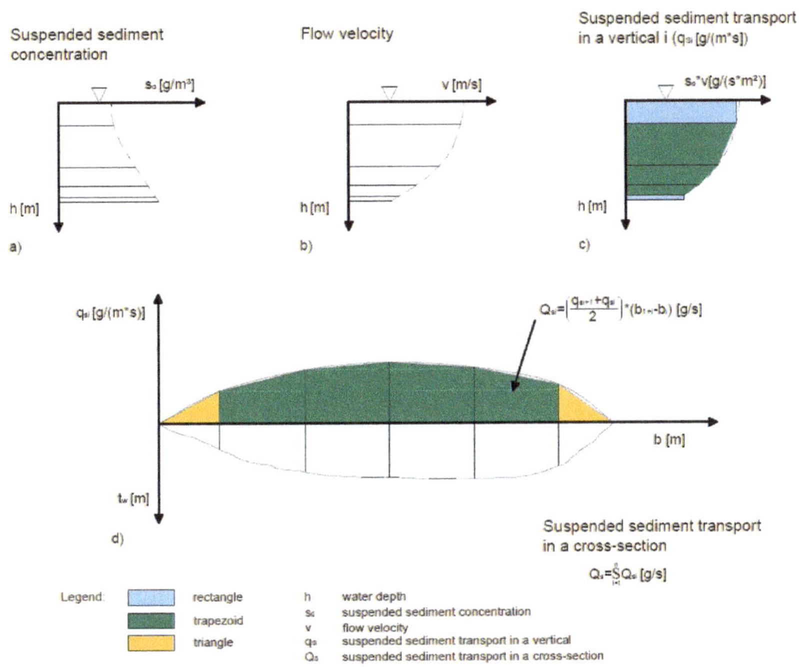

Figure 2. Calculation of the suspended sediment load, based on [20,21]. (**a**) shows the suspended sediment concentration in a vertical, (**b**) shows the flow velocity in a vertical, (**c**) shows the suspended sediment transport in a vertical and its calculation, (**d**) shows the suspended sediment transport in a cross-section and its calculation. h is the water depth, s_0 is the suspended sediment concentration, v is the flow velocity, q_{si} is the suspended sediment transport in a vertical, Q_s is the suspended sediment transport in a cross-section.

2.4. Indirect Measurement of Suspended Load

Sediment concentrations were as well determined indirectly, using in situ turbidity measurement. The turbidity of water expresses its reduced transparency, which is caused by particles in water, by scattering or absorbing light rays passing through water. The instruments used in daily use are based on the laws of nephelometry (nephelometers). Nephelometric measurement measures a 90° scattering of light in the visible or infrared range. Wedges measuring optical reflection measure light rays scattered in 140–165° in the infrared range. Such devices are also suitable for analyzing a sample of small volume (a few cm^3). The dispersion of light depends on the size of the particles in the sample, their color, and the shape of the particles [22]. The advantage of this method is that the probe can be fixed in a suitable place and its operation can be automated, which greatly increases the temporal resolution of sediment data. The disadvantage of the method is that turbidity depends on the size, composition, color, and shape of the sediment. The disadvantage of installation is that the accuracy can be greatly reduced by biofilm formation on the probe, so it is necessary to ensure continuous cleaning. Being an indirect method, calibration is required to convert turbidity into sediment concentration.

When using direct methods, data are generated only if the sampling team is physically present on site, takes the samples, and performs the necessary laboratory and processing work. The temporal resolution of data obtained by direct methods should not be arbitrarily low, as they depend on the available human resources and measurement conditions. Indirect methods are based on the use of a physical relationship, so each requires a calibration between the measured and the parameter being sought. An increased number of samplings is required until the calibration is established, after which it is sufficient to check the correctness of the calibration set up by sampling on a regular basis. However, the need for routine calibration is expected to diminish over time [8]. Indirect methods do not give information specific to the entire cross-section, only point information. A separate calibration is required for cross-sectional extension of point information. Another advantage of the indirect methods is that they can be remotely controlled, so they can provide data without on-site presence, at almost any temporal resolution.

To continuously detect turbidity, an OTT Hydrolab HL7 multiparameter probe was used. The parameters measured by the probe can be selected at the time of purchase, and the manufacturer offers a total of 13 sensors for the probe. The probe operated in this specific research measures turbidity, α-chlorophyll, and water temperature. The turbidity sensor can measure turbidity in the range of 0–3000 NTU. Measurement is carried out according to ISO 7027 using light with a wavelength of 880 nm (infrared). Since the measurement accuracy of optical turbidity meters is significantly impaired by the biofilm formed on the sensor, the probe is equipped with a central brush that keeps the instrument clean at all times.

The operation of the turbidity probe and the handheld turbidity meter differ based on the standards used. The probe measures optical reflections of infrared light according to ISO 7027, while the handheld turbidity meter measures 90° dispersion according to EPA 180.1. The analyzed sample is also different, the turbidity probe is in situ, while the handheld turbidity meter analyzes a disturbed sample several times.

The multiparameter probe is fixed and built as a remote registration station. The station was installed on the sand-bed alluvial reach of the Danube River in the ~1479.6 rkm section of the river on a floating pontoon next to the left bank (Figure 3) in January 2020. The support structure, consisting of a protective tube and an instrument box, is mounted on the downstream edge of the pontoon. The placement on the float ensures that the measuring point is always at a depth of ~1.5 m. The subsurface placement provides greater protection from driftwood. The probe is located in a protective basket at the bottom of the protective tube attached to the pontoon. The station consists of a probe, a data logger, additional electronics, and the supporting structure. The probe measures the parameters every hour and sends them to the data logger located in the instrument box, which transmits the data to the LDVWD, where the measured data can be seen with minimal delay. Thanks to the low power consumption of the devices, the installed 50 W solar panel and the 22 Ah buffer battery supply the system with power without problems.

The turbidity measured by the probe is checked by comparative measurements regularly, but at almost every flood wave. During the regular inspection, a sample is taken next to the probe, and then the turbidity of the sample taken is measured also with a hand-held turbidity meter, as described in Section 2.2.

Before carrying out further calculations, the adequacy of the values recorded by the probe was checked. According to the user information, the probe must be calibrated before the first use and at regular intervals. The calibration standard series consists of seven samples (0.1–4000 NTU), of which at least five are required for the Danube River (0.1–1000 NTU).

Figure 3. Automatic turbidity monitoring station (Danube, 1479.6 rkm, left bank, authors' photo).

The probe must therefore be calibrated to improve its measurement accuracy. In order to monitor and improve the measurement accuracy of the probe without the calibration series, water samples are regularly taken from the probe and the turbidity of the sample taken is measured with a handheld turbidimeter. The turbidity of a sample is measured three times and a sample is taken twice from next to the probe.

The correct functioning of the chlorophyll sensor has to be checked with campaign-like measurements as well. During the occasional sampling, the correct registration of chlorophyll content and water temperature is checked, and additional chemical parameters (pH, specific conductivity, dissolved oxygen content, and saturation) are also measured. These parameters are measured and checked for purposes not relevant to this study.

In the used method, the turbidities registered by the installed probe were corrected based on the control measurements made with the handheld turbidity meter. Based on the reliable relationship between the turbidity and the sediment concentration, the corrected turbidity was converted into a near-bank suspended sediment concentration. Using the relationships established based on the simultaneous suspended sediment measurements and near-bank turbidity measurements, the sediment yield along the section was calculated by using a relationship between it and the near-bank sediment concentration (near-bank SSC in Figure 4). A flowchart of the method used is shown in Figure 4.

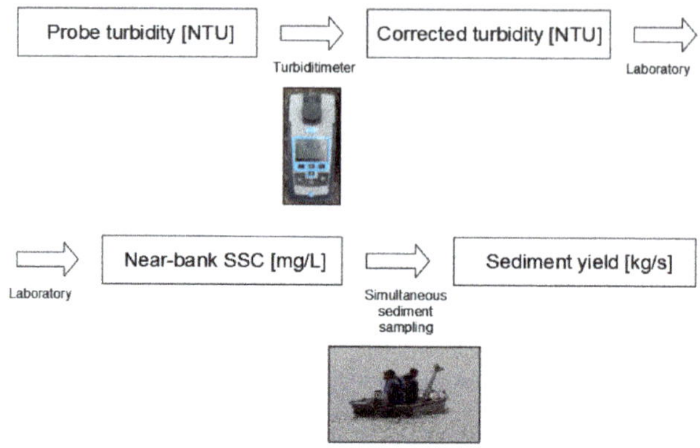

Figure 4. Determination of the suspended sediment load based on turbidity—the applied method.

3. Results

3.1. Turbidity-Suspended Sediment Concentration Relationship

One of the essential steps of this method is to establish the relationship between the turbidity and the suspended sediment concentration. This relationship was set up based on more than 1000 measurements between 2018 and 2022 that were carried out in the frame of regular monitoring at the LDVWA. The measurements were done for four sections on the alluvial reach of the Danube (Dunaújváros, Dombori, Baja, and Mohács). Based on the results, the relationship between the turbidity and the suspended sediment concentration is very strong for each location (R^2 = 0.92–0.99), and overall, as well (R^2 = 0.95); furthermore, it was determined that the relationship in those four cases was insensitive for the location. Accepting these results, the turbidity measured by the handheld turbidimeter can be converted into suspended sediment concentration in a reliable way [23]. The suspended sediment concentration as the function of the turbidity can be seen in Figure 5.

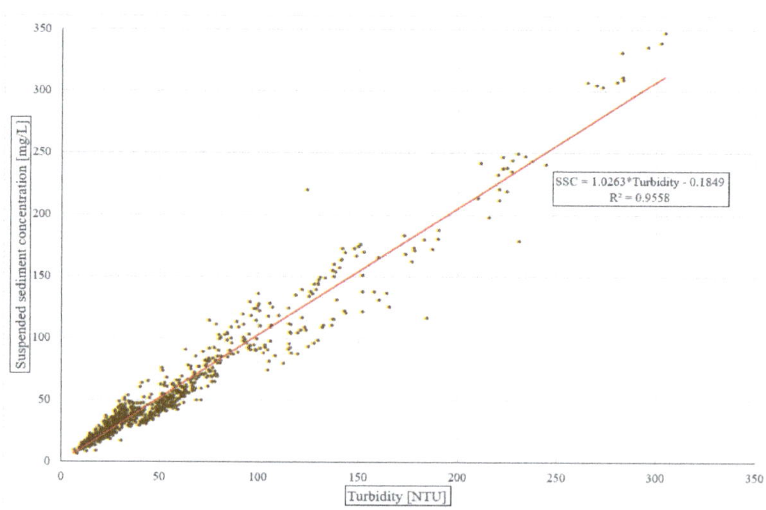

Figure 5. Turbidity versus suspended sediment concentration.

3.2. Accuracy of Probe-Recorded Turbidites

Since the installation, 28 control measurements have been carried out, of which 23 have been turbidity comparisons. Based on the measurement results, it can be seen that under

50 NTU turbidimeter turbidity the relative accuracy of the probe is low, as it measures 50–125% of the turbidimeter turbidity. The turbidity range below 50 NTU is in case of low discharge conditions with a low amount of suspended sediment, as well. Because of this, a larger relative error in this range will not cause a significant error in absolute terms. In the turbidity range above 50 NTU, the probe measures 60–90% of the handheld turbidity. The difference between the turbidity measured by the turbidimeter and the probe may be because of the difference between the measurement methods. The difference can also be caused by the fact that the probe is an in situ device, while the turbidimeter analyzes a small volume of sample which is repeatedly disturbed.

In Figure 6, the turbidity measured with the turbidimeter is plotted as a function of the turbidity measured by the probe. Based on the equation of the regression line, the turbidity measured by the probe can be corrected.

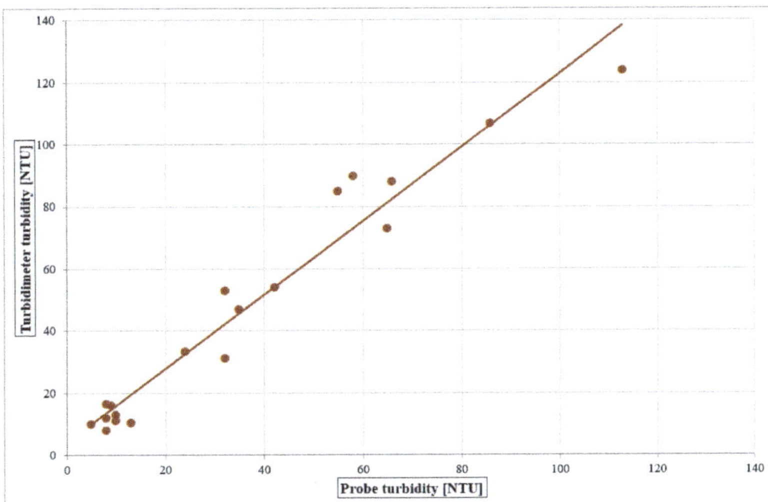

Figure 6. Turbidimeter turbidity versus probe turbidity relationship.

The correlation between the turbidity meter and the turbidity probe turbidities can be expressed as follows:

$$\text{Turbidimeter turbidity} = 1.1893 * \text{Probe turbidity} + 3.9477 \quad (4)$$

with a correlation coefficient squared $R^2 = 0.95$, which can be considered a rather good correlation.

3.3. *Water Level versus Suspended Sediment Concentration*

In order to find a relationship between water level (changes) and turbidity, the relationship between the two parameters in the time series was sought. The recorded turbidity as a function of water level was given for the floods listed in Table 1.

Several conclusions can be drawn based on the water level and sediment concentration time series, as well as the peak water level and sediment concentration. It is assumed that sediment concentration (similarly to discharge) peaks after the water level. It can be observed in the time series that the water level and sediment concentration do not always peak at the same time. Comparing the water levels and the sediment concentrations, it can be concluded that floods with approximately the same peak water level do not peak with the same sediment concentration. It can be also observed that peak values do not follow each other consistently: the peak water level of 400 cm had a peak concentration of 60 mg/L, but two weeks later the next flood peaked with a lower, 350 cm water level and a higher, 100 mg/L sediment concentration. The examined floods also include ones where there has been no increase in concentration during the entire duration of the flood.

Table 1. Attributes of the analyzed flood waves [23].

Period	Peak Water Level [cm]	Peak Discharge [m^3/s]	Peak Turbidity [NTU]
February 2020	639	4450	150
June–July 2020	574	4000	50
June–July 2020	483	3350	25
June–July 2020	400	2800	30
July–September 2021	668	4700	230
July–September 2021	540	3750	70
July–September 2021	496	3450	55
July–September 2021	397	2800	60

Figure 7 shows the sediment concentration as a function of water level for three periods from Table 1. The three biggest floods peaked with water levels between 570 and 670 cm, but the peak sediment concentrations are not proportional to the peak water level, as they were between 50 and 230 mg/L. In addition, in the case of the two largest flood waves, a hysteresis of sediment concentrations (similar to discharge hysteresis) can be observed. The shape of the curves for each flood wave are different. The hysteresis curve of the 2020 February flood wave has the shape of an asymmetrical 8. The next flood wave, which occurred in June and July 2020, does not have an exact shape nor a significant rise in the suspended sediment concentration, while the flood wave of June and July 2021 has a shape of an open 8. The higher suspended sediment concentrations occurred at the falling limb in the case of these flood waves.

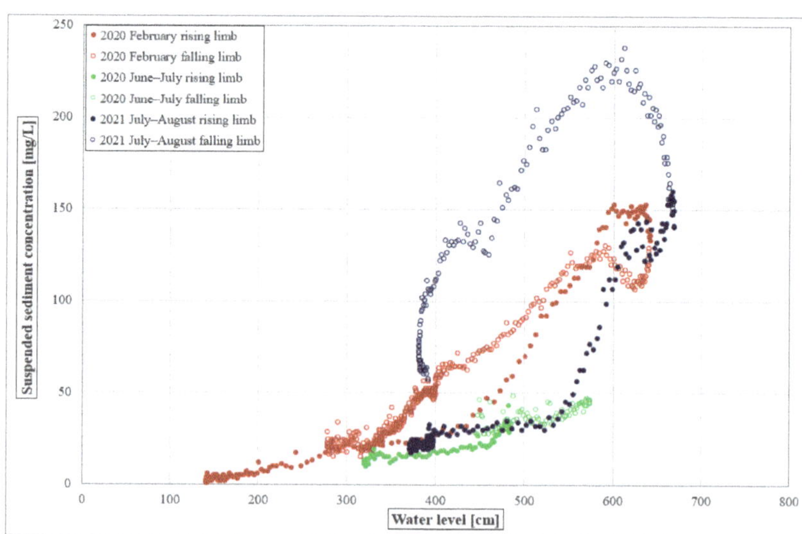

Figure 7. Suspended sediment concentration as a function of water level in the case of three flood waves, with the separation of rising and falling limbs, based on [23].

According to [24], the main factors influencing sediment hysteresis in the case of this particular gauging station of the Danube River may be the magnitude and sequence of events, PSD, and sediment source. As in the discussed analysis, these latter two factors were not investigated, and at the moment, no explanation is sought for the described hysteresis phenomena. However, sediment dynamics are complex, and the hysteresis patterns may be linked to many other factors as well.

3.4. Discharge and Sediment Sampling Measurement

The results of the discharge and sediment sampling measurements for the establishment of a relationship between turbidities and the total suspended load can be seen in Table 2. The measurements were performed by LDVWD and UPS FWS with standard methods (pump sampling and ADCP discharge measurement) in the relatively low water period of the years 2021–2022, when only a few medium-sized flood waves happened (Figure 7).

Table 2. Discharge and sediment yield.

Date	Water Level [cm]	Discharge [m^3/s]	Sediment Yield [kg/s]	Hydrological Condition	Performer
11 February 2020	546	3722	445	Falling	LDVWD
13 May 2020	171	1633	35	Stagnating	LDVWD
12 June 2020	219	1914	30	Rising	LDVWD
23 June 2020	490	3500	161	Rising	LDVWD
29 June 2020	422	2620	194	Falling	UPS FWS
8 July 2020	444	3098	113	Falling	LDVWD
10 August 2020	580	3979	345	Falling	LDVWD
4 February 2021	513	3713	265	Falling	LDVWD
6 May 2021	278	2223	62	Rising	LDVWD
25 May 2021	523	3587	136	Falling	LDVWD
7 September 2021	442	2900	300	Falling	UPS FWS

In order to provide a picture of the water regime of the Danube River, the hydrograph of the Baja gauging station for the investigated period is provided, with indication of the measurements (Figure 8).

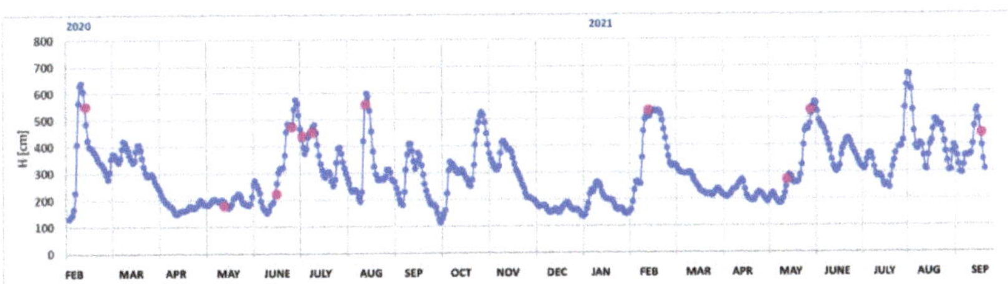

Figure 8. Hydrograph of the Baja station in the investigated period. Water level time series is marked with blue and measurement days are indicated as purple points.

The suspended sediment yield rating curve (suspended sediment yield as the function of the discharge) based on the data above is shown in Figure 8. The fit of the regression to calculate the suspended sediment yield based on the discharge is medium. The correlation coefficient squared is $R^2 = 0.59$ for the exponential curve with Equation (5).

$$\text{Suspended sediment yield} = 11.149 * e^{0.0009*\text{Discharge}} \tag{5}$$

As a comparison, the suspended sediment yield rating curve based on the old data (1951–1965) is also shown in Figure 9. The correlation coefficient squared is $R^2 = 0.76$, which indicates a good relationship between the discharge and the suspended sediment yield.

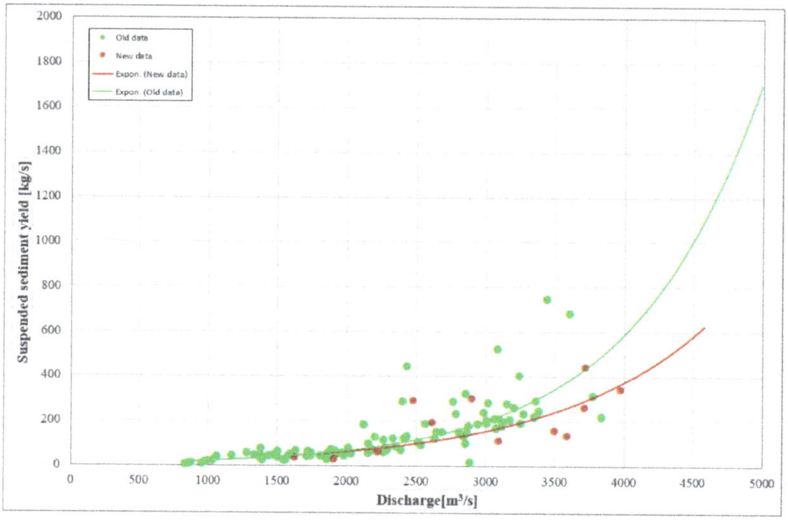

Figure 9. Suspended sediment yield rating curve for the old (1951–1965) and the new (2020–2021) measurements.

3.5. Near-Bank Suspended Sediment Concentration versus Suspended Sediment Yield

In the final step of the analysis, a relationship between the near-bank suspended sediment concentration and suspended sediment yield was sought. The near-bank suspended sediment concentration was calculated based on the probe turbidity corrected by the probe-turbidimeter relationship.

Since the installation of the probe, 11 sediment yield measurements have been carried out. Based on the results so far, there is a usable relationship between the two quantities. Based on the literature available [12,15,25], two straight lines were fitted to the measurement results, as the slope of the regression line is significantly decreasing around a sediment concentration of 80 mg/L. The fit of the regression line in the lower range is very strong, $R^2 = 0.90$. Due to the low number of measurements (three), the goodness of the fit of the regression line in the higher range is approximate. This approximate fit is very strong as well, $R^2 = 0.85$. The regression lines fit well for the rising limb, falling limb, and for the stagnating conditions, as well. The relationship between the near-bank suspended sediment concentration and the suspended sediment yield based on the findings of this study is shown in Figure 10, while the comparison between the correlations with discharge for the old and new data is shown in Figure 9.

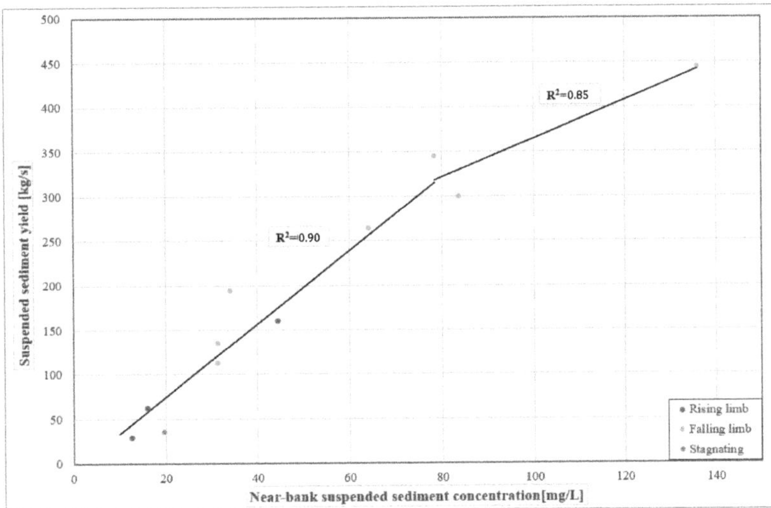

Figure 10. Suspended sediment yield as a function of near-bank suspended sediment concentration.

4. Discussion

4.1. Water Level versus Sediment Concentration

Based on the analysis of the simultaneous water level and sediment concentration data, it can be concluded that an unequivocal relationship cannot be established between the water level (or discharge) and the sediment concentration. The equivocal relationship can be traced back to a number of reasons, which may be more on a watershed scale. The sediment concentration during a flood wave may depend on the factors triggering it (snowmelt, precipitation). The concentration may also depend on which subbasin the flood wave comes from and the characteristics of this subbasin. In addition to the above, the amount of suspended sediment is also influenced by the level of the flood wave: the flood wave can descend at such a high level that it washes out the silt deposited in the side-branches or by reaching the floodplain, and the deposited silt may be washed into the bed as well. The arriving flood may also increase the amount of sediment being transported by tearing up the bed armor formed between floods. The succession of the floods may also affect the sediment concentration, since the first flood wave can wash out the deposited silt from the side-branches or tear up the bed armor (thus increasing the sediment concentration); subsequent flood waves can no longer do this. Because of these, the sediment regime of a river can usually not be described with the traditional relationship ($G = f(Q)$) due to its complexity [26].

There is an amount of international literature on the direction of hysteresis of water level and suspended sediment concentration (or discharge and turbidity) [27–29]. In the research presented in [27], simultaneous flow and turbidity data of several flood waves were examined in medium-sized river basins, where several hysteresis samples (clockwise, counter-clockwise, eight-shaped, and complex) were distinguished, and the different hysteresis patterns were explained by the factors triggering the flood wave, the number of sediment sources, the distance of sediment sources from the examined cross-section, and the mobility of individual sediment sources.

4.2. Suspended Sediment Yield Rating Curve

The accuracy of the suspended sediment rating curves must be checked, regardless of the regression coefficients. The goodness of them was checked by calculating the difference between the measured value and the value calculated by the usage of the curve equation.

In the case of the 2020–2021 measurement campaign, the average difference between them was 52% in relative terms and 88 kg/s in absolute terms. In half of the cases (6 out of 11), the relative difference was above 50%, and in five cases, the difference was equal to or above 100 kg/s.

By analyzing the historical data, it can be concluded that based on the regression coefficient ($R^2 = 0.76$) or the average relative and the average absolute difference (31%, 46 kg/s), this relationship is much more accurate than the one based on the recent measurement. However, the calculation error in the discharge range above 2000 m^3/s is higher with an average value of 72 kg/s. About 22% of the differences (13 out of 58) in this range are above 100 kg/s.

Based on these, the accuracy of the suspended sediment yield rating curve is not adequate enough to perform accurate calculations about the annual sediment regime.

4.3. Near-Bank Suspended Sediment Concentration versus Suspended Sediment Yield

Two regression lines were fitted for the simultaneous data of the near-bank suspended sediment concentration and the suspended sediment yield. The accuracy of the relationship between the near-bank suspended sediment concentration and the suspended sediment yield (R^2 = 0.85–0.90, Figure 9) is better than the suspended sediment yield rating curve based on the discharge either in the case of historical or new data (R^2 = 0.42–0.57, Figure 8).

The breakpoint of the slope between the two regression lines will be more accurately marked by increasing the number of measurements. In the article [15] published in 2020, the authors defined the breakpoint between the regression lines at around 30 mg/L suspended sediment concentration for the Danube reach between Sződliget and Ráckeve. There is a relatively long distance between these stations and Baja (130–190 km), the PSD of the transported sediment is rather different, and the difference between the breakpoints (breakpoint of Baja is at 80 mg/L) can also be caused by the different geometries of the two cross-sections.

The results so far provide a good starting point; however, 11 measurements are not yet enough to establish reliable relationships covering all ranges. In the future, the measurements will continue in order to refine the relationships. After the establishment of an acceptable relationship which covers all the ranges, this integrated method will be compared with the sediment yield rating curve method by determining and comparing the annual and flood wave sediment regimes. Further tests will be carried out about the suitability of the selected gauging station (Baja), as during the measurements made so far, the water level of the Danube River did not exceed the point of overflowing the banks of the main riverbed (400 m wide), so the effects of the floodplain (7 km wide) on the sediment transport in the case of the integrated method could not yet be detected.

Overall, the prospect of large-scale applications of proven suspended sediment surrogate technologies is a revolutionary concept in fluvial sediment monitoring technologies with important future benefits, providing for safer, more frequent and consistent, arguably more accurate, and ultimately less expensive fluvial sediment data collection [8], which underlines the importance of the presented research.

Author Contributions: Conceptualization, L.V; methodology, L.V.; formal analysis, L.V. and E.A.T.; resources, L.V. and E.A.T.; writing-original draft preparation: L.V. and E.A.T.; writing review and editing, L.V. and E.A.T.; visualization, L.V. and E.A.T.; supervision, L.V. and E.A.T. All authors have read and agreed to the published version of the manuscripts.

Funding: This research was funded by General Directorate of Water Management, H-1253 Budapest (probe), Lower Danube Valley Water Directorate, H-6500 Baja (sediment samplings) and the University of Public Service, Faculty of Water Sciences, H-6500 Baja (sediment samplings). The station was designed by the General Directorate of Water Management and was built by the Lower Danube Valley Water Directorate and VITUKI HUNGARY Kft, H-1097 Budapest. Measurements of UPS FWS were carried out in frame of a research contract in the HOMER project and during higher education teaching of field measurement techniques. The funding from General Directorate of Water Management and Lower Danube Valley Water Directorate were institutional.

Institutional Review Board Statement: Not applicable.

Informed Consent Statement: Not applicable.

Data Availability Statement: Data used for this article can be requested via e-mail from the authors.

Acknowledgments: The discharge measurements and sediment sampling were performed by the measurement team of the LDVWD and UPS FWS, respectively.

Conflicts of Interest: The authors declare no conflict of interest.

References

1. Griffiths, R.E.; David, J.T. Importance of measuring discharge and sediment transport in lesser tributaries when closing sediment budgets. *Geomorphology* **2017**, *269*, 59–73. [CrossRef]
2. Common Implementation Strategy for the Water Framework Directive—Integrated Sediment Management Guidelines and Good Practices in the Context of the Water Framework Directive. Available online: https://environment.ec.europa.eu/system/files/2022-09/CISdocumentsedimentfinalTO_BE_PUBLISHED_1430554724.pdf (accessed on 15 March 2023).
3. Baranya, S.; Józsa, J.; Török, G.; Kondor, G.T.; Ficsor, J.; Mohácsiné, S.G.; Habersack, H.; Haimann, M.; Riegler, A.; Liedermann, M.; et al. A Duna hordalékvizsgálatai a SEDDON osztrák-magyar együttműködési project keretében. *J. Hungar. Hydrol. Soc.* **2015**, *95*, 41–46, (in Hungarian with English abstract).
4. Habersack, H.; Baranya, S.; Holubova, K.; Vartolomei, F.; Skiba, H.; Schwarz, U.; Krapesch, M.; Gmeiner, P.; Haimann, M. *Danube Sediment Management Guidance, Output 6.1 of the Interreg Danube Transnational Project Danube Sediment*; European Commission: Vienna, Austria, 2019.
5. Habersack, H.; Baranya, S.; Holubova, K.; Vartolomei, F.; Skiba, H.; Schwarz, U.; Krapesch, M.; Gmeriner, P.; Haimann, M. *Sediment Manual for Stakeholders. Interreg Danube Transnational Programme Danube Sediment Project*; Institute of Hydraulic Engineering and River Research: Vienna, Austria, 2019.
6. Brasington, J.; Richards, K. Turbidity and suspended sediment dynamics in small catchments in the Nepal Middle Hills. *Hydrol. Process.* **2000**, *14*, 2559–2574. [CrossRef]
7. A Lenzi, M.; Marchi, L. Suspended sediment load during floods in a small stream of the Dolomites (northeastern Italy). *Catena* **2000**, *39*, 267–282. [CrossRef]
8. Gray, J.R.; Gartner, J.W. Technological advances in suspended-sediment surrogate monitoring. *Water Resour. Res.* **2009**, *45*, W00D29. [CrossRef]
9. Navratil, O.; Esteves, M.; Legout, C.; Gratiot, N.; Nemery, J.; Willmore, S.; Grangeon, T. Global uncertainty analysis of suspended sediment monitoring using turbidimeter in a small mountainous river catchment. *J. Hydrol.* **2011**, *398*, 246–259. [CrossRef]
10. Tamás, E.A.; Ficsor, J. Questions in the quantitative analysis of sediment load—Example of three major rivers in Hungary. *E3S Web Conf.* **2018**, *40*, 04023. [CrossRef]
11. Walling, D.E. The changing sediment loads of the world's rivers. *Ann. Wars. Univ. Life Sci. SGGW Land Reclam.* **2008**, *39*, 3–20. [CrossRef]
12. Habersack, H.; Liedermann, M.; Trirrhart, M.; Haimann, M.; Kreisler, A. Innovative approaches in Sediment Transport Monitoring and analysis concept. In Proceedings of the 35th IAHR World Congress, Chengdu, China, 8–13 September 2013.
13. Kutai, R. *Pontbeli Zavarosságmérésen Alapuló Hordalékhozam Becslő Eljárás Implementálása a Dunára, Tudományos Diákköri Konferencia*; Budapest University of Technology and Economics: Budapest, Hungary, 2014; (In Hungarian with English abstract).
14. Pomázi, F.; Baranya, S.; Török, G. Nagy folyók lebegtetett hordalékvándorlásának új vizsgálati módszerei 1—A továbbfejlesztett hordalékmonitoring módszertan bemutatása. *J. Hungar. Hydrol. Soc.* **2020**, *100*, 37–47, (In Hungarian with English abstract).
15. Pomázi, F.; Baranya, S. Nagy folyók lebegtetett hordalékvándorlásának új vizsgálati módszerei 2—Közvetlen és közvetett lebegtetett hordalékmérési eljárások összehasonlító vizsgálata. *J. Hungar. Hydrol. Soc.* **2020**, *100*, 64–73. (In Hungarian with English abstract)
16. *ME-10-231-20:2009*; Felszíni vizek lebegtetett hordalékának mérése szivattyús mintavevővel. Környezetvédelmi és Vízügyi Minisztérium: Budapest, Hungary, 2008. (In Hungarian)
17. *Laboratory Investigation of Suspended-Sediment Samplers*; Interageny Report No. 5; Hydraulics Laboratory, Iowa University: Iowa, IA, USA, 1941.
18. Gray, J.R.; Landers, M.N. *Measuring Suspended Sediment. Comprehensive Water Quality and Purification*; US Geological Survey: Reston, VA, USA, 2014; pp. 159–204. [CrossRef]
19. *ME-10-231-17:2009*; Felszíni Vizek Vízhozamának Mérése ADCP Berendezéssel. Környezetvédelmi és Vízügyi Minisztérium: Budapest, Hungary, 2008. (In Hungarian)
20. *VMS 251/8-81*; Hidrológiai Mérési Adatok Elsődleges Feldolgozása: A Lebegtetett Hordalék Adatai, OVH Vízügyi Szabványosítási és Egységesítési Központ. MSZH Kiadó: Budapest, Hungary, 1981. (In Hungarian)
21. Haimann, M.; Gmeiner, P.; Liedermann, M.; Aigner, J.; Kreisler, A.; Riegler, A.; Blamauer, B.; Baranya, S.; Török, G.T.; Ficsor, J.; et al. *Hordalékvizsgálati Kézikönyv*; Research report; SEDDON Project; 2014. (In Hungarian with English abstract)
22. Sutherland, T.; Lane, P.; Amos, C.; Downing, J. The calibration of optical backscatter sensors for suspended sediment of varying darkness levels. *Mar. Geol.* **2000**, *162*, 587–597. [CrossRef]

23. Vas, L.T. A Duna lebegtetett hordaléktöménységének monitoringja fix telepítésű zavarosságmérő műszer alkalmazásával. *J. Hungar. Hydrol. Soc.* **2022**, *102*, 14–25. (In Hungarian with English abstract)
24. Malutta, S.; Kobiyama, M.; Chaffe, P.L.B.; Bonumá, B.N. Hysteresis analysis to quantify and qualify the sediment dynamics: State of the art. *Water Sci. Technol.* **2020**, *81*, 2471–2487. [CrossRef] [PubMed]
25. Haimann, M.; Liedermann, M.; Lalk, P.; Habersack, H. An integrated suspended sediment transport monitoring and analysis concept. *Int. J. Sediment Res.* **2014**, *29*, 135–148. [CrossRef]
26. Lohani, A.K.; Goel, N.K.; Bhatia, K.K.S. Deriving stage–discharge–sediment concentration relationships using fuzzy logic. *Hydrol. Sci. J.* **2007**, *52*, 793–807. [CrossRef]
27. Lannergård, E.E.; Fölster, J.; Futter, M.N. Turbidity-discharge hysteresis in a meso-scale catchment: The importance of intermediate scale events. *Hydrol. Process.* **2021**, *35*, e14435. [CrossRef]
28. Lewis, D.J.; Kenneth, W.T.; Randy, A.; Dahlgren, J.N. *Turbidity and Total Suspended Solid Concentration Dynamics in Streamflow from California Oak Woodland Watersheds 1*; Pacific Southwest Research Station: Albany, CA, USA, 2002.
29. Wymore, A.S.; Leon, M.C.; Shanley, J.B.; McDowell, W. Hysteretic Response of Solutes and Turbidity at the Event Scale Across Forested Tropical Montane Watersheds. *Front. Earth Sci.* **2019**, *7*, 126. [CrossRef]

applied sciences

MDPI

Article

Bedload Sediment Transport Estimation in Sand-Bed Rivers Comparing Traditional Methods and Surrogate Technologies

Philipe Ratton [1,*], Tobias Bernward Bleninger [1], Rodrigo Bahia Pereira [2] and Fábio Veríssimo Gonçalves [2]

[1] Post-Graduate Program on Water Resources and Environmental Engineering (PPGERHA), Federal University of Paraná (UFPR), Campus Centro Politécnico, Curitiba 81530-000, PR, Brazil
[2] Post-Graduate Program on Environmental Technology (PPGTA), Federal University of Mato Grosso do Sul (UFMS), Campo Grande 79070-900, MS, Brazil
* Correspondence: philipe.ratton@gmail.com

Abstract: Bedload sediment transport in rivers can cause impacts, such as bed erosion/deposition, sandbank formation and changes in flow capacity. Bedload sampling techniques have limitations related to spatial and temporal resolution. These constraints are more relevant in rivers with dunes and high sediment transport. This paper presents a comparison between bedload transport rates estimated with direct and indirect methods in a river with sand dunes. The case study area is a stretch of the Taquari River, in Brazil. Surveys were carried out on three consecutive days, during a flood season. A SonTek M9-ADCP with HydroSurveyor capabilities activated was used to simultaneously measure bathymetry and water velocities throughout a river reach, and also to perform moving-bed tests at six verticals along a predefined cross-section. A mechanical trap (Helley–Smith) was used to collect bedload samples at the same time and positions where the moving-bed tests were performed. Sediment transport was calculated and compared following different approaches: (1) ADCP-BT (Bottom Tracking); (2) modified ISSDOTv2 method (dune tracking); (3) HelleySmith mechanical trap; (4) and five empirical equations. The results showed good agreement between the methodologies, indicating the potential of using ADCPs for hydro sedimentological studies due to the advantages of integrating bathymetry, flow velocity and bedload data.

Keywords: bedload; sediment transport; ADCP; dunes; riverbed geomorphology

Citation: Ratton, P.; Bleninger, T.B.; Pereira, R.B.; Gonçalves, F.V. Bedload Sediment Transport Estimation in Sand-Bed Rivers Comparing Traditional Methods and Surrogate Technologies. *Appl. Sci.* **2023**, *13*, 5. https://doi.org/10.3390/app13010005

Academic Editors: Gordon Gilja, Manousos Valyrakis, Panagiotis Michalis, Thomas Pahtz and Oral Yagci

Received: 3 November 2022
Revised: 8 December 2022
Accepted: 15 December 2022
Published: 20 December 2022

1. Introduction

Bedload sediment transport in rivers can interfere with hydro-morphodynamic processes, such as bed erosion/deposition, sandbank formation and changes in flow capacity. When these processes become excessive and out of balance, they might affect economic activities, such as waterway navigation, harbor installation and operation, hydroelectric power generation and water supply. It is important to understand the hydro sedimentological behavior, quantifying bedload transport, in order to establish an effective management of sediments and water resources [1].

Several bedload sampling techniques are cited in the literature. Conventional sampling procedures are based on mechanical traps deployed on the river bottom to collect sediment samples at different positions along a cross-section, during a certain time interval [2]. However, these techniques present limited spatial and temporal resolutions and high uncertainties [3]. Conventional sampling is technically difficult, time-consuming, expensive and dangerous during floods. These constraints are even more relevant in rivers with dunes and high sediment transport, as the position of the sampler on the river bottom and the heterogeneous bedload transport affect the measurements [4,5].

Surrogate technologies, such as acoustic Doppler current profilers (ADCP), provide faster and safer measurements with better resolution [6–8]. ADCPs were initially developed with the aim of computing water velocities and discharges. However, recent research indicates the ADCP's potential to estimate bedload rates through a feature called bottom tracking (BT) [8–10].

Bottom tracking consists of sending acoustic pulses from slanted ADCP beams that reflect on the river bottom and return to the sensor. The Doppler shift in the frequency of the return pulse is related to the boat's velocity (when the instrument is mounted on a moving boat). In rivers with a moving-bed (i.e., sediment particles in motion near the bottom), static measurements (named moving-bed tests) allow for computing spatially averaged bed particle velocities. Thereafter, bedload transport rates can be estimated from the product between moving-bed velocities, thickness of the bedload layer and sediment concentration in the bedload layer. This method is herein referred to as ADCP-BT.

Bedload may consist of fine (clays, silts and sand) to coarse (gravels and cobbles) sediment particles. The entrainment and transport of gravel beds are normally more intermittent than sand beds, due to their larger and typically less sorted particle size, providing a more spatially and temporally heterogeneous transport. Nevertheless, the supposed more homogeneous transport in sand-bed rivers tends to be restricted and local due to the transport of bed material as bedforms (ripples and dunes), which increases the spatial-temporal heterogeneity of bedload rates [11].

Additionally, to the application of ADCPs to estimate moving-bed velocities and hence bedload transport rates, the instrument with activated HydroSurveyor capabilities (in the case of SonTek M9-ADCP model) allows the collection of synchronized data of bathymetry and flow velocities, where not only the vertical beam, but also four slanted beams are used to obtain depth information. After detailed and consecutive surveys along a river reach, it is possible to analyze longitudinal bathymetric profiles, and map aggradation and degradation areas. This approach is known as dune tracking, since it evaluates dune migration rates, and has been successfully employed to estimate bedload transport [5,12,13].

Abraham et al. [14] developed a method using the dune tracking approach, which was named ISSDOTv2 (Integrated Section Surface Difference Over Time, Version 2). This method computes bedload based on the difference in time-sequenced three-dimensional bathymetric data. Scour volumes calculated from the difference plots of bed level are related to the average bedload transport due to the dune's movement. A systemic bias inherent in the methodology is related to dune migration not being captured due to insufficient temporal sampling frequency, which tends to underestimate bedload transport as the time interval between consecutive surveys increases. Shelley et al. [15] implemented a procedure to correct scour volumes and bedload estimates using information about dune characteristics (wavelength, height and celerity) and the time interval between the surveys.

Currently, there is no standard procedure for the identification and discrimination of geometric bedform characteristics. Many researchers apply different methodologies, obtaining variable results, which confirms the present lack of a consistent and nonarbitrary quantitative descriptor of bed morphology. With the aim to fulfill this gap, Gutierrez et al. [16] proposed the application of robust spline filters and continuous wavelet transforms to discriminate these morphodynamic features, allowing for a quantitative recognition of bedform hierarchies. This methodology was applied in this work to reduce subjectivity in establishing the parameter values needed in the ISSDOTv2 correction term (proposed by Shelley et al. [15]).

The present paper shows a comparison between different direct and indirect methods applied to estimate bedload transport rates in a river with sand dunes, during a flood season. Two approaches were carried out using an ADCP: (1) moving-bed tests using the bottom tracking feature; and (2) dune tracking. Moreover, a Helley-Smith (HS) mechanical trap was used to collect bedload samples at six verticals along a predefined cross-section. The field measurements were carried out during one campaign (31 January 2018 to 2 February 2018), on a daily basis, obtaining pairs of data sets for each methodology. Additionally, five empirical transport equations were used with hydraulic and sedimentometric parameters averaged for the river cross-section: Einstein [17], Einstein-Brown [18], Kalinske [19], Meyer-

Peter-Müller [20] and Van Rijn [21]. Bedload rates were compared to evaluate the accuracy of all methodologies.

2. Study Site

The Taquari River drains a catchment area of 65,000 km^2 and is located in the midwestern region of Brazil, in the Upper Paraguay Basin, whose watershed area is approximately 600,000 km^2, spreads across three countries (Brazil, Bolivia and Paraguay) and includes Pantanal, one of the world's largest freshwater wetlands (140,000 km^2 [22]). Known for its biodiversity, Pantanal is considered a World Natural Heritage site by UNESCO, reinforcing the importance of better understanding sediment dynamics in the region.

The study site is located in the upper reach of the Taquari River (18°31′36.26″S, 54°43′56.90″W), in the city of Coxim, Brazil. In Figure 1, P1 is the monitored cross-section within the surveyed river reach, P2 is a control point at the Coxim River and P3 is a stream gauging station called Coxim (code 66870000, in the Brazilian National Water and Sanitation Agency) at the Taquari River. Historical data available at P3 include water level, water discharge and cross-section bathymetric profiles.

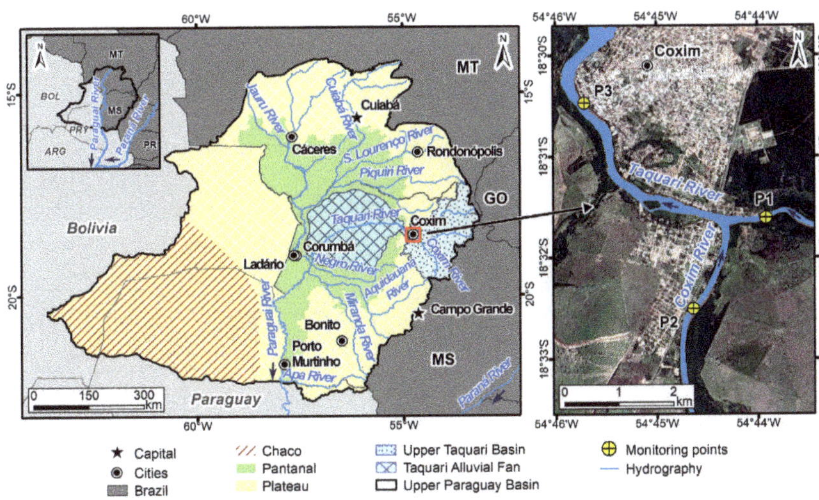

(a) (b)

Figure 1. (**a**) Upper Paraguay Basin, Pantanal wetland and study site at Taquari River in Coxim, Brazil; (**b**) points of measurement in Coxim: P1 and P3 at Taquari River, P2 at Coxim River.

Sediment transport (in suspension and bedload) in the study area is high during floods, increasing the morphology dynamics and frequently altering the bedforms geometries. At the Coxim gauging station (P3, located 4.5 km downstream of the study site and after the confluence with the Coxim River), the drainage area is 27,600 km^2 and the annual average discharge is 350 m^3/s. Floods normally occur during the summer (December–March), with maximum discharges higher than 1600 m^3/s at extreme events. During the drought season (July–September), flow is near 250 m^3/s. The annual average rainfall in the region is 1230 mm/year, and the monthly average rainfall during the summer is 200 mm/month.

At the monitored cross-section (P1), during the field campaign, in a flood season, the river width was approximately 110 m, mean water depth was near 2.0 m, mean flow velocity was 1.2 m/s and the water level slope was 15–25 cm/km. The riverbed is mobile, and bedload is formed by fine and medium sand (125 μm < d < 500 μm). Along the 1.0 km surveyed reach, upstream of the confluence of the Taquari and Coxim Rivers, the morphology is dynamic, with the interaction of bars, ripples and dunes of varying sizes.

3. Materials and Methods

3.1. Field Measurements

Hydrodynamics, bedload transport and morphological changes were monitored using an acoustic Doppler current profiler (SonTek M9-ADCP) on 3 consecutive days in the first trimester of 2018. An external real-time kinematic global positioning system (RTK-GPS) was coupled with the ADCP to improve vessel positional precision (subdecimetric in horizontal). The RTK system used was formed by two Leica 1200 GX GNSS dual frequency GPS receivers (L1 and L2) and a Pacific Crest PDL radio. The GPS mobile station was attached to the ADCP through a metal pole, directly above the ADCP sensor. The GPS base station was fixed at a random point on the right margin (approximately 15 m away from the riverbank), tracked throughout the field campaign.

The M9-ADCP was applied to measure water discharges at cross-section P1, following procedures regarding quality control [23]. Static ADCP moving-bed tests were also performed at 6 verticals locations along the cross-section P1 (Figure 2a), to evaluate the average velocity of sediment particles moving close to the riverbed. These results were used to estimate bedload transport rates.

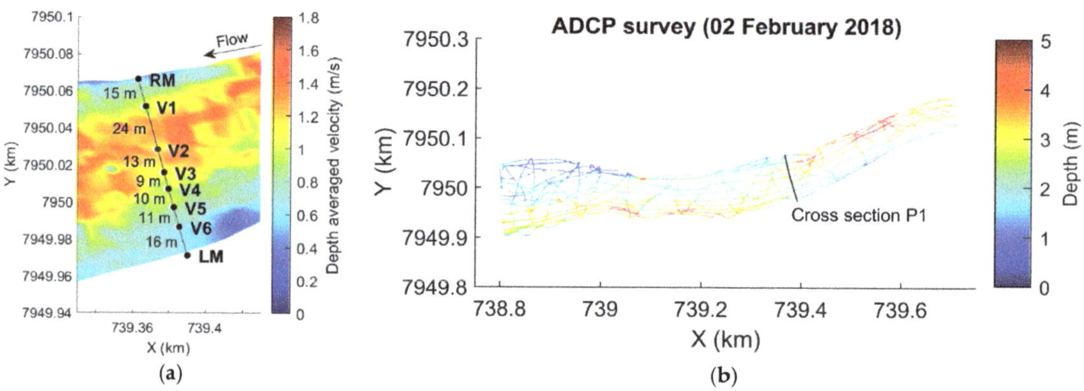

Figure 2. (**a**) Cross-section P1 and verticals of static measurements; (**b**) depth points measured during the ADCP bathymetric survey. Flow occurs from right to left.

In addition to discharge measurements and moving-bed tests, the ADCP was used for two bathymetric surveys (integrated with simultaneous flow velocity collection), using the HydroSurveyor software capabilities [24]. The surveyed reach was approximately 1 km long (380 m upstream and 620 m downstream of the cross-section P1) (Figure 2b). Each bathymetric survey lasted 3 h and the time interval between them was near 24 h. The collection frequency was 1 Hz, with 5 depth points measured per second (from the ADCP vertical beam and its 4 slanted beams). During the survey, navigation was conducted with longitudinal, transversal and oblique track lines, trying to cover the biggest possible area of the river. At regions of higher river width, e.g., near the confluence with the Coxim River (downstream), the transversal distance between longitudinal track lines reached 30 m. However, near the region where the modified ISSDOT method was applied, the transversal distance was below 20 m.

The distances between consecutive depth points (1 Hz frequency) measured with the ADCP varied according to the boat speed and local depth. Since maximum depths were approximately 5–6 m and boat speed did not exceed 20 km/h (normally less than 15 km/h), the distance between successive measured points was less than one meter (usually some decimeters), with overlapping data due to the 5-beam scheme. The width of each surveyed "swath" varied mainly between 1 m and 3 m.

For logistic reasons, it was not possible to conduct all types of surveys (water discharges, moving-bed tests, bathymetry, flow velocities, and bedload samples) on the same days. Table 1 and Figure 3 summarize the temporal resolution and the equipment used.

Table 1. Surveys carried out during the field campaign.

Survey	Equipment	31 January 2018	1 February 2018	2 February 2018
Water discharges	ADCP + RTK	X	X	X
Moving-bed tests	ADCP + RTK	X	X	
Bathymetry	ADCP + RTK		X	X
Flow velocities field	ADCP + RTK		X	X
Bedload samples	Helley-Smith	X	X	
Bed material samples	USBM-54	X		

"X"—indicates the days when the measurements were made.

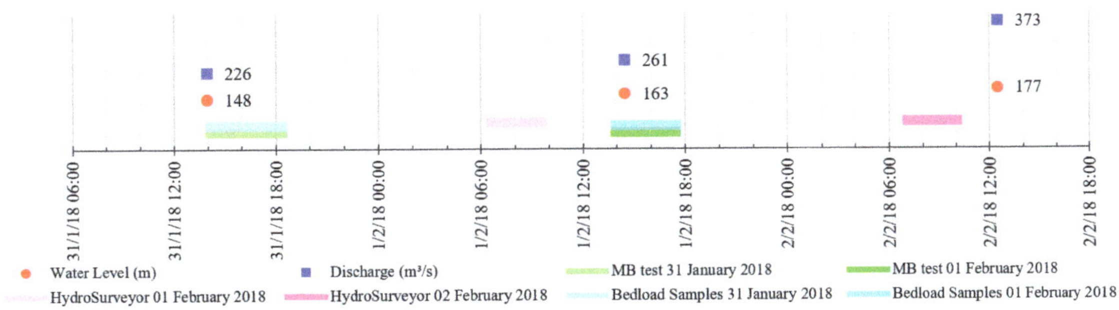

Figure 3. Temporal resolution of field surveys.

3.2. Digital Elevation Model

After post-processing the bathymetric data (eliminating outliers and smoothing the water level), an underwater digital elevation model (DEM) was generated for each survey day (Figure 4). For this purpose, the processed bathymetry was interpolated in a curvilinear grid with an average resolution of 1.4 m, using methods available on the QuickIn module of the Delft3D software [25]: grid cell averaging, triangular interpolation and internal diffusion. The two DEMs (1 February 2018 and 2 February 2018) were used to investigate and analyze bedforms along the river reach and to extract longitudinal bed elevation profiles (application of the modified ISSDOTv2 method). The justification for the choice of the grid and Delft3D software capabilities was due to subsequent morphodynamical simulations (not shown in this paper).

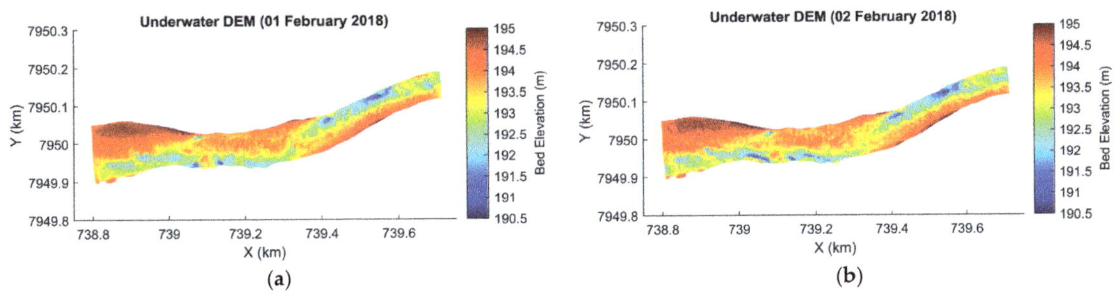

Figure 4. Underwater DEMs generated for (**a**) 1 February 2018 and (**b**) 2 February 2018.

3.3. Bed Material and Bedload Samples

Bed material and bedload samples were collected with an USBM-54 sampler and with a Helley-Smith (HS) mechanical trap, respectively, on two consecutive days. The samplers were handled from an anchored boat at the same 6 verticals at the cross-section P1 where moving-bed tests were performed. The distances between verticals varied from 9 m to 25 m and the duration of each HS sampling was 2 min. Samples were analyzed in a laboratory using laser diffraction technology, and particle size distribution curves were obtained.

In order to reduce the uncertainties associated with the HS sampling, four samples were collected in each vertical. In total, 48 bedload samples were obtained, 24 on each measurement day. Unfortunately, it was not possible to use a video camera to monitor the inlet of sediment into the HS sampler nozzle due to the high turbidity of the river, caused by the high concentration of suspended sediments in the water column. Complementary measurements performed with a USD-49 sampler indicated an average suspended sediment concentration (SSC) of 190 mg/L at 20% of depth and 310 mg/L at 80% of depth, calculated following the filtration method [26].

Therefore, despite the caution during bedload sampling, e.g., slowly lowering and raising the sampler through the water column, it was not possible to monitor the entrainment of sediments into the HS trap. This constraint, thus, hampered the assessment of a potential decrease in the quality and reliability of the bedload samples.

After drying and weighing all 48 sediment samples, unit bedload transport rates at each vertical were computed as the average of 4 daily samples. The total bedload transport at the cross-section was calculated by integrating the unit rates with the increment widths of each vertical.

A method proposed by Frings and Vollmer [3] was applied to estimate the uncertainties related to HS samplings. The uncertainty was determined based on parameters, such as the number of subsections along the cross-section (n_s = 6), the number of samples per subsection (n_m = 4), the temporal-spatial variation of bedload transport (z_1 = 0.15 and z_2 = 8) and the proportion of the total cross-section in which transport occurs (k = 1). From these values, a relative uncertainty of 63% was calculated for HS estimates of total bedload transport. Details about the method can be found in the aforementioned reference.

3.4. Moving-Bed Tests (ADCP-BT)

The ADCP-BT method, also called the kinematic model, computes bedload transport rate based on Equation (1), where v_{MB} is the average moving-bed velocity, δ_b is the thickness of the bedload active layer and c_b is the sediment concentration in the bedload layer [27–29].

$$q_{b,\ ADCP} = v_{MB} \cdot \delta_b \cdot c_b \tag{1}$$

Moving-bed tests (MB) were performed with ADCP to measure v_{MB}, using the bottom tracking feature. Measurements were carried out at the same verticals where HS samples were collected. Tests lasted 5 min.

The ADCP-BT method was applied following two different approaches: (1) based on Einstein [17], hereafter called ADCP-BTe; and (2) based on Van Rijn [27], hereafter called ADCP-BTvr. In the ADCP-BTe method, the thickness of the bedload layer (δ_{b_e}) in each vertical was adopted as twice the median diameter (d_{50}) of HS samples (Equation (2) [17]) and the sediment concentration in the bedload layer (c_{b_e}) followed Equation (3), both classical formulae from the literature [1]. In the ADCP-BTvr method, δ_{b_vr} and c_{b_vr} were computed based on the formulations proposed by Van Rijn [27], (Equations (4)–(8)). Density was adopted as $\rho_s = 2650\ \text{kg/m}^3$ and porosity as $p = 0.4$ (usual values for sand [1]).

$$\delta_{b\_e} = 2 \times d_{50} \tag{2}$$

$$c_{b\_e} = (1 - p) \times \rho_s \tag{3}$$

$$\delta_{b\_vr} = 0.3 \times d_{50} \times D_*^{0.7} \times T^{0.5} \tag{4}$$

$$D_* = d_{50} \times \left[\frac{(s-1)g}{v^2}\right]^{\frac{1}{3}} \quad (5)$$

$$T = \frac{(u'_*)^2 - (u_{*cr})^2}{(u_{*cr})^2} \quad (6)$$

$$u'_* = \frac{\overline{u}}{5.75 \times \log\left(\frac{12h}{3d_{90}}\right)} \quad (7)$$

$$\frac{c_{b\_vr}}{c_0} = 0.18 \times \frac{T}{D_*} \quad (8)$$

where ρ_s is the density of sand (kg/m^3); p is the porosity of sand (dimensionless); D_* is the dimensionless particle parameter; T is the transport stage parameter (or excess shear parameter) (dimensionless); v is the kinematic viscosity of water (m^2/s); u'_* is the bed shear velocity related to grains (m/s); u_{*cr} is the critical bed shear velocity (m/s) according to Shields; $\overline{u}$ is the depth-averaged flow velocity (m/s); h is the water depth (m); d_{90} represents the grain size (m) of the superficial bed sediment for which 90% of the particles are lower in weight; c_0 is the maximum (bed) concentration (=0.65).

Due to the arrangement of the equipment on the boat, it was not possible to track exactly the same point (in each vertical) with the ADCP and the Helley-Smith sampler. This contributes to the uncertainties related to the sampling procedures and the variability of the results. However, the distance between both devices was kept the shortest as possible, not exceeding 3 m. The results of MB tests were used to estimate unit bedload transport rates at the 6 verticals and then integrate them along the cross-section.

3.5. Empirical Formulae

Five empirical formulae (Table 2) were applied to estimate bedload transport at Taquari River cross-section: Kalinske [19], Meyer-Peter-Müller [20], Einstein [17], Einstein-Brown [18] and Van Rijn [21]. The equations presented at the end of Table 2 are general formulae used to compute some of the parameters needed in the empirical transport equations. From now on, the mentioned methods are referred to by the following acronyms, respectively: K, MPM, E, EB and VR.

Table 2. Empirical formulae applied to estimate bedload transport rates.

Reference	Equations
Kalinske [19]	$\frac{q_b}{u_* \times d} = 10 \times \Psi'^2$
Meyer-Peter-Müller [20]	$\phi = (4 \times \Psi' - 0.188)^{3/2}$
Einstein [17]	$\Psi_E = \frac{(s-1) \times d}{R\prime \times I}$ $R\prime = (ks/ks\prime)^{3/2} \times R\phi \ vs \ \Psi_E \rightarrow see\ chart\ in\ the\ reference$
Einstein-Brown [18]	$\phi = 40 \times \Psi\prime^3$
Van Rijn [21]	$q_b = 0.015 \times v \times h \times (d/h)^{1.2} \times M_e^{1.5}$ $M_e = \frac{v - v_{cr}}{\sqrt{(s-1) \times g \times d}}$
General formulae	$\phi = \frac{q_b}{d^{3/2} \times \sqrt{(s-1) \times g}}$ $\Psi = \frac{\tau_0}{(s-1) \times \rho \times g \times d}$ $\Psi' = \frac{\tau'_0}{(s-1) \times \rho \times g \times d} = (ks/ks')^{3/2} \times \Psi = \frac{1}{\Psi_E}$ $q_{b1} = q_b \times \rho_s$ $ks = \frac{1}{n}$ $\tau_0 = \rho \times g \times R \times I$

All the equations developed by these researchers use information about hydraulic parameters and bed material characteristics to evaluate bedload transport capacity. The values applied in the equations are space and time averaged, showing a more general and simplified approach than other methods that account for local effects. The presented formulae were selected for this study because they have been found to provide accurate bedload transport estimates in large sand rivers with dunes.

In Table 2, q_b is the unit bedload transport rate in volume ($m^3/s{\cdot}m$); q_{b1} is the unit bedload transport rate in mass ($kg/s{\cdot}m$); τ_0 is the bed shear stress (N/m^2); u_* is the shear velocity, also called friction velocity (m/s); Ψ is the shields parameter, also called flow parameter (dimensionless); Ψ' is the flow parameter related to grains, not influenced by bedforms roughness (dimensionless); Ψ_E is the flow parameter used by Einstein, inverse of Ψ' (dimensionless); ϕ is called transport parameter (dimensionless); the apostrophe $'$ refers to the influence of grains, (disregarding bedforms); g is the acceleration due to gravity (m/s^2); d is the sediment diameter (m); ρ is the density of water (kg/m^3); ρ_s is the density of sediments (kg/m^3); I is the water surface slope of the study reach (m/m); h is the flow depth (m); R is the hydraulic radius (m) of the cross-section (ratio of the cross-sectional area to the wetted perimeter); M_e is a mobility parameter which represents the excess mobility of sediments (dimensionless); s is the specific gravity of sediments (dimensionless); v is the depth-averaged flow velocity (m/s); v_{cr} is the critical water velocity for currents based on shields (initiation of motion) (m/s); ks is a roughness coefficient of the bed ($m^{1/3}/s$), equivalent to the inverse of Manning's n roughness coefficient ($s{\cdot}m^{-1/3}$).

Kalinske [19] developed a simple theory of bedload transport where the main driver was particle velocity. He assumed that particle velocity adapted instantaneously to fluid velocity (linear relationship), which is the main shortcoming of his approach because of the different processes of particle motion (sliding, rolling and saltating). In his model, bedload transport rate is computed from the volume of particles in motion (per unit streambed area) and their mean velocity.

The Meyer-Peter-Müller formula is normally restricted to 0.4 to 28.6 mm uniform sediments or mixtures in steep rivers with high rates of bedload transport [4,30]. Despite the fact that the data upon which the formula was based were obtained in flows with little or no suspended load, the MPM equation is commonly employed to predict bedload transport in a wide range of fluvial systems, even for flows with appreciable suspended loads.

For Einstein [17], bedload transport reflects the random nature of particle paths and exchanges between the bed and stream. In his approach, the main driver of bedload transport is the number of sediment particles entrained. Although he used probabilistic concepts, his bedload equation is deterministic (and in good agreement with the MPM equation). The concepts of his work are in use today, e.g., bedload transport as an intermittent process, dimensional analysis, probabilities of entrainment and deposition, influence of turbulence on incipient motion [31,32].

The Einstein-Brown formula is a modification developed by H. Rouse, M.C. Boyer and E.M. Laursen of a formula by Einstein [33], presented in Brown [18]. The equation was based on flume data with well-sorted sediments (d = 0.3–28.6 mm). The Einstein and the Einstein-Brown formulae are based on the concept that the fluid forces on the bed sediment fluctuate randomly due to the turbulence of the flow. They consider that a small amount of sediment moves randomly in jumps or steps even when the mean hydraulic forces at the bed are very small, resulting in continuous values of bedload at low transport rates [34].

The Van Rijn [21] formula was developed for particle sizes ranging between 0.2 and 2 mm. The bedload transport model for steady flow proposed by the author is a parameterization of a detailed grain saltation model representing the basic forces acting on a bedload particle. The Van Rijn model is appropriate for rivers transporting sandy sediments in conditions of subcritical flow.

3.6. Modified ISSDOTv2 Method

In rivers with bedforms, dune tracking is considered an effective way for estimating bedload transport. In this case, successive surveys were performed along a longitudinal bathymetric profile and, based on sediment and bedform characteristics, unit bedload was calculated [5,12,35,36].

The integrated section surface difference over time method (ISSDOTv2) uses the scour volumes computed from the difference plots of time-sequenced bathymetric data to estimate bedload transport [14,15,37]. This method is represented by Equation (9), which is mathematically equivalent and dimensionally homogeneous to Equation (10), developed by Richardson et al. [38]. The main advantage of the ISSDOTv2 method is that Equation (9) is applied over a longitudinal profile, while Equation (10) requires averaged wave heights and velocities.

$$q_{b,\mathrm{ISSDOTv2}} = \frac{(1-p) \times \rho_s \times V}{1.82 \times \Delta t_{bat} \times N_d} \tag{9}$$

$$q_b = \frac{(1-p) \times v_d \times H_d \times \rho_s}{2} \tag{10}$$

where $q_{b,\mathrm{ISSDOTv2}}$ is the bed-material load moving in the sand wave computed by the ISSDOTv2 method (in kg/s); ρ_s is the density of sand (=2650 kg/m^3); p is the porosity of sand (adopted as 0.4); 1.82 is a constant that accounts for the nontriangular shape of dunes; N_d is the number of dunes along the longitudinal profile; V is the scoured volume (m^3) during the time interval Δt_{bat} between the bathymetric surveys (s); v_d is the dune speed (celerity, in m/s); and H_d is the dune height (m).

The condition for applying the methodology is that the scoured volumes are equal to (or very close to) the depositional volumes. In rivers, unlike laboratory experiments, this condition is rarely observed, since flow is unsteady and the interaction between turbulence and high shear stresses results in the suspension of bed-material. Some of these suspended sediment particles might travel downstream and deposit on a different dune from which they originated, thus, adding depositional volume to another dune. Abraham et al. [37] recommend scour-deposition ratios between 0.8 and 1.2 (20% variation); otherwise, the profile is considered unsuitable for the method.

Furthermore, a systemic bias inherent in the methodology underestimates the bedload rates. This error is related to the dune celerity and to the time interval between the surveys. In the process of subtracting longitudinal bathymetric profiles, some scour volumes are not computed (triangles C and D in Figure 5). The larger the time interval and the dune celerity, the greater the not computed volumes and the lower the calculated bedload. Shelley et al. [15] proposed a procedure to correct this systemic bias, represented by Equation (11), allowing the application of the ISSDOTv2 method with large time intervals.

$$q_{b,\mathrm{ISSDOTv2},corr} = q_{b,\mathrm{ISSDOTv2}} + \frac{v_d^2 \times \Delta t_{bat} \times H_d}{2 \times \lambda} \tag{11}$$

where $q_{b,\mathrm{ISSDOTv2},corr}$ is the corrected dune transport rate (kg/s); and λ is the dune wavelength (m).

Originally, the ISSDOTv2 method was developed to be applied over longitudinal bathymetric profiles surveyed with multibeam echosounders. The higher spatial resolution of the depth points measured with multibeam yields high accuracy in the interpolated digital elevation models (DEMs), thus, improving the quality of bedload estimates. In the present study, the lower resolution resulting from the ADCP survey (using only 5 beams) increases the uncertainties in the DEMs' construction and in the subsequent profiles' extraction. This modified approach, regarding the difference in bathymetric data collection and its spatial resolution, motivated the authors to test a modified ISSDOTv2 method, hereafter called mISSDOTv2.

The mISSDOTv2 method was applied in the central region of the study area (Figure 6), in a reach approximately 300 m long and with 85 longitudinal profiles spaced approximately 1.15 m apart. This area comprises the cross-section P1. Only 33 profiles were selected (39%) because they presented scour-deposition ratios within the range of 20% (0.8 to 1.2). Bedload rate was calculated for each selected profile and subsequently corrected according to dune geometric characteristics obtained from wavelet-spline analysis and to the time interval between the bathymetries ($\Delta t_{bat} = 24$ h). The results were extrapolated to the region where profiles were discarded. The total bedload transport at the cross-section was computed by integrating all longitudinal swaths.

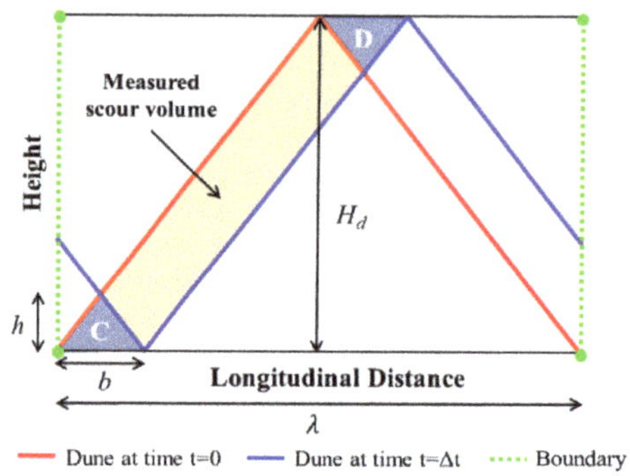

Figure 5. Schematic diagram showing the systemic bias due to the not-computed scour volumes (adapted from [15]).

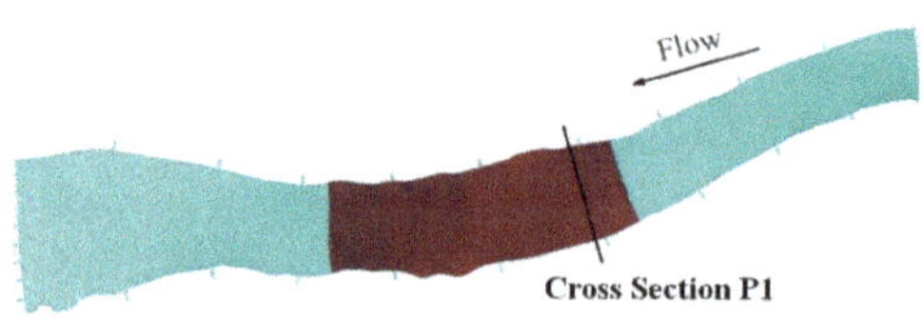

Figure 6. River reach where the modified ISSDOTv2 method was applied (red area).

3.7. Spline Filters and Wavelet Transforms

Since bedforms can be interpreted as harmonic variations, they can be decomposed through mathematical analysis for waves or oscillatory systems. Available techniques include spectral analysis, moving average smoothing techniques, fractal theory and logistic regression, among others.

In this work, a methodology based on the use of robust spline filters and continuous wavelet transforms to discriminate bedform geometry was applied [16]. This technique overcomes the limitations of the Fourier transform used in spectral analysis, as it can be applied to nonstationary, intermittent, aperiodic, discontinuous, nonlinear and three-dimensional processes, such as riverbed morphology.

The algorithm of the spline filter uses the penalized least squares method and smoothes uniformly sampled data through the discrete cosine transform. The wavelet function used is the Morlet function, which can capture frequencies in the ripple scale (wavelength less than 60 cm), although the measurement resolution is lower.

In this methodology, continuous wavelet analysis is performed on the original signal (bathymetric longitudinal profile: elevation or depth), generating a global wavelet spectrum that provides information about bedform wavelengths with greater frequency and intensity. Then, the spline filter is applied sequentially to the original signal and its derivatives, using the transformed wavelet analysis with the characteristic lengths obtained from the global spectrum to decompose and hierarchize the bedform profile into three categories.

Usually, the first hierarchy corresponds to ripples or small dunes (wavelengths in the order of 5 m), the second hierarchy is associated with medium (10 m) or large (100 m) dunes and the third hierarchy comprises mega-dunes or bars. The sum of the three hierarchies' signals results in the original signal. More details about the methodology can be found in Gutierrez et al. [16,39] and Torrence and Compo [40].

4. Results

4.1. Discharges, Flow Velocity, Grain Size and DEM

During the field campaign (31 January 2018, 1 February 2018 and 2 February 2018), measurements were carried out at cross-section P1 (Figure 7, Table 3). On these days, discharges were 226 m^3/s, 261 m^3/s and 373 m^3/s, respectively. The mean flow velocity varied from 1.1 m/s to 1.3 m/s, mean water depth varied from 2.0 m to 2.3 m and cross-section width was approximately 110 m. Bedload consists of fine sand (125 µm < d < 250 µm) and medium sand (250 µm < d < 500 µm). Figure 8 shows depth-averaged velocities mapped along the river reach after ADCP surveys on two consecutive days.

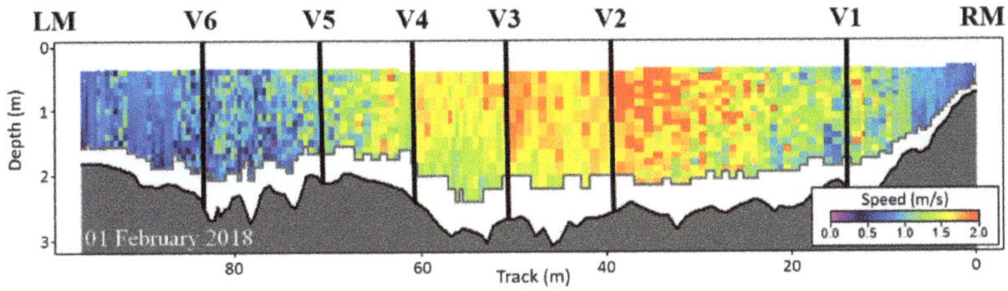

Figure 7. Flow velocity measured with ADCP at cross-section P1 on 1 February 2018. Vertical black lines are the positions of moving-bed tests and where bedload samples were collected.

Table 3. Measured and estimated hydraulic parameters at Taquari River, cross-section P1.

Parameters	31 January 2018	1 February 2018	2 February 2018
Water Level (cm)	148	163	177
Discharge (m^3/s)	226	261	373
Mean Flow Velocity (m/s)	1.13	1.18	1.33
Cross-Section Area (m^2)	200.3	221.4	280.4
Width (m)	101.3	109.4	120.8
Mean Depth (m)	1.98	2.02	2.32
Hydraulic Radius (m)	1.85	1.81	2.30
Water Level Slope (cm/km)	15	25	25
d_{50} (mm) (bedload)	0.268	0.274	-
d_{90} (mm) (bedload)	0.397	0.406	-
d_{50} (mm) (bed material)	0.302	-	-
d_{90} (mm) (bed material)	0.418	-	-
Transport Stage	11.4	12.5	15.7
Manning's n ($s/m^{1/3}$)	0.016	0.020	0.021

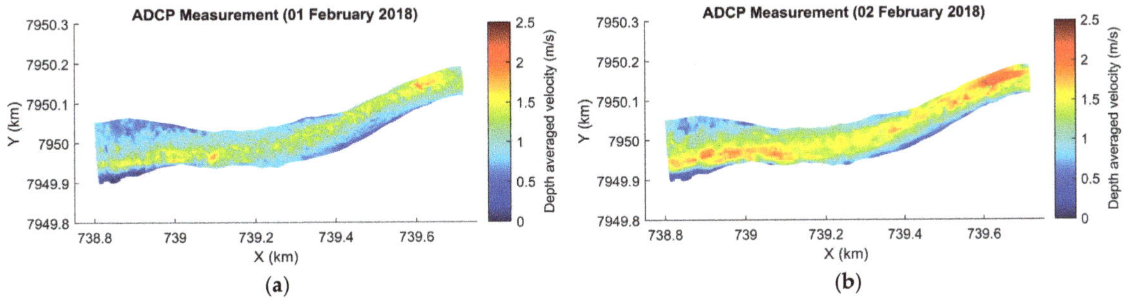

(a) **(b)**

Figure 8. Depth-averaged velocities measured with ADCP and interpolated along the river reach. (**a**) 1 February 2018; (**b**) 2 February 2018.

Figure 9a illustrates a DEM (1 February 2018) of the region where the modified ISSDOTv2 method was applied. By the plot difference of the two DEMs (2 February 2018–1

February 2018), the areas with aggradation and degradation during the time interval between the surveys were identified (Figure 9b). Analyzing the vertical variation at all grid points, 96% of the points presented erosion/deposition heights in the range ±50 cm, and 81% in the range ±25 cm. The distribution of the morphological variation was reasonably symmetrical, with 48% of the points undergoing aggradation and 52% undergoing degradation.

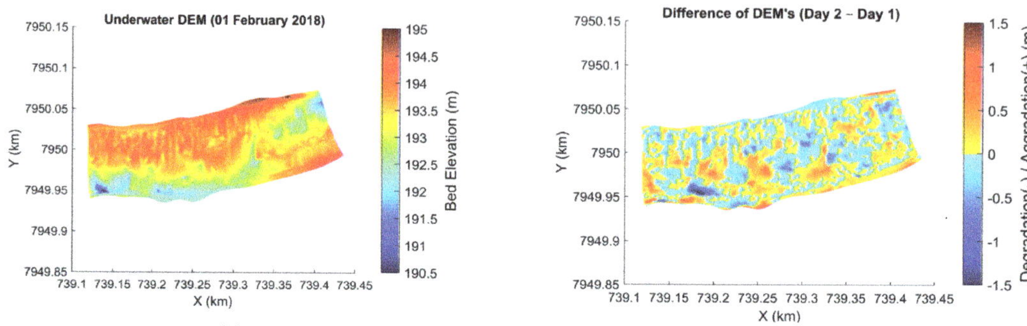

(**a**) (**b**)

Figure 9. (**a**) Underwater DEM (1 February 2018) of the region where the modified ISSDOTv2 method was applied; (**b**) difference of DEMs showing regions of aggradation (+) and degradation (−) between the surveys.

4.2. Unit Bedload Transport

Unit bedload rates were computed at the six verticals along the cross-section P1, where HS samples were collected, and MB tests were performed with ADCP. Despite the highest resolution in water velocities obtained with SmartPulseHD (automatically variable frequency: 1.0 MHz and/or 3.0 MHz), the moving-bed velocities measured with this feature were very similar to those obtained with the two manual configuration schemes (1.0 MHz and 3.0 MHz fixed frequencies). The average standard deviation of measurements performed on 31 January 2018 (Figure 10), for example, was 0.019 m/s. On that day, the highest standard deviation happened in Vertical 3 (Std Dev = 0.044 m/s).

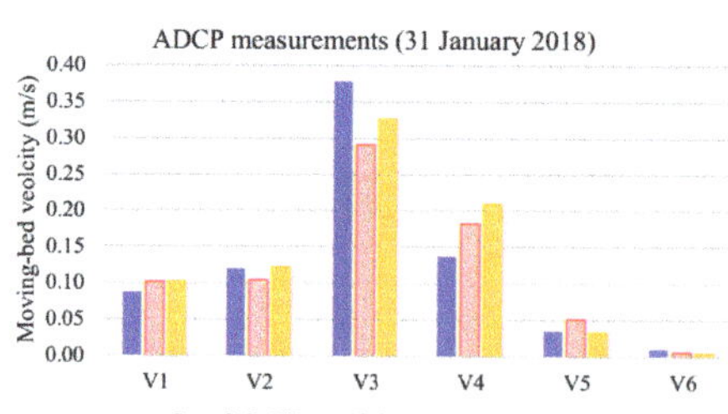

Figure 10. Moving-bed velocities measured with ADCP on 31 January 2018 at the six verticals of cross-section P1, in Taquari River.

Figure 11 presents the unit bedload rates estimated with HS versus the depth-averaged water velocities measured at each vertical. The low correlation of the plotted data (R = 0.29 for 31 January 2018 and R = 0.44 for 1 February 2018) reinforces the high uncertainties associated with the HS samplings in sand-bed rivers.

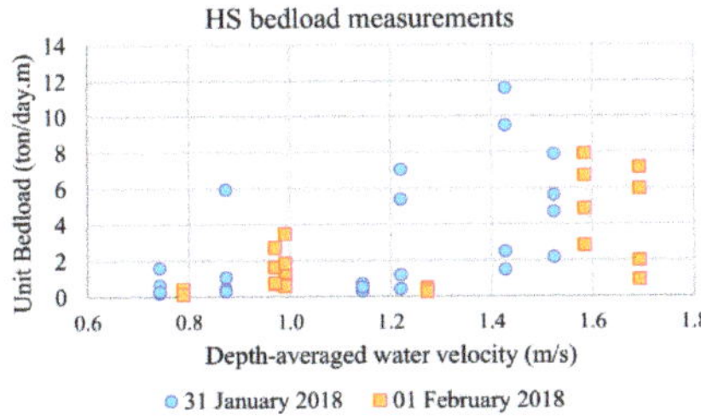

Figure 11. Unit bedload transport rates estimated with Helley-Smith sampler versus depth-averaged water velocities.

Comparing the unit bedload rates based on ADCP-BTe and ADCP-BTvr (Figure 12), there is no trend where one method yields higher bedload rates than the other. For example, at some verticals, the ADCP-BTe unit rates were higher than the ADCP-BTvr estimates, but at other verticals, the opposite happened. This suggests that a compensation of differences exists, influenced by the necessary parameters to apply Equation (1) in each vertical: the thickness of the bedload layer (δ_b) and the sediment concentration in the bedload layer (c_b). In general, the estimates of c_{b_vr} were lower than c_{b_e} (which was constant), with an average ratio of 35%. On the other hand, estimates of δ_{b_vr} were in average twice of δ_{b_e}, with higher deviations observed at the verticals where flow velocities and bedload transport were more intense.

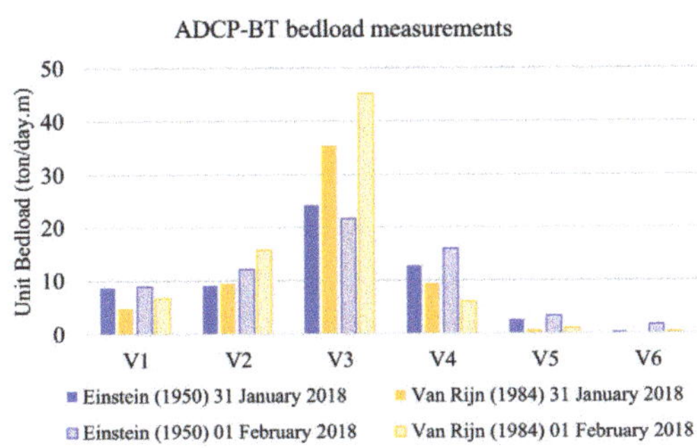

Figure 12. Unit bedload transport rates estimated with ADCP-BT, following Einstein [17] and Van Rijn [27] formulations to compute the parameters of Equation (1).

4.3. *Bedform Characteristics and mISSDOTv2*

After correcting and interpolating the bathymetric data on the curvilinear grid, longitudinal bed profiles were extracted from the underwater DEM. Figure 13 shows a pair of longitudinal bed profiles (at grid position N = 26) of 1 February 2018 (blue) and 2 February 2018 (red). This pair of profiles met the selection criteria for use in the mISSDOTv2 method, i.e., presented erosion-deposition ratios within the range [0.8–1.2]. Inconsistencies observed in some profiles might be related to bathymetric data interpolations on the grid, which is a constraint of the modified ISSDOTv2 in comparison to the original methodology based on multibeam surveys.

Despite the feasibility of identifying bedforms along the illustrated profiles, it is challenging to determine by visual inspection the height, wavelength and celerity of the dunes. Intending to decrease subjectiveness of personal interpretation while determining geometries and dimensions, the methodology developed by Gutierrez et al. [16,39] was applied.

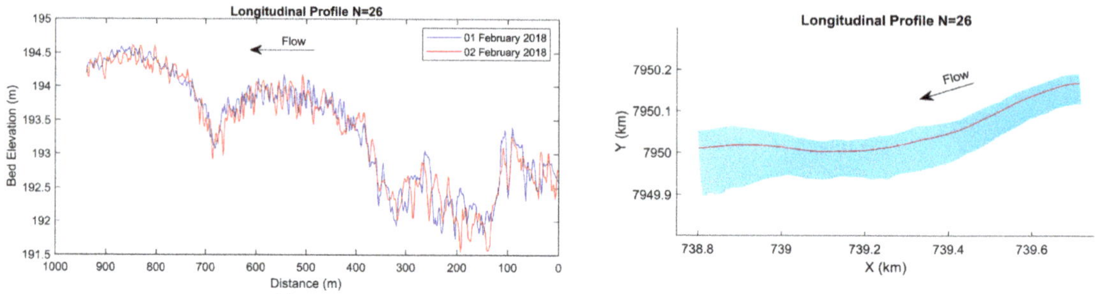

(a) (b)

Figure 13. Pair of longitudinal bed profiles (at N = 26) measured at Taquari River. (**a**) Elevation view; (**b**) plan view (grid).

The first wavelet analysis was performed on the original signal (longitudinal bed elevation profile) and generated a global wavelet spectrum indicating the most frequent bedform wavelengths. Figure 14 presents the first step results for the profile N = 26 (1 February 2018). The wavelet power spectrum contours show the spatial distribution of wavelengths (λ), with a 95% confidence interval, and the cone of influence (dashed blue line) discriminates reliable from spurious local power spectrum results. The global wavelet power spectrum indicates the main wavelength frequencies found at the analyzed profile (peaks located to the right of the dotted red line correspond to wavelengths at 95% of the confidence level). Characteristic wavelengths for the presented profile are: 14 m, 32 m, 100 m, 168 m and 387 m.

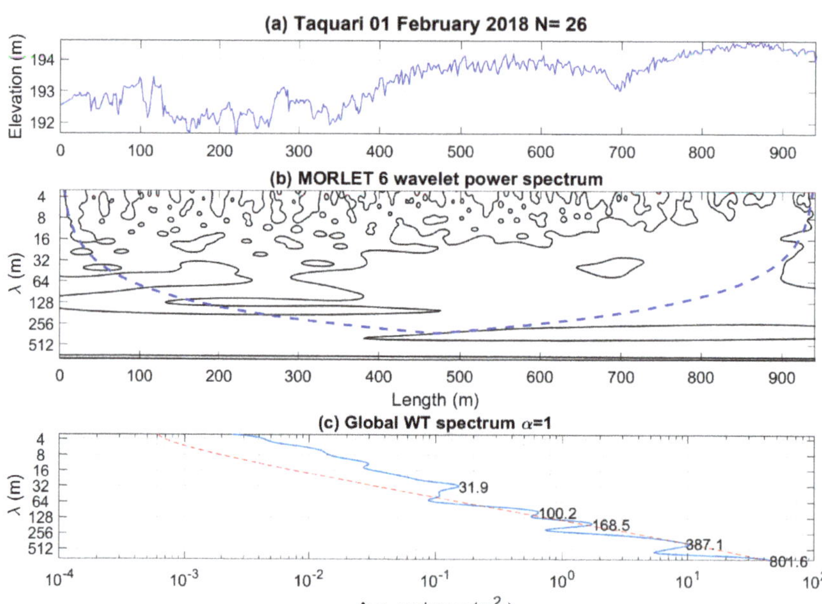

Figure 14. Wavelet analysis for profile N = 26 on 1 February 2018. Flow occurs from left to right. (**a**) Longitudinal bed elevation; (**b**) spatial distribution of wavelengths (λ), through wavelet power spectrum contours using the Morlet function; (**c**) global wavelet power spectrum, indicating the main wavelength frequencies.

On the second step, longitudinal bedforms were decomposed and hierarchized based on the wavelengths previously found. Three hierarchies were defined: the first referring to small dunes; the second, associated with medium or large dunes; and the third hierarchy, comprising mega-dunes or bars. Figure 15 illustrates the decomposition of the longitudinal profile N = 26 (1 February 2018).

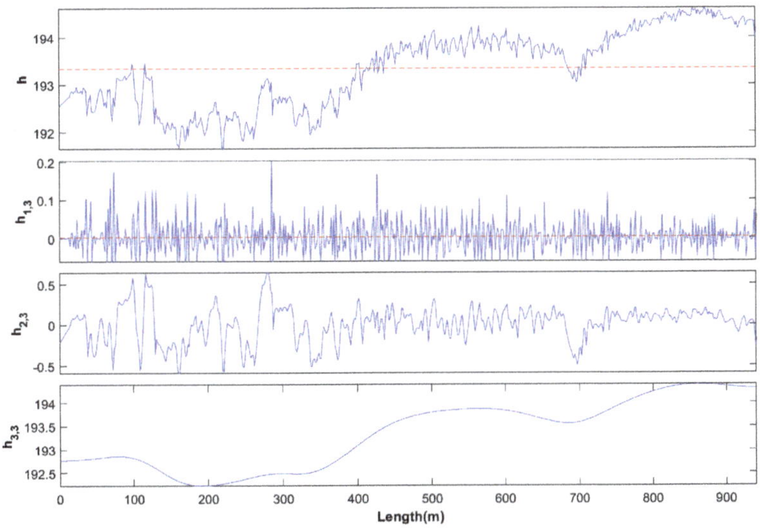

Figure 15. Wavelet-spline discrimination for longitudinal bedform profile N = 26 on 1 February 2018, with heights and wavelengths of the 3 hierarchies, where h represents the original signal, i.e., the measured bed elevations (m); $h_{1,3}$ is the first hierarchy (small dunes); $h_{2,3}$ is the second hierarchy (medium and large dunes); and $h_{3,3}$ is the third hierarchy (bars). Flow occurs from left to right. Values are in meters.

The results from the wavelet transforms provided information to analytically quantify bedform dimensions so that the mISSDOTv2 method could be applied with the correction term (Equation (11)). Nevertheless, it was necessary to visually assess the dune celerity for each selected longitudinal profile pair, computing the ratio of the distance between the dune's crests to the time interval of 24 h. As there were many dunes with varying sizes and speeds over a single profile path, it was necessary to average the individual estimates to establish a unique dune celerity value for use in Equation (11). Thus, it was not possible to use the same objectivity as in the wavelet-spline methodology.

The second hierarchy ($h_{2,3}$) resulting from the wavelet-spline discrimination was adopted as the bedform reference to evaluate the systemic loss of area and correct the dune transport rate using Equation (11). For each selected profile, one numeric value was determined for each parameter: dune wavelength (λ), dune height (H_d) and dune celerity (v_d). The variations observed between the profiles were: $30 < \lambda < 80$ (m); $0.2 < H_d < 0.8$ (m); $1 < v_d < 20$ (m/day). The average area loss of the selected profiles was 35% of the total corrected dune transport rate.

4.4. Total Bedload Transport

The total bedload transport is the mass (or volume) of bedload that passes through a cross-section during a defined time interval. Figure 16a shows the total bedload transport at cross-section P1 computed based on Helley-Smith (HS) samples, moving-bed tests (ADCP-BTe and ADCP-BTvr), the modified ISSDOTv2 method and five empirical equations (K, MPM, E, EB and VR). The predictions were created for the three consecutive survey days. The result of the mISSDOTv2 method (776 tons/day), however, represents the mean bedload transport between the last two days (1 February 2018 and 2 February 2018), since it uses longitudinal profiles from two DEMs generated with bathymetric data measured under two distinct hydraulic conditions.

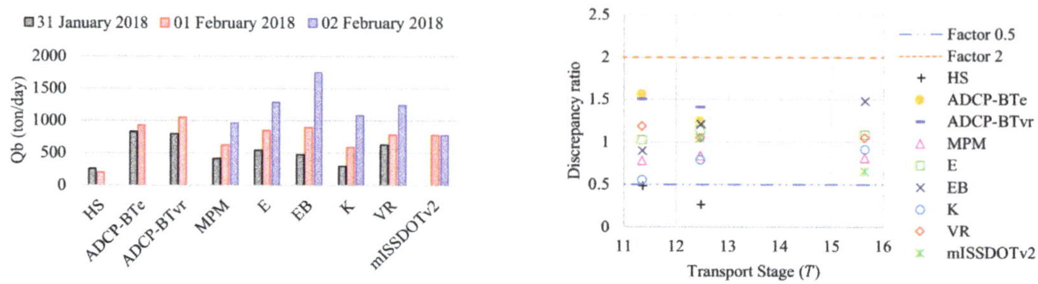

Figure 16. (**a**) Total bedload estimates at cross-section P1; (**b**) discrepancy ratios of total bedload estimates.

The average total bedload transport rates of all methods are: 528 ton/day (31 January 2018), 742 ton/day (1 February 2018) and 1182 ton/day (2 February 2018) (Figure 16a). The accuracy of the methods was evaluated in terms of a discrepancy ratio, defined as the ratio of the computed total bedload of each method and the average of all methods. Figure 16b shows the discrepancy ratios as a function of the transport stage (Equation (6)). Several authors consider the methods to give consistent sediment transport rates when the discrepancy ratio is between 0.5 and 2 [4,27,30,41–43]. This condition was met for all estimates on the three days, except for the HS measurements, which were probably underestimated.

The lowest sediment transport rates were estimated by HS on the first two days (this method was not applied on the third day) and by the MPM equation on the third day (disregarding the mISSDOTv2 method, which represents the mean bedload between the last two days).

The maximal sediment transport rates on the first two days were accomplished with ADCP-BTe and ADCP-BTvr (which were not applied on the third day). Despite the variations in unit bedload rates computed with these two approaches, the total bedload transport was similar: (A) 31 January 2018–829 tons/day with ADCP-BTe and 794 tons/day with ADCP-BTvr; (B) 1 February 2018–926 tons/day with ADCP-BTe and 1046 tons/day with ADCP-BTvr.

On 2 February 2018, the maximal rate was predicted by the EB equation. When the Shields parameter (Ψ) is high, the EB equation begins to consider a portion of the suspended (fine) sediment as bedload (on last day, Shields parameter was calculated as $\Psi = 1.3$). This is the main reason for the higher bedload computed on that day, when the discharge severely increased (+43% in relation to the previous day).

5. Discussion

The presence of a sampling device on the river bottom alters the pattern of the flow and sediment transport in its vicinity [44]. The Helley-Smith sampler was developed to perform collections in environments carrying coarse sediments; however, when used in sand-bed rivers (d < 0.5 mm), the samples can be overestimated [45]. For example, samplings with HS at the East Fork River near Pinedale, Wyoming, USA, where particle sizes varied between 0.25 mm and 0.5 mm, were overestimated by 150% [46].

Studies have shown that the effectiveness of HS for bedload sampling is influenced by mesh size [47,48], nozzle wall thickness [49], flow turbulence and bedform dynamics [44]. The position of the device on the river bottom (above or below the bedload layer; and aligned or not with the flow) can cause oversampling or undersampling of bedload rates [50,51]. In rivers with bedforms, the spatial variability of bedload transport is a source of uncertainty for the physical samples, since the bedload rate at a specific cross-section tends to decrease temporally from a maximum near the crest to a minimum in the trough as dunes migrate downstream [52].

Even for pressure-difference samplers with hydraulic efficiency near 100%, the sampling efficiency is affected by errors and uncertainties as mentioned above [53]. Therefore, caution is needed when considering the difficulties of obtaining accurate measurements with mechanical traps, especially when recognizing the drawbacks of the HS sampler when used in sand-bed rivers. Although it was not possible to identify the specific factors that caused the lower rates of the HS estimates in the Taquari River, the magnitude of the computed total transport was useful for comparison with the other methods.

Considering the surrogate method ADCP-BT, it is interesting to point out that Villard et al. [8] have also estimated bedload transport with ADCP (bottom-tracking), testing the Einstein [17] and Van Rijn [27] approaches. After a linear functional analysis comparing bedload estimates from the ADCP-BT method, the Helley-Smith sampler and the Van Rijn formulae [27], they concluded that there was no improvement in either relation, when modeled values (based on Van Rijn [27]) of bedload layer thickness (δ_b) and bedload concentration (c_b) were used.

A possible reason for the high bedload rates estimated with ADCP-BTe is related to c_b determination using Equation (3) with $p = 0.4$, which is equivalent to adopting the maximum possible concentration in the bedload layer. Actually, the expected porosity should be higher than 0.4, even with coarser bed particles. The ADCP-BTvr, in turn, considers a bed particle's saltation height (δ_b) that, although does exist in sand-bed streams subjected to bed shear velocities, is very small, in the order of a few diameters.

Latosinski et al. [1] assessed bedload rates in a large sand-bed river (Paraná river, in Argentina) using ADCP-BT, combining theoretical parameters and apparent velocity of bed-particles from ADCP. Comparing ADCP-BTe and ADCP-BTvr, they concluded that there is indeed a compensation of differences between the methodologies, in such a way that the ADCP-BTe approach yields better results, even though it lacks a clear physical meaning.

From Villard et al. [8] and Latosinski et al. [1] findings, the adoption of constant values of δ_b and c_b (based on Einstein [17]) seem to be quite appropriate to explore the ADCP-BT capability. The results from the present work at Taquari River endorse this statement, despite the lack of a physically based foundation in the definition of variables.

Regarding the complementary use of the ADCP sensor (exploiting its 5 beams) for collecting bathymetric data should be recognized for its lower spatial resolution in comparison to a multibeam echosounder survey. While errors in depth measurements and RTK-GPS positioning contribute to errors in estimated bed elevations, the interpolation step to create the DEMs is considered to be the dominant source of error [54]. Nonetheless, the presented study intended to test a simpler, cheaper and easier approach to evaluate the capability to estimate bedload transport using a modified ISSDOTv2 method.

The ADCP application (with HydroSurveyor capabilities activated to measure bathymetry) shows great potential, despite the spatial resolution constraints. However, since the bathymetry interpolation applied to create DEMs increases noise and uncertainties in the longitudinal profiles, the application of mISSDOTv2, although promising, must be carried out with caution.

During the field campaign, it was not possible to repeat the bathymetric survey more than once, preventing the establishment of a linear relationship between lost area and time interval to evaluate the computed corrected values [15] in the modified ISSDOTv2 method. During the second bathymetry, the pilot of the boat tried to follow the original track lines, maintaining the boat speed as constant as possible, but in general it was not possible to cover the same exact paths. Then, the visual inspection of longitudinal profile pairs (before interpolation) and the recognition of equivalent dunes at different times were hampered by the limited time-spatial resolution of the ADCP bathymetric surveys.

Two pieces of evidence were considered to evaluate if bedforms traveled less than 50% of their wavelength in the time interval between the consecutive surveys (24 h). The first evidence was based on dune-tracking measurements performed with an ADCP in the Paraná River [55], where bed material and flow velocities are similar to those found at the Taquari River, which resulted in dune velocities of approximately 1 m/day. The second piece of evidence was based on the results of a computational morphological modeling of the study site with the Delft3D software [25]. Simulations were performed, testing different sediment transport formulations and providing longitudinal bed profiles that pointed out that dunes traveled less than 50% of their wavelength in 24 h (Figure 17).

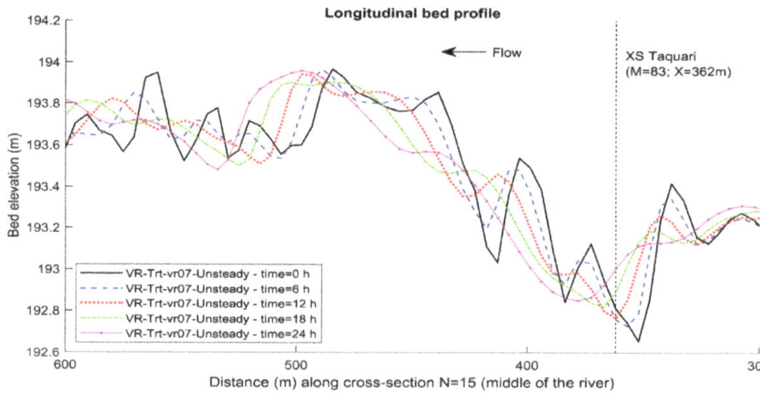

Figure 17. Temporal variation of a longitudinal bed profile from Delft3D simulations using Van Rijn formulations.

Despite these efforts, it was not possible to guarantee that dunes did not travel more than their wavelength in the 24-h interval between the ADCP surveys. This is the main uncertainty related to the application of the mISSODTv2 in the presented research. Nevertheless, the authors considered it useful to present the results to illustrate the potential of the methodology, and to warn about the necessary precautions. A major piece of advice is to perform several transects along a fixed longitudinal path before the bathymetric surveys, with the aim of recognizing bedform features at different time steps. From the analysis of the distance traveled by these bedforms, the proper time interval between the bathymetric surveys should be chosen, considering the equipment used, the extension of the area, and the local sand wave velocities for a specific hydrologic condition.

It is worth mentioning that the bedload empirical formulations applied in this paper (Table 2) predict the maximum bedload transport rate that a flow in equilibrium can produce, given the local hydraulic and sedimentological conditions. This transport capacity may not be equal to the actual transport if the channel is undergoing aggradation or degradation processes. Since it is not trivial to determine when this capacity is reached, the use of field-measured data to check the validity of bedload transport equations (and vice versa) should be carried out carefully.

Table 4 summarizes the main advantages and disadvantages of the methods analyzed in this paper. As seen, it is not possible to choose one method as the most efficient and recommended of all. Each one has its own constraints and is more suitable for specific conditions concerning economic costs, hydraulic/sedimentologic characteristics, equipment availability, field staff capacity, desired spatial-temporal resolution and hydrological regime.

Table 4. Advantages and disadvantages of methods applied to estimate bedload sediment transport in sand-bed rivers with bedforms.

Method	Advantages	Disadvantages
Helley-Smith Sampler	• Physical sampling allowing bed material analysis. • Cheaper instrument.	• Limited spatial resolution. • Limited temporal resolution. • Intrusive method, causing bed disturbance. • High uncertainties related to the sampler (hydraulic) efficiency and the sampling efficiency. • Time-consuming and dangerous during floods. • The representativeness of the sample is affected by the presence of bedforms. • Originally developed for environments carrying coarse sediments.
ADCP-BT	• Faster and safer measurements. • Higher spatial and temporal resolutions compared to physical sampling. • Nonintrusive method. • Simultaneous measurement of flow velocity at each vertical point. • Results are useful to correct water discharges estimates based on the bottom-tracking feature.	• The measured average moving-bed velocity is influenced by the ADCP frequency and the acoustic pulse length. • The measured average moving-bed velocity is influenced by bedload particle grain size. • Uncertainties related to the determination of the parameters: thickness of the bedload active layer and sediment concentration in the bedload layer. • The representativeness of the sample is affected by the presence of bedforms. • Specialized (and expensive) equipment and trained staff are needed.
Modified ISSDOTv2	• Nonintrusive method. • Useful in rivers with bedforms. • Simultaneous measurement of flow velocities throughout the river reach. • The method is applied over longitudinal bathymetric profiles. • Bedload estimates tend to be more accurate, since they consider a wide area and the spatial variability of bedload transport.	• There must be bedforms (especially dunes). • A systemic bias related to dune migration not captured in different time-sequenced 3D bathymetric data must be corrected. • Necessity to determine the geometric bedforms characteristics and celerity to correct the scour volumes and bedload estimates. • Specialized (and expensive) equipment and trained staff are needed. • Require detailed and consecutive surveys along a river reach (time-consuming). • Difficulties in consecutively tracking the same boat paths in large rivers. • Uncertainties related to the bathymetry resolution (spatial and temporal). • Uncertainties related to the interpolation applied to create the digital elevation models (DEMs). • Requires that the scoured volumes be equal to the depositional volumes (steady regime).

Table 4. *Cont.*

Method	Advantages	Disadvantages
Empirical Formulae	• Easy to implement. • Wide use and discussion in the literature.	• Field measurements are necessary for the determination of hydraulic and sedimentometric parameters. • Parameters are space-time averaged for the river cross-section, not accounting for local effects. • Each equation is developed for specific site conditions. • The formulations predict the maximum bedload transport rate that a flow in equilibrium can produce.

6. Conclusions

Field measurements were carried out during a flood season on a sand-bed river with high sediment transport (bedload and suspended load), when transport stages were between 11 and 16. A SonTek M9-ADCP was employed for different purposes during the field campaign: (a) measure discharges at the main cross-section; (b) perform moving-bed tests at six verticals along the cross-section; and (c) simultaneously map bathymetry and flow velocities throughout the 1.0-km length of the study reach.

Bedload sediment transport rates were estimated using the following four approaches: (1) physical sampling, using the Helley-Smith mechanical trap; (2) moving-bed tests with ADCP (bottom tracking feature); (3) dune tracking, based on a modified ISSDOTv2 method; and (4) five empirical equations (Einstein [17]; Einstein-Brown [18]; Kalinske [19]; Meyer-Peter and Müller [20]; Van Rijn [21]). Results were consistent, with discrepancy ratios between 0.5 and 2, except for the HS measurements that underestimated the bedload transport.

Data collected with ADCP may be used for purposes other than those used in the present work. Williams et al. [56], for instance, employed dense observations of depths and flow velocities acquired from an ADCP to calibrate and validate a hydrodynamic model implemented for a braided river. In this regard, the data collected at the Taquari River were also used in three-dimensional hydrosedimentological modeling with Delft3D software [25], the results of which are presented in the author's thesis [57].

The simultaneously collected and spatially distributed depths and flow velocities are extremely useful for mapping hydrodynamic and sedimentological behaviors and increasing our understanding of local physical processes. The application of these data-points with empirical formulations, such as Van Rijn [21], allows the analysis of bedload transport distribution throughout the monitored area, overcoming the limitations inherent to cross-section averaged estimates. Although this procedure was not shown in this paper, some preliminary results indicate good agreement with the total bedload rates presented here [57].

The calculation of moving-bed particle velocity from the difference in bottom-tracking and RTK-GNSS-derived velocities, for the case when the boat is in motion, is a routine not yet available, neither on RiverSurveyor software nor on HydroSurveyor software. The development of a tool integrated into HydroSurveyor that encompasses this issue would be a great contribution to mapping bedload velocities and their spatial variability throughout a river reach. It was suggested to the ADCP manufacturer.

Another promising approach to estimate bedload rates in rivers with bedforms is the acoustic mapping velocimetry (AMV) method [58]. The technique combines components and processing protocols from two contemporary nonintrusive approaches: acoustic and image-based. The bedform mapping is conducted with acoustic surveys (preferably multibeam-echo soundings), while the estimation of the velocity of the bedforms is obtained with processing techniques pertaining to image-based velocimetry. The bedload rates are then computed by applying Equation (10) [38], where dune heights can be calculated manually and individually for each dune or by using an automated technique [59].

Author Contributions: Conceptualization, P.R., T.B.B., R.B.P. and F.V.G.; methodology, P.R., T.B.B., R.B.P. and F.V.G.; formal analysis, P.R.; investigation, P.R., T.B.B., R.B.P. and F.V.G.; resources, T.B.B. and F.V.G.; data curation, P.R. and R.B.P.; writing—original draft preparation, P.R.; writing—review and editing, P.R., T.B.B. and R.B.P.; supervision and project administration, T.B.B. and F.V.G.; funding acquisition, T.B.B. and F.V.G. All authors have read and agreed to the published version of the manuscript.

Funding: This research was partially funded by the following: (1) CNPq (Conselho Nacional de Pesquisa—National Research Council) productivity stipend (no. 8786885193878624, call: CNPq no. 12/2017); (2) CAPES (Coordenação de Aperfeiçoamento de Pessoal de Nível Superior—Coordination for the Improvement of Higher Education Personnel Brasil)—Finance Code 001; (3) FINEP (Financiadora de Estudos e Projetos—Financier of Studies and Projects)—MORHIS Project (01.13.0455.00); (4) UFMS (Universidade Federal do Mato Grosso do Sul—Federal University of Mato Grosso do Sul).

Institutional Review Board Statement: Not applicable.

Informed Consent Statement: Not applicable.

Data Availability Statement: All data that support the findings of this study are available from the corresponding author upon reasonable request.

Acknowledgments: We would like to cordially thank the Federal University of Mato Grosso do Sul (UFMS) and the Federal University of Paraná (UFPR) for the logistic support and financing during field surveys. The support and assistance provided by other researchers from UFMS and UFPR during field activities was also critical to the successful data collection. We would also like to thank the labs (LAMIR, LABEAM, LAQUA and HEROS) for analyzing the sediment samples and the Technological Institute of Transport and Infrastructure (ITTI-UFPR) for the support during the research.

Conflicts of Interest: The authors declare no conflict of interest. The funders had no role in the design of the study; in the collection, analyses, or interpretation of data; in the writing of the manuscript; or in the decision to publish the results.

References

1. Latosinski, F.G.; Szupiany, R.N.; Guerrero, M.; Amsler, M.L.; Vionnet, C. The ADCP's Bottom Track Capability for Bedload Prediction: Evidence on Method Reliability from Sandy River Applications. *Flow Meas. Instrum.* **2017**, *54*, 124–135. [CrossRef]
2. Edwards, T.K.; Glysson, G.D. *Field Methods for Measurement of Fluvial Sediment*; Techniques of water-resources investigations of the United States Geological Survey; U.S. Geological Survey: Denver, CO, USA, 1999; ISBN 978-0-607-89738-8.
3. Frings, R.M.; Vollmer, S. Guidelines for Sampling Bedload Transport with Minimum Uncertainty. *Sedimentology* **2017**, *64*, 1630–1645. [CrossRef]
4. Claude, N.; Rodrigues, S.; Bustillo, V.; Bréhéret, J.-G.; Macaire, J.-J.; Jugé, P. Estimating Bedload Transport in a Large Sand–Gravel Bed River from Direct Sampling, Dune Tracking and Empirical Formulas. *Geomorphology* **2012**, *179*, 40–57. [CrossRef]
5. Gaeuman, G.; Jacobson, R.B. Field Assessment of Alternative Bed-Load Transport Estimators. *J. Hydraul. Eng.* **2007**, *133*, 1319–1328. [CrossRef]
6. Kostaschuk, R.; Villard, P.; Best, J. Measuring Velocity and Shear Stress over Dunes with Acoustic Doppler Profiler. *J. Hydraul. Eng.* **2004**, *130*, 932–936. [CrossRef]
7. Rennie, C.D.; Millar, R.G.; Church, M.A. Measurement of Bed Load Velocity Using an Acoustic Doppler Current Profiler. *J. Hydraul. Eng.* **2002**, *128*, 473–483. [CrossRef]
8. Villard, P.; Church, M.; Kostaschuk, R. Estimating Bedload in Sand-Bed Channels Using Bottom Tracking from an Acoustic Doppler Profiler. *Fluv. Sedimentol. VII* **2005**, *35*, 197–209. [CrossRef]
9. Rennie, C.D.; Millar, R.G. Deconvolution Technique to Separate Signal from Noise in Gravel Bedload Velocity Data. *J. Hydraul. Eng.* **2007**, *133*, 845–856. [CrossRef]
10. Yorozuya, A.; Kanno, Y.; Fukami, K.; Okada, S. Bed-Load Discharge Measurement by ADCP in Actual Rivers. In *Proceedings of the River Flow*; Bundesanstalt für Wasserbau: Braunschweig, Germany, 2010; pp. 1687–1692.
11. Rennie, C.D.; Vericat, D.; Williams, R.D.; Brasington, J.; Hicks, M. Calibration of Acoustic Doppler Current Profiler Apparent Bedload Velocity to Bedload Transport Rate. In *Gravel-Bed Rivers*; Tsutsumi, D., Laronne, J.B., Eds.; John Wiley & Sons, Ltd.: Chichester, UK, 2017; pp. 209–233. ISBN 978-1-118-97143-7.

12. Simons, D.B.; Richardson, E.V.; Nordin, C.F. Bedload Equation for Ripples and Dunes. *Geol. Surv. Prof. Pap. 462 H* **1965**, 1–9. [CrossRef]
13. Villard, P.V.; Church, M. Dunes and Associated Sand Transport in a Tidally Influenced Sand-Bed Channel: Fraser River, British Columbia. *Can. J. Earth Sci.* **2003**, *40*, 115–130. [CrossRef]
14. Abraham, D.; Kuhnle, R.A.; Odgaard, A.J. Validation of Bed-Load Transport Measurements with Time-Sequenced Bathymetric Data. *J. Hydraul. Eng.* **2011**, *137*, 723–728. [CrossRef]
15. Shelley, J.; Abraham, D.; McAlpin, T. Removing Systemic Bias in Bed-Load Transport Measurements in Large Sand-Bed Rivers. *J. Hydraul. Eng.* **2013**, *139*, 1107–1111. [CrossRef]
16. Gutierrez, R.R.; Abad, J.D.; Parsons, D.R.; Best, J.L. Discrimination of Bed Form Scales Using Robust Spline Filters and Wavelet Transforms: Methods and Application to Synthetic Signals and Bed Forms of the Río Paraná, Argentina: BED FORM SCALE DISCRIMINATION. *J. Geophys. Res. Earth Surf.* **2013**, *118*, 1400–1418. [CrossRef]
17. Einstein, H.A. *The Bed-Load Function for Sediment Transportation in Open Channel Flows*; Technical Bulletins; United States Department of Agriculture, Economic Research Service: Washington, DC, USA, 1950.
18. Brown, C.B. Chapter XII Engineering Hydraulics. In *Sediment Transportation*; John Wiley and Sons: New York, NY, USA, 1950.
19. Kalinske, A.A. Movement of Sediment as Bed Load in Rivers. *Trans. AGU* **1947**, *28*, 615. [CrossRef]
20. Meyer-Peter, E.; Müller, R. Formulas for Bed-Load Transport. *IAHSR 2nd Meet. Stockh. Append. 2* **1948**, 39–64.
21. van Rijn, L.C. Unified View of Sediment Transport by Currents and Waves. I: Initiation of Motion, Bed Roughness, and Bed-Load Transport. *J. Hydraul. Eng.* **2007**, *133*, 649–667. [CrossRef]
22. Hamilton, S.K.; Sippel, S.J.; Melack, J.M. Inundation Patterns in the Pantanal Wetland of South America Determined from Passive Microwave Remote Sensing. *Arch. Für Hydrobiol.* **1996**, 1–23. [CrossRef]
23. Mueller, D.S.; Wagner, C.R. Measuring Discharge with Acoustic Doppler Current Profilers from a Moving Boat. *Geol. Surv. Tech. Methods 3A 22* **2009**. [CrossRef]
24. SonTek. *HydroSurveyor User's Manual, Software Version 1.5, Firmware Version 3.80*; SonTek—A Xylem Brand: San Diego, CA, USA, 2014.
25. Deltares. *Delft3D-FLOW, User Manual, Hydro-Morphodynamics*, Version: 3.15; Deltares: Delft, The Netherlands, 30 August 2018; 672p.
26. APHA—American Public Health Association. *Standard Methods for the Examination of Water and Wastewater*, 23rd ed.; Bridgewater, L.L., Baird, R.B., Eaton, A.D., Rice, E.W., Eds.; American Public Health Association: Washington, DC, USA, 2017; ISBN 978-0-87553-287-5.
27. van Rijn, L.C. Sediment Transport, Part I: Bed Load Transport. *J. Hydraul. Eng.* **1984**, *110*, 1431–1456. [CrossRef]
28. van Rijn, L.C. *Principles of Sediment Transport in Rivers, Estuaries and Coastal Seas*; Aqua Publications: Amsterdam, The Netherlands, 1993; ISBN 978-90-800356-2-1.
29. Rennie, C.D.; Villard, P.V. Site Specificity of Bed Load Measurement Using an Acoustic Doppler Current Profiler. *J. Geophys. Res.* **2004**, *109*, F03003. [CrossRef]
30. Batalla, R.J. Evaluating Bed-Material Transport Equations Using Field Measurements in a Sandy Gravel-Bed Stream, Arbúcies River, NE Spain. *Earth Surf. Process. Landf.* **1997**, *22*, 121–130. [CrossRef]
31. Ancey, C. Bedload Transport: A Walk between Randomness and Determinism. Part 1. The State of the Art. *J. Hydraul. Res.* **2020**, *58*, 1–17. [CrossRef]
32. Ancey, C. Bedload Transport: A Walk between Randomness and Determinism. Part 2. Challenges and Prospects. *J. Hydraul. Res.* **2020**, *58*, 18–33. [CrossRef]
33. Einstein, H.A. Formulas for the Transportation of Bed Load. *Trans. ASCE* **1942**, *107*, 561–573. [CrossRef]
34. Vanoni, V.A. *Sedimentation Engineering. Manuals and Reports on Engineering Practice No. 54*; American Society of Civil Engineers—ASCE: New York, NY, USA, 1975.
35. Fredsoe, J. Unsteady Flow in Straight Alluvial Streams. Part 2. Transition from Dunes to Plane Bed. *J. Fluid Mech.* **1981**, *102*, 431–453. [CrossRef]
36. Hubbell, D.W. Apparatus and Techniques for Measuring Bedload. *Geol. Surv. Water Supply Pap. 1748* **1964**, 74. [CrossRef]
37. Abraham, D.; McAlpin, T.; May, D.; Pratt, T.; Shelley, J. Update on ISSDOTv2 Method for Measuring Bed-Load Transport with Time Sequenced Bathymetric Data. In Proceedings of the 3rd Joint Federal Interagency Conference on Sedimentation and Hydrologic Modeling, SEDHY 2015, Reno, NE, USA, 19–23 April 2015.
38. Richardson, E.V.; Simons, D.B.; Posakony, G.J. *Sonic Depth Sounder for Laboratory and Field Use*; U.S. Dept. of the Interior, Geological Survey: Whashington, DC, USA, 1961; p. 7.
39. Gutierrez, R.R.; Mallma, J.A.; Núñez-González, F.; Link, O.; Abad, J.D. Bedforms-ATM, an Open Source Software to Analyze the Scale-Based Hierarchies and Dimensionality of Natural Bed Forms. *SoftwareX* **2018**, *7*, 184–189. [CrossRef]
40. Torrence, C.; Compo, G.P. A Practical Guide to Wavelet Analysis. *Bull. Am. Meteor. Soc.* **1998**, *79*, 61–78. [CrossRef]
41. Haddadchi, A.; Omid, M.H.; Dehghani, A.A. Bedload Equation Analysis Using Bed Load-Material Grain Size. *J. Hydrol. Hydromech.* **2013**, *61*, 241–249. [CrossRef]
42. López, R.; Vericat, D.; Batalla, R.J. Evaluation of Bed Load Transport Formulae in a Large Regulated Gravel Bed River: The Lower Ebro (NE Iberian Peninsula). *J. Hydrol.* **2014**, *510*, 164–181. [CrossRef]
43. Yang, C.T. The Movement of Sediment in Rivers. *Geophys. Surv.* **1977**, *3*, 39–68. [CrossRef]

44. Gomez, B. Bedload Transport. *Earth Sci. Rev.* **1991**, *31*, 89–132. [CrossRef]
45. Helley, E.J.; Smith, W. Development and Calibration of a Pressure-Difference Bedload Sampler. *Open File Rep.* **1971**. [CrossRef]
46. Emmett, W.W. *A Field Calibration of the Sediment-Trapping Characteristics of the Helley-Smith Bed-Load Sampler*; Studies of bedload transport in river channels; U.S. Govt. Print. Off.: Washington, DC, USA, 1980; p. 44.
47. Beschta, R.L. Increased Bag Size Improves Helley-Smith Bed Load Sampler for Use in Streams with High Sand and Organic Matter Transport. In Proceedings of the Erosion and Sediment Transport Measurement, IAHS Publ. no. 133, Florence, Italy, 22–26 June 1981; pp. 17–25.
48. Druffel, L.; Emmett, W.W.; Schneider, V.R. *Laboratory Hydraulic Calibration of the Helley-Smith Bedload Sediment Sampler, Open-File Report 76-752*; Department of the Interior, Geological Survey: Bay St. Louis, MI, USA, 1976; p. 71.
49. Pitlick, J. Variability of Bed Load Measurement. *Water Resour. Res.* **1988**, *24*, 173–177. [CrossRef]
50. Gaudet, J.M.; Roy, A.G.; Best, J.L. Effect of Orientation and Size of Helley-Smith Sampler on Its Efficiency. *J. Hydraul. Eng.* **1994**, *120*, 758–766. [CrossRef]
51. Gaweesh, M.T.K.; van Rijn, L.C. Bed-Load Sampling in Sand-Bed Rivers. *J. Hydraul. Eng.* **1994**, *120*, 1364–1384. [CrossRef]
52. Childers, D. Field Comparisons of Six Pressure-Difference Bedload Samplers in High-Energy Flow. In *USGS Water Resources Investigations, Report 92-4068*; U.S. Geological Survey: Vancouver, WA, USA, 1999; p. 71.
53. Bunte, K.; Klema, M.; Hogan, T.; Thornton, C. Testing the Hydraulic Efficiency of Pressure Difference Samplers While Varying Mesh Size and Type. *Tech. Comm. Fed. Interag. Sediment. Proj.* **2017**, 117.
54. Williams, R.D.; Rennie, C.D.; Brasington, J.; Hicks, D.M.; Vericat, D. Linking the Spatial Distribution of Bed Load Transport to Morphological Change during High-Flow Events in a Shallow Braided River: Spatially Distributed Bedload Transport. *J. Geophys. Res. Earth Surf.* **2015**, *120*, 604–622. [CrossRef]
55. Gamaro, P.E.M.; Maldonado, L.H.; Castro, J.L. Aplicação Do Método Das Dunas Para Determinação Da Descarga de Fundo No Rio Paraná. In *Proceedings of the XI Encontro Nacional de Engenharia de Sedimentos*; Associação Brasileira de Recursos Hídricos (ABRH): João Pessoa-PB, Brazil, 2014; pp. 1–14.
56. Williams, R.D.; Brasington, J.; Hicks, M.; Measures, R.; Rennie, C.D.; Vericat, D. Hydraulic Validation of Two-Dimensional Simulations of Braided River Flow with Spatially Continuous ADcp Data: TWO-DIMENSIONAL SIMULATION OF BRAIDED RIVER FLOW. *Water Resour. Res.* **2013**, *49*, 5183–5205. [CrossRef]
57. Ratton, P. Mapeamento e Modelagem 3D do Transporte de Sedimentos do Leito em Rios com Dunas. Ph.D. Thesis, Federal University of Paraná (UFPR), Postgraduate Program in Water Resources and Environmental Engineering (PPGERHA), Curitiba, PR, Brazil, 2020.
58. Muste, M.; Baranya, S.; Tsubaki, R.; Kim, D.; Ho, H.; Tsai, H.; Law, D. Acoustic Mapping Velocimetry. *Water Resour. Res.* **2016**, *52*, 4132–4150. [CrossRef]
59. van der Mark, C.F.; Blom, A.; Hulscher, S.J.M.H. Quantification of Variability in Bedform Geometry. *J. Geophys. Res.* **2008**, *113*, F03020. [CrossRef]

applied sciences

MDPI

Article

Estimation of Changes in Sediment Transport along the Free-Flowing Middle Danube River Reach

Dejana Đorđević [1,*], Enikő Anna Tamás [2], Ljubiša Mihajlović [3], Csaba Abonyi [4], Aleksandar Vujanović [5] and Béla Kalocsa [6]

1 Faculty of Civil Engineering, University of Belgrade, 11000 Belgrade, Serbia
2 Faculty of Water Sciences, University of Public Service, H-6500 Baja, Hungary; tamas.eniko.anna@uni-nke.hu
3 Directorate for Inland Waterways, Ministry of Construction, Transport and Infrastructure, 11000 Belgrade, Serbia; ljmihajlovic@plovput.rs
4 Lower Danube Valley Water Authority, H-6500 Baja, Hungary; abonyics@gmail.com
5 Saobraćajni Institut CIP d.o.o., 11000 Belgrade, Serbia; acavujanovic@yahoo.com
6 Palustris Co., H-6500 Baja, Hungary; kalocsa.bela@gmail.com
* Correspondence: dejana@grf.bg.ac.rs

Abstract: The subject of this study is an approximately 300 km long Middle Danube River reach that spans from river kilometer (rkm) 1581 in Hungary to (rkm) 1255 in Serbia. The observed drying of floodplains in Hungary some thirty years ago initiated the hydrological studies. However, problems related to the navigation route maintenance of the Danube River and those in the water supply of irrigation and drainage canal networks are now present in the whole free-flowing middle Danube region. The study aims at investigating the correlation between the observed water level decrease and recorded incision of the river bed at gauging stations and the indirect estimation of the long-term sediment transport along the sand-bed reach based on the surveillance cross-sections' data collected during regular monitoring surveys on the navigable Danube. It starts with hydrological analyses of the 70-year-long time series of water level and discharge yearly data and continues with morphological and correlation studies. It ends with the estimation of sediment transport. The decreasing trend in water levels and the increasing trend in cross-sectional areas are persistent. There is a linear correlation between the two. Longitudinal changes in sediment transport indicate the existence of both degrading and aggrading riverbed reaches.

Keywords: low water levels; incision of riverbeds; sand bed; surveillance cross-sections; indirect estimation of sediment transport

Citation: Đorđević, D.; Tamás, E.A.; Mihajlović, L.; Abonyi, C.; Vujanović, A.; Kalocsa, B. Estimation of Changes in Sediment Transport along the Free-Flowing Middle Danube River Reach. *Appl. Sci.* **2023**, *13*, 10513. https://doi.org/10.3390/app131810513

Academic Editor: Gordon Gilja

Received: 16 August 2023
Revised: 11 September 2023
Accepted: 14 September 2023
Published: 21 September 2023

1. Introduction

Large alluvial rivers carry large discharges. As such, they are valuable water resources that are used for multiple purposes (e.g., drinking and industrial water supplies, irrigation, electricity production in hydropower plants, and navigation). At the same time, they can carry significant amounts of sediments that play a role in shaping the riverbed and its planform. Since water is the carrying fluid, sediment dynamics depend on hydrodynamics, which are now modified by different human interventions and uses [1]. In the long run, morphological changes can significantly affect the hydrological cycle of surface and groundwaters through permanent decreases or increases in water levels in the channel and the groundwater table, which, in turn, can produce various imbalances in the environment [2]. This is why regularly monitoring water levels (and/or discharges), on the one hand, and river sediment, on the other, are essentially important. However, some large rivers, like the Danube River, run through several countries, and along some stretches, they are boundary rivers. While the fact that they are transboundary rivers has advantages in the flow of goods and services, the fact that they are also boundary rivers may cause problems in crisis times, especially regarding sediment monitoring. This is the case with the downstream part of the free-flowing Middle Danube River reach (from Bezdan to Novi Sad in Serbia, a part of which is a boundary with Croatia). There has been no sediment monitoring at all along this stretch of the river, not only during the crisis in the 1990ies [3]. The situation is similar to bed load monitoring in the Hungarian reach, where the bed load is sampled only in the gravel bed stretches at the most upstream gauging stations (GSs) of the country, while data from suspended sediment sampling, which is more frequent than bed load sampling, are of questionable quality because (1) the sample is usually not related to flood events when the suspended sediment concentrations are the highest, (2) the cross-section is not usually surveyed during the measurement campaign and (3) the discharge is not usually measured simultaneously with suspended sediment sampling. In contrast, water level monitoring was not and is not susceptible to the crisis to the same extent as the sediment monitoring, and there are only a few gaps in the stage time series.

Owing to the relationship between changes in the water level (stages) and morphological changes in the riverbed (either the incision or aggradation of the channel), trends in the time series of the stage data might be an indicator of trends in riverbed changes. Riverbed deformation results from an imbalance in the sediment transport capacity of the flow and sediment supply. Since water stage monitoring is simpler and cheaper than bathymetric surveys and sediment monitoring, analyzing its trends is a convenient initial step in the assessment of the average multiannual change in sediment transport.

The first sign which initiated analyses of long-term trends in the water stage and discharge data in the free-flowing Middle Danube River reach was an alarming drying of floodplain forests in the Gemenc area in Hungary in the late 1980ies and early 1990ies. The first analyses covered the time series from 1901 to 1992 at 17 GS along the Hungarian reach and 8 GS downstream of Budapest, including Dunaújváros at river kilometer (rkm) 1581, Dunaföldvár (rkm 1560), Paks (rkm 1531), Baja (rkm 1479) and Mohács (rkm 1447) [4]. The results have shown a decreasing trend of water levels at all these gauging stations. However, there was no attempt to correlate it with the suspected incision of the riverbed, most probably initiated by river training activities, mostly carried out in the second half of the 19th century and the first half of the 20th. Recently, Tamás et al. [5]. extended hydrological analyses to the entire free-flowing Middle Danube River reach (from Dunaújváros—rkm 1581 in Hungary to Novi Sad—rkm 1255 in Serbia) following the reports on water intake problems at different locations, out of which the Bezdan pumping station which supplies the Danube-Tisza-Danube irrigation and drainage canal network in Serbia [6]; and decreasing inundation frequencies in the nature reserve Kopački rit on the right bank floodplain in Croatia [7] were outstanding. The analysis covered a 70-year-long period from 1951 to 2019. During the first 30 years of this period, many river training works were completed along this reach to facilitate navigation, thus contributing to setting up one of the ten Transeuropean transportation corridors—the Danube-Rhine corridor. Despite the obligation of all Danubian countries to regularly monitor the river channel for operational purposes, there were only a few attempts to analyse the sediment budget along the river e.g., [8] until the Interreg project "Danube sediment" [9]. The project focussed on the adjustment of the sediment monitoring methodologies and the data in the Danube River basin.

This paper aims to present a comprehensive study of the sand bed free-flowing Middle Danube River reach, with hydrological and morphological analyses and the indirect estimation of the average multiannual sediment transport along the study reach. Given the fact that the suspended sediment monitoring on the Hungarian sub-reach is relatively scarce in space and time [3] and that there was no sediment monitoring at all on the Serbian side [3], the estimation of the sediment transport is, for now, based on analyses of the bathymetric data of surveillance cross-sections from the Directorate for Inland Waterways in Serbia and the Lower Danube-Valley Water Authority in Hungary. This procedure is standard in estimating reservoir sedimentation processes, e.g., the reservoir sedimentation at Shumburit Dam in Ethiopia [10]. A similar analysis was performed along the gravel bed reach of the Danube River between rkm 1800 and rkm 1826 for the period

1992–2001 by Rákóczi and Sass [11]. Martin and Ham [12] demonstrated the advantages of a morphological method based on available field surveys in evaluating the performance of different bed load transport formulae over larger spatial and temporal scales (decadal-scale) on a 65 km long gravel-bed reach of the Fraser River. Tang et al. [13] used bathymetric data collected in two years bounding 30 years (1977 and 2006) to analyze riverbed deformation, i.e., cumulative volume change in the channel of the largest Mississippi Delta distributary—the Atchafalaya River. Analysis of the bathymetric data revealed, among other facts, a discrepancy between the transport capacity of the flow and the inflow of sediments [13].

Sediment transport can also be successfully estimated from dune-tracking methods based on bathymetric surveys of channel longitudinal profiles recorded with a certain time shift [14,15]. Recently, Gaeuman and Jacobson [16] suggested a modified method based on routing local erosion and deposition volumes. It is applicable to channels with a complex bedform morphology when it is difficult to delineate individual bedforms. However, the conventional dune-tracking method and its modifications require on-purpose bathymetric surveys. This is why it is not possible to use these methods for the estimation of sediment transport in less monitored rivers like the free-flowing Danube River reach, as on-purpose surveys have not been performed there.

The research questions to which the answers are sought in this paper are as follows:

(1) Are the previously observed decreasing trends in water levels stagnating? Did they revise, or are they continuing after 30 years from the initial studies?
(2) Is there a correlation between the observed minimum water level and the recorded incision of the riverbed at gauging stations?
(3) What are the long-term sediment transport characteristics along the investigated reach?

The paper is organized as follows. After the description of the study reach, the methods of hydrological and morphological analyses are presented, and the procedure for the estimation of the sediment transport is described. These are followed by a presentation of the results of homogeneity tests and trend analyses of water level time series and discharges, correlation analyses between the cross-sectional area change at GSs and the corresponding lowering of the water level, and the estimation of the average multiannual sediment transport within the sand bed reach with no tributaries. The relationships between all three phenomena are discussed in the fourth section. The key conclusions of the study are provided in the last section. They include an answer showing that the time series of water levels are inhomogeneous along the entire reach while the time series of discharges are homogeneous. It is found that the trend of lowering water levels and the deepening of the riverbed continues at the same rate and that there is a linear correlation between minimum water levels at GSs and percentage increases in the corresponding cross-sectional areas. At the upstream end of the sand bed reach, the 25-year sediment transport rate is positive, indicating the aggradation of the riverbed. In the remaining 86% of the investigated reaches, the 25-year sediment transport rate is negative, indicating the degradation of the riverbed.

2. Description of the Study Reach

The subject of this study is an approximately 300 km long reach spanning from rkm 1581 in Dunaújváros, Hungary, to rkm 1255 in Novi Sad, Serbia (Figure 1), and it belongs to the Middle Danube. The reach is situated between the Gabcikovo (rkm ~1800) and Iron Gate I (~rkm 1000) dams and hydropower plants. The Danube River is alluvial downstream of ~rkm 1600, although the transition from gravel bed to sand bed is around Paks in Hungary (rkm 1531). Since the Danube is a free-flowing river along this reach, in the course of creating the Danube waterway, now part of the European TEN-T network, there was extensive river-training work along the reach in the past. This included cutting off mostly overdeveloped meander bends and the installation of different stone structures after the 1870s (e.g., groynes and longitudinal stone toes). These river-training measures were known for increasing flow energy and sediment transport capacities, and they resulted in deepening erosion in the riverbed. Additionally, there were intensive dredging activities in the upstream part of the study reach between Dunaújváros (rkm 1581) and Paks (rkm 1531) in the second half of the 20th century until the beginning of the 2000s, which provoked further morphological changes in the river channel.

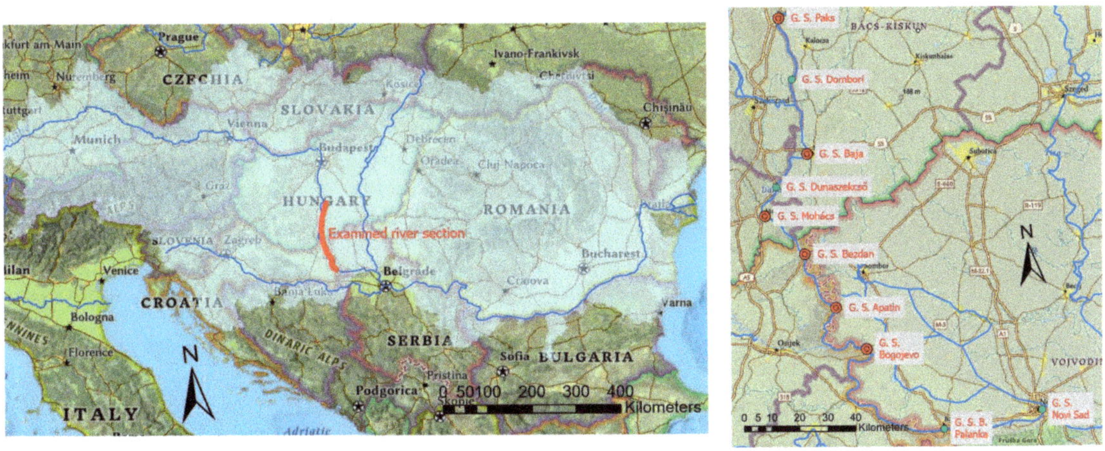

(a) (b)

Figure 1. (**a**) Map of Central Europe with the Danube River Basin and the Danube River (the investigated section is highlighted in red). (**b**) Detailed map of the investigated reach; the gauging stations used in the detailed analysis in the present study are marked with red dots.

To the best of the authors' knowledge, no comprehensive analysis of the changes in the water regime or, at the same time, the morphological parameters of the entire free-flowing alluvial reach of the middle Danube River has been performed to date. To fill this gap, available time series of the water level and discharge data for the entire free-flowing alluvial reach of the Danube River and the morphological evolution of cross-sections at corresponding gauging stations (GSs) are analyzed in the present study. Moreover, the digital database of surveillance cross-sections from the state institutions authorized for hydrographic surveys along the international waterway is used to assess variations in the long-term sediment transport along the reach.

3. Materials and Methods

3.1. Methods of Statistical Analysis of Hydrological Data

In the first phase of the hydrological study, analyses were performed using time series of water levels and discharges for the 70-year-long period between 1950 and 2019 for the entire free-flowing alluvial reach of the River Danube downstream of Budapest, Hungary, and upstream of Slankamen, Serbia (~300 km). This was considered necessary, as the older studies published on this topic used different periods, older datasets, and different/shorter river reaches, e.g., in [17]. The analysis included the official water level and discharge data of the Hungarian Hydrological Forecasting Service and the Hydrometeorological Service of the Republic of Serbia for a total of ten gauging stations (Figure 1b): Dunaújváros (rkm 1581), Dunaföldvár (rkm 1560), Paks (rkm 1531), Baja (rkm 1479) and Mohács (rkm 1447) in Hungary, and Bezdan (rkm 1425), Apatin (rkm 1402), Bogojevo (rkm 1367), Bačka Palanka (rkm 1299), and Novi Sad (rkm 1255) in Serbia, similar to Tamás et al. [5].

The first step in this analysis was a test of the homogeneity of the time series. Computations were made in XLStat (Addinsoft, 2023) [18]. A two-tailed, two-sample Kolmogorov–Smirnov test at a confidence level of $\alpha = 0.05$ was used. The homogeneity test was first performed for values of the yearly characteristics of water levels and discharges for all stations. Consequently, the trend analysis was performed for the annual characteristic values to find trend parameters. Given the fact that there are no large tributaries between the upstream end of the reach and Apatin, it is presumed that discharges upstream of the Drava River confluence, which is downstream of Apatin, do not change significantly. That is why annual discharge data were analyzed only at two gauging stations, Baja and Bezdan, which are located upstream of the confluence of the Drava River. The discharges for the two stations were also tested for normality using XLStat [19].

For further analyses, as the intention was to connect water level changes with morphology, the two stations at the upstream end of the reach (Dunaújváros and Dunaföldvár) were excluded, as the bed material of the river there contains a relatively large portion of gravel, and according to previous studies [20] dredging played a role in the lowering of the water levels on this part of the reach. The two stations at the downstream end of the reach (Bačka Palanka and Novi Sad) were also excluded as the backwater effect and sediment deposition upstream of the Iron Gate I Dam (~rkm 1000) can influence water levels there [14]. The trend of the yearly characteristic values of water level data at the six remaining stations (Paks, Baja, Mohács, Bezdan, Apatin, and Bogojevo) was examined. All trend analyses were performed using linear regression with the least squares (LSQ) method in MS Excel.

3.2. Bathymetric Data Collection

Bathymetric data that were used as a basis for the estimation of the sediment transport were obtained from the databases of the Directorate for Inland Waterways—Plovput in Serbia, the former Environmental and Water Management Research Institute (VITUKI), and the Lower-Danube-Valley Water Directorate (LDVWD) in Hungary. These included 23 surveillance cross-sections in Hungary at an average distance of approximately 2000 m and 46 surveillance cross-sections in Serbia at an average distance of approximately 1000 m. All cross-sections were distributed along the sand bed reach, spanning from Baja (rkm 1479 + 000) to the proximity of the confluence of the Drava and Danube Rivers (rkm 1387 + 200) (Figure 2).

Plovput is the state institution responsible for hydrographic surveys on international inland navigable waterways in Serbia. Plovput's typical cross-section single-beam surveying equipment includes the following:

- Survey Echo Sounder with a depth measurement precision of 1 cm +/− 0.1% depth;
- DGPTS-RTK positioning with a measurement precision of +/−20 cm;
- Data-processing and post-processing applications.

VITUKI was a coordinator for hydrographic surveys on the Danube River until 2012, and the LDVWD is in charge of river management and navigation routes on the Danube reach downstream of rkm 1560 + 000. The hydrographic survey equipment of VITUKI included a single-beam ultrasonic echosounder with an accuracy of 2 cm and a total station on the shore. The accuracy of the total station was less than 2 mm at 200 m in the robotic survey mode. The equipment of the LDVWD consisted of a single-beam ultrasonic echosounder and an RTK GNSS receiver. The accuracy of both instruments was 2 cm.

The echosounder used to survey cross-sections in Serbia was mounted on a 5 m long boat, specifically equipped for this activity. All collected data (which contained the x, y, and z coordinates of the measured points) are in the State Geographical Coordinate System. Before the commencement of a survey, the coordinates of boundary points were entered into the software. Survey lines ran along predefined cross-sections. Data on location and depth (x, y, z) were transferred to the software for hydrographic surveys. The software synchronized the data constantly. Thus, the boat position was supplied in real time. The speed of sound in water was entered according to the information provided by an SVP (sound velocity profiler) device. A differential GPS station was mounted at the reference point with known geographic coordinates. A station was connected to the boat with the radio signal, sending information on differential correction and providing the required accuracy for the performed survey. Depth information was obtained by using the time necessary for ultrasound waves to travel from the echosounder to the river bottom and back. Data on depth and location were synchronized in real time, and information was stored in ASCII format in the form of x, y (position), and z (depth) coordinates. After completing the survey, quality control was performed, spikes were removed, and data were stored in the database with cleaned x, y, and z coordinates for each cross-section.

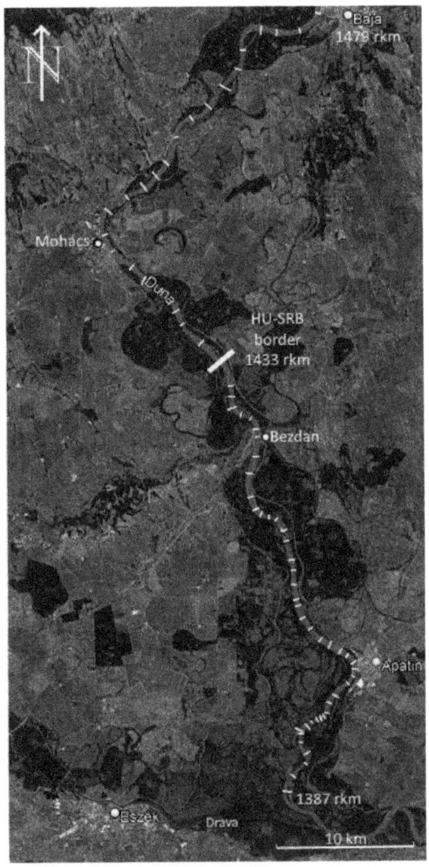

Figure 2. Investigated river reach with indicated positions of surveillance cross-sections (yellow marks) on the Hungarian and Serbian sides. The white line presents the border. The average distance between the cross-sections in Hungary is approximately 2000 m and in Serbia 1000 m.

VITUKI surveyed cross-sections with a special motor boat. The total station on the occupied stance on the shore followed the active signal, coming from the echosounder beacon-receiver placed on the same rod with fixed height and without horizontal offset. The processing result was a 3D point (x, y, and z coordinate) list.

The LDVWD surveyed the bed with satellite positioning (RTK GNSS) and single-beam ultrasonic depth measurement equipment based on a specific cross-section plan prepared in advance. To convert the satellite positioning coordinates into a Cartesian spatial coordinate system, original positioning in the global system and the officially accepted Field Transformation Procedure were used.

3.3. *Analysis of Morphological Changes*

Since Hungary and Serbia use different surveying systems, it was necessary to make all data comparable. After the conformance of the data in the two countries, the cross-sections were visualized and processed using the AutoCAD software (Map 3D 2011) and Plovput's in-house AutoCAD routines.

To establish a standard for comparisons along the stretch being studied, the next step involved defining a reference. To achieve this, Étiage Navigable (EN) was selected, which is not a reference *level* but a reference *water surface elevation profile* along the reach, belonging to a relatively low discharge and hydraulically steady state. EN was originally determined by the Danube Commission in 2012 for each gauging station. This requires that each cross-section has its own reference level. However, as EN is significantly below the bankfull stage, it was decided to raise it by 2 m. Once this was accomplished, the area of each cross-section below the newly established EN + 2 m was calculated.

The correlation between the decrease in water levels and the incision of the riverbed was analyzed using cross-sectional areas of gauging stations along the sand bed reach upstream of the Drava River confluence. The analysis was conducted for the selected years during the 60-year-long period (years that had hydrographic surveys performed in both countries). In total, eight different years were found, starting from 1975, and the first available hydrographic survey data in Serbia from 1964 were also included in the analysis of the GSs on the Serbian side. The rate of cross-sectional area change was analyzed first, followed by a correlation between the percentages of the cross-sectional increase and minimum water levels. The percentage of the cross-sectional increase was expressed with reference to the first year with available bathymetric data. All trend and correlation analyses were performed using linear regression with the LSQ method in MS Excel.

3.4. Indirect Method for Estimation of the Sediment Transport

Sediment transport was inferred indirectly from the 69 surveillance cross-sections from Baja to the confluence of the Drava River. The study is limited to this shortened sand bed reach, with no large tributaries with massive sediment inputs. Surveillance cross-sections are from the databases of the three previously mentioned institutions. The procedure for defining the cross-sectional area is described in the previous subsection. The estimation of the sediment transport should be based on data from the two years when hydrographic surveys were made in both countries. Since there was no overlap in the last 25 years, we decided to use data from surveys in two consecutive years in the two countries. These were datasets collected in 1996 and 2021 by Plovput and datasets collected in 1997 and 2022 by VITUKI and LDVWD, respectively. The datasets were used for the estimation of 25-year sediment transport changes along Serbian and Hungarian reaches. The areas of 69 cross-sections were used to calculate channel volumes between each pair of cross-sections along the study reach (*sr*):

$$\forall_{sr,i}(t) = 0.5(A_i(t) + A_{i+1}(t))\ \Delta x, \quad i = 1,\ 68 \tag{1}$$

The long-term sediment transport between two cross-sections is thus

$$Q_{s,i} = \frac{\forall_{sr,i}(t_1) - \forall_{sr,i}(t_2)}{\Delta t}\ \rho_s(1-\lambda) = \frac{\Delta\forall_{sr,i}}{\Delta t}\ \rho_s(1-\lambda), \quad i = 1,\ 68 \tag{2}$$

where ρ_s is sediment particle density and λ is a porosity of the bed material.

4. Results

4.1. Results of the Homogeneity Tests for the Hydrological Data

Homogeneity tests for water level yearly characteristics were performed by Tamás et al. in 2021 [5], and the results are presented in Figure 3a–c only for the Baja GS, as all gauging stations along the investigated reach showed very similar behavior. Both minimum and mean water stages and levels were highly inhomogeneous, with large D-statistic values (>0.3) and very low probabilities (p) (less than 0.01; at some GSs, these were not even traceable; see Table 1). For the sake of the completeness of the results and the explanation for why the evaluation of the sediment transport was limited to one part of the investigated reach, the D-statistics and p-values are taken from [5] and presented again in Table 1.

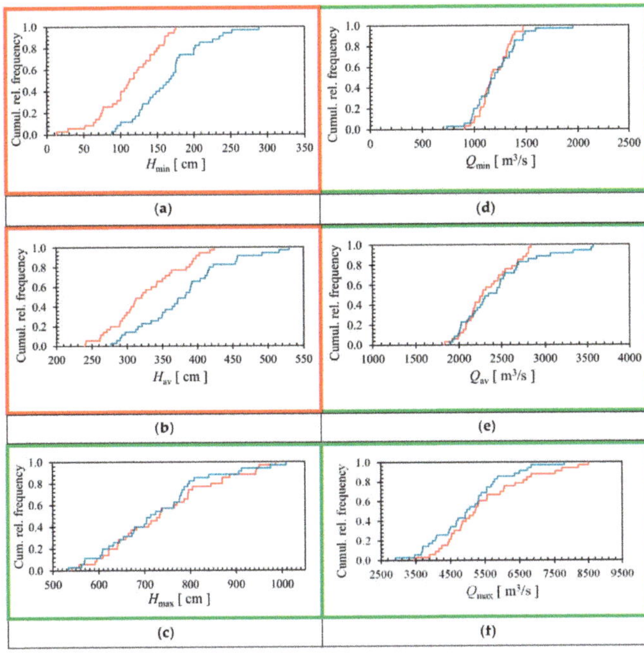

Figure 3. Results of the two-tailed (α = 0.05), two-sample Kolmogorov–Smirnov test of the homogeneity of the time series of (**a–c**) water stages and (**d,e**) discharges at the Baja GS. The first row of panels (**a,d**) refers to the water stage and discharge minima, respectively; the second row (**b,e**) refers to the mean values, respectively; and the third row (**c,f**) refers to the water stage and discharge maxima, respectively. It is readily apparent that the time series of water stage minima and means are non-homogeneous (red frames), while all discharge data and water stage maxima are homogeneous (green frames). Blue lines—data for the first 35 years (1950–1984). Red lines—data for the second time interval of 35 years (1985–2019).

Table 1. Results of the homogeneity test of water level yearly characteristics [5].

Gauging Station	Minima			Means			Maxima		
	D	*p*	Homog.	*D*	*p*	Homog.	*D*	*p*	Homog.
Dunaújváros	1.000	<0.0001	NO	0.943	<0.0001	NO	0.343	0.033	NO
Dunaföldvár	0.971	<0.0001	NO	0.857	<0.0001	NO	0.257	0.197	YES
Paks	0.714	<0.0001	NO	0.543	<0.0001	NO	0.159	0.837	YES
Baja	0.457	0.001	NO	0.400	0.007	NO	0.114	0.976	YES
Mohács	0.457	0.001	NO	0.429	0.003	NO	0.086	1.000	YES
Bezdan	0.371	0.016	NO	0.371	0.016	NO	0.114	0.976	YES
Apatin	0.429	0.003	NO	0.371	0.016	NO	0.114	0.976	YES
Bogojevo	0.429	0.003	NO	0.429	0.003	NO	0.171	0.683	YES
Bačka Palanka	0.257	0.255	YES	0.357	0.038	NO	0.121	0.976	YES
Novi Sad	0.229	0.320	YES	0.171	0.683	YES	0.114	0.976	YES

The homogeneity test was also performed for discharges, and the results are presented in Table 2, as in [5], but only for the two gauging stations, as was explained in Section 3.1. They are graphically presented only for the Baja GS (Figure 3d–f), as they are very similar for all other GSs. Unlike the water level data, the yearly characteristic discharges at the investigated gauging stations are homogeneous for all three characteristics—minima, means, and maxima (all *D*-statistic values are less than 0.2, and values of probability

(p) are much larger than 0.70). In the same study ([5]), it was shown that the annual discharge minima, means, and maxima showed no trend at the downstream end of the study reach over the last 70 years.

Table 2. Results of the homogeneity test of yearly discharge characteristics (taken from [5]).

Gauging Station	Minima			Means			Maxima		
	D	p	Homogen	D	p	Homogen	D	p	Homogen
Baja	0.139	0.896	YES	0.143	0.879	YES	0.190	0.569	YES
Bezdan	0.143	0.867	YES	0.171	0.683	YES	0.114	0.976	YES

The results of the discharge normality test are visualized in Figure 4. Although the frequency of the measured data (green columns) shows slight asymmetry, the test undoubtedly indicates that all three yearly discharge characteristics follow normal distributions at all four gauging stations.

4.2. Results of the Hydrological Data Trend Analysis

The results of the trend analysis of the annual characteristic values of water levels are presented in Figure 5, based on data from [5]. The data show a linear decreasing trend at all gauging stations; the trendline slope is negative along the reach except in Novi Sad at the downstream end, where the free-flowing Danube enters the Iron Gate I reservoir. It is readily apparent that the rate of decrease in minimum water levels decreases in the downstream direction. The decreasing rate is more than three times greater on the gravel bed (approximately 3.9 cm/year) than on the sand bed sub-reach, where the average decreasing rate upstream of the confluence of the Drava River is 1.1 cm/year. Along the sand bed reach between Baja and Bogojevo (just upstream of the Drava River confluence), it almost linearly decreases from 1.3 cm/year to 0.9 cm/year. Based on these data, the total change in water level minima (ΔZ) for the analyzed 70-year period ranges from 1.45 m at the beginning of the sand bed reach (Paks) to approximately 0.60 m at Bogojevo (Figure 6). Between Mohács and Apatin, the total drop in water levels changes almost linearly from 0.76 m to 0.70 m. The corresponding average yearly water level decrease (δZ) is around 2 cm/year at Paks, 1.3 cm/year at Baja, and approximately 1 cm/year between Mohács and Apatin. Downstream of the Drava River confluence, it is less than 0.9 cm/year.

Variation of trends for the mean values on the gravel-bed sub-reach is almost the same as for minima, while that on the sand-bed reach is similar, except the rates are higher. Since means include the influence of water level maxima, any anomaly in the water level maxima reflects on them. It is visible at Mohács (Figure 5), where the rate of decrease is approximately 25% greater than expected (0.0146) had not there been this anomaly in the water level maxima.

The annual characteristic values of discharges on the reach between Baja and Apatin did not change significantly during the investigated period, as annual maxima show a slightly increasing trend, while mean values and minima fluctuate around almost constant values (Q_{min} around 1100 m^3/s and Q_{mean} around 2300 m^3/s) [5].

4.3. Correlation between the Minimum Water Level and Cross-Sectional Area Changes at GSs

The evolution of cross-sections at four GSs on the sand bed reach from Baja to Apatin is presented in Figure 7. Visual inspection of the cross-sections shows that they are more or less stable starting in 1985 (dark blue line). The most stable cross-section is in Baja (Figure 7a), where the amplitude of changes is the smallest. The cross-section in Apatin is the most unstable one, having the largest amplitudes in bed level changes (Figure 7d). The analysis of the cross-sectional area change compared with the first survey supports these visual observations (Table 3). This revealed that the Apatin cross-section became stable after 2004 upon additional river training work. Thus, 2004 was used as the reference year for this GS. It is readily apparent that the rate of change (value of the parameter, a) increases in the downstream direction.

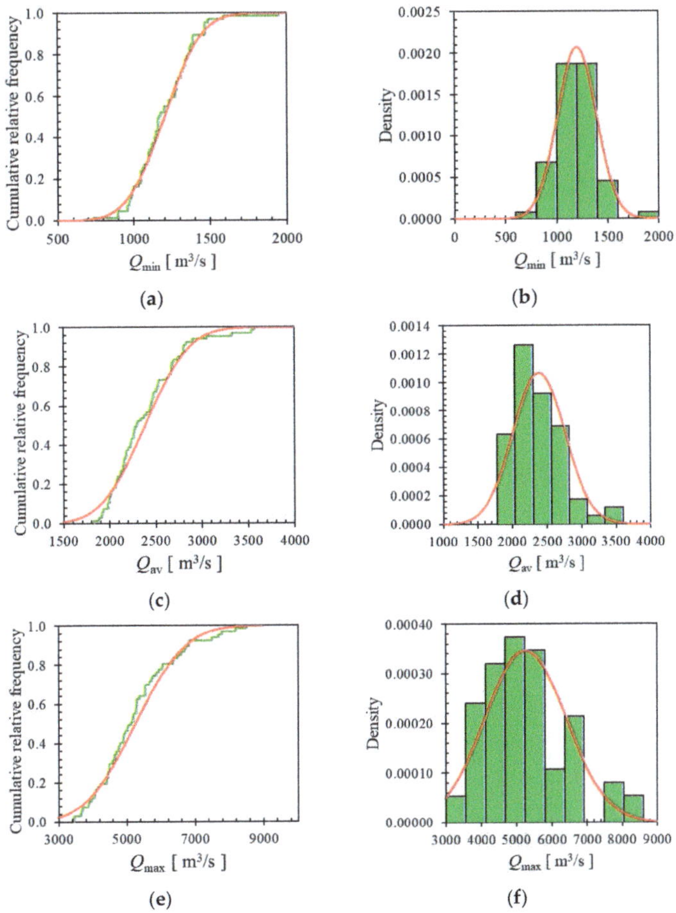

Figure 4. Normality test for discharges (**a**,**b**) minima, (**c**,**d**) means and (**e**,**f**) maxima at Baja GS. Green colour is used for empirical distributions and red for thenormal distribution in the Cumulative distribution functions (CDFs) and Probability density functions (PDFs) alike. The three characteristic values belong to normally distributed populations with the following statistics: (**a**,**b**) N (1204, 192.87), (**c**,**d**) N (2395, 374.15), and (**e**,**f**) N (5272, 1153).

Table 3. Change trends of the cross-sectional area with reference to the first survey, $A_Y/A_{ref} - 100 = a\text{Year} - b$; the reference years ($ref$) are 1975 for GSs in Hungary and 1964 for GSs in Serbia, except for Apatin, where 2004 is taken as the reference.

Gauging Station	River km	a	b
Baja	1478.70	0.08	152.36
Mohács	1446.90	0.24	489.29
Bezdan	1425.59	0.59	1172.1
Apatin	1401.90	1.92	−3890.6

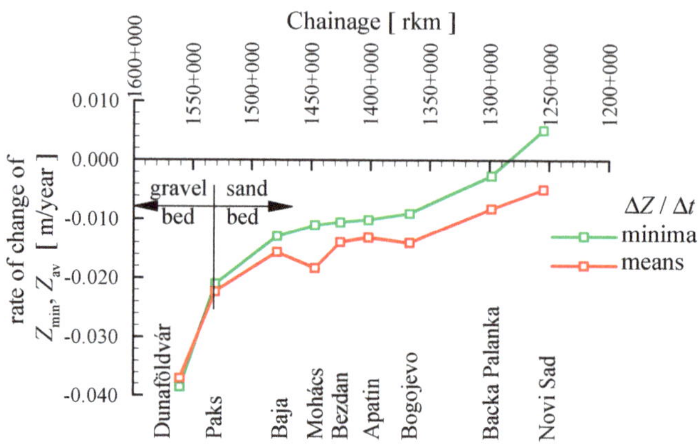

Figure 5. Variations in the trends of changes in characteristic yearly water level values—minimum and mean along the free-flowing Middle Danube River reach.

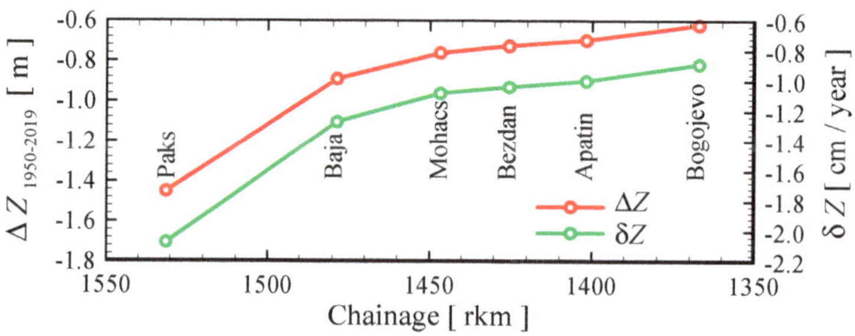

Figure 6. The total (ΔZ) and yearly (δZ) change in water level minima based on the yearly time series trends based on rkm.

The results of the analysis of rates of change in the cross-sectional area between the two successive surveys ($\Delta A/\Delta t$) at the four considered GSs are presented in Figure 8. They confirm the visual observations. The lowest rates are at Baja GS (Figure 8a). They are less than 20 m^2/year or 0.75% of the cross-sectional area below EN + 2 m, except between 2004 and 2007 (23.5 m^2/year) when there was a large flood wave in the spring of 2006 with very long overbank flow durations along the entire Danube River course. The most prominent morphological changes at the Mohács GS happened between the first three surveys—an aggradation of approximately 50 m^2/year (1975–1985) and a degradation of around 65 m^2/year (1985–1988) (Figure 8b). These changes correspond to 2% and 2.6% of the cross-sectional area each. After that, $\Delta A/\Delta t$ did not exceed 30 m^2/year, except, again, between 2004 and 2007, when it was around 35 m^2/year. Changes at Bezdan GS (Figure 8c) were less than 10 m^2/year before 1997, i.e., less than 0.5% of the cross-sectional area. Between the following two surveys, the cross-section aggraded and degraded for equal amounts of approximately 45 m^2/year or 2% of A. After the large flood wave of 2006, the changes in the cross-sectional area increased, and between the last two surveys, they exceeded 20 m^2/year, or 0.8% of A. Finally, Figure 8d quantitatively confirms that the GS in Apatin is the most unstable in the reach. This section is one of 17 critical sectors for inland navigation in Serbia. The cross-sectional area aggraded at a rate of 137 m^2/year between 1985 and 1988, which is 6.9% of A. The aggradation/degradation between other surveys was always greater than 50 m^2/year, or more than 3% of A.

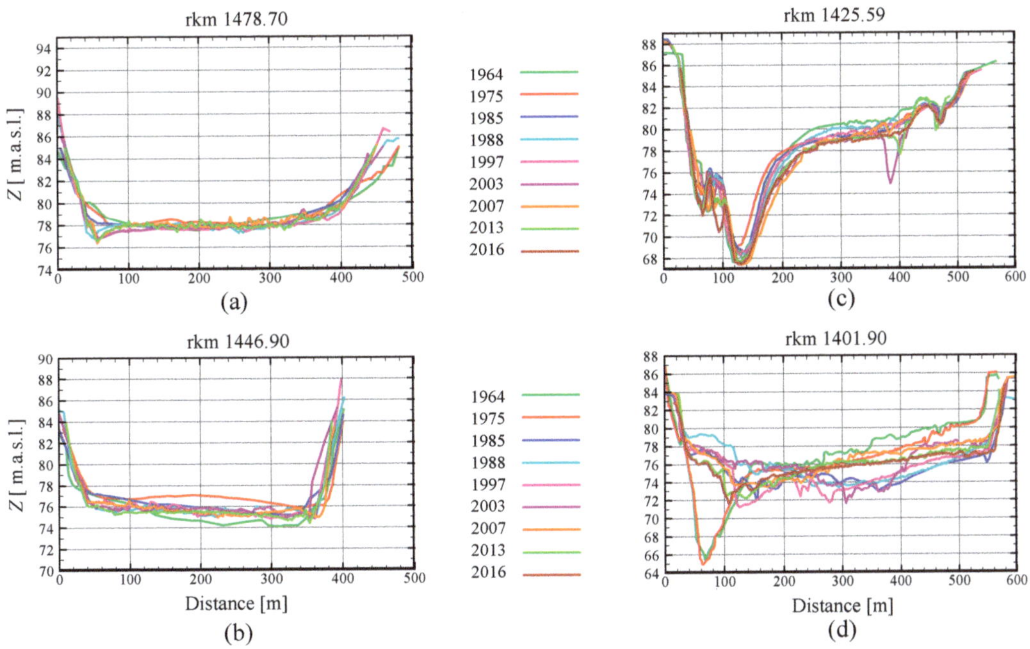

Figure 7. Evolution of cross-sections at GSs in (**a**) Baja, (**b**) Mohács, (**c**) Bezdan, and (**d**) Apatin.

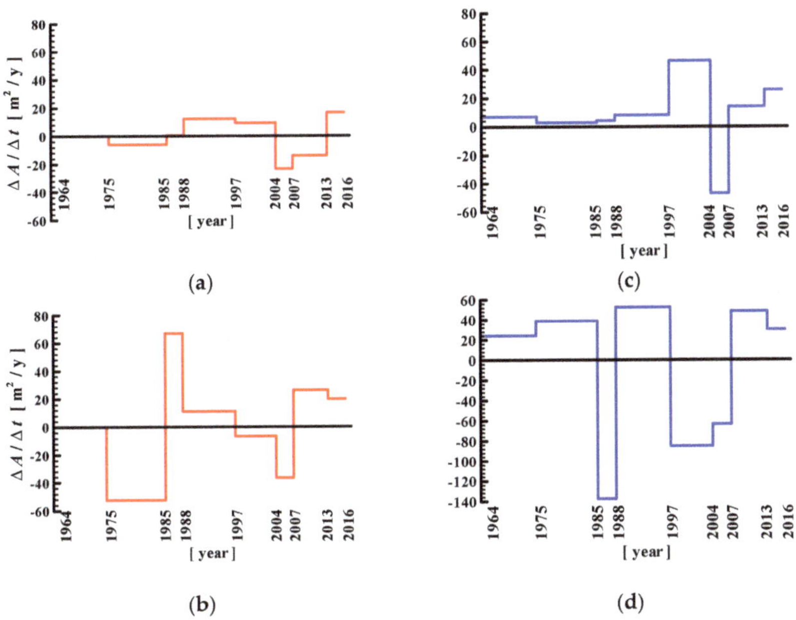

Figure 8. Rate of change in the cross-sectional area in (**a**) Baja; (**b**) Mohács; (**c**) Bezdan; and (**d**) Apatin according to available bathymetric surveys. Red color is used for GSs on the Hungarian, and blue color for GSs on the Serbian side.

Finally, the correlation between the percentages of the observed cross-sectional area increase (morphological variable) and the minimum water level (hydrological variable) is presented in Figure 9. The cross-sectional area increase is calculated compared with the reference year, i.e., the first year with available bathymetric data, as indicated in the caption of Table 3. There is a linear correlation between the two variables, and the decrease in the low water level follows the incision of the riverbed. Values of regression coefficients are provided in Table 4. Figure 9 confirms what was already presented in Figures 5 and 6, i.e., that the highest rate of water level decrease is in Baja (rkm 1478.70) and that Z_{min} decreases at a rate of 0.12 m^2 with a percentage increase in the cross-sectional area, with a coefficient of determination of $R^2 > 0.6$ (Table 4). The rate of Z_{min} decrease is almost the same as in Mohács (rkm 1446.8) and Bezdan (rkm 1425.59). It is approximately 0.03. However, the coefficient of determination in Bezdan ($R^2 = 0.62$) is significantly greater than that in Mohács ($R^2 \approx 0.09$). The trend in Apatin (rkm 1401.90) decreases for the entire period (1964–2016), i.e., the slope of the trendline is positive but very small ($a = 0.004$). Moreover, the coefficient of determination is very low ($R^2 \approx 0.02$). These values altogether indicate that there is no correlation between the morphological and hydrological variables at this particular GS.

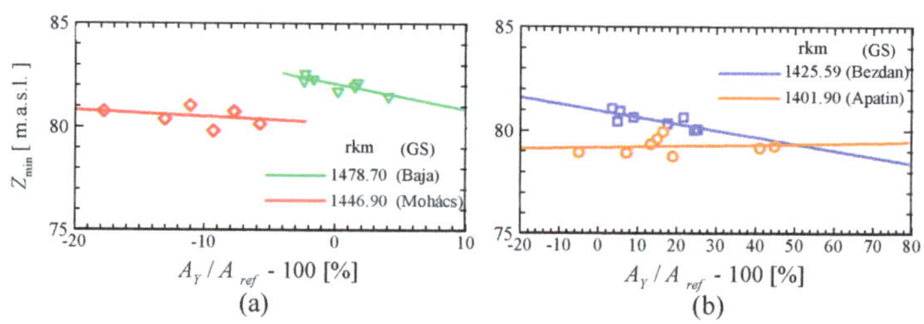

Figure 9. Correlation between the minimum water level decrease and the percentage of cross-sectional area increase. The percentage of cross-sectional area increase in a given year (A_Y) is calculated in relation to the cross-sectional area in the reference year (A_{ref}). The reference year on the Serbian side is 1964, and on the Hungarian side, it is 1975. Correlations for GSs on (**a**) Hungarian and (**b**) Serbian sides.

Table 4. Correlation coefficients in the linear regression equation, $Z_{min} = a\,(A_Y/A_{ref} - 100) + b$, where the reference years (*ref*) are 1975 for GSs in Hungary and 1964 for GSs in Serbia.

Gauging Station	River km	a	B	R^2
Baja	1478.70	−0.124	8.21	0.64
Mohács	1446.90	−0.031	8.02	0.09
Bezdan	1425.59	−0.032	8.10	0.62
Apatin	1401.90	0.004	7.92	0.02

4.4. Estimation of the Sediment Transport

Variations in the cross-sectional area (A) along the study reach in the two consecutive years 1996/1997 and 2021/2022 (as explained in Section 3.4) are presented in Figure 10. Generally, the cross-sectional area (A) increased in 52 cross-sections (75%), while it decreased in the remaining 17 cross-sections (25%). The longest stretch with the observed decrease in A is between Baja and Mohács (11 cross-sections). The percentage decrease reduces in the flow direction from 21% just downstream of Baja to 1% just upstream of Mohács. The remaining six cross-sections are on the Serbian reach. The percentage reduction is generally less than 3%. However, there are two locations with 12 and 17.5% reductions (just upstream of Bezdan and some 20 km upstream of the Drava River confluence, respectively). The stretch with the most pronounced increase in A is between Bezdan and Apatin, with a maximum enlargement of A of approximately 30%.

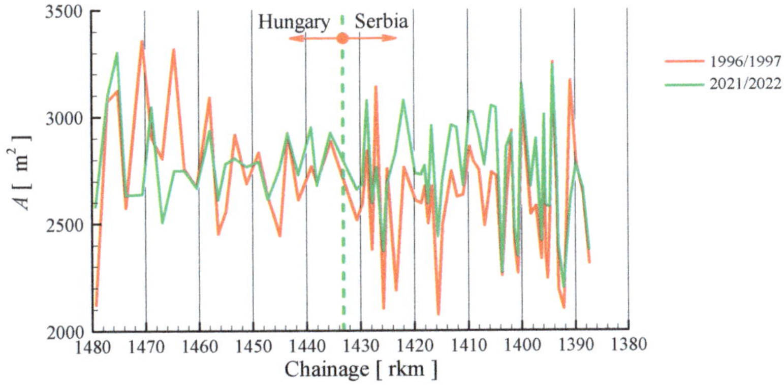

Figure 10. Longitudinal profiles of the main-channel cross-sectional area in 1996/1997 and 2021/2022.

These longitudinal profiles of A were used to estimate the 25-year long-term sediment transport, presented in Figure 11. The largest multiannual sediment transport rate (Q_s) (around $65{\cdot}10^3$ t/year) is on the Hungarian side between Baja and Mohács, where the aggradation of the riverbed is evident (Figure 10). The most intensive degradation of the riverbed is downstream of Bezdan and resulted from a multiannual sediment transport rate of approximately $45{\cdot}10^3$ t/year. Generally, downstream of Mohács, the sediment transport rate is negative, meaning there is a continuous uptake of sediments from the riverbed, i.e., the incision of the riverbed. The uptake rates are mostly greater than $10{\cdot}10^3$ t/year, and between some surveillance cross-sections, they reach approximately $25{\cdot}10^3$ t/year.

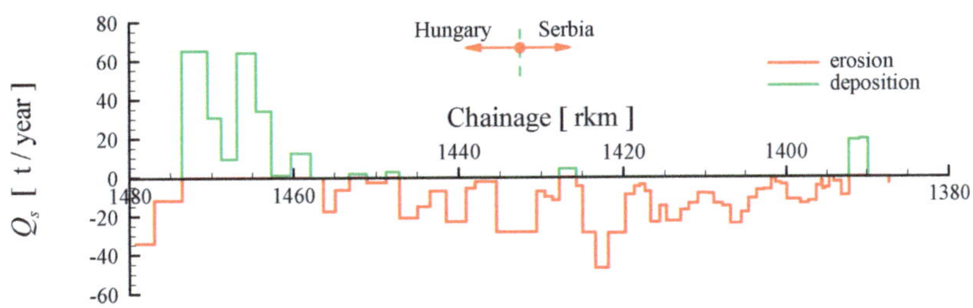

Figure 11. Twenty-five-year long-term changes in sediment transport with sediment deposition and erosion patterns.

5. Discussion

The minima and means of water levels are inhomogeneous along the entire reach except at Bačka Palanka and Novi Sad, where the Danube approaches and enters the Iron Gate Reservoir and where the effects of the backwaters are or could be felt (Table 1). The non-homogeneity of water levels results from their strong relationship with riverbed changes, which are the consequence of intensive river training work along this stretch of the river. On the other hand, the yearly maxima are homogeneous, as the shape of the main-channel riverbed at high flows merely influences the water level, and it is supposed that the riverbed incision mainly affects the riverbed of the main channel and not the floodplains. Unlike water level data, the yearly characteristic discharges at the investigated gauging stations are homogeneous for all three characteristics—minima, means, and maxima. The homogeneity of the discharge data (Figure 3d–f) can be explained by the fact that the shape of the riverbed does not affect water quantity, which arrives from upstream, and the fact that there are no large tributaries downstream of Szob (rkm 1700).

The trend analysis of a 70-year long time series for the annual characteristic water level values of the entire free-flowing Middle Danube reach [5] confirmed the decreasing trends in all three characteristics (minima, means, and maxima) found in [4,17] along the Hungarian stretch until 1992 and 2005, respectively. Inspecting the variation in the rates of change of Z_{min} and Z_{av} along the studied reach reveals that rates along the gravel bed reach are up to three times larger than those along the sand bed reach. Between Mohács and Bogojevo, the $\Delta Z/\Delta t$ is almost the same for Z_{min}, with an average value of 1.1 cm/year. Downstream of the Drava River confluence (in Bačka Palanka) and at the entrance to the Iron Gate I reservoir (in Novi Sad), the decreasing rate rapidly reduces because of the deposition of sediments brought by the Drava River and the backwater effects of the Iron Gate I Dam. Moreover, under the influence of Iron Gate I backwaters, the incision terminates and changes into aggradation, an usual physical process where a river enters a reservoir, lake, or sea. Thus, this is a positive answer to the first research question.

The second research question about the correlation between the observed minimum water levels and the recorded incision of the riverbed resulted from the fact that the water level decrease is usually a consequence of a channel incision. The riverbed incision is a continuous process in rivers with numerous cut-offs. Rectification of the river course by cutting-off meanders and disconnection of narrower side channels along anabranching river reaches was a river engineering technique applied in the 19th century aiming at facilitating navigation, flood control and agricultural production along European rivers that played a role in demographic, economic, cultural and urban development, such as the Rhine and Danube Rivers. Dredging and groyne construction might amplify the incision process, both of which happened in the second half of the 20th century along the Middle Danube. The comparison between the longitudinal profiles of the sand bed free-flowing Middle Danube River reach in 1996/1997 and 2021/2022 (Figure 12) shows that the riverbed incision continued after 2013 when a report on long-term morphological development was published [21] (this was approximately 110 years after the start of cutting-off work on the Middle Danube). Long-term analysis of longitudinal riverbed profiles of the Upper Rhine River has shown, at one location, a drop in thalweg as large as 7 m 120 years after Johann Tulla's work [22]. Thus, a recorded maximum thalweg drop of approximately 6 m in a bend upstream of the Drava River confluence (Figure 12) agrees with observations in other rectified rivers.

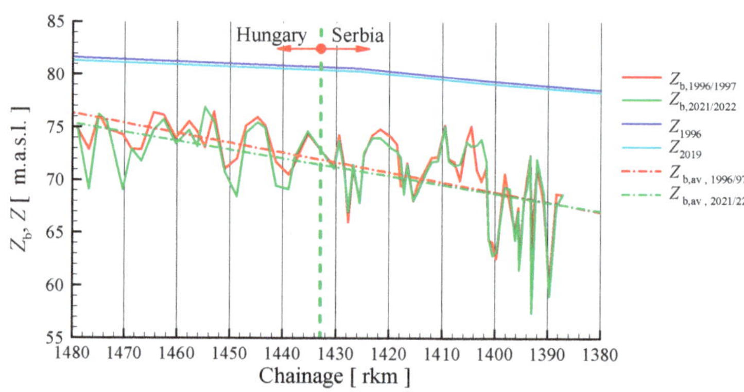

Figure 12. Actual and averaged longitudinal profiles of the Danube River with corresponding water surface profiles in 1996/1997 and 2021/2022.

In addition to actual longitudinal riverbed profiles, large-scale bed slopes at the beginning and end of the considered time interval are drawn in Figure 12. These readily demonstrate that the large-scale bed slope of the study reach is decreasing, i.e., that the river channel is still responding to changes from 150 and 60 years ago. The average bed slope reduced from 0.094‰ to 0.083‰.

To the best of the authors' knowledge, there is no information on the correlation between the minimum water level and the incision of the channel at GSs on the Danube River.

In this study, a linear relationship between the Z_{min} and the corresponding percentage of the cross-sectional area increase due to a channel incision after the reference year of observation was found at three out of four considered gauging stations. The exception is the Apatin GS, where the channel dynamically changes with the growth and washing of a large sand bar. This location is one of the critical sectors for navigation along the Serbian/Croatian border reach. The rates of Z_{min} lowering decrease in the downstream direction, which is consistent with the results of the trend analysis in Figures 5 and 6 and in [6].

Since the Danube River channel on the investigated reach was predominantly of the meandering type in the 19th century, it was rectified with numerous cut-offs (Figure 4.1.12 in [21] (p. 42)). Only on the Hungarian side were there more than 20 cut-offs between Dunaföldvár and the Hungarian–Serbian border [23]. Thus, an alternating degradation and aggradation pattern could be expected. This pattern is further amplified by the construction of river-training structures starting in 1876 and dredging in the second half of the 20th century. The answer to the third research question, i.e., what are the long-term sediment transport characteristics along the investigated reach, is sought through the indirect estimation of the sediment transport rate between Baja and the Drava River confluence, as there are no large tributaries along the reach that could affect riverbed morphology. Both figures—Figure 10 (with the raw data) and Figure 11 (with the results of the sediment transport estimation)—show an expected pattern. The stretch between Baja and a location some distance upstream of Mohács aggrades (the main-channel cross-sectional area reduces—Figure 10 and positive values of Q_s in Figure 11, which are more than four times greater than the negative ones along the reach exposed to degradation). It is expected since it is located downstream of a stretch with numerous cut-offs and intensive dredging in the second half of the 20th century (the area close to Dunaföldvár). The short river reach around Baja can be considered stable (Figures 7a and 8a), possibly owing to the very wide floodplain (~6 km), where large discharges with large velocities (high energy level) can spread. This area is known as Gemenc from ~1503 to ~1469 rkm. Downstream of Gemenc, the river channel is narrow until it reaches Mohács. Close to Mohács, a reach with numerous cut-offs and river-training structures built to aid navigation, it starts again (see the map on the right in Figure 4.1.12 in [21] (p. 42)). As such, it continuously degrades with a mean 25-year long-term sediment transport of approximately 10 t/year. The most intensive washing of the riverbed material during the analyzed period (around 50 t/year) is throughout the length of 5 km downstream of Bezdan (even the translation of the name of this small town—bottomless—suggests that there is a permanent deepening of the riverbed at this location). This is logical, as Bezdan is at the upstream end of the 25 km rectified river channel. The sediment transport rate ceases toward the Drava River confluence. Another quantitative drop in the washing rate is close to Apatin, where the sediment transport drops from 13.5 t/year to 3 t/year on average.

During the investigated 250-year-long time interval, several large floods with peaks exceeding 7000 m^3/s happened (in the summer of 2002, the spring of 2006, the winter of 2010, and the biggest one in the summer of 2013). Among them, the one with undoubtedly the most prominent effect on the morphological changes was that of the spring of 2006, when there were 48 days with discharge larger than the bankfull discharge of 4500 m^3/s and 5 days with a discharge larger than 7500 m^3/s (see Figure 8, block VI, for the time interval between 2004 and 2007, when there was a significant deepening of the cross-sections along the entire reach). Since large amounts of sediment are carried during extreme flood events, it would be interesting to see whether and how it affected the sediment transport rate along the study reach and, consequently, sedimentation and erosion patterns. As was already mentioned, there were only two almost-coinciding years with bathymetric surveys in both countries at the disposal for the analysis—1996/1997 and 2021/2022. Thus, it would even be useful to analyze asynchronous data to infer to what extent floods alleviate or deteriorate the incision of the riverbed in the main channel and how much time is needed after the flood event to reach dynamic equilibrium. From the standpoint of this analysis, it would be advisable if bathymetric surveys were performed along the entire reach after each large flood wave.

6. Conclusions

A comprehensive hydrological–morphological study of the 300 km long free-flowing Middle Danube River reach between rkm 1581 (Dunaújváros), Hungary, and rkm 1255 (Novi Sad), Serbia, led to the following conclusions:

1. The data series of minimum and average water levels are inhomogeneous and show a continuous decreasing trend. They are both the consequence of the permanent riverbed changes initiated by the rectification of the meandering river channel in the 19th century, and they amplified with the consequent construction of groynes and longitudinal stone toes since 1876 and intensive dredging after the 1950s.
2. The rate and type of cross-sectional area change (an increase or a decrease) depend on the location of the cross-section along the river course. The slowest decrease is downstream of reaches with cut-offs and intensive dredging. More intensive changes in the cross-sectional area are at the upstream ends of rectified main-channel reaches.
3. The correlation between low water levels and the percentage increase in the main channel cross-sectional area at official gauging stations in the last 50 years is linear. Thus, the hypothesis about the continuous lowering of water levels caused by the incision of the main channel is correct. The large-scale longitudinal riverbed slope decreased in the last 25 years by 0.01‰. A maximum water level decrease of approximately 1 m is on the upstream end of the study reach, and it reduces when approaching the Drava River confluence.
4. In the absence of systematic sediment transport monitoring, regularly surveyed cross-sections of the navigable river can be used to estimate the long-term averaged sediment transport, i.e., to infer the average rates of aggradation and degradation of the channel along the reach.
5. Along the sand bed reach of the free-flowing Middle Danube with no large tributaries, the maximum value of the positive sediment transport in the last 25 years is approximately 30% greater than the maximum absolute value of the negative transport.
6. The maximum absolute value of the negative transport is near Bezdan, at the upstream end of another rectified river sub-reach. The sediment transport ceases toward the confluence of the Drava River.
7. In 86% of the study reach, the long-term sediment transport for the studied period is negative. Thus, sediment balance changes also prove that decreasing water levels are a good indicator of an ongoing severe riverbed erosion process.

Author Contributions: Conceptualization, D.Đ. and E.A.T.; methodology, E.A.T., D.Đ. and B.K.; investigation, D.Đ., E.A.T., L.M., C.A., A.V. and B.K.; resources, D.Đ., E.A.T., L.M., C.A., A.V. and B.K.; data curation, A.V., C.A., L.M., E.A.T. and D.Đ.; writing—original draft preparation, D.Đ. and E.A.T.; writing—review and editing, E.A.T. and D.Đ.; visualization, D.Đ., E.A.T., L.M., C.A., A.V. and B.K.; supervision, D.Đ., E.A.T., L.M., C.A., A.V. and B.K. All authors have read and agreed to the published version of the manuscript.

Funding: This research received no external funding.

Institutional Review Board Statement: Not applicable.

Informed Consent Statement: Not applicable.

Data Availability Statement: Data used for this research can be requested via email from the authors.

Acknowledgments: The authors are grateful for the provision of data from the official institutions of both countries: the Hungarian Hydrological Forecasting Service; the Directorate for Inland Waterways; the Ministry of Construction, Transport, and Infrastructure—Plovput in Serbia; and the Lower-Danube-Valley Water Directorate (LDVWD) in Hungary.

Conflicts of Interest: The authors declare no conflict of interest.

References

1. Garcia, M. (Ed.) *Sedimentation Engineering: Processes, Measurements, Modeling, and Practice*; American Society of Civil Engineers: Reston, VA, USA, 2008.
2. Plavšić, J. *Engineering Hydrology*, 1st ed.; Univerzitet u Beogradu—Građevinski Fakultet: Belgrade, Serbia, 2019. (In Serbian)
3. Sediment Monitoring in the Danube River, Budapest University of Technology and Economics University of Natural Resources and Life Sciences, Vienna, with Contributions by the Project Partners, 2019, Interreg, Danube Transnational Programme, Danube Sediment. Available online: https://www.interreg-danube.eu/uploads/media/approved_project_output/0001/27/65948979 2a6c2b58c4e322ac8c609943565c3095.pdf (accessed on 3 March 2023).
4. Kalocsa, B.; Zsuffa, I. A Duna magyar szakaszának vízállásváltozásai. *Hidrológiai Közlöny* **1997**, *77*, 183–192.
5. Tamás, E.A.; Đorđević, D.; Kalocsa, B.; Vujanović, A. Hydrological indicators of the riverbed incision along the free-flowing Danube River reach from Budapest to Slankamen relevant for the lateral connectivity between the river channel and floodplains. In Proceedings of the 43rd IAD Conference Rivers and Floodplains in the Anthropocene—Upcoming Challenges in the Danube River Basin, Neuburg, Germany, 9–11 June 2021; pp. 62–69.
6. Savić, R.; Bezdan, A. Uticaj promena nivoa dunava na mogućnost zahvatanja vode u OKM HS DTD. *Zb. Rad. Građevinskog Fak.* **2009**, *18*, 61–71.
7. Tadić, L.; Bonacci, O.; Dedić, T. Dynamics of the Kopački Rit (Croatia) wetland floodplain water regime. *Environ. Earth Sci.* **2014**, *71*, 3559–3570. [CrossRef]
8. Schwarz, U.; Babić-Mladenović, M.; Bondar, C.; Gergov, G.; Holubova, K.; Modev, S.; Rákóczi, L.; Rast, G.; Steindl, J.; Sorin, T.; et al. *Assessment of the Balance and Management of Sediments of the Danube Waterway: Current Status, Problems and Recommendations for Action*; Working Paper for the Danube River Basin; World Wide Fund for Nature: Vienna, Austria, 2008; pp. 1–60.
9. Sediment Balance Assessment for the Danube, Water Research Institute & Project Partners (BME, BOKU, OVF, NARW, NIHWM, LfU, NIMH, EAEMDR, HRVODE, IzVRS, TUM, JCI, PLOVPUT). 2020. Available online: https://www.interreg-danube.eu/uploads/media/approved_project_output/0001/39/d7f9e88e194b7dcea22b51235d653c50d358b7ae.pdf (accessed on 3 March 2023).
10. Endalew, L.; Mulu, A. Estimation of reservoir sedimentation using bathymetry survey at Shumburit earth dam, East Gojjam zone Amhara region, Ethiopia. *Heliyon* **2022**, *8*, e11819. [CrossRef] [PubMed]
11. Rákóczi, L.; Sass, J. A Főmeder és az Ágrendszerek Medermorfológiai-, Üledék- és Hordalékviszonyainak Alakulása. Vituki RT Water Management Scientific Research Joint Stock Company. Available online: http://www.szigetkoz.biz/monitoring/MTA200 1/sass.htm (accessed on 3 March 2023).
12. Martin, Y.; Ham, D. Testing bedload transport formulae using morphologic transport estimates and field data: Lower Fraser River, British Columbia. *Earth Surf. Process. Landf.* **2005**, *30*, 1265–1282. [CrossRef]
13. Tang, M.; Jun Xu, Y.; Xu, W.; Wang, B.; Cheng, H. Three-decadal erosion and deposition of channel bed in the Lower Atchafalaya River, the largest distributary of the Mississippi River. *Geomorphology* **2021**, *380*, 107638. [CrossRef]
14. Babic Mladenovic, M.; Kolarov, V.; Damjanovic, V. Sediment regime of the Danube River in Serbia. *Int. J. Sediment Res.* **2013**, *28*, 470–485. [CrossRef]
15. Lisimenka, A.; Kubicki, A. Bedload transport in the Vistula River mouth derived from dune migration rates, southern Baltic Sea. *Oceanologia* **2019**, *61*, 384–394. [CrossRef]
16. Gaeuman, D.; Jacobson, R.B. Field Assessment of Alternative Bed-Load Transport Estimators. *J. Hydraul. Eng.* **2007**, *113*, 1319–1328. [CrossRef]
17. Goda, L.; Kalocsa, B.; Tamás, E.A. River Bed Erosion on the Hungarian Section of the Danube. *J. Environ. Sci. Sustain. Soc.* **2007**, *1*, 47–54. [CrossRef]
18. Addinsoft. *XLSTAT Statistical and Data Analysis Solution*; Addinsoft: New York, NY, USA, 2021. Available online: https://www.xlstat.com (accessed on 13 March 2023).
19. Zsuffa, I. *Műszaki Hidrológia [Technical Hydrology]*; Műegyetemi Kiadó: Budapest, Hungary, 1996.
20. Rákóczi, L. Sediment Regime of the River Danube (1956–1985). In *Hydrological Processes of the Danube River Basin*; Brilly, M., Ed.; Springer: Dordrecht, Germany, 2010. [CrossRef]
21. Long-Term Morphological Development of the Danube in Relation to the Sediment Balance, 7 Water Research Institute & Project Partners (BME, BOKU, OVF, NARW, NIHWM, LfU, NIMH, EAEMDR, HRVODE, IzVRS, TUM, JCI, PLOVPUT). 2020. Available online: https://www.interreg-danube.eu/uploads/media/approved_project_output/0001/39/6299a3c18cf54d8ef7e0215084 4e27358da8038d.pdf (accessed on 3 March 2023).
22. Schmitt, L. Navigation, Hydropower, Sediment Legacies, Functional Restoration and Transboundary Management of the Upper Rhine, Seminar@IWG-WB—31 March 2023. Available online: https://www.wb.iwg.kit.edu/research.php?tab=%5B2434%5D#tabpanel-2434 (accessed on 2 August 2023).
23. Ihrig, D. (Ed.) *A Magyar Vízszabályozás Története (The History of Hungarian Water Regulation)*; OVH: Budapest, Hungary, 1973; 408p.

applied sciences

Article

Upstream and Downstream Changes in the Channel Width and Sinuosity Due to Dam Construction in Tropical Rivers: The Case of Colombia

Diana C. Alvarado, Andrés Vargas-Luna * and Juan Diego Giraldo-Osorio

Civil Engineering Department, Engineering School, Pontificia Universidad Javeriana, Cra. 7a # 40-62, Bogotá 110231, Colombia; diana_alvarado@javeriana.edu.co (D.C.A.); j.giraldoo@javeriana.edu.co (J.D.G.-O.)
* Correspondence: avargasl@javeriana.edu.co

Abstract: In Colombia, most of the energy is produced by using water resources. However, the morphological impact of damming has not been thoroughly studied yet. Therefore, upstream and downstream changes in the channel width and sinuosity along the river due to the Betania, Prado, Salvajina, and Urrá I Dams, four of the oldest hydroelectric projects, were estimated. These changes were reported by using aerial photographs and satellite images to compare the river before and after dam construction. The analysis was complemented by including hydrological trends and geological characteristics of the areas to evaluate their relevance on the impacts on channel morphology. It was shown that factors such as valley confinement and the bank's composition are key to determining the magnitude of the impact downstream of the dam. Upstream of the dam, contrastingly, the influence of the reservoir geometry controls the magnitude of the morphological changes, marking the boundaries of affected areas. The impacts of dam construction on river morphology vary notably, but including the geological characteristics of the river reach can be useful to improve predictions of the channel morphology response. The proposed methodology can be used to identify biotic compensation measures for new projects, a task that is not well defined in several countries.

Keywords: downstream and upstream changes; morphological evolution; river morphology; erodibility; valley confinement

Citation: Alvarado, D.C.; Vargas-Luna, A.; Giraldo-Osorio, J.D. Upstream and Downstream Changes in the Channel Width and Sinuosity Due to Dam Construction in Tropical Rivers: The Case of Colombia. *Appl. Sci.* **2023**, *13*, 7109. https://doi.org/10.3390/app13127109

Academic Editors: Gordon Gilja, Manousos Valyrakis, Panagiotis Michalis, Thomas Pahtz and Oral Yagci

Received: 18 April 2023
Revised: 19 May 2023
Accepted: 29 May 2023
Published: 14 June 2023

1. Introduction

The construction of dams is an anthropic practice that provides economic benefits for our society but has also caused significant changes to riverine landscapes worldwide. Around 800,000 dams have been built around the world with different objectives: irrigation, flow control, navigation, and hydropower, among others [1]. Inevitably, dam construction and operation have immediate and long-term effects on river morphology [2–4] both upstream [5,6] and downstream [7,8] of those structures. Thus, predicting the effects of dams is key to maintaining biodiversity and ecosystem services [9,10].

Upstream of dams, the aggradation processes in the alluvium along reservoir reaches are dominant due to the reduction in flow velocities, responsible for sediment transport, and the backwater effects following impoundment [11–16]. Bed aggradation in these regions generates a water level rise, increasing the flood frequency and the vulnerability of the existent infrastructure, imposing an additional risk to human settlements located in the surroundings and considerable challenges for managers [17,18]. Therefore, predicting the extent of the morphological changes in the upstream area of a reservoir can benefit flood risk estimations. Morphological changes upstream of dams are evaluated in varied ways by measuring different morphological parameters. For example, by using aerial photographs, Liro [6] determined changes due to the Czorstyn reservoir at a control distance of 2.2 km, reporting an increase in bar areas and the river channel, and changes in bank erosion due to flooding, highlighting the importance of the material present in the confining valley.

Evans et al. [11] also used aerial photographs to define the length of the upstream influence area (4 km) based on increased bar and sand areas. However, no previous studies have considered the importance of geology in upstream areas, except for Escudero et al. [5], who made an approximation at a regional scale. There is also a lack of studies dealing with the changes in the channel width in these areas.

The downstream morphological long-term changes depend on site-specific drivers, such as the altered flow and sediment regimes, and geological controls [19]. Clear water released from reservoirs is expected to increase bed degradation and erosion rates in the downstream reaches [20,21]. Nevertheless, this trend is not universal and varies with the distance from the dam [22], ranging from tens to hundreds of kilometers [23,24]. This variation may be dominated by the lithological and structural controls of the downstream valley that determine channel geometry [25] and the interaction between the main channel and its floodplains [26]. Including the surface geological settings in morphological analyses may contribute to developing more effective methods to predict such changes, but this practice has not yet been addressed.

Upstream and downstream morphological evolution is commonly expressed with changes in variables such as the channel width, water depth, slope, and sediment characteristics through time. Moreover, a few studies have shown that damming can even alter bar patterns [27] and river planforms [28]. With a few studies carried out in the tropics, the morphological changes reported in the literature are mainly collected from temperate and Mediterranean environments. The impacts due to dam construction and operation in tropical rivers may have different magnitudes than those observed in other regions because of the differences observed in [29]: (1) the degree of discharge alteration, (2) sediment loads and dynamics, and (3) the nature and pace of post-disturbance adjustments. Considering the ecological value of tropical systems, it is important to perform studies in these regions to assess similitudes and differences in the morphological response. Morphological effects due to dam construction are commonly reported based on hydrological series of discharges and sediment loads, and historical bathymetries at limited gauging stations [7,23,24,28,30–32]. Recent advances in geographic information systems (GIS) have resulted in a wider use of GIS tools for estimating morphological changes in rivers due to damming [4,11,33]. However, these applications have also been focused on specific areas without considering the analysis of geological records. In this paper, GIS tools are used to analyze post-dam trends and to quantify the evolution of the channel width and sinuosity over time by correlating the magnitude of those changes with the hydrological and geological settings of each river reach.

The river width is adopted in this research as the reference parameter to establish the magnitude of the morphological changes, considering its key relevance in channel geometry and the river character [34–37]. In addition to the channel width, sinuosity is also included in this work to consider possible planform changes. The channel width and sinuosity are obtained using all the information available for the study sites, including their planform from remote sensing (satellite images, aerial photographs), hydrology (gauging stations), and geology. The geology was analyzed from a regional to a local scale, considering confinement, composition (formations), and erodibility of the banks to understand the magnitude of the morphological changes that each river reach experiences after dam construction.

Compensatory mitigation is intended to look for a way to repay the negative effects of development projects [38]. The definition of biodiversity offsets focuses on restoring the altered functions of the intervened ecosystem [39]. This analysis is important to establish the extent of the alteration in the ecosystem, defining a clear application of this work. Considering that our interest is to identify the elements that are relevant to delimiting the expected morphological changes due to dam construction in a simplified manner, other important factors, such as land use changes, are not included.

This paper aims at identifying and discussing the changes in the channel width and sinuosity upstream and downstream of four large dams located in Colombia (South

America), considering this region as an appropriate representative of tropical systems. The work also explores the integration of hydrological and geological information in the analysis to establish a way to identify the areas that are more susceptible to those changes. Our results can help in setting the extent of the affectations and the biotic compensation areas due to new projects to be considered in the definition of biodiversity offsets.

2. Materials and Methods

2.1. Study Sites

The selected reservoirs are located in the Colombian Caribbean hydrographic basin, which is irrigated by the main rivers of the country and houses around 85% of its population [40,41]. This study considers dams constructed across three rivers of this hydrographic basin: the Magdalena River, the Cauca River, and the Sinú River. The dams built across the selected rivers are located on the fall line, with an upstream bedrock reach and a downstream alluvial bed. Anthropogenic interventions in these areas have generated severe changes in land use, which have resulted in increased sediment loads [42,43]. However, the morphological impacts of damming in this hydrographic area have not yet been studied in detail; therefore, understanding the morphological trends and the magnitude of their changes due to dam construction is a useful task in the future planning and management of water resources.

Spatial and temporal criteria were considered to define the study sites. The spatial criterion was limited by the fact that in this work, we use free satellite images with a lowest spatial resolution of 30 m; therefore, river reaches upstream and downstream of each dam must have a minimum width of 60 m. Considering the temporal resolution and to be able to identify morphological changes, images before and after the construction of the selected dam are required, with a minimum operating time of 5 years. The dams that satisfy the selection criteria are presented in Table 1 and Figure 1. In the case of Betania and Prado Dams, both structures affect the Magdalena River and are separated by a river reach of, approximately, 178 km (see Table 1).

Table 1. Main characteristics of the dams selected for the study.

Dam	Year Starting Operation	Height (m)	Reservoir Volume (10^6 m^3)	Reservoir Area (m^2)	Downstream Initial Channel Width (m)	Rivers Upstream of Reservoir	Rivers Downstream of Reservoir
Prado	1972	75	1010	125,400	302	Prado, Cunday	Magdalena
Salvajina	1985	148	764	203,100	84	Cauca	Cauca
Betania	1986	170	1971	700,000	236	Magdalena, Yaguara	Magdalena
Urrá I	2000	100	1740	740,000	112	Sinú, Verde	Sinú

At the dates of construction of the Prado, Salvajina, and Betania projects (see Table 1), environmental impact studies were not mandatory according to Colombian regulations. Therefore, existing documents [44–46] only include technical recommendations, excluding the analysis of ecological and morphological impacts on the intervened river reaches. The Urrá I project was conceived after the installation of environmental law in the country [47]; however, the possible morphological effects due to dam construction were not addressed in its environmental impact study.

2.2. Data

2.2.1. Average Monthly Precipitation Data within the Basins

The precipitation data within the basins were obtained from the total precipitation product of ERA5-Land. ERA5-Land is a reanalysis that provides a consistent evolution of land variables over several decades using a spatial resolution of 10 km. This reanalysis produces data that go back several decades in time, providing an accurate description of the climate of the past [48]. This last feature was particularly useful in our research, since the precipitation station network was limited during the Betania project pre-dam period (1960–1984). Seeking coherence in the analysis, the same product was used for the Salvajina (pre-dam period 1976–1983) and Urrá I (pre-dam period 1991–1998) projects, even though for those dates, the precipitation station network had improved.

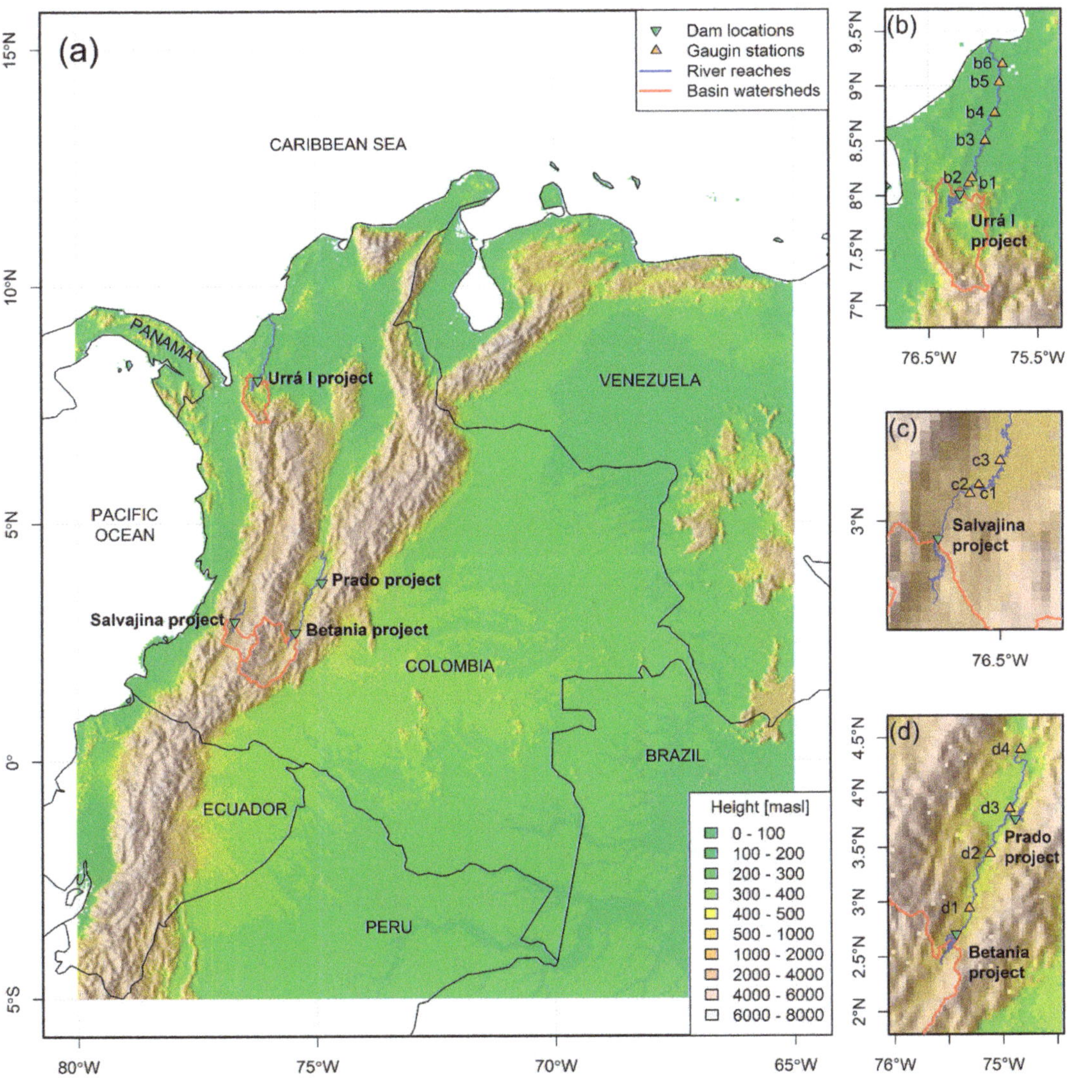

Figure 1. Location of the selected dams in Colombia: (**a**) general location, (**b**) Urrá I project, (**c**) Salvajina project, and (**d**) Betania and Prado projects (see Table 2 for information about selected gauging stations).

Table 2. Gauging stations selected for the study.

Dam	Station Name	ID	Coordinates Long.	Lat.	Elevation (masl)	Period (Years) Pre-Dam	Post-Dam	Downstream Distance (km)
Urrá I	El Toro	b1	−76.13	8.11	100	1991–1998	2002–2010	26
	Carrizola	b2	−76.10	8.16	55	1993–1998	2002–2020	40
	Nueva Colombia	b3	−75.98	8.50	20	1991–1998	2002–2020	112
	Montería	b4	−75.89	8.75	15	1963–1998	2002–2020	162
	Sabana Nueva	b5	−75.85	9.03	9	1979–1998	2002–2020	205
	Palma Central	b6	−75.82	9.20	6	1991–1998	2002–2010	230
Salvajina	La Balsa	c1	−76.60	3.09	987	1976–1983	1987–2020	27
	Tablanca	c2	−76.57	3.12	980	1978–1983	1987–2010	36
	La Bolsa	c3	−76.50	3.20	964	1967–1983	1987–2016	75
Prado and Betania	Puente Santander	d1	−75.31	2.94	431	1960–1984	1988–2019	42
	Angostura	d2	−75.12	3.44	345	1975–1984	1988–2020	119
	Purificación	d3	−74.94	3.85	306	1960–1984	1988–2019	182
	Nariño	d4	−74.84	4.39	277	1978–1984	1988–2019	273

Using the monthly total precipitation product from ERA5-Land, the average precipitation for each month was calculated within the basins that supply the reservoirs. So, each basin had a monthly precipitation series between 1950 and 2020. For precipitation change analysis, the pre-dam and post-dam periods were forced to coincide with the selected periods in the discharge stations closest to the dams, since these are assumed to be the stations whose records are more affected (see Section 2.2.2 and Table 2).

2.2.2. Discharge and Sediment Data

River width changes are directly related to the flow regime; the discharge information was provided free of charge by the *Instituto de Hidrología, Meteorología y Estudios Ambientales* (IDEAM; http://dhime.ideam.gov.co/atencionciudadano/; accessed on 30 April 2023) and the *Corporación Autonoma Regional del Valle del Cauca* (CVC; https://www.cvc.gov.co/; accessed on 28 February 2023). Gauging stations were chosen for their location in the main channel and recorded time before and after dam construction (see Table 2). In this table, identificatory numbers of each station (ID) are the same as in Figure 1. Gauging stations were only selected downstream of the dams because upstream stations do not have information prior to dam construction.

The hydrological regime in the influence area of each selected dam was analyzed for selecting the images in the summer or with similar summer discharges, avoiding years under the influence of El Niño–Southern Oscillation (ENSO) phenomena extreme phases (El Niño or La Niña years), in which the river can exhibit abrupt width changes. After dam construction and according to the design criteria, the natural discharge regime of a river can be maintained, decreased, or increased throughout the year [49–51]. As the spatial information was taken in the summer season, it was necessary to estimate the magnitude of change with statistical tests. We used Student's *t*-test [52,53] to evaluate the mean change, and Fisher's [54] *F*-test to evaluate the variance change, for the pre- and post-dam periods. Although the literature describes a general tendency due to dam operation (increased discharges during the dry period, medium discharges maintained, and decreased high discharges [51,55,56]), the influence of a dam is also observed in annual values. Therefore, minimum, medium, and maximum annual flow time series were built, and the differences (in percentage) between the pre-dam and post-dam periods for all stations were evaluated. The Kolmogorov–Smirnov [57,58] and Mann–Whitney [59] non-parametric tests were performed to evaluate changes in the distribution before and after dam construction.

In Colombia, only a few gauging stations count with a record of suspended sediment transport. Nevertheless, we also included suspended sediment measurements from available reports and theses in our analysis [60–64]; a summary of the reduction trend in this parameter is shown in Table 3. The gathered data allowed us to establish an approximate distance downstream of the dams at which the suspended sediment load was almost fully recovered.

Table 3. Suspended sediment reduction downstream of the selected dams.

Project	Closest Station to the Dam			Farthest Station to the Dam		
	Station Name	Downstream Distance (km)	Reduction in Suspended Sediment (%)	Station Name	Downstream Distance (km)	Reduction in Suspended Sediment (%)
Prado and Betania	Puente Santander	42	70–73	Nariño	279	9
Salvajina	Salvajina	1	75	Juanchito	134	30
Urrá I	Pasacaballos	6	81	Montería	175	40

2.2.3. Remote Sensing Data

To define the morphological changes in the selected river reaches, it is important to gather information before and after dam construction. For this purpose, three different types of spatial information were collected: aerial photographs, spectral images, and radar images. For the dams built between 1972 and 1986 (i.e., Prado, Salvajina, and Betania), the acquisition of aerial photographs supplied by the *Instituto Geográfico Agustín Codazzi* (IGAC; https://www.igac.gov.co/; accessed on 15 April 2023) was required. A total of 85 aerial photographs were purchased to characterize the pre-dam morphology (see Appendix B). Aerial photographs are acquired following flight lines that coincide with the river alignment, but several flights of the same year or nearly a year (not more than 5 years) may be needed. In the case of the Betania and Prado projects, the predominant spatial information is from 1969 (most photographs), so this year was used as a general reference. For the Salvajina project, a similar behavior was observed for the year 1961.

The satellite images chosen for this study were obtained from (1) Landsat missions provided by the United States Geological Survey (USGS; https://earthexplorer.usgs.gov/; accessed on 30 March 2023), with a spatial resolution of 30 m (Landsat 4–5 TM, between 1982 and 2012; Landsat 7 ETM+, between 1999 and 2003; and Landsat 8, from 2013 to the present), and (2) Aster images provided by the National Aeronautics and Space Administration (NASA; https://earthdata.nasa.gov/; accessed on 30 March 2023), with a spatial resolution between 10 and 30 m, available from 2000 to the present. In areas with high cloudiness, Sentinel 1-GDR radar images from the European Spatial Agency (ESA; https://scihub.copernicus.eu/; accessed on 30 March 2023), with 10 m spatial resolution, were used.

Spatial uncertainty in measurements of the river width depends on adequate delimitation of the water body borders [65]. These uncertainties appear when channels are digitized from aerial photographs and satellite images with vegetation, clouds, and shadows. The images used in this study are of a different type and year; therefore, river alignments may exhibit considerable changes. These errors are, however, not easy to estimate; therefore, only the uncertainty in the spatial resolution of each image and how this affects the river width estimation [66] were considered. Because most images considered in this study were Landsat images with a 30 m pixel size, all spatial information was worked on with this resolution to avoid error propagation. From our selected sites, the Cauca River is most sensitive to this aspect, with a width of 86 m (36% of possible variation), followed by the Sinú River with a width of 112 m (26.7% of possible variation). As the Magdalena River is the widest (channel width from 236 m to 302 m), it exhibits less uncertainty, from 10% to 13%.

2.3. Morphological Parameter Estimation

Aerial photographs and satellite images were integrated into GRASS [67] for estimating morphological characteristics. The methodology used in this study for processing the images is presented in Figure 2. The aerial photos were georeferenced and the river channels digitized with a maximum root mean square error (RMSE) of 10 m with the settings of the GRASS GIS (see Figure 2a).

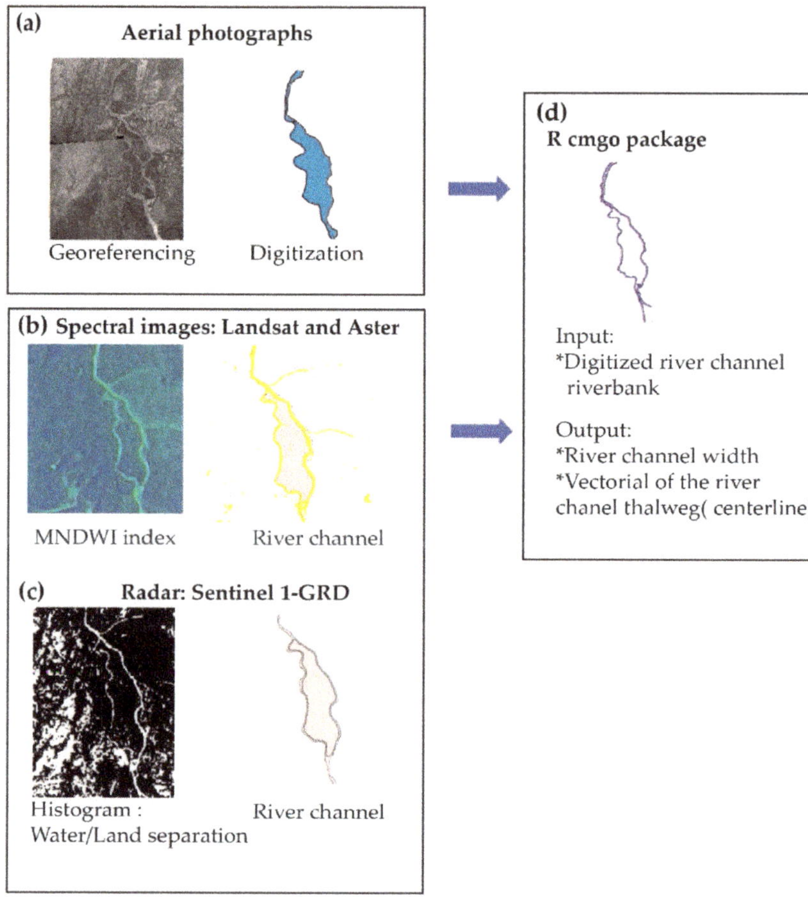

Figure 2. Methodology for obtaining the morphological characteristics of rivers: (**a**) georeferencing and digitization of aerial photographs, (**b**) delimitation of the river channel in spectral images, (**c**) delimitation of the river channel in radar images, and (**d**) morphological parameter assessment using the R package cmgo.

Satellite images with a spectral range were processed to identify the river channel width. Based on their spectral characteristics, different indices have been developed for the extraction of water bodies [68]. Considering the geomorphology of the study areas (i.e., high mountain ranges and piedmont plains), the Modified Normalized Difference Water Index (MNDWI; see Figure 2b) was chosen, defined by:

$$MNDWI = \frac{GREEN - MIR}{GREEN + MIR} \quad (1)$$

where the MIR and $GREEN$ bands must be selected for the different Landsat and Aster images since the band numbers change among missions. All water body extraction methods add additional bodies to the main channel, such as tributaries and effluents, swamps, clouds, and rock bodies, such as outcrops, which need to be removed or cleaned, leaving only one channel. In Colombia, the cloud cover is high, especially in the mountainous areas of all selected dams. Although the extraction of water bodies with the MNDWI, in general, is for values greater than 0 in the cloudy areas, the selection value may vary between −0.2 and 0, so a constant threshold check is necessary for each image.

In the Urrá I project sector, due to its high cloudiness, a Sentinel 1-GRD image was used for 2020 (Figure 2c) and corrected by following the procedures proposed by Filipponi (2019) [69], a methodology available in the free Sentinel Applications Platform (SNAP) tool (https://step.esa.int/main/download/snap-download/; accessed on 30 October 2022). The image was converted from digital pixel values to calibrated SAR backscatter values, which were analyzed with the histogram setting of SNAP, using the low return signal behavior of open water [70], allowing us to differentiate water and land. This information was processed in a similar way to the Landsat and Aster images, only leaving the main channel of the river.

After obtaining the channel using the methods previously described, the morphological parameters were calculated with the R package cmgo (see Figure 2d) [71]. This open-source code package is versatile and easy to use. The package gives channel metrics such as width, sinuosity, and slope. The R package cmgo requires separating the channel banks on the left and right sides, information that is used for calculating the width, centerline, and slope of the channel. Considering the purpose of this study, only the width and centerline were used in the analysis.

The downstream and upstream changes for each dam were measured using sections selected based on the tributaries, the changes in relief, and the river's slope (verification realized with an SRTM DEM obtained from the United States Geological Service (USGS; https://earthexplorer.usgs.gov/; accessed on 15 October 2022)). The channel width of each section was averaged and compared during the temporal window for each case.

2.4. Geological Information

Considering that one of our hypotheses was that morphological changes are conditioned by the geological characteristics and relief of the river reach, this information was gathered for each study site. The geological information was relevant to define the erodibility of the banks of the analyzed river reaches. This information was obtained from the *Servicio Geológico Colombiano* (SGC [72]) that allows free access to the Colombian geology through a viewer (http://srvags.sgc.gov.co/JSViewer/Atlas_Geologico_colombiano_2015; accessed on 30 September 2022) and geological reports from different areas (https://recordcenter.sgc.gov.co; accessed on 30 September 2022).

From more resistant (less erodible) to more erodible, the materials can be listed in the following order: metamorphic rocks, igneous rocks and limestone, and sedimentary rocks. Erodibility varies according to grain size and intercalation for the latter, in which the more erodible materials are the unconsolidated ones, such as alluvium, colluvium, and recent-origin fans [73,74]. In this study, only the outcropping chronostratigraphic units (CUs) close to the river were considered, and a detailed description of these units can be found in Appendix A.

To determine the erodibility in the riverbanks of each study reach, we regionally applied the Global Erodibility Index [75], which is based on the subjacent rock type (see Appendix A). This index varies from 1.0 (the hardest rocks) to 3.2 (unconsolidated sediments that do not suffer the effects of pressure and temperature).

Relief changes are reflected in the valley confinement, which is related to the geological and geomorphological environment: high relief areas exhibit high confinement, wide valleys and plains offer low confinement, and the transition zone (between high and low zones) defines the medium confinement regions. The confinement controls the mobility of a river channel, affecting the morphological changes [76]. These transitional changes (from a confined valley to a valley with floodplains) were classified in three confinement degrees: low confinement (where the width of the floodplain or vast valleys is larger than 20 km), medium confinement (with narrow valleys, in which their width varies between 3 and 20 km), and highly confined (extremely narrow valleys with a width less than 3 km).

3. Geological Context of Each Study Site

For each study site, the geology was evaluated upstream and downstream of the dam from a regional to a local scale based on the information available from geological maps, topography, hydrography, and various studies. A geological conceptual map of each study site was produced.

3.1. The Urrá I Project

The Sinú River basin is in the Western Andes Mountain Range. The tributaries come from the Paramillo massif (Verde and Sinú Rivers; see Figure 3), feeding the reservoir of the Urrá I Dam. This area has high humidity due to the interaction between the NE trade winds and the Choco jet current from the west, causing high rainfall in the upper area of the Sinú River basin and the nearby regions [77]. The quaternary alluvium (Q-al) is the chronostratigraphic unit (CU) that mainly interacts with the Sinú River in this vast valley, except between 118 km and 156 km, where the e3e4-Sm CU appears [78]. The Betanci swamp and the Grande swamp complex stand out in the basin for their characteristics and water interchange with this river (see Figure 3).

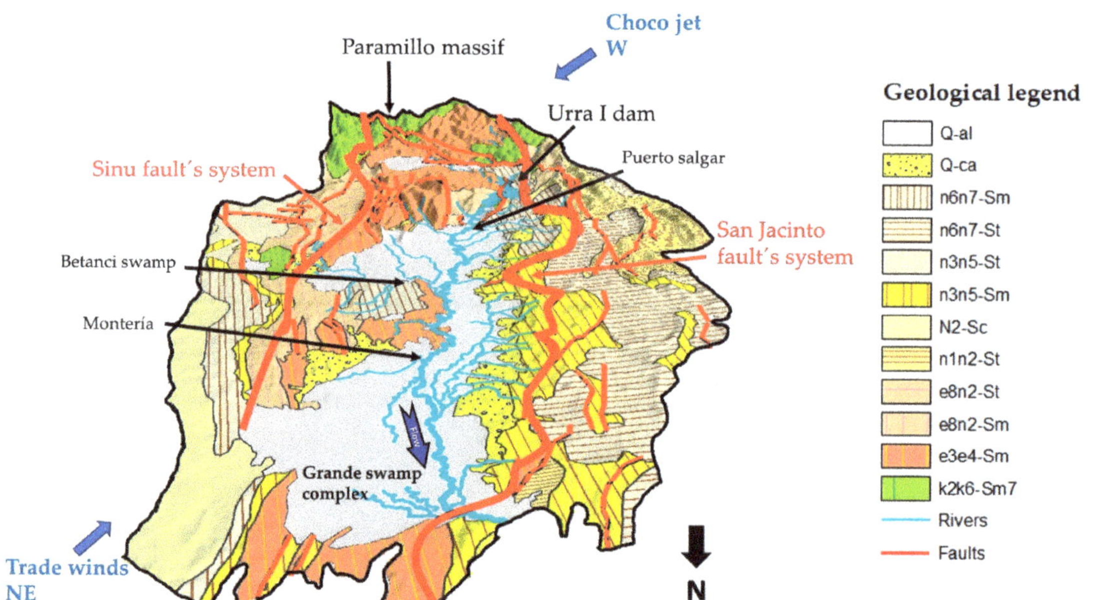

Figure 3. Geological conceptual map of the Urrá I Dam area (geology from the SGC and DEM from the SRTM mission/2000).

3.2. The Salvajina Project

The Cauca River basin is located between the Central and Western Andes Mountain Ranges; its valley is the inter-Andean graben [79] and is limited by the Romeral and Cauca fault systems (see Figure 4). There are volcanic-type formations in the upper part of the basin or high mountain: lavas, tuffs, and basalts [72]. Downstream of the Salvajina Dam, the river goes through a transition from a high mountain to an extensive alluvial valley following a parallel course and close to the Cauca fault system, near the site of Timba.

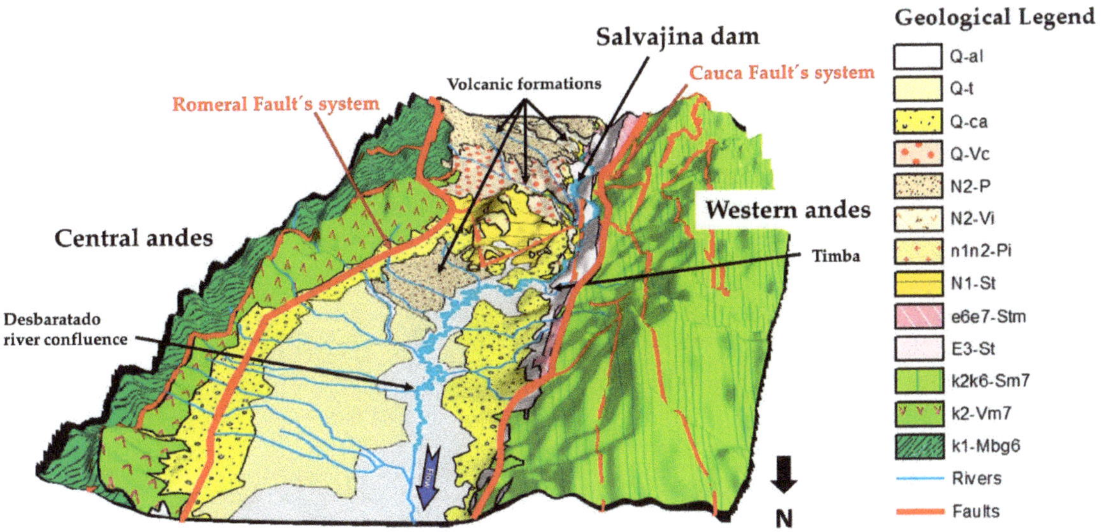

Figure 4. Geological conceptual map of the Salvajina Dam area (geology from the SGC and DEM from the SRTM mission/2000).

Between Timba and the confluence of the Desbaratado River, the planform exhibits high sinuosity. From this confluence, the river stops being sinuous when it is near Cali, and the river's course tends toward the Cauca fault system and the western mountain range. These changes in sinuosity and the river course respond to the underlying geological structure. The river is confined by the terraces (Q-t CU, see Figure 4) coming from the central mountain range (this cordillera is 1400 m higher than the western one).

3.3. The Betania and Prado Projects

The Magdalena River basin is located between the Central and Eastern Andean Mountains Ranges; the Garzón Suaza and Chusma fault systems limit this basin (see Figure 5). Betania is the first dam that affects the Magdalena River, and downstream 178 km is an affluent (Prado River) that is dammed for the Prado Dam. The basin is divided into two sub-basins, Neiva and Girardot, separated by a valley or transition zone [80,81]; a lithologic formation n4n6-Sc CU (Honda group) is the base of the basin and overlying different kinds of quaternary materials, such as volcanic, colluvial, terraces, and alluvial.

In the Neiva sub-basin from 0 km to 90 km, it is possible to identify the Foehn-effect-generated Tatacoa Desert [82,83]. The repercussions of this effect are present throughout the year (poor vegetation, little soil weathering). The climate of this area is classified as a tropical wet and dry or savanna climate, with the driest month having a precipitation of less than 60 mm during the dry summer [84].

The transition to the Girardot sub-basin occurs in a narrow valley from 90 km to 158 km, which ends at the Pata and Natagaima highs [85]. In the Girardot sub-basin from 158 km to 244 km, the Magdalena River follows a north-south trend closer to the Eastern Andes and Prado fault (Garzón Suaza fault system). From 178 km to 244 km, quaternary terraces and the Guamo fan are predominant on the left bank, confining the Magdalena River against the Eastern Mountain Range and rectifying the river.

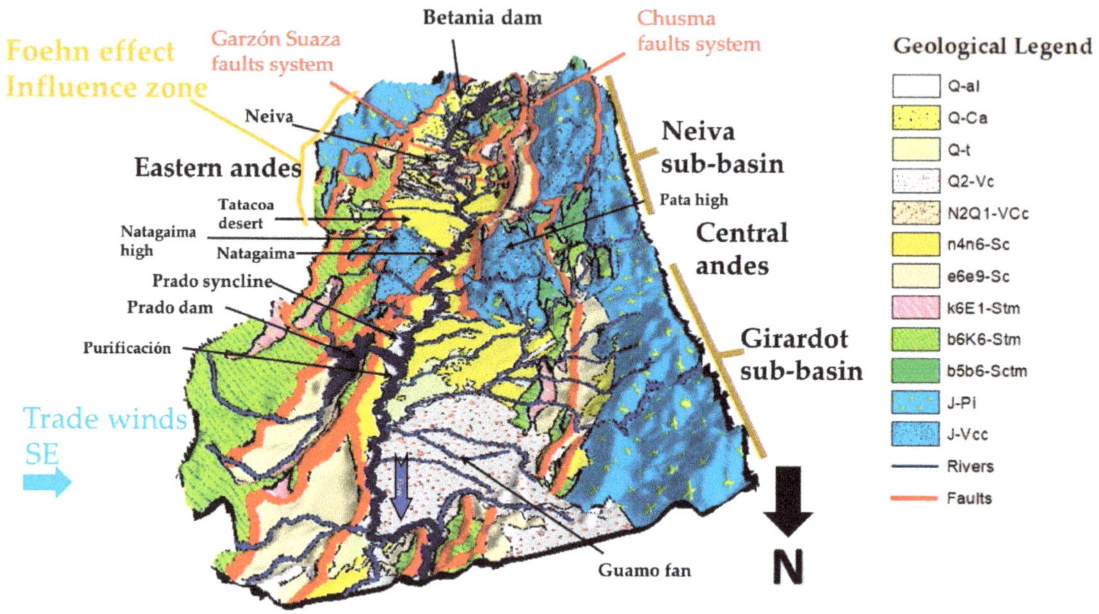

Figure 5. Geological conceptual map of the Betania–Prado Dam area (geology from the Servicio Geológico Colombiano and DEM from the SRTM mission/2000).

4. Results

4.1. Hydrology

For each discharge time series at the available gauging stations (see Table 2), a 3-year period was removed and was not considered in the later analysis: from 1 year earlier to 1 year after the start of dam operation.

4.1.1. Hydrological Regime

The hydrological regime for the pre- and post-dam periods was evaluated at each flow gauging station. In addition, we calculated the changes in monthly precipitation within the basins of each project in order to identify whether the changes in discharge are influenced by the precipitation change within the basins.

In Figure 6, the hydrological regime for both periods in two gauging stations downstream of each dam is shown. In Appendix C, the test results (Student's t-test and Fisher's F-test) applied on the average monthly discharge and precipitation within the basins in the pre-dam and post-dam periods are summarized. According to the statistical results presented in Appendix C, no significant changes in the mean or variance values of the monthly precipitation within the basins were observed, except for some specific months. Therefore, those precipitation changes do not explain the monthly mean flow changes in the downstream gauging stations.

The hydrological effect due to dam construction is similar in the four selected study sites: increased low discharges during the dry period and decreased high discharges during the rainy period (see Figure 6). Downstream of the Urrá I Dam (see Figure 6a and Table 4), the differences between the pre- and post-dam periods occur in the means during the dry period, while differences in the variances are more significant in the rainy season.

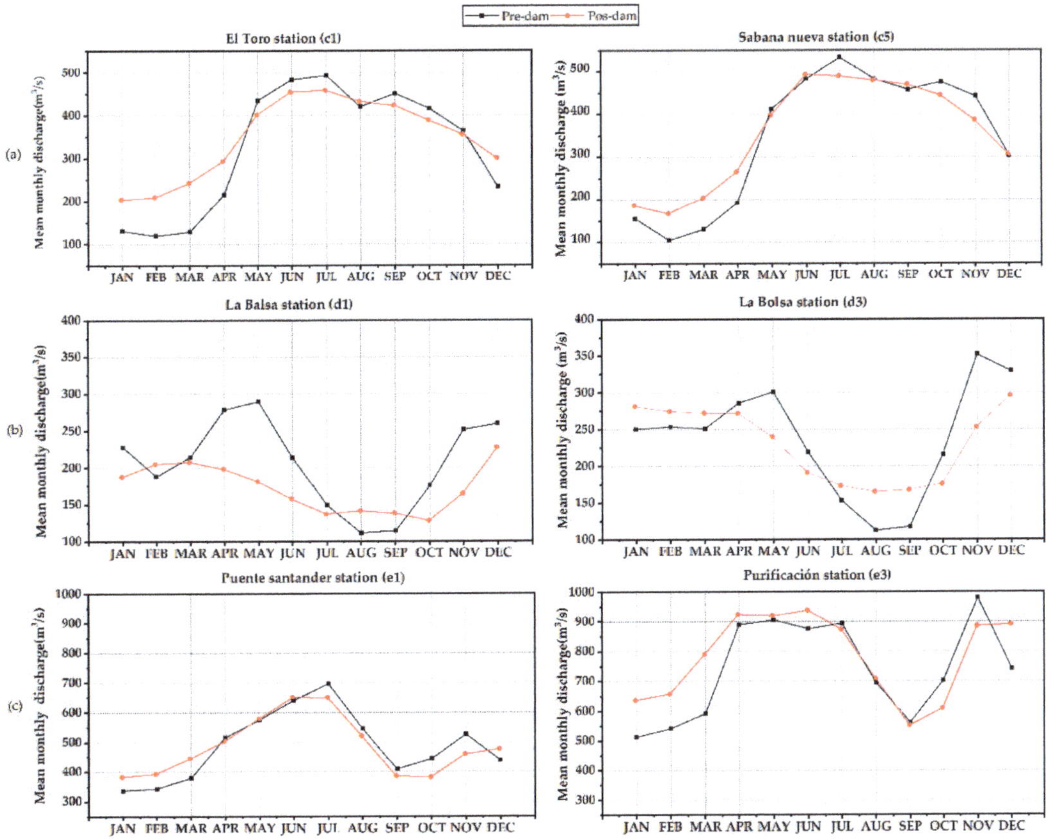

Figure 6. Mean monthly discharge in downstream stations for the pre-dam (black line) and post-dam (red line) periods. Left and right columns show the nearest and the furthest station to the dam, respectively. (**a**) Urrá I project, (**b**) Salvajina project, and (**c**) Betania and Prado projects.

Downstream of the Salvajina Dam (see Figure 6b and Table 4), the discharges decrease significantly in the rainy period compared to the other dams; this is reflected in the lower values (*t*-test) before and after dam construction. We also observed a considerable increase in discharges during the dry season, from August to September (see Figure 6b).

The hydrological effects due to the Betania Dam construction are reflected by the discharge series at the Puente Santander and Angostura stations, while the hydrological effects observed downstream of the Purificación station reflect the combined effects of the Betania and Prado Dams. It was impossible to identify the affectation of only the Prado Dam, which started operation in 1972, since the gauging station records are only available from 1975, so the combined effect of both dams was considered. In the area affected only by the Betania Dam (until the Angostura station), attenuated discharges are observed during the rainy season and increased discharges are present during the dry period (January to April). The discharges during the same dry period at the Purificación station were higher in the post-dam period than the ones observed before dam construction. The percentages of the difference between the low discharges pre- and post-dam in this station were larger than those observed in the gauging stations that are only influenced by the Betania Dam. Of the three studied areas, this one has the slightest changes in the mean and variance, which implies a lower impact of these dams on the hydrological regime.

Table 4. Upstream changes for each dam in Section 1 (influenced zone) and Section 2 (non-influenced zone).

Dam	Reach	Coordinates					Characteristics		
		Initial		Ending					
		Lat.	Lon.	Lat.	Lon.	L (km)	Period	W (m)	S
					Verde River				
Urrá I	1	7.85	−76.29	7.83	−76.13	8	Pre Pos	81 373	1.3 1.1
	2	7.83	−76.13	7.80	−76.33	5	Pre Pos	60 56	1.0 1.1
					Sinú River				
	1	7.84	−76.29	7.83	−76.25	6.7	Pre Pos	56 203	1.2 0.9
	2	7.83	−76.25	7.79	−76.24	6.3	Pre Pos	55 62	1.0 1.0
					Cauca River				
Salvajina	1	2.77	−76.71	2.75	−76.70	2.7	Pre Pos	33 92	1.06 1.05
	2	2.75	−76.70	2.71	−76.74	5	Pre Pos	44 44	1.15 1.15
					Yaguará River				
Betania	1	2.69	−75.51	2.64	−75.52	2.5	Pre Pos	71 269	1.1 1.1
	2	2.64	−75.52	2.63	−75.5	3.5	Pre Pos	55 47	1.0 1.0
					Magdalena River				
	1	2.56	−75.49	2.53	−75.5	6.5	Pre Pos	156 342	1.4 1.1
	2	2.53	−75.5	2.47	−75.56	9.5	Pre Pos	122 140	1.3 1.3

Note: Lat.: latitude; Lon.: longitude; L: length; W: river width average; S: river sinuosity average; pre: pre-dam; pos: post-dam.

4.1.2. Annual Discharges

The changes in the annual discharge (minimum, mean, and maximum annual flow) were evaluated by calculating the difference (as a percentage) between the pre- and post-dam periods (see Figure 7). The Kolmogorov–Smirnov and Mann–Whitney non-parametric tests were also applied to test the hypothesis of no difference at a 5% significance level between the distributions of the annual discharge values from the pre- and post-dam periods. The results of these tests are presented in Appendix C. The general effect of the studied dams in all areas was to increase the minimum annual discharges and decrease the maximum annual discharges. For the studied reservoirs, dam operation focuses on retaining the water excess during the wet period, decreasing the peak discharges to release it later in a controlled way during the dry period, increasing the minimum discharges for maintaining water consumption. At the same time, the medium annual discharges tend to remain constant.

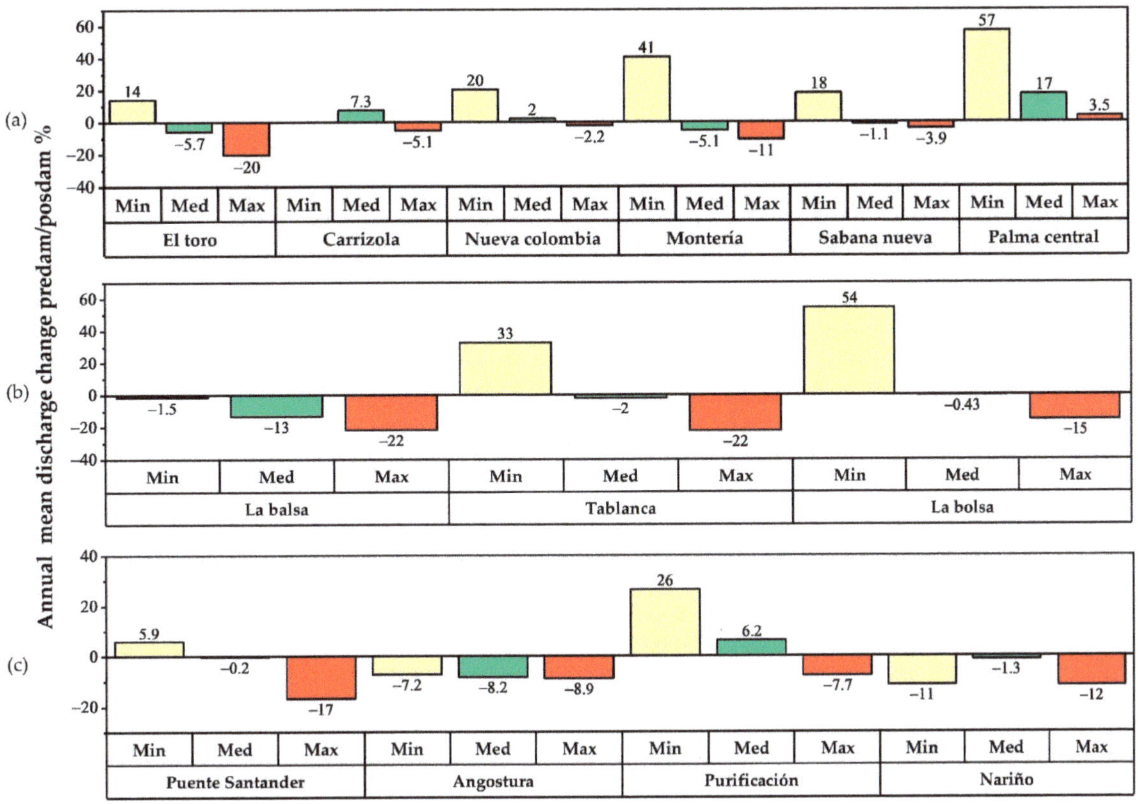

Figure 7. Changes in annual discharges between pre-dam and post-dam periods as a percentage at downstream stations of the (**a**) Urrá I Dam, (**b**) Salvajina Dam, and (**c**) Betania and Prado Dams.

Downstream of the Urrá I Dam (see Figure 7a), all stations exhibit an increase in minimum annual discharges and show a different trend in medium annual discharges. In the Palma central station, all annual differences in discharges are positive, indicating that there is no influence of the dam on this station but, rather, an increase in discharges due to the tributaries. Downstream of the Salvajina Dam, the maximum and medium annual discharges decrease and the minimum annual discharges increase up to 54% (see Figure 7b). After dam construction, the distribution changed for the minimum annual discharges at the La Bolsa station and the maximum annual discharges at the La Balsa station. Downstream of the Betania Dam, the minimum discharges do not have a single trend (see Figure 7c). The maximum and medium annual discharges decreased after dam construction, except at the Purificación station (influenced by the Betania and Prado Dams), where the minimum and medium annual discharges increased. Only at the Purificación station in the post-dam period did the distribution of the minimum annual discharges change.

4.2. Erodibility Index

Based on the geological information, bank material was classified and ranked by using the Global Erodibility Index (EI) according to the recommendations found in the literature [75]. In this work, the EI demonstrated its applicability and its capacity to be applied in morphodynamic modeling. This classification is shown with the morphological changes described in the next section.

4.3. Morphological Changes

4.3.1. Downstream Morphological Changes

The channel width and sinuosity per section were calculated for all years, considering the pre- and post-dam periods. The values were averaged by river reach, and to identify the effects due to dam construction, the differences in the channel width and sinuosity were presented as a percentage, considering the pre-dam condition. A river reach is defined by the affluence of tributaries or changes in the river morphology. These values are schematically arranged in plots that include the confinement degree, the chronostratigraphic units, the erodibility index, and the main tributaries downstream of the dam. In the next sections, the main changes in the channel morphology of each selected dam are presented.

- Morphological changes downstream of the Urrá I Dam

The spatial information available for this study site is distributed as follows: pre-dam years (1991 and 1998) and post-dam years (2007, 2018, and 2020). The percentage of change in the channel width and sinuosity referring to the pre-dam year of 1991 for this study site as a function of the downstream distance is presented in Figure 8. This figure emphasizes that downstream of the Urrá I Dam, the general trend of the channel width after dam construction was to increase (see Figure 8f). After the construction of the dam, the Sinú River increased its width by 17% on average; meanwhile, sinuosity only decreased to −1.31% on average. This general trend is consistent with the increase in low discharges and the sediment retention in the reservoir. However, these changes in the channel width are not correlated with the modifications to channel sinuosity. Figure 8e allows identifying areas with the changing channel width that almost maintain a constant sinuosity. These areas coincide with the location of the Betanci stream affluence and the Grande complex swamps (Figure 8d).

In the high mountains, the river dynamic is driven by changes in lithology, because of the less thickness of the quaternary alluvium. This can be seen in the first 5 km downstream of the dam (see Figure 8), where one of the banks is dominated by rock, so the change in the channel width is less than that observed in quaternary-alluvium dominant areas.

In Figure 8, it is also observed that in the year 2007, the channel width exhibited the maximum observed values (close to an increase of 20%). Areas with this high percentage of increase in the channel width are concordant with more erodible zones, where the quaternary alluvium is on both sides of the channel (see Figure 8b,c). In 2020 (the last year of comparison), some reaches had a reduced channel width in comparison with the maximum channel width observed during the study period; this negative trend indicates bed erosion and channel incision, a behavior that can be clearly identified near the city of Montería. The morphological changes drastically reduced 219 km downstream of the dam, near the Cotoco site; therefore, this length can be assumed as the extent of the morphological impact due to dam construction.

- Morphological changes downstream of the Salvajina Dam

For this study site, the pre-dam years are 1961 (obtained from aerial photographs) and 1984, while the post-dam years are 2007, 2019, and 2020. The percentage of change in the channel width and sinuosity referring to the pre-dam year of 1991 for this study site as a function of the downstream distance is presented in Figure 9.

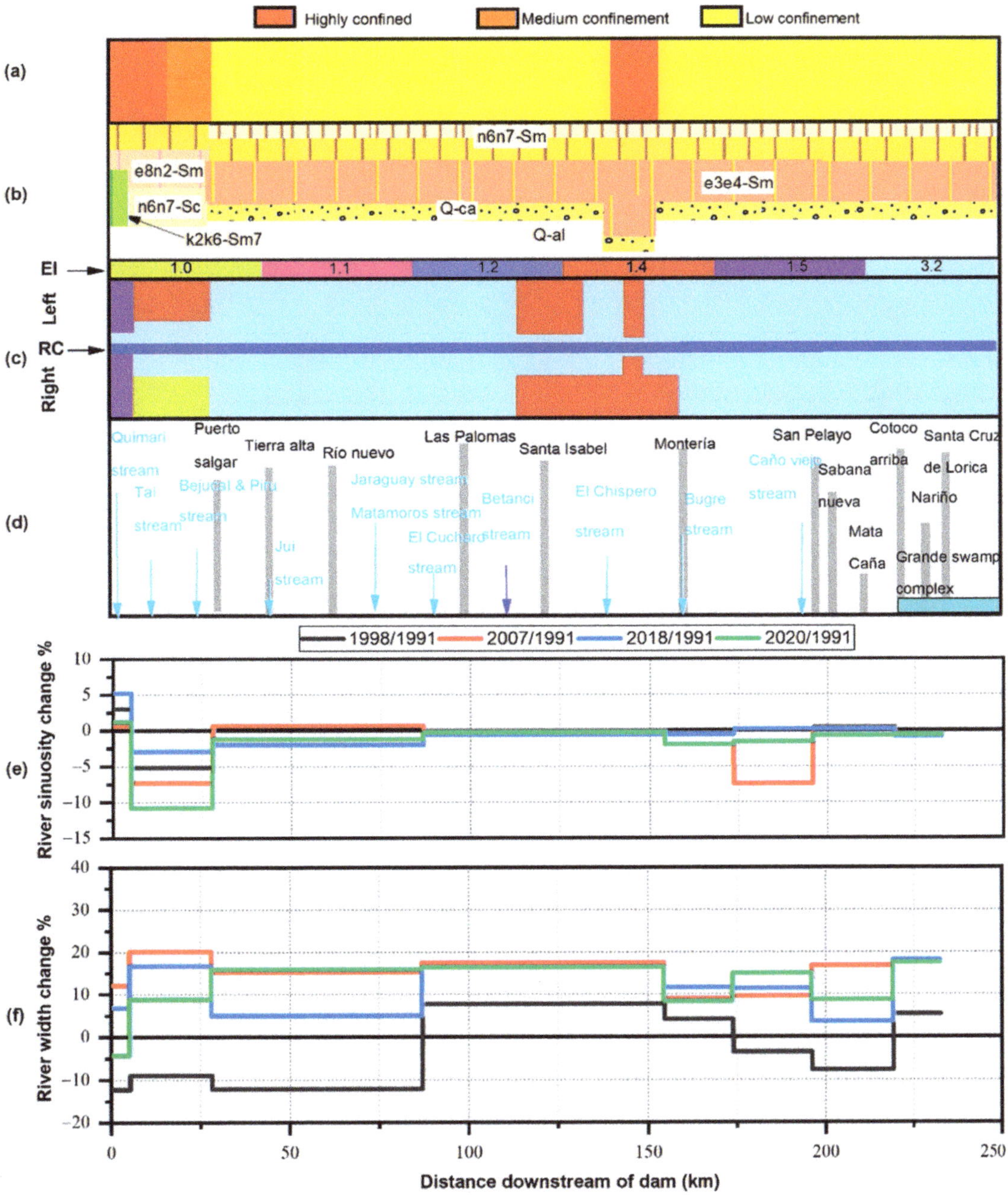

Figure 8. Morphological changes downstream of the Urrá I Dam: (**a**) confinement grade; (**b**) geological units near the Sinú River; (**c**) Erodibility Index (EI) on the left and right sides of the river channel (RC); (**d**) sites of importance near the Sinú River, including affluent rivers and streams; (**e**) river sinuosity change as a percentage with respect to the pre-dam year 1991; and (**f**) river width change as a percentage with respect to the pre-dam year 1991.

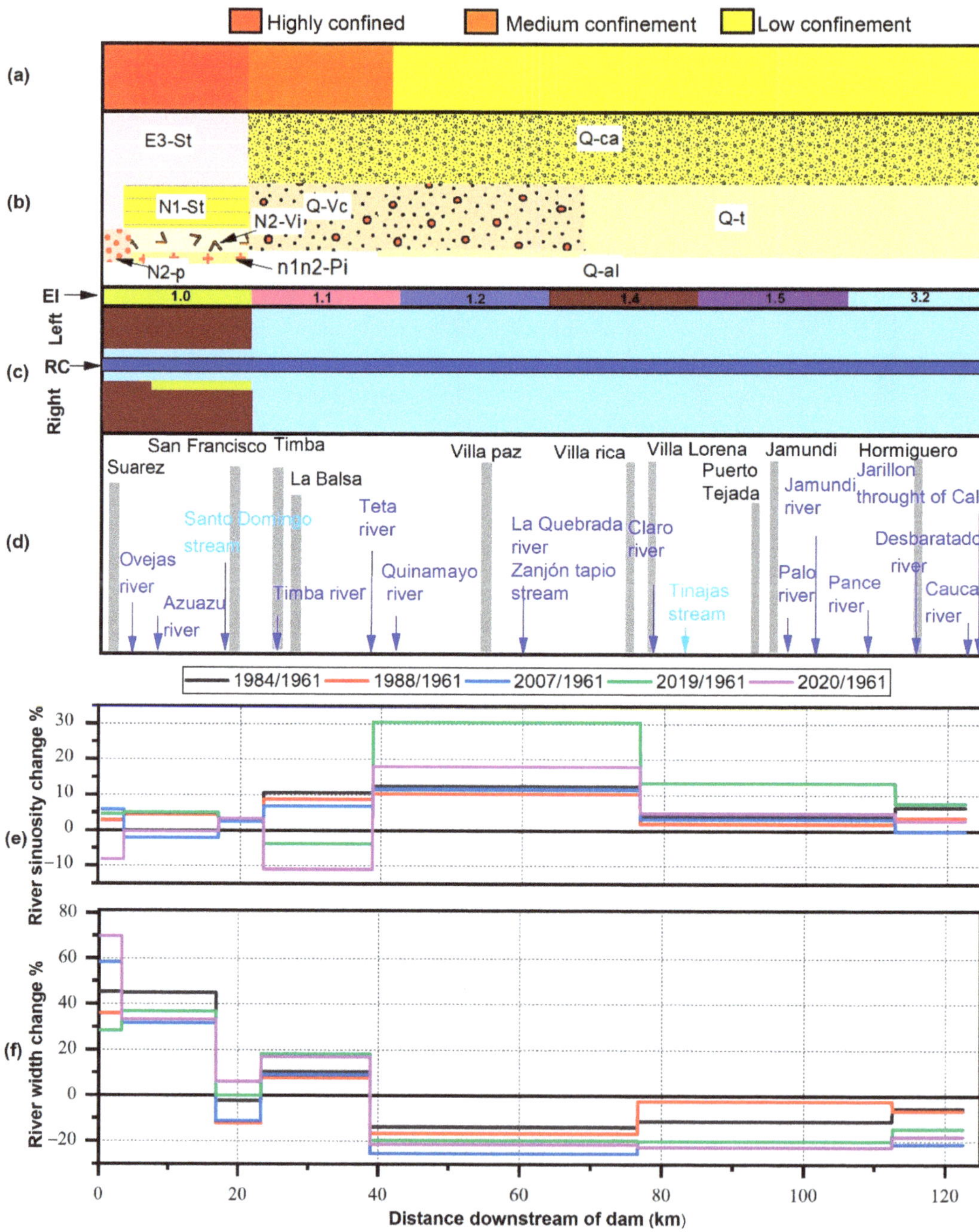

Figure 9. Morphological changes downstream of the Salvajina Dam: (**a**) confinement grade; (**b**) geological units near the Cauca River; (**c**) Erodibility Index (EI) on the left and right banks of the river channel (RC); (**d**) sites near the Cauca River, including affluent rivers and streams; (**e**) river sinuosity change as a percentage with respect to the pre-dam year 1961; and (**f**) river width change as a percentage with respect to the pre-dam year 1961.

Figure 9 shows that downstream of this dam, the morphological changes had two contrasting behaviors, observed in the high mountain area and the alluvial valley. In the high mountain area (between 0 km and 39 km) with high-to-medium confinement, the tendency of the channel was to increase its width, with some changes in sinuosity, during the post-dam period, with a short reach in which the channel width decreased between 1984 and 2019. The most recent year, 2020, presented an increase in the channel width in most areas. The first 17 km of the Cauca River downstream of the Salvajina Dam exhibit a similar behavior to that reported for the Sinú River: the channel width and sinuosity respond to the bank composition; fewer changes are observed in areas with low erodibility, and vice versa. The transition from high mountain to alluvial valley (see Figures 4 and 9a,b) occurs between 17 km and 23 km, where the channel width increases but sinuosity remains constant. From 23 km to 39 km, the channel width increases, which coincides with the transition zone toward the vast valley, where medium confinement is found with quaternary materials (Q- Vc and Q-ca CUs). In the alluvial valley area, from 39 km to 123 km, low confinement and quaternary lithology are observed. In this zone, the Erodibility Index is the highest for the studied river reach (with values up to 3.2). However, there are different quaternary alluvium types (Q-Vc, Q-Ca, Qt, and Q-al) and the river is sinuous. The general tendency of the channel width is to decrease, increasing channel sinuosity. This increase in sinuosity implies a decrease in the river slope, which is reflected in a reduction in the erosive capacity.

The quaternary terraces (Q-t CU) dominate on the right side of the valley, between 70 km and 123 km, confining the river to the left side or pulling it toward the Western Andes. The river stops being sinuous exactly in confluence with the Desbaratado River, approximately 118 km downstream of the dam. From 118 km to 123 km, the channel width and sinuosity did not significantly change in the post-dam period. This narrowing trend of the river channel combined with an increase in discharges during the dry period may indicate the existence of an incision process in this area. This aspect is contrary to the expected increase in bank erosion processes, which emphasizes the relevance of the composition of the alluvial valley materials. In general, before the construction of the dam, the Cauca River width increased by 3.5 m (4.33%) and its sinuosity by 0.1 (6.5%). During the post-dam period, the channel width decreased by −0.17 m (−0.21%) and the sinuosity by 0.09 (4.74%). The morphological changes drastically reduced 118 km downstream of the dam, near the Desbaratado River; therefore, this length can be assumed as the extent of the morphological impact due to dam construction.

- Morphological changes downstream of the Betania and Prado Dams

The pre-dam years are 1969 (obtained from aerial photographs) and 1984 for the Betania and Prado Dams, respectively, while the post-dam years are 1988, 1998, 2015, and 2020. The percentage of change in the channel width and sinuosity referring to the pre-dam year of 1969 for this case as a function of the downstream distance is presented in Figure 10.

The Magdalena River crosses different confinements and lithologies, separated into three areas according to the area's geology: the Neiva sub-basin, the Girardot sub-basin, and the transition zone between them (valley). In the Neiva sub-basin, from 0 km to 90 km, the narrowing trend of the river was kept in the pre-dam and post-dam periods without important changes in sinuosity. After dam construction, the decrease in the channel width intensified. There is, however, no general trend, because the banks present contrasting lithologies. Two areas are highlighted with unique geological characteristics:

- Between 14–18 km, there is an erodible volcanic lithology (EI = 1.5) on a high-confinement area. Then, an increase in the channel width was observed in the post-dam period.
- From 18 to 25 km, the channel width was reduced by about −62% between pre- and post-dam periods. This is a Magdalena River transition area from high- to low-confinement. The transition area is close to where the Frío River flows into the Magdalena River.

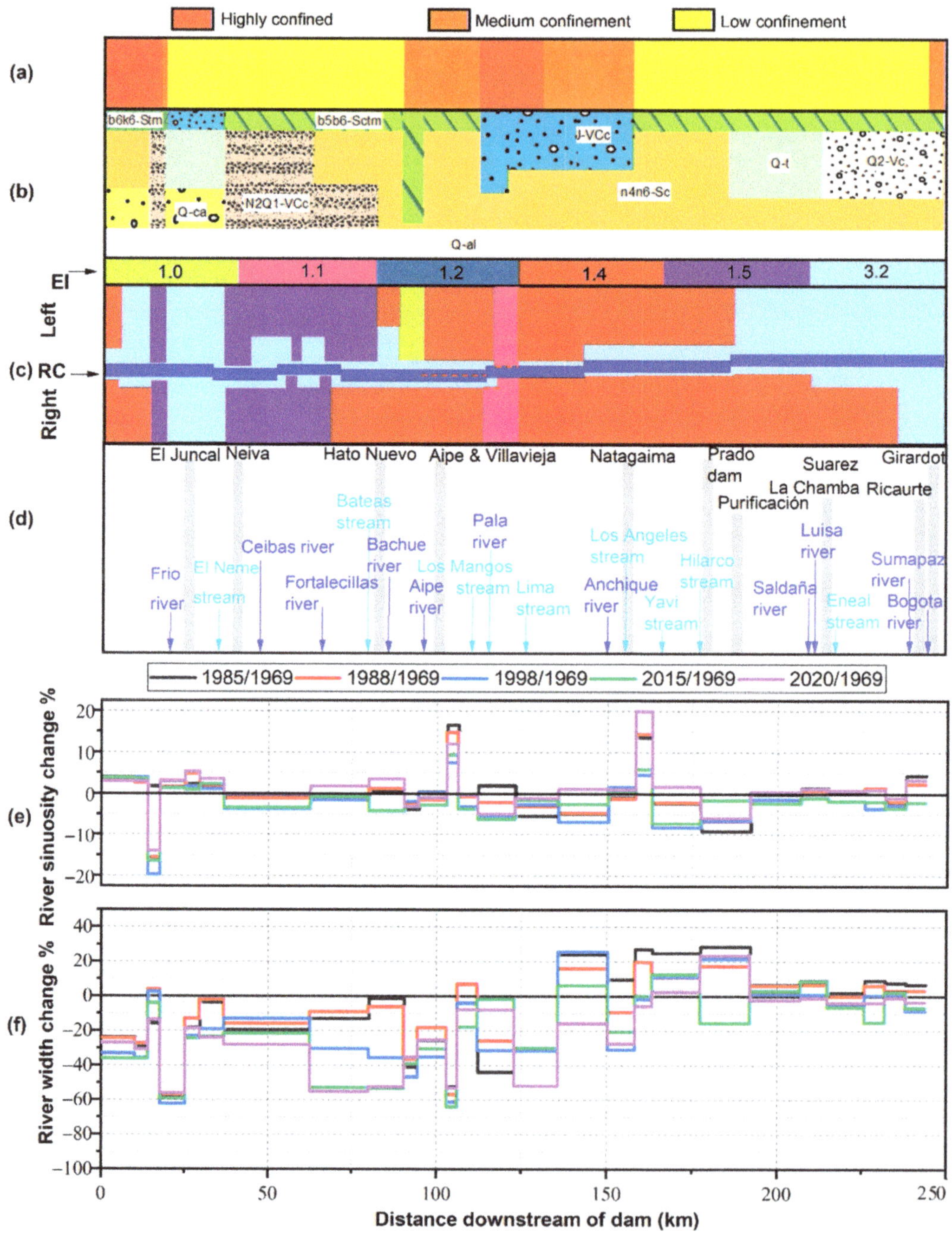

Figure 10. Morphological changes downstream of the Betania and Prado Dams: (**a**) confinement grade; (**b**) geological units near the Magdalena River; (**c**) Erodibility Index (EI) on the left and right sides of the river channel (RC) and faults (----); (**d**) sites near the Magdalena River, including affluent rivers and streams; (**e**) river sinuosity change as a percentage with respect to the pre-dam year 1969; and (**f**) river width change as a percentage with respect to the pre-dam year 1969.

The transition zone, a valley with medium-to-high confinement, between 90 km and 158 km presents a thinner alluvial quaternary deposit. The general trend of the channel width is to decrease. That decreasing trend was enhanced in the post-dam period. Some sections are highlighted: For example, the river reach located between 103 km and 106 km exhibits the lowest channel width change throughout the Magdalena River in the post-dam period (−64%) and a maximum sinuosity of 16%. This reach corresponds to the confluence zone of four streams into the Magdalena River. Between 112 km and 123 km, the channel width increases due to the presence of a fault zone (sheared rocks), which makes this zone erodible.

In the Girardot sub-basin, with low confinement from 158 km to 244, there was a channel-widening trend during the pre-dam period. After dam construction, a general narrowing trend is observed until 192 km downstream of the dam. This trend is primarily associated with bars that decrease its channel width, delimited by the n4n6-SC CU formation. From 192 km to 244 km, quaternary formations, such as Q-t, alluvial terraces, and the Q2-Vc CU Guamo fan appear, which confine the river, generating an almost straight planform. In this area, there are no significant changes in sinuosity and width, but precisely between 215 km to 226 km, the channel width shows no change during the pre-dam period. The post-dam period maintains a value of −6% for all years, defining the extent of the morphological impact due to dam construction, which occurs until 215 km downstream of the dam (near the city of Suárez).

In this study site, it was not possible to differentiate the changes due to each dam (Betania and Prado), because the gauging station installation date does not coincide with the starting of the operation of the Prado Dam. The hydrological regime in this area presents increased discharges during the dry period, but those are not reflected in the observed changes in the channel width. This aspect also shows the relevance of the resistance of the riverbanks, which coincide with more consolidated formations.

4.3.2. Upstream Morphological Changes

The morphological changes upstream of the selected dams are presented schematically in a similar way to that used for the downstream areas; however, these plots do not include the tributaries of the main river. The morphological changes upstream of the selected dams showed similar trends, and the obtained results are summarized in Table 4, while the schematic plots are included in Appendix D. To identify the extent of the morphological effects upstream of the reservoir, two reaches were adopted: (1) the length of the channel until the influence of the reservoir was identified and (2) the length of the channel upstream of the previous one in which the effect of the reservoir vanished.

The morphological changes in the reach upstream of the reservoir (identified by reach 1 in Table 4) are related to the increased water levels in the reservoir with the correlated increase in the channel width. The magnitude of the increase in the channel width varies among the study sites, but it is possible to emphasize that upstream changes in the channel width are 1 order of magnitude larger than those observed downstream of the dam.

Upstream of the Urrá I Dam, an increase in the channel width of 600% and 562% was observed along the Verde and Sinú Rivers, respectively—a situation that occurs in zones with high confinement and medium erodibility but that are affected by the presence of faults that increase instability. Upstream of the Betania Dam, the Yagura River has low confinement and high erodibility and presents an increase in the channel width of up to 456%. Contrastingly, the zone of the Magdalena River with medium confinement and medium erodibility exhibits an increase up to 167%. Upstream of the Salvajina Dam, the channel width increases up to 123% in zones with high confinement and medium erodibility. The sinuosity, however, remains almost constant for the cases of Salvajina and Betania Dams, while in the Urrá I Dam, it presents high variability.

Considering the results presented in Table 4 and the plots presented in Appendix D, it is possible to identify that the magnitude of the morphological changes upstream of the studied sites depends on the hydrological variability of the water levels in the reservoir and the geometrical settings of the reservoir vase, such as the confinement degree and the valley lateral slope. The geological characteristics of the forming materials seem to have an influence on the morphological changes as high erodibility generates an increase in the channel width. The coincidence of the channel course with fault chains modifies the behavior of certain zones from low to high erodibility. Due to the changing morphology, the zones identified as reach 1 in each study site should be considered in risk reduction activities because of the increased water levels and channel widths.

5. Discussion

5.1. Downstream Dam Effects

Although geology is considered essential in influencing river morphology, a few studies have considered the riverbank's composition in the study of the morphological effects of dams. It is more common to use the geological information to define the right place to build a dam, analyzing, for instance, the rock type and geological structure [86]. Studies dealing with the morphological effects of dams use geologic information to obtain a context of the area of interest with lithological descriptions at a regional scale, but this information is not included in the morphodynamic analysis. Geology determines the spatial variation of riverbank materials (formations) and their associated resistance [73]. Grant et al. [30], for example, distinguished the importance of geology as a primary control, determining valley and landscape configurations and creating a relationship between sediment flows and post-dam changes. It is important to mention that due to the Colombian topography, the selected dams are built on the fall line (with an upstream bedrock reach and a downstream alluvial bed). Therefore, our analyses focused on this high–low intervention type.

Downstream of a dam, it is expected to find narrower channels through time [2,33]; however, a few authors have reported other trends during the post-dam period. Williams and Wolman [8] reported contrasting trends based on data from 21 dams in the United States. They explained these different responses due to the resistance of the alluvial bank materials, identified through field campaigns. In the Snake River, Nelson et al. [87] reported wider channel widths after dam construction, associated with large tributary floods. The results of our study suggest that including the geological framework and confinement degree of the selected area may clarify the possible channel width response, allowing identification of the more susceptible areas. This information is valuable in defining the extent of the biotic compensation measures in environmental impact studies. In mountainous areas (regions with high-to-medium confinement) near a dam, there is a general trend to have increased widths, even though the thickness of the alluvial quaternary area is thinner than in the alluvial valley. Moreover, in some sectors, the riverbanks can be in direct contact with the geological formations. Just downstream of the dams of the selected projects, a high-confinement zone is found. However, the high mountain area downstream of the Betania Dam (Magdalena River) is composed of more erodible materials (feldespathic sandstones and conglomerates rich in lithics). These low-resistance riverbanks explain the wider channels observed after dam construction (see Figure 10). In the other two cases (Salvajina and Urrá I projects), narrower channels are present, an aspect that can be attributed to incision processes induced by high-resistance banks (see Figures 8 and 9).

In contrast, in the alluvial valley region (area with low confinement), quaternary formations predominate, and it is there where the trend may vary. The valley composition of the selected sites determines the channel width response. The Sinú and Cauca River valleys present the quaternary alluvium in both riverbanks; however, downstream of the Salvajina project, a graben structure (i.e., distinct escarpment on each riverbank caused by downward displacement) composed of terraces and colluvial exists. These material sequences generate channel confinement, a configuration that is not present downstream of the Urrá I project. This difference in the geological structure generates channel narrowing in the Cauca River valley and channel widening in the Sinú River valley (see Figures 8 and 9). Downstream of the Betania project, the Magdalena River exhibits decreased channel widths through the valley up to 160 km, where the alluvial material interacts with consolidate rocks. The increased presence of unconsolidated materials and the end of a confined area, at 160 km downstream of the dam, determine the wider channels in that zone (see Figure 10). In Figure 11, three moments with similar discharges of a sector of the Magdalena River in which the channel width reduced by more than 50% are presented. Besides channel width reduction, in this figure, it is possible to identify how the dynamics of the reach considerably reduced, while maintaining a low-impacted flow regime (see the Purificación station regime in Figure 6).

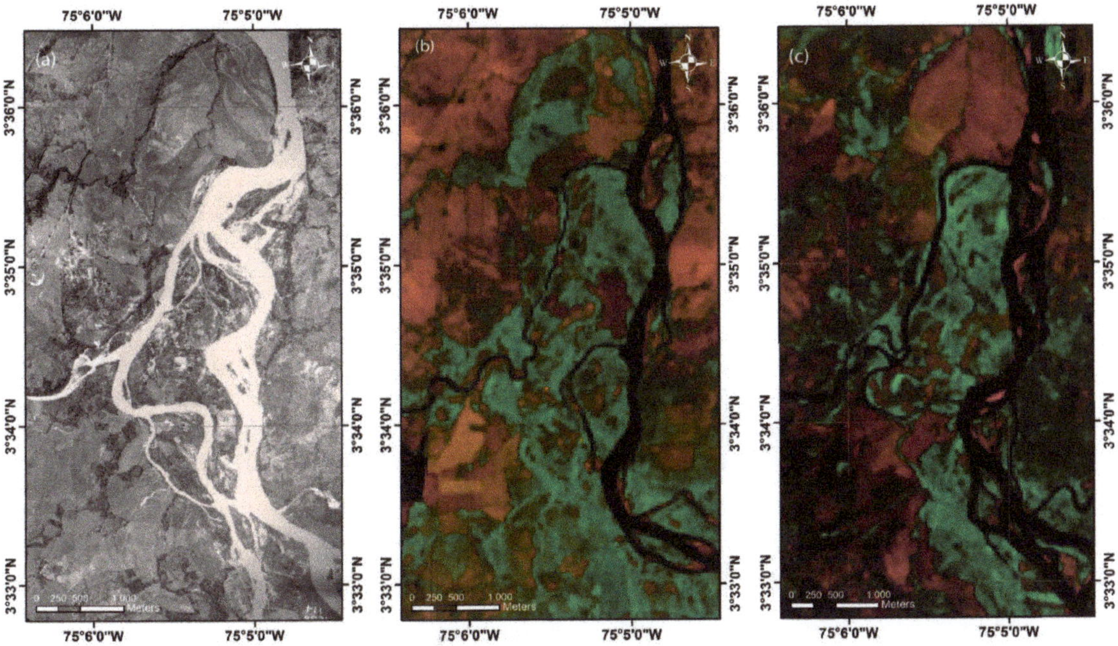

Figure 11. Downstream changes in the Betania Dam in the sector near Natagaima: (**a**) aerial photograph of March 1969 (Q = 233 m^3/s), (**b**) RGB composite 743 of March 1988 (Q = 326 m^3/s), and (**c**) RGB composite 754 of March 2020 (Q = 350 m^3/s).

Generally, in morphological studies downstream of dams, the information about the river width is measured at selected cross sections [7,23,24,28,30–32], but the morphological impacts due to dams cannot be analyzed in short distances [87,88]. Therefore, the use of GIS tools, as the one selected for this study (R package cmgo), allows covering long distances (until 240 km downstream in our case) to identify the real extent of the impacts. The geological information at a regional scale is compatible with the spatial resolution of free available imagery and DEMs, allowing identification of areas more susceptible to change and defining the extent of morphological alterations [76,89].

5.2. Upstream Dam Effects

Studies addressing the quantification of morphological changes upstream of dams are scarce and geographically limited. These studies have mainly developed conceptual models to identify the upstream effects of dams; among these, studies by Liro [12] and Lu et al. [16] were mainly based on the relationship between deposition and erosion. Specifically, Liro [12] described the effects of morphological changes on flooding. In general, morphological studies summarize upstream effects as bed aggradation and changes in the river profile (e.g., Csiki and Rhoads [90]). Our results suggest that reservoir operation and its geometrical characteristics, such as the valley confinement degree and slope, determine the magnitude and extent of the changes in channel morphology upstream of dams. Therefore, the geology of the forming materials seems to have an influence on the morphological long-term changes: wider and less sinuous channels were observed in zones with high erodibility as quaternary materials and fault zones (see Figures A1–A3 in Appendix D). However, we also showed that the change in the channel width in the upstream zone is 1 order of magnitude larger than in the downstream areas (see Table 4 and Appendix D).

In Figure 12, an example of these changes in the Magdalena River upstream of the Betania Dam is presented. This figure emphasizes that the geomorphology of the dam area is of paramount importance for determining long-term morphological effects. The drastic reduction in the reservoir levels observed in this figure between 1988 and 2020 allows us to see the important planform changes recorded in this area. Our methodology, and the morphological studies of these areas in general, is found to be highly relevant in flood risk reduction analyses.

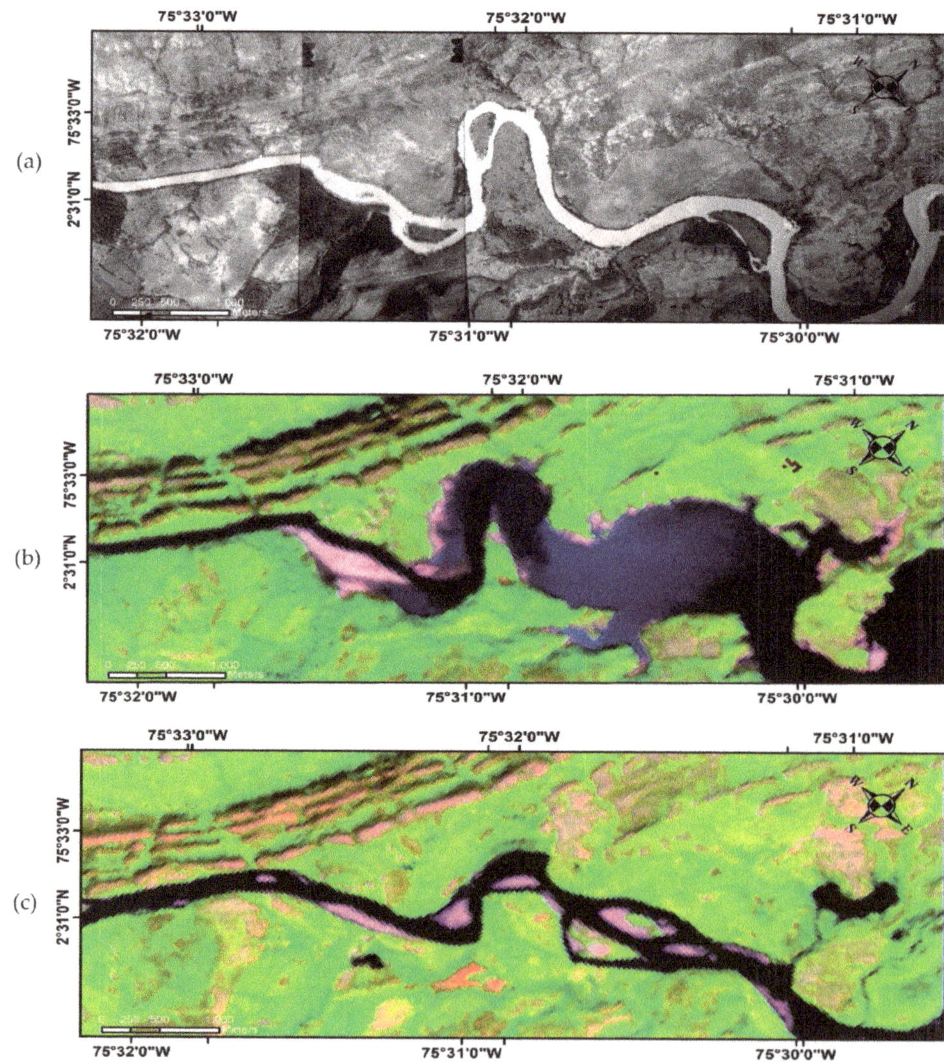

Figure 12. Upstream changes due to the Betania Dam across the Magdalena River: (**a**) aerial photograph of 1969, (**b**) RGB composite 743 of 1988, and (**c**) RGB composite 754 of 2020.

5.3. Limitations

It is important to mention that our study addresses the downstream morphological impacts of dams at the reach scale, which can be useful in regional analyses; however, our approach is motivated but limited at the same time by information scarcity. Detailed data on sediment transport characteristics, land use, and land cover changes, as well as the channel shape and slope evolution, would have allowed us to reach more quantitative results. Additionally, the Global Erodibility Index (EI) used in this work was adjusted to a global scale, but the Colombian geological formations are slightly different. Field visits and laboratory tests may reinforce the application of the EI in the Colombian region. We assumed that all quaternary formations (unconsolidated materials) have the same EI, neglecting the sediment characteristics. Therefore, the quaternary formations must be better described for identifying lithologies and evaluating fluvial erodibility in morphology [28,91,92].

6. Conclusions

The magnitudes and trends of upstream and downstream morphological changes due to dam construction vary widely. In this contribution, the relationship between the geology of each river reach and the channel width and sinuosity response was addressed. For that purpose, the geology of the areas close to the river channel was correlated to the riverbank resistance through the Erodibility Index (EI). The selected dams are built on the fall line of Colombian rivers, considered an appropriate representative of tropical systems.

Although the hydrological alteration reported in the tropical rivers considered in this contribution is relatively low, the observed morphological changes are of considerable magnitude. The magnitude of these morphological changes can be related to the high sedimentological alteration, considering that none of the studied dams includes a sediment bypass. Therefore, the reduction in both the suspended and bed loads is almost complete downstream of the dam.

Downstream of dams, the geomorphological settings define the degree of confinement and riverbank resistance, key characteristics of defining the morphological trends and future behavior of river channels. Therefore, including the geological framework of the selected area may clarify the possible morphological response of the channel width, allowing identification of the more susceptible zones. This delimitation may be valuable in setting the study area for biotic compensation measures and environmental impact studies.

Upstream of dams, the geology of the forming materials also influences the morphological changes. The geometrical characteristics of the reservoir tail (valley confinement degree and transversal slope) and its operation determine the magnitude and extent of the changes in channel morphology.

Despite the high sediment transport rates of Colombian rivers (and tropical systems in general), which allow quick recovery of sediment loads, the morphological impacts downstream of dams can reach considerable lengths. This fact demonstrates the relevance of conducting morphological studies before dam construction, including the geological characteristics of the underlying materials, to consider the possible negative effects of such interventions.

The use of the geological information about the area of interest and its relationship with the Erodibility Index (EI) used in this research to estimate the riverbank resistance demonstrated its usefulness. This approach can also be used in the calibration of the erodibility coefficients required in morphodynamic modeling exercises to predict morphological changes upstream and downstream of dams.

Author Contributions: D.C.A. collected the data, performed calculations, analyzed the data, and wrote the main text of the manuscript. D.C.A., J.D.G.-O. and A.V.-L. prepared the layout of the research work. A.V.-L. supervised the overall activity and the progress of the manuscript construction. J.D.G.-O. and A.V.-L. commented on the manuscript and improved the language. All authors have read and agreed to the published version of the manuscript.

Funding: The authors express their gratitude to Pontificia Universidad Javeriana, which supported this research through the APC fundind (VRI support to promote the scientific paper writing). Also, Pontificia Universidad Javeriana gave financial support to purchase the aerial photos.

Institutional Review Board Statement: Not applicable.

Informed Consent Statement: Not applicable.

Data Availability Statement: Not applicable.

Acknowledgments: The authors thank the *Corporación del Valle del Cauca* (CVC) for supplying information about the hydrological stations downstream of the Salvajina Dam. The Pontificia Universidad Javeriana is also acknowledged for financial support to purchase the aerial photos used in this research.

Conflicts of Interest: The authors declare no conflict of interest.

Appendix A

Chronostratigraphic units' description:

The chronostratigraphic units' description was obtained from the *Servicio Geológico Colombiano* [45], and the unit coding is presented here:

k2k6 − Sm7

Age + Lithology + Terrain *

Turonian − Maastrichtiense + Sedimentary marine + 7

* Terrain: Geological provinces of Colombia [72].

- **Urrá I dam area:**

Q-al: Alluvial deposits and floodplains
Q-ca: Alluvial fans and colluvial deposits
n6n7-Sm: Intercalations of mudstones, calcareous arenites, and coarse-grained–conglomeratic quartz sandstone
n6n7-St: Conglomerate intercalations, medium–conglomerate lithic–sublithic arenites, calcareous arenites, and mudstones
n3n5-Sm: Fine–coarse-grained lithic and feldspathic sandstones with glauconite and calcareous shale intercalation with mudstones
n1n2-St: Medium-grained lithic–sublithic sandstones interstratified with mudstones, locally peat, charcoal, and sandy conglomerate lenses
e8n2-Sm: Limestone, sandstone, and calcareous mudstone
e8n2-St: Ferruginous fine-grained–conglomeratic sandstone intercalation with calcareous siltstone and coal
e3e4-Sm: Grain-decreasing lithic sandstones to conglomeratic–fine-grained sandstone intercalation with mudstones and micritic limestones
k2k6-Sm7: Claystones, muddy and lithic sandstones, cherts, limestone, tuffs and agglomerates, local cataclassites

- **Salvajina dam area:**

Q-al: Alluvial deposits and floodplains
Q-ca: Alluvial fans and colluvial deposits
Q-t: Alluvial terraces
Q-Vc: Volcanoclastic flows made up of pyroclasts and epiclasts of andesitic composition
N2-P: Ash tuffs and lapilli intercalation with lavas of andesitic composition
N2-Vi: Lavas and ignimbrites of andesitic composition
n1n2-Pi: Quartzodiorites, tonalites, diorites, and granodiorites

N1-St: Claystones, siltstones, sandstones, and conglomerates
e6e7-Stm: Conglomerates, lithic muddy sandstones, quartz sandstones, siltstones, claystones, and coals
E3-St: Fine-grained quartz sandstones to conglomerate intercalation with conglomerates, mudstones, and coals
k2-Vm7: Tholeiitic basalts, dolerites, picrites, basics tuff, and volcanic breccias
k2k6-Sm7: Claystones, muddy and lithic sandstones, cherts, limestone, tuffs and agglomerates, local cataclassites
K1-Mbg6: Amphibolite schists and quartzites

- **Betania and Prado projects:**

Q-al: Alluvial deposits and floodplains
Q-ca: Alluvial fans and colluvial deposits
Q-t: Alluvial terraces
Q2-Vc: Volcanoclastic flows constituted of pyroclasts and epiclasts of andesitic and dacitic composition
N2Q1-VCc: Felspathic sandstones, claystones, and conglomerates rich in lithics of andesites and dacites
n4n6-Sc: Lithic sandstones with intercalation of gray–green claystones and conglomerates
e6e9-Sc: Intercalations of red beds of conglomerates, lithic conglomeratic sandstones, and claystones
k6E1-Stm: Red claystones with intercalations of fine quartz sandstones, coal beds in the basin
b6k6-Stm: Intercalations of shales, limestones, sandstones, cherts, and phosphorites
b5b6-Sctm: Red conglomerates, feldesphatic sandstones and claystones, quartz sandstones, limestones and shales; in the top, ferruginous quartz and glauconites sandstones
J-Pi: Monzogranite from Algeciras and Stock of Ibagué
J-VCc: Tuffs, agglomerates, and lavas, occasionally intercalations of red beds of lithic sandstones and siltstones

Global Erodibility Index (EI):

1.0—acid plutonic rocks, metamorphic rocks, and carbonate sedimentary rocks
1.1—acid volcanic rocks
1.2—mixed sedimentary rocks
1.4—basic volcanic rocks; siliciclastic rocks of fine–medium grain sizes
1.5—basic plutonic rocks, siliciclastic rocks of high variability in grain sizes
3.2—unconsolidated sediments

Appendix B

Table A1. Images used in the study for the selected dams.

Project	Aerial Photographs		Satellite Images		
	Flight	Date	Name	Mission	Date
Betania	C1431	12 September 1971			
	C1866	1 March 1973	LT05_L1TP_008058_19850322	Landsat 5	22 March 1985
	C1343	9 December 1970	LT04_L1TP_008058_19880322	Landsat 4	22 March 1988
	C1518	11 September 1974	LT05_L1TP_008058_19981004	Landsat 5	4 October 1998
	C1564	5 March 1975	LC08_L1TP_008058_20151019	Landsat 8	19 October 2015
	C1256	22 March 1969	LC08_L1TP_008058_20200322	Landsat 8	22 March 2020
	C1470	17 November 1973			

Table A1. *Cont.*

Project	Aerial Photographs		Satellite Images		
	Flight	Date	Name	Mission	Date
Prado	C1256	22 March 1969	LT05_L1TP_008057_19850322	Landsat 5	22 March 1985
	M1393	24 February 1966	LT04_L1TP_008057_19880322	Landsat 4	22 March 1988
	M152	29 March 1957	LT05_L1TP_008057_19981004	Landsat 5	4 October 1988
	M153	30 March 1957	LC08_L71TP_008057_20151019	Landsat 8	19 October 2015
			LC08_L1TP_008057_20200322	Landsat 8	22 March 2020
Urrá I	No flights		LT05_L1TP_010054_19910321	Landsat 5	21 March 1991
			LT05_L1TP_010054_19980103	Landsat 5	3 January 1998
			AST_L1T_00301132007153521	Aster	13 January 2007
			LC08_L1TP_010054_20181228	Landsat 8	28 December 2018
			S1B_IW_GRDH_1SDV_20200 111T104920	Sentinel 1B	11 January 2020
Salvajina	C1421	16 March 1972	LT05_L1TP_009057_19840902	Landsat 5	2 September 1984
	C1518	11 September 1974	LT04_L1TP_009058_19880516	Landsat 4	16 May 1988
	M1098	14 August 1961	AST_L1T_00301292007153659	Aster	29 January 2007
	M1467	14 August 1961	LC08_L1TP_009058_20190818	Landsat 8	18 August 2019

Appendix C

Table A2. Comparison of monthly discharges in pre-dam and post-dam periods at each flow gauging station. The values are the p-values of Student´s t-test and Fisher´s F-test. The p-values below $\alpha = 0.05$ are bolded.

ID	TEST	January	February	March	April	May	June	July	August	September	October	November	December
b1	T	**0.01**	**0.00**	**0.00**	0.13	0.57	0.66	0.91	0.57	0.90	0.63	0.90	0.25
	F	0.91	0.74	0.74	0.89	0.39	**0.02**	0.39	**0.02**	0.34	0.37	**0.02**	0.31
b2	T	**0.01**	**0.01**	**0.00**	0.15	0.55	0.67	0.68	0.84	0.61	0.69	0.91	0.21
	F	0.92	0.66	0.77	0.96	0.44	**0.02**	**0.03**	0.23	0.09	0.35	**0.00**	0.26
b3	T	0.10	0.06	**0.02**	0.09	0.53	0.87	0.36	0.81	0.68	0.48	**0.04**	0.68
	F	0.72	0.64	0.41	0.81	0.98	0.63	0.54	0.18	0.53	**0.00**	0.82	0.46
b4	T	0.53	**0.00**	**0.00**	**0.01**	0.76	0.70	0.12	0.25	0.11	**0.00**	**0.00**	0.49
	F	**0.04**	0.39	0.15	0.43	0.97	0.69	0.32	0.37	0.28	**0.00**	0.28	0.18
b5	T	0.63	**0.04**	**0.00**	**0.00**	0.49	0.62	0.87	0.62	0.82	0.06	**0.04**	0.59
	F	0.56	**0.01**	0.76	**0.02**	0.47	0.97	0.72	**0.02**	**0.14**	**0.02**	0.60	**0.05**
b6	T	**0.00**	**0.02**	**0.00**	**0.01**	0.25	**0.02**	0.24	0.36	0.30	0.90	0.93	0.17
	F	0.95	0.42	0.20	0.98	0.53	**0.03**	0.13	0.47	0.44	0.19	0.70	0.43
c1	T	0.71	0.43	0.85	**0.00**	**0.00**	**0.03**	0.75	0.05	0.08	0.16	**0.02**	0.48
	F	0.94	0.96	0.35	0.19	0.25	0.60	0.48	**0.00**	0.41	0.26	0.73	0.19
c2	T	0.47	0.23	0.75	0.19	**0.02**	0.13	0.35	**0.00**	**0.01**	0.98	0.42	0.63
	F	0.31	0.36	0.67	0.74	0.34	0.69	0.23	**0.02**	0.45	0.92	0.51	**0.03**
c3	T	0.41	0.56	0.56	0.66	**0.02**	0.18	0.2	**0.00**	**0.00**	0.05	0.00	0.45
	F	0.76	0.50	0.83	0.93	0.65	**0.00**	0.26	0.50	0.07	*0.03*	0.60	0.38
d1	T	0.25	0.29	0.15	0.82	0.96	0.65	0.27	0.50	0.51	**0.02**	0.07	0.44
	F	0.89	0.81	0.42	0.25	0.07	0.76	**0.02**	0.59	0.37	0.08	0.14	0.01
d2	T	0.95	0.59	0.82	0.15	0.11	0.38	0.44	0.72	0.06	**0.00**	0.28	0.88
	F	0.15	0.98	0.86	0.71	0.73	0.73	0.53	0.92	0.69	0.82	0.66	0.29
d3	T	0.07	0.12	**0.01**	0.66	0.80	0.30	0.70	0.61	0.78	**0.04**	0.24	0.08
	F	0.90	0.91	0.32	0.20	0.47	0.34	0.08	0.72	0.55	0.18	0.42	0.06
d4	T	0.77	0.99	0.18	0.52	0.32	0.70	0.31	0.47	0.08	**0.01**	**0.00**	0.29
	F	0.88	0.47	0.30	0.73	0.62	0.73	0.50	0.99	0.71	**0.02**	0.68	0.08

Table A3. Comparison of monthly precipitation in pre-dam and post-dam periods at each project basin. The values are the *p*-values of Student´s *t*-test and Fisher´s *F*-test. The *p*-values below α = 0.05 are bolded.

BASIN	TEST	January	February	March	April	May	June	July	August	September	October	November	December
Betania	T	0.33	0.96	0.80	0.64	0.58	0.69	**0.01**	0.53	0.44	0.35	0.73	**0.05**
	F	0.74	0.15	0.39	0.68	0.10	0.25	0.27	0.07	0.78	0.49	0.07	0.82
Salvajina	T	0.36	0.66	0.11	0.25	0.89	0.65	**0.00**	0.26	0.46	0.06	0.10	0.40
	F	0.07	0.34	0.65	0.90	0.70	0.08	**0.01**	0.10	0.78	**0.01**	0.42	0.39
Urrá I	T	0.95	0.11	0.48	0.88	0.30	0.48	0.15	0.54	**0.00**	0.78	0.22	0.61
	F	0.42	0.29	0.49	**0.02**	0.80	0.97	0.93	0.51	0.46	0.27	0.50	0.43

Table A4. Comparison of downstream gauging station annual flow. The flow values are the average of minimum (MIN), median (MED), and maximum (MAX) annual discharges at each station. In addition, the *p*-values of Kolmogorov–Smirnov (KS) and Mann–Whitney (MW) tests are shown. The *p*-values below α = 0.05 are bolded.

Project	Station	Annual Value	Average Discharge (m^3/s) Pre-Dam	Post-Dam	KS *p*-Value	MW *p*-Value
Urrá I	El Toro	Min	71	81	0.06	0.16
		Med	348	328	0.96	0.48
		Max	953	759	0.06	0.04
	Carrizola	Min	79	117	0.49	0.24
		Med	329	353	0.70	0.92
		Max	846	803	0.88	0.72
	Nueva Colombia	Min	74	89	**0.01**	**0.05**
		Med	347	354	0.91	0.98
		Max	820	802	0.44	0.70
	Montería	Min	69	97	**0.01**	**0.02**
		Med	376	357	0.90	0.57
		Max	876	781	**0.01**	**0.01**
	Sabana Nueva	Min	78	92	**0.04**	0.22
		Med	352	348	0.80	0.87
		Max	714	686	0.76	0.89
	Palma Central	Min	72	113	**0.00**	**0.01**
		Med	352	412	**0.07**	**0.05**
		Max	620	642	0.86	0.42
Salvajina	La Balsa	Min	67	66	0.798	0.592
		Med	204	177	0.983	0.817
		Max	542	424	0.144	0.023
	Tablanca	Min	58	77	0.858	0.654
		Med	198	194	0.44	0.196
		Max	576	447	0.277	0.793
	La Bolsa	Min	68	105	**0.008**	**0.02**
		Med	233	232	0.458	0.313
		Max	635	537	0.212	0.062
Betania and Prado	Puente Santander	Min	153	162	0.798	0.592
		Med	488	487	0.983	0.817
		Max	1796	1496	0.144	**0.023**
	Angostura	Min	235	218	0.858	0.654
		Med	657	603	0.44	0.196
		Max	2007	1829	0.277	0.793
	Purificación	Min	216	273	**0.008**	**0.02**
		Med	744	790	0.458	0.313
		Max	2591	2392	0.212	0.062
	Nariño	Min	456	404	0.625	0.486
		Med	1153	1138	0.977	0.986
		Max	3563	3147	0.279	0.093

Appendix D

Schematic plots showing upstream morphological changes due to the reservoir construction in each study site.

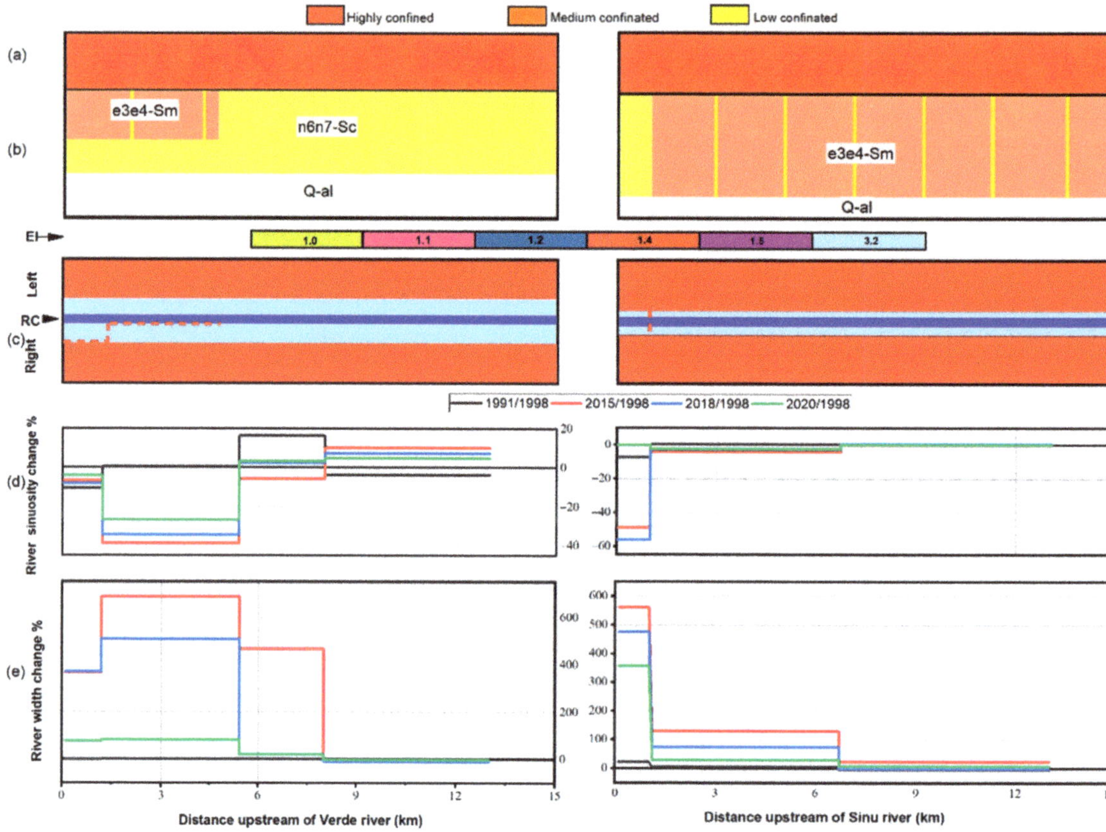

Figure A1. Morphological changes upstream of the Urrá I Dam across Verde and Sinú Rivers: (**a**) confinement grade, (**b**) geological units near the rivers, (**c**) Erodibility Index (EI) on the left and right sides of the river channel (RC) and faults (----), (**d**) river sinuosity change as a percentage with respect to the pre-dam year 1998, and (**e**) river width change as a percentage with respect to the pre-dam year 1998.

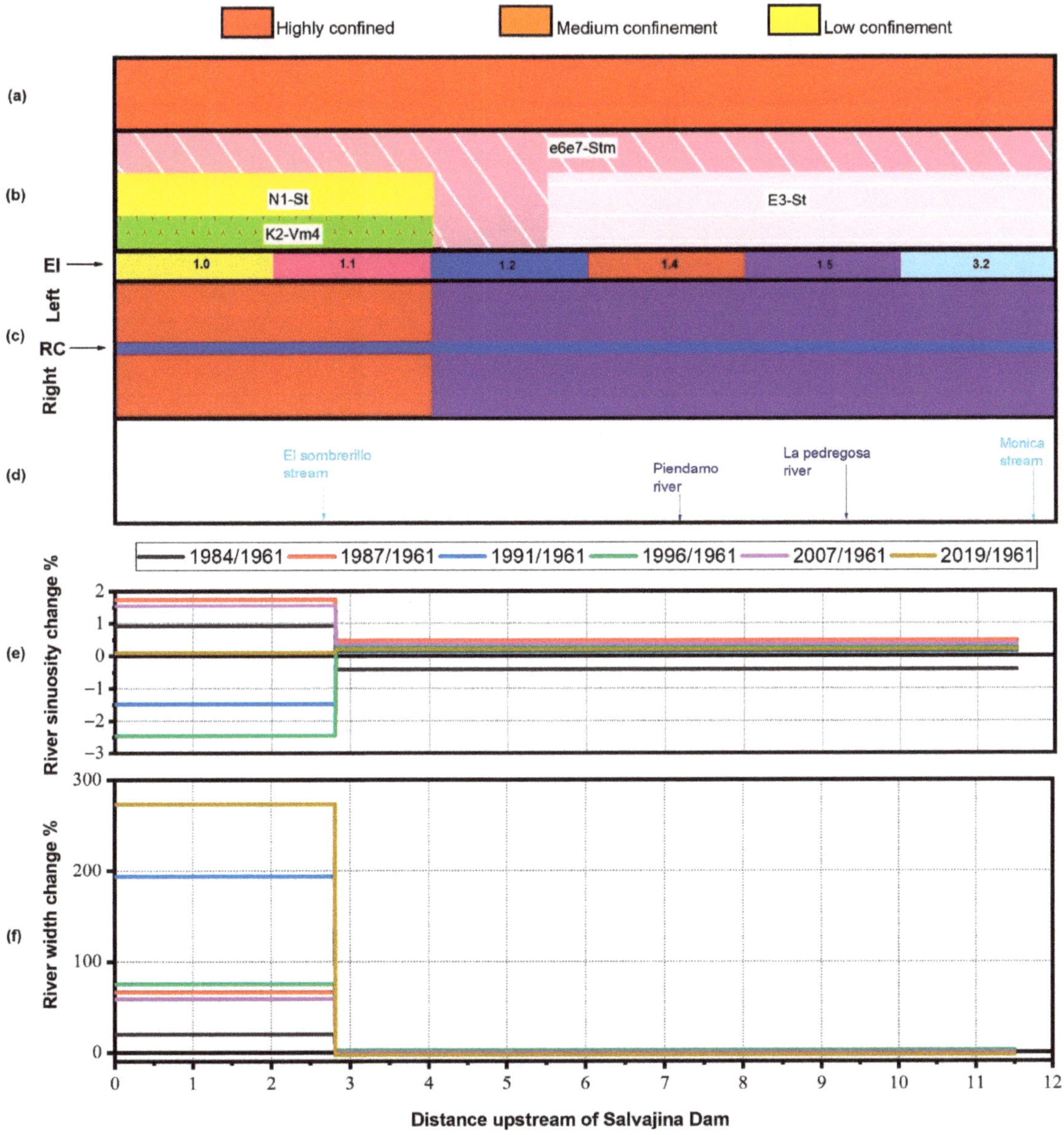

Figure A2. Morphological changes upstream of the Salvajina Dam: (**a**) confinement grade, (**b**) geological units near the Cauca River, (**c**) Erodibility Index (EI) on the left and right sides of the river channel (RC), (**d**) sites near the river, (**e**) river sinuosity change as a percentage with respect to the pre-dam year 1961, and (**f**) river width change as a percentage with respect to the pre-dam year 1961.

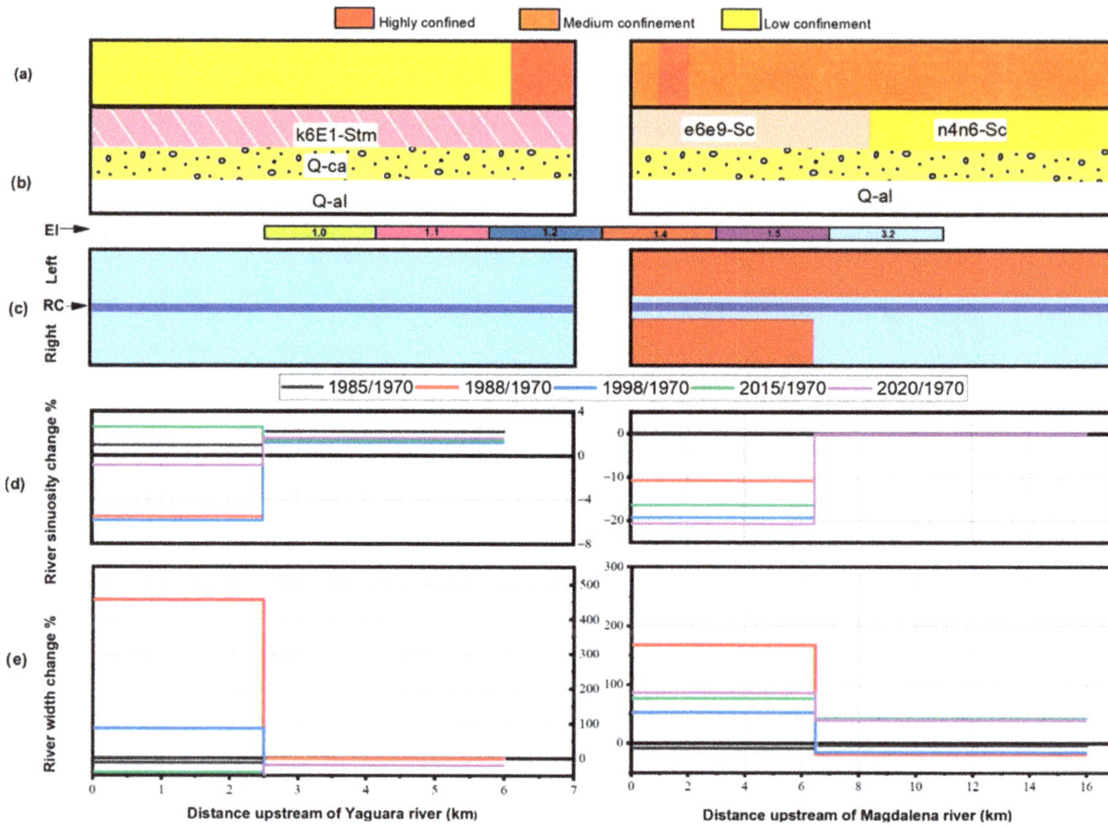

Figure A3. Morphological changes upstream of the Betania Dam across the Yaguara River and the Magdalena River: (**a**) confinement grade, (**b**) geological units near the rivers, (**c**) Erodibility Index (EI) on the left and right sides of the river channel (RC), (**d**) river sinuosity change as a percentage with respect to the pre-dam year 1970, and (**e**) river width change as a percentage with respect to the pre-dam year 1970.

References

1. Friedl, G.; Wüest, A. Disrupting Biogeochemical Cycles—Consequences of Damming. *Aquat. Sci.* **2002**, *64*, 55–65. [CrossRef]
2. Csiki, S.; Rhoads, B.L. Hydraulic and Geomorphological Effects of Run-of-River Dams. *Prog. Phys. Geogr. Earth Environ.* **2010**, *34*, 755–780. [CrossRef]
3. Gordon, E.; Meentemeyer, R.K. Effects of Dam Operation and Land Use on Stream Channel Morphology and Riparian Vegetation. *Geomorphology* **2006**, *82*, 412–429. [CrossRef]
4. Skalak, K.J.; Benthem, A.J.; Schenk, E.R.; Hupp, C.R.; Galloway, J.M.; Nustad, R.A.; Wiche, G.J. Large Dams and Alluvial Rivers in the Anthropocene: The Impacts of the Garrison and Oahe Dams on the Upper Missouri River. *Anthropocene* **2013**, *2*, 51–64. [CrossRef]
5. Baena-Escudero, R.; Guerrero-Amador, I.C.; Rinaldi, M.; González-Sayago, A. Hydrological and Geomorphic Effects Upstream of the Cantillana Dam along the Guadalquivir River (Southern Spain). *Geomorphology* **2021**, *388*, 107786. [CrossRef]
6. Liro, M. Development of Sediment Slug Upstream from the Czorsztyn Reservoir (Southern Poland) and Its Interaction with River Morphology. *Geomorphology* **2016**, *253*, 225–238. [CrossRef]
7. Brandt, S.A. Classification of Geomorphological Effects Downstream of Dams. *CATENA* **2000**, *40*, 375–401. [CrossRef]
8. Williams, G.P.; Wolman, M.G. *Downstream Effects of Dams on Alluvial Rivers*; United States Geological Survey (USGS): Washington, DC, USA, 1984; pp. 1–94.
9. Liu, X.; Wu, R.; Chen, X.; Zhou, Y.; Yang, L.; Ouyang, S.; Wu, X. Effects of Dams and Their Environmental Impacts on the Genetic Diversity and Connectivity of Freshwater Mussel Populations in Poyang Lake Basin, China. *Freshw. Biol.* **2020**, *65*, 264–277. [CrossRef]

10. Stanley, E.H. Chapter 15—Understanding Ecosystem Effects of Dams. In *Fundamentals of Ecosystem Science*, 2nd ed.; Weathers, K.C., Strayer, D.L., Likens, G.E., Eds.; Academic Press: Cambridge, MA, USA, 2021; pp. 287–291, ISBN 978-0-12-812762-9.
11. Evans, J.E.; Huxley, J.M.; Vincent, R.K. Upstream Channel Changes following Dam Construction and Removal Using a GIS/Remote Sensing Approach. *J. Am. Water Resour. Assoc.* **2007**, *43*, 683–697. [CrossRef]
12. Liro, M. Conceptual Model for Assessing the Channel Changes Upstream Form Dam Reservoir. *Quaest. Geogr.* **2014**, *33*, 61–74. [CrossRef]
13. Liro, M. Dam Reservoir Backwater as a Field-Scale Laboratory of Human-Induced Changes in River Biogeomorphology: A Review Focused on Gravel-Bed Rivers. *Sci. Total Environ.* **2019**, *651*, 2899–2912. [CrossRef] [PubMed]
14. Maddock, T., Jr. Behaviour of Channels Upstream from Reservoirs. *Int. Assoc. Sci. Hydrol. Publ.* **1966**, *71*, 812–823.
15. Petts, G.E. Long-Term Consequences of Upstream Impoundment. *Environ. Conserv.* **1980**, *7*, 325–332. [CrossRef]
16. Lu, Y.; Zuo, L.; Ji, R.; Liu, H. Deposition and Erosion in the Fluctuating Backwater Reach of the Three Gorges Project after Upstream Reservoir Adjustment. *Int. J. Sediment Res.* **2010**, *25*, 64–80. [CrossRef]
17. Grant, G.E. The Geomorphic Response of Gravel-Bed Rivers to Dams: Perspectives and Prospects. In *Gravel-Bed Rivers*; John Wiley & Sons, Ltd.: Hoboken, NJ, USA, 2012; pp. 165–181, ISBN 978-1-119-95249-7.
18. Kondolf, G.M.; Gao, Y.; Annandale, G.W.; Morris, G.L.; Jiang, E.; Zhang, J.; Cao, Y.; Carling, P.; Fu, K.; Guo, Q.; et al. Sustainable Sediment Management in Reservoirs and Regulated Rivers: Experiences from Five Continents. *Earths Future* **2014**, *2*, 256–280. [CrossRef]
19. Graf, W.L. Downstream Hydrologic and Geomorphic Effects of Large Dams on American Rivers. *Geomorphology* **2006**, *79*, 336–360. [CrossRef]
20. Surian, N.; Cisotto, A. Channel Adjustments, Bedload Transport and Sediment Sources in a Gravel-Bed River, Brenta River, Italy. *Earth Surf. Process. Landf.* **2007**, *32*, 1641–1656. [CrossRef]
21. Surian, N.; Rinaldi, M. Morphological Response to River Engineering and Management in Alluvial Channels in Italy. *Geomorphology* **2003**, *50*, 307–326. [CrossRef]
22. Lu, X.X.; Siew, R.Y. Water Discharge and Sediment Flux Changes over the Past Decades in the Lower Mekong River: Possible Impacts of the Chinese Dams. *Hydrol. Earth Syst. Sci.* **2006**, *10*, 181–195. [CrossRef]
23. Hadley, R.F.; Emmett, W.W. Channel Changes Downstream from a Dam. *JAWRA J. Am. Water Resour. Assoc.* **1998**, *34*, 629–637. [CrossRef]
24. Yang, S.L.; Milliman, J.D.; Xu, K.H.; Deng, B.; Zhang, X.Y.; Luo, X.X. Downstream Sedimentary and Geomorphic Impacts of the Three Gorges Dam on the Yangtze River. *Earth-Sci. Rev.* **2014**, *138*, 469–486. [CrossRef]
25. Rennie, C.D.; Church, M.; Venditti, J.G. Rock Control of River Geometry: The Fraser Canyons. *J. Geophys. Res. Earth Surf.* **2018**, *123*, 1860–1878. [CrossRef]
26. Jain, V.; Fryirs, K.; Brierley, G. Where Do Floodplains Begin? The Role of Total Stream Power and Longitudinal Profile Form on Floodplain Initiation Processes. *GSA Bull.* **2008**, *120*, 127–141. [CrossRef]
27. Słowik, M.; Dezső, J.; Marciniak, A.; Tóth, G.; Kovács, J. Evolution of River Planforms Downstream of Dams: Effect of Dam Construction or Earlier Human-Induced Changes? *Earth Surf. Process. Landf.* **2018**, *43*, 2045–2063. [CrossRef]
28. Jiongxin, X. Channel Pattern Change Downstream from a Reservoir: An Example of Wandering Braided Rivers. *Geomorphology* **1996**, *15*, 147–158. [CrossRef]
29. Chong, X.Y.; Vericat, D.; Batalla, R.J.; Teo, F.Y.; Lee, K.S.P.; Gibbins, C.N. A Review of the Impacts of Dams on the Hydromorphology of Tropical Rivers. *Sci. Total Environ.* **2021**, *794*, 148686. [CrossRef]
30. Grant, G.E.; Schmidt, J.C.; Lewis, S.L. A Geological Framework for Interpreting Downstream Effects of Dams on Rivers. In *A Peculiar River*; American Geophysical Union (AGU): Washington, DC, USA, 2003; pp. 203–219, ISBN 978-1-118-66573-2.
31. Smith, N.D.; Morozova, G.S.; Pérez-Arlucea, M.; Gibling, M.R. Dam-Induced and Natural Channel Changes in the Saskatchewan River below the E.B. Campbell Dam, Canada. *Geomorphology* **2016**, *269*, 186–202. [CrossRef]
32. Takahashi, M.; Nakamura, F. Impacts of Dam-Regulated Flows on Channel Morphology and Riparian Vegetation: A Longitudinal Analysis of Satsunai River, Japan. *Landsc. Ecol. Eng.* **2011**, *7*, 65–77. [CrossRef]
33. Cluett, L.J.; Radford, B.T.M. Downstream Morphological Change in Response to Dam Construction: A GIS Based Approach for the Lower Ord River, Western Australia. *Water Sci. Technol.* **2003**, *48*, 1–8. [CrossRef]
34. Leopold, L.B.; Maddock, T., Jr. *The Hydraulic Geometry of Stream Channels and Some Physiographic Implications*; United States Geological Survey: Washington, DC, USA, 1953; pp. 1–57.
35. Leopold, L.B.; Wolman, M.G. *River Channel Patterns: Braided, Meandering, and Straight*; United States Geological Survey: Washington, DC, USA, 1957; pp. 39–85.
36. Booker, D.J. Predicting Wetted Width in Any River at Any Discharge. *Earth Surf. Process. Landf.* **2010**, *35*, 828–841. [CrossRef]
37. Vargas-Luna, A.; Crosato, A.; Byishimo, P.; Uijttewaal, W.S.J. Impact of Flow Variability and Sediment Characteristics on Channel Width Evolution in Laboratory Streams. *J. Hydraul. Res.* **2019**, *57*, 51–61. [CrossRef]
38. Gardner, T.A.; Von Hase, A.; Brownlie, S.; Ekstrom, J.M.M.; Pilgrim, J.D.; Savy, C.E.; Stephens, R.T.T.; Treweek, J.; Ussher, G.T.; Ward, G.; et al. Biodiversity Offsets and the Challenge of Achieving No Net Loss. *Conserv. Biol. J. Soc. Conserv. Biol.* **2013**, *27*, 1254–1264. [CrossRef] [PubMed]

39. Harman, W.; Starr, R.; Carter, M.; Tweedy, K.; Clemmons, M.; Suggs, K.; Miller, C. *A Function-Based Framework for Stream Assessment and Restoration Projects*; United States Environmental Protection Agency (EPA), Office of Wetlands, Oceans and Watersheds: Washington, DC, USA, 2012; p. 344.
40. Baez-Villanueva, O.M.; Zambrano-Bigiarini, M.; Ribbe, L.; Nauditt, A.; Giraldo-Osorio, J.D.; Thinh, N.X. Temporal and Spatial Evaluation of Satellite Rainfall Estimates over Different Regions in Latin-America. *Atmos. Res.* **2018**, *213*, 34–50. [CrossRef]
41. Giraldo-Osorio, J.D.; Trujillo-Osorio, D.E.; Baez-Villanueva, O.M. Analysis of ENSO-Driven Variability, and Long-Term Changes, of Extreme Precipitation Indices in Colombia, Using the Satellite Rainfall Estimates CHIRPS. *Water* **2022**, *14*, 1733. [CrossRef]
42. Restrepo, J.D.; Kettner, A.J.; Syvitski, J.P.M. Recent Deforestation Causes Rapid Increase in River Sediment Load in the Colombian Andes. *Anthropocene* **2015**, *10*, 13–28. [CrossRef]
43. Restrepo, J.D.; Escobar, H.A. Sediment Load Trends in the Magdalena River Basin (1980–2010): Anthropogenic and Climate-Induced Causes. *Geomorphology* **2018**, *302*, 76–91. [CrossRef]
44. Gutiérrez Rodríguez, H. *Central Hidroeléctrica del río Prado: Nota Preliminar*; Instituto de Aprovechamiento de Aguas y Fomento Eléctrico—Servicio Geológico Colombiano (SGC): Bogotá, Colombia, 1959; p. 15.
45. ICEL. *Propuesta para la Ejecución de la Central Hidroeléctrica de Betania*; Instituto Colombiano de Energía Eléctrica (ICEL)—Ministerio de Minas y Energía: Bogotá, Colombia, 1978; p. 66.
46. Mazuera González, Ó.E. *Proyecto de Regulación del río Cauca*; Corporacion Autonoma Regional del Valle del Cauca (CVC): Cali, Colombia, 1972; p. 219.
47. Congreso de Colombia. *Fundamentos de la Política Ambiental Colombiana*; República de Colombia: Bogotá, Colombia, 1993; p. 47.
48. Muñoz-Sabater, J. ERA5-Land Monthly Averaged Data from 1950 to Present. Copernicus Climate Change Service (C3S) Climate Data Store (CDS) 2019. Available online: https://cds.climate.copernicus.eu/cdsapp#!/dataset/reanalysis-era5-land-monthly-means (accessed on 15 May 2023).
49. Zuo, Q.; Liang, S. Effects of Dams on River Flow Regime Based on IHA/RVA. *Proc. IAHS* **2015**, *368*, 275–280. [CrossRef]
50. Šarauskienė, D.; Adžgauskas, G.; Kriaučiūnienė, J.; Jakimavičius, D. Analysis of Hydrologic Regime Changes Caused by Small Hydropower Plants in Lowland Rivers. *Water* **2021**, *13*, 1961. [CrossRef]
51. Song, X.; Zhuang, Y.; Wang, X.; Li, E.; Zhang, Y.; Lu, X.; Yang, J.; Liu, X. Analysis of Hydrologic Regime Changes Caused by Dams in China. *J. Hydrol. Eng.* **2020**, *25*, 05020003. [CrossRef]
52. Meitzen, K.M. Stream Flow Changes across North Carolina (USA) 1955–2012 with Implications for Environmental Flow Management. *Geomorphology* **2016**, *252*, 171–184. [CrossRef]
53. Yaghmaei, H.; Sadeghi, S.H.; Moradi, H.; Gholamalifard, M. Effect of Dam Operation on Monthly and Annual Trends of Flow Discharge in the Qom Rood Watershed, Iran. *J. Hydrol.* **2018**, *557*, 254–264. [CrossRef]
54. Dahmen, E.R.; Hall, M.J. *Screening of Hydrological Data: Tests for Stationarity and Relative Consistency*; ILRI Publication; International Institute for Land Reclamation and Improvement: Wageningen, The Netherlands, 1990; ISBN 978-90-70754-23-5.
55. Mathias Kondolf, G.; Batalla, R.J. Chapter 11 Hydrological Effects of Dams and Water Diversions on Rivers of Mediterranean-Climate Regions: Examples from California. In *Developments in Earth Surface Processes*; Catchment Dynamics and River Processes; Garcia, C., Batalla, R.J., Eds.; Elsevier: Amsterdam, The Netherlands, 2005; Volume 7, pp. 197–211.
56. Zhang, W.; Mu, S.; Zhang, Y.; Chen, K. Seasonal and Interannual Variations of Flow Discharge from Pearl River into Sea. *Water Sci. Eng.* **2012**, *5*, 399–409. [CrossRef]
57. Ali, R.; Ismael, A.; Heryansyah, A.; Nawaz, N. Long Term Historic Changes in the Flow of Lesser Zab River, Iraq. *Hydrology* **2019**, *6*, 22. [CrossRef]
58. Cunnane, C. *Statistical Distributions for Flood Frequency Analysis*; Operational Hydrology Report (OHR); Secretariat of the World Meteorological Organization: Geneva, Switzerland, 1989; ISBN 978-92-63-10718-3.
59. Da Silva, R.M.; Santos, C.A.G.; Moreira, M.; Corte-Real, J.; Silva, V.C.L.; Medeiros, I.C. Rainfall and River Flow Trends Using Mann–Kendall and Sen's Slope Estimator Statistical Tests in the Cobres River Basin. *Nat. Hazards* **2015**, *77*, 1205–1221. [CrossRef]
60. Barajas Ibañez, H.; Leiva Fonseca, P.A. Evaluación del Cambio Morfológico y Sedimentológico del Cauce del Río Sinú Aguas Abajo de la presa del Embalse Urrá Asociados a su Construcción y Operación. Bachelor's Thesis, Universidad de La Salle, Bogotá, Colombia, 2016.
61. Arandia Suárez, A. Estimación de los Cambios Hidrológicos, Sedimentológicos y Morfológicos en ríos Aluviales Intervenidos por Presas de Gran Altura. Master's Thesis, Universidad de Los Andes, Bogotá, Colombia, 2018.
62. Laverde Mesa, L. Evaluación del Impacto de los Embalses por Retención de Sedimentos Sobre la Morfología del Cauce del Río Magdalena Mediante el Análisis de Tramos Representativos en el Comportamiento Sedimentológico Entre el Embalse de Betania y el Municipio de Regidor. Bachelor's Thesis, Universidad de La Salle, Bogotá, Colombia, 2016.
63. Ramírez Callejas, C.A.; Bocanegra Vinasco, R.A.; Sandoval García, M.C. La carga sedimentaria en el río Cauca en su alto valle geográfico. *Ing. Recur. Nat. Ambiente* **2009**, *8*, 68–82.

64. CVC; Universidad del Valle. *Caracterización del río Cauca y Tributarios: Tramo Salvajina—La Virginia*; Universidad del Valle: Santiago de Cali, Colombia; Corporación Autónoma Regional del Valle del Cauca (CVC): Santiago de Cali, Colombia, 2001; p. 116.
65. Donovan, M.; Belmont, P.; Notebaert, B.; Coombs, T.; Larson, P.; Souffront, M. Accounting for Uncertainty in Remotely-Sensed Measurements of River Planform Change. *Earth-Sci. Rev.* **2019**, *193*, 220–236. [CrossRef]
66. Pogson, M.; Smith, P. Effect of Spatial Data Resolution on Uncertainty. *Environ. Model. Softw.* **2015**, *63*, 87–96. [CrossRef]
67. Bruno, L.O. GRASS: A Free and Open Source Solution for Hydrographic Body Analysis. *Nativa* **2017**, *5*, 24–30. [CrossRef]
68. Xu, H. Modification of Normalised Difference Water Index (NDWI) to Enhance Open Water Features in Remotely Sensed Imagery. *Int. J. Remote Sens.* **2006**, *27*, 3025–3033. [CrossRef]
69. Filipponi, F. Sentinel-1 GRD Preprocessing Workflow. *Proceedings* **2019**, *18*, 11. [CrossRef]
70. Zhang, W.; Hu, B.; Brown, G.S. Automatic Surface Water Mapping Using Polarimetric SAR Data for Long-Term Change Detection. *Water* **2020**, *12*, 872. [CrossRef]
71. Golly, A.; Turowski, J.M. Deriving Principal Channel Metrics from Bank and Long-Profile Geometry with the R Package Cmgo. *Earth Surf. Dyn.* **2017**, *5*, 557–570. [CrossRef]
72. Gómez Tapias, J.; Montes Ramírez, N.E.; Nivia Guevara, Á.; Diederix, H. Mapa Geológico de Colombia 2015. Available online: https://www2.sgc.gov.co/MGC/Paginas/mgc_1M2015.aspx (accessed on 10 May 2023).
73. Zondervan, J.R.; Stokes, M.; Boulton, S.J.; Telfer, M.W.; Mather, A.E. Rock Strength and Structural Controls on Fluvial Erodibility: Implications for Drainage Divide Mobility in a Collisional Mountain Belt. *Earth Planet. Sci. Lett.* **2020**, *538*, 116221. [CrossRef]
74. Sklar, L.S.; Dietrich, W.E. Sediment and Rock Strength Controls on River Incision into Bedrock. *Geology* **2001**, *29*, 1087–1090. [CrossRef]
75. Moosdorf, N.; Cohen, S.; von Hagke, C. A Global Erodibility Index to Represent Sediment Production Potential of Different Rock Types. *Appl. Geogr.* **2018**, *101*, 36–44. [CrossRef]
76. Fryirs, K.A.; Wheaton, J.M.; Brierley, G.J. An Approach for Measuring Confinement and Assessing the Influence of Valley Setting on River Forms and Processes. *Earth Surf. Process. Landf.* **2016**, *41*, 701–710. [CrossRef]
77. Poveda, G.; Mesa, O.J. On the Existence of Lloró (the Rainiest Locality on Earth): Enhanced Ocean-Land-Atmosphere Interaction by a Low-Level Jet. *Geophys. Res. Lett.* **2000**, *27*, 1675–1678. [CrossRef]
78. Quintero Camacho, W. *Informe Preliminar de Áreas de Interés Regional para Encontrar Yacimientos de Hidrocarburos no Convencionales (Shale Gas), a Partir de Información Geofísica (Magnetometría y Gravimetría) en la Cuenca del Sinú—San Jacinto*; Servicio Geológico Colombiano (SGC): Bogotá, Colombia, 2016; p. 18.
79. Barbosa Camacho, G. *Memoria Explicativa: Mapa Geológico del Departamento del Cauca*; Instituto de Investigación e Información Geocientífica, Minero-Ambiental y Nuclear (INGEOMINAS)—Servicio Geológico Colombiano (SGC): Santiago de Cali, Colombia, 2003; p. 221.
80. Núñez Tello, A.; Ferreira Vesga, P.; Rodríguez, M.Á. *Levantamiento Geológico de la Plancha 323, Neiva, Huila*; Instituto de Investigación e Información Geocientífica, Minero-Ambiental y Nuclear (INGEOMINAS)—Servicio Geológico Colombiano (SGC): Bogotá, Colombia, 2002; p. 94.
81. Mojica, J.; Franco, R. Estructura y evolución tectónica del valle medio y superior del Magdalena, Colombia. *Geol. Colomb.* **1990**, *17*, 41–64.
82. Villarroael, C.A.; Setoguchi, T.; Brieva, J.; Macía, C. Geology of the La Tatacoa "Desert" (Huila, Colombia): Precisions on the Stratigraphy of the Honda Group, the Evolution of the "Pata High" and the Presence of the La Venta Fauna. *Mem. Fac. Sci. Kyoto Univ. Ser. Geol. Mineral.* **1996**, *58*, 41–66.
83. Armenta Porras, G.E. Análisis detallado del Efecto Foehn generado por la Cordillera Oriental en el Alto Magdalena (Huila y Tolima). Master's Thesis, Universidad Nacional de Colombia, Bogotá, Colombia, 2013.
84. Dill, H.G.; Andrei, B.; Sorin-Ionut, B.; Kristian, U.; Jorge, G.T.; Daniel, B.; Thomas, C. The "Badland Trilogy" of the Desierto de La Tatacoa, Upper Magdalena Valley, Colombia, a Result of Geodynamics and Climate: With a Review of Badland Landscapes. *CATENA* **2020**, *194*, 104696. [CrossRef]
85. Kammer, A.; Piraquive Bermúdez, A. Evidencias sedimentológicas y estructurales para un origen paleógeno de la Falla de Chusma, Valle Superior del Magdalena, borde occidental de la sub-cuenca de Neiva. *Geol. Colomb.* **2013**, *38*, 43–64.
86. Barjasteh, A. Influence of Geological Structure on Dam Behavior and Case Studies. In *Dam Engineering*; Mudanya Üniversitesi: Bursa, Turkey, 2019; p. 106. ISBN 978-1-78985-480-0.
87. Nelson, N.C.; Erwin, S.O.; Schmidt, J.C. Spatial and Temporal Patterns in Channel Change on the Snake River Downstream from Jackson Lake Dam, Wyoming. *Geomorphology* **2013**, *200*, 132–142. [CrossRef]
88. Adib, A.; Foladfar, H.; Roozy, A. Role of Construction of Large Dams on River Morphology (Case Study: The Karkheh Dam in Iran). *Arab. J. Geosci.* **2016**, *9*, 661. [CrossRef]
89. O'Brien, G.R.; Wheaton, J.M.; Fryirs, K.; Macfarlane, W.W.; Brierley, G.; Whitehead, K.; Gilbert, J.; Volk, C. Mapping Valley Bottom Confinement at the Network Scale. *Earth Surf. Process. Landf.* **2019**, *44*, 1828–1845. [CrossRef]
90. Csiki, S.J.; Rhoads, B.L. Influence of Four Run-of-River Dams on Channel Morphology and Sediment Characteristics in Illinois, USA. *Geomorphology* **2014**, *206*, 215–229. [CrossRef]

91. Turowski, J.M. Alluvial Cover Controlling the Width, Slope and Sinuosity of Bedrock Channels. *Earth Surf. Dyn.* **2018**, *6*, 29–48. [CrossRef]
92. Schumm, S.A.; Dumont, J.F.; Holbrook, J.M. *Active Tectonics and Alluvial Rivers*; Cambridge University Press: New York, NY, USA, 2002; ISBN 978-0-521-89058-8.

applied sciences

MDPI

Article

Development of a Two-Dimensional Hybrid Sediment-Transport Model

Yaoxin Zhang [1,*], Mohammad Al-Hamdan [1,2,3] and Daniel Wren [4]

1 National Center for Computational Hydroscience and Engineering, University of Mississippi, Oxford, MS 38655, USA; yzhang@olemiss.edu
2 Department of Civil Engineering, University of Mississippi, University, MS 38677, USA; mzalhamd@olemiss.edu
3 Department of Geology and Geological Engineering, University of Mississippi, University, MS 38677, USA
4 National Sedimentation Laboratory (NSL), USDA Agricultural Research Service (ARS), Oxford, MS 38655, USA; daniel.wren@usda.gov
* Correspondence: yzhang@ncche.olemiss.edu or yaoxin@olemiss.edu; Tel.: +1-6629158972

Abstract: This paper presents the development of a two-dimensional hydrodynamic sediment transport model using the finite volume method based on a collocated unstructured hybrid-mesh system consisting of triangular and quadrilateral cells. The model is a single-phase nonequilibrium sediment-transport model for nonuniform and noncohesive sediments in unsteady turbulent flows that considers multiple sediment-transport processes such as deposition, erosion, transport, and bed sorting. This model features a hybrid unstructured mesh system for easy mesh generation in complex domains. To avoid interpolation from vertices in conventional unstructured models, this model adopted a second-order accurate edge-gradient evaluation method to consider the mesh irregularities based on Taylor's series expansion. In addition, the multipoint momentum interpolation corrections were integrated to avoid possible nonphysical oscillations during the wetting-and-drying process, common in unsteady sediment transport problems, to ensure both numerical stability and numerical accuracy. The developed sediment transport model was validated by a benchmark degradation case for the erosion process with armoring effects, a benchmark aggradation case for the deposition process, and a naturally meandering river for long-term unsteady sediment-transport processes. Finally, the model was successfully applied to simulate sediment transport in a reservoir that was significantly affected by typhoon events.

Keywords: deposition; erosion; unsteady; nonequilibrium

Citation: Zhang, Y.; Al-Hamdan, M.; Wren, D. Development of a Two-Dimensional Hybrid Sediment-Transport Model. *Appl. Sci.* **2023**, *13*, 4940. https://doi.org/10.3390/app13084940

Academic Editors: Gordon Gilja, Manousos Valyrakis, Panagiotis Michalis, Thomas Pahtz and Oral Yagci

Received: 19 March 2023
Revised: 5 April 2023
Accepted: 11 April 2023
Published: 14 April 2023

1. Introduction

Sediment transport, defined as sediments driven by water and moving with the water, is one of the most important processes when studying morphological and environmental problems. Sediment transport may result in sediment depositions in lakes, reservoirs [1,2] and coastal wetlands [3,4]; erosion of riverbanks [5,6], coastlines [7,8], and downstream of dams [9,10]; local scour downstream of hydraulic structures [11,12]; gully erosion [13,14]; channel evolution [15,16]; adsorption/de-adsorption and resuspension [17], etc.

In addition to conventional physical models, with the advancement of computer technology and numerical methods, numerical models have become powerful tools to study sediment transport in rivers, lakes, reservoirs, and coastal regions. In natural rivers, sediment in the water can be described as a two-phase flow [18,19], in which the sediment phase and the water phase are governed by the continuity and the momentum equations and interact with each other all the time [20,21]. In the conventional single-phase flow method, sediment transport is described as a diffusion phenomenon using the momentum equation and the transport equations: the suspended load moves in the form of a suspension in the water column, while the bed load moves by sliding, rolling, or saltation along or near the bed. Despite the advancement of the two-phase flow method [21–23] in recent years, the single-phase flow method [20,24–26] is still dominant in sediment transport modeling due to its relative simplicity and higher computational efficiency.

Sediment transport models may be categorized according to the dimension of study domains (one dimensional (1D), two dimensional (2D), or three dimensional (3D)), flow conditions (steady or unsteady), uniformity of sediment particles (single-sized or multiple-sized), transport mode of sediment (equilibrium or nonequilibrium), and the cohesion of sediment materials (noncohesive or cohesive). In natural rivers, nonuniform and noncohesive sediments are dominant, and the flows are unsteady and turbulent. Moving sediments interact with the channel and can both change channel geometry and be influenced by channel geometry. The dynamic interactions between flow and sediments imply that global equilibrium sediment transport (the sediment transport rate equals the transport capacity) is rare. Local transport may be temporarily in equilibrium with the flow, however, nonequilibrium sediment transport is much more common in natural channels [18]. The 1D sediment transport models (e.g., [25]; CCHE1D [27]; MIKE 11 [28]; HEC-RAS [29]; and GSTAR-1D [30]) were developed for averaged solutions with high computing efficiency for long-term simulations in large-scaled domains, while 3D models (e.g., CCHE3D [31]; Delft3D [32]) are for highly-sophisticated solutions of short-term simulations in local small domains. The 2D models (e.g., [19]; CCHE2D [24]; SRH-2D [26]; [33]) are in between 1D and 3D models with respect to computing efficiency and numerical accuracy. Theoretically, sediment transport is a 3D phenomenon and 3D models should be used. However, in practice for different problems, model selection needs to balance computing efficiency and numerical accuracy.

Unlike hydrodynamic flow models, there are more uncertainties and challenges with the sediment transport models since numerous semiempirical and empirical formulas have been developed, covering all aspects of the sediment transport process, such as the sediment settling velocity, sediment incipient motion, mobile bed roughness, critical shear stress, suspended-load transport capacity, and bed-load transport capacity. A good review of some of those formulas can be found in [20]. Most of those formulas were derived based on physical laws using experimental data under steady flow conditions, and application of them to truly unsteady sediment transport has been a concern [21]. Each formula has its own advantages, limitations, and application ranges. A sediment transport model cannot accommodate all the available formulas, which would make the model difficult to use. The same model cannot use multiple approaches in most cases since different formulas often yield significantly different simulation results according to the authors' experiences.

In the review of sediment transport models by Papanicolaou et al. [34], model limitations were identified, including turbulence calculations, entrainment formulas established under uniform flow conditions, fractional (nonuniform) sediment transport calculations, evaluation of sediment dispersion and diffusion coefficient, soil contributions from banks, hill slopes, and floodplains, and two-phase flow modeling. Most of these problems have not been resolved. Basically, sediment transport modeling needs to address the unsteady and nonequilibrium natures of sediment transport [26]. On the other hand, the mesh system (structured or unstructured) determines the numerical methods in sediment transport models. Despite their advantages and disadvantages, both structured and unstructured meshes have been widely used. For example, the CCHE2D model [19,24] and the Delta3D model [32] were based on structured meshes, while SRH-2D [26] was based on unstructured polygon meshes. Compared to structured meshes, unstructured meshes are more suitable and flexible for geometrically complex domains with complicated boundaries. Among unstructured meshes, triangle meshes, and quadrilateral meshes are the most popular and easiest to generate [35]. In addition, structured (quadrilateral) meshes can be shared by both structured models and unstructured models in many cases.

In this study, an alternative single-phase 2D depth-integrated sediment transport model for turbulent flows was developed. The model aims to simulate unsteady, nonequilibrium, and nonuniform (fractional) sediment transport processes for noncohesive sediment materials on both the laboratory scale and the field scale. The model features (1) an unstructured hybrid-mesh system consisting of either triangle cells or quadrilateral cells or mixed triangle and quadrilateral cells for geometrically-complex domains with high adaptivity; (2) an edge gradient evaluation method with second-order accuracy designed for mesh irregularity and nonuniformity; and, (3) a multipoint momentum interpolation correction method to remove possible nonphysical oscillations in the wetting-and-drying process. The development details are presented in the Section 2 (Numerical Model) and Section 3 (Numerical Method). Selected examples and applications will demonstrate and validate the proposed sediment-transport model.

2. Numerical Model

2.1. Flow Model

The flow model is the backbone of the sediment-transport model. The proposed sediment-transport model is based on a 2D depth-integrated hydrodynamic-flow model developed by Zhang et al. [36] where the continuity equation and the momentum equations for unsteady turbulent flow are:

$$\frac{\partial h}{\partial t} + \frac{\partial (hu)}{\partial x} + \frac{\partial (hv)}{\partial y} = 0 \tag{1}$$

$$\frac{\partial (hu)}{\partial t} + \frac{\partial (hu^2)}{\partial x} + \frac{\partial (huv)}{\partial y} = -gh\frac{\partial \eta}{\partial x} + \frac{1}{\rho}[\frac{\partial (h\tau_{xx})}{\partial x} + \frac{\partial (h\tau_{xy})}{\partial y}] + \frac{(\tau_{wx} - \tau_{bx})}{\rho} + f_{cor}hv \tag{2}$$

$$\frac{\partial (hv)}{\partial t} + \frac{\partial (huv)}{\partial x} + \frac{\partial (hv^2)}{\partial y} = -gh\frac{\partial \eta}{\partial y} + \frac{1}{\rho}[\frac{\partial (h\tau_{yx})}{\partial x} + \frac{\partial (h\tau_{yy})}{\partial y}] + \frac{(\tau_{wy} - \tau_{by})}{\rho} - f_{cor}hu \tag{3}$$

where t represents time (s); u and v are depth-integrated velocity (m/s) components in the x and y directions, respectively; η is the water surface elevation (m); ρ is the water density (kg/m^3); $h = \eta - z_b$ is the local water depth (m) and z_b is the bed elevation (m); g is the gravitational acceleration (m/s^2); f_{cor} is the Coriolis parameter; τ_{bx} and τ_{by} are shear stresses (Pa) on the bed surface and were calculated as follows:

$$\frac{\tau_{b(x,y)}}{\rho} = \frac{gn^2}{h^{1/3}}\sqrt{u^2 + v^2}(u, v) \tag{4}$$

n is Manning's roughness (m$^{-1/3}$s); τ_{wx}, τ_{wy} are surface wind shear stresses (Pa):

$$(\tau_{wx}, \tau_{wy}) = \rho_{air}c_{fa}\sqrt{U_w^2 + V_w^2}(U_w, V_w) \tag{5}$$

where c_{fa} is the friction coefficient at the water surface and U_w and V_w are wind velocity (m/s); and, $\tau_{xx}, \tau_{xy}, \tau_{yx}$, and τ_{yy} are the depth-integrated Reynolds stresses (Pa) including both viscous and turbulent effects and approximated based on the Bousssinesq assumption:

$$\tau_{xx} = 2\rho(\nu + \nu_t)\frac{\partial u}{\partial x} \tag{6a}$$

$$\tau_{xy} = \tau_{yx} = \rho(\nu + \nu_t)(\frac{\partial u}{\partial y} + \frac{\partial v}{\partial x}) \tag{6b}$$

$$\tau_{yy} = 2\rho(\nu_t + \nu)\frac{\partial v}{\partial y} \tag{6c}$$

where a mixing-length model [21] was adopted to calculate the eddy viscosity ν_t (m^2/s).

2.2. Sediment-Transport Model

In this study, a single-phase sediment-transport model was developed, mainly consisting of the suspended-load transport equation, the bed-load transport equation, and the bed-change equation.

2.2.1. Suspended-Load Transport

For suspended loads, the governing equation for the kth size class reads [19]

$$\frac{\partial(hC_k/\beta_{sk})}{\partial t} + \frac{\partial(huC_k)}{\partial x} + \frac{\partial(hvC_k)}{\partial y} = \frac{\partial}{\partial x}[h(\varepsilon_s \frac{\partial(C_k)}{\partial x} + D_{sxk})] + \frac{\partial}{\partial y}[h(\varepsilon_s \frac{\partial(C_k)}{\partial y} + D_{syk})] + \alpha\omega_{sk}(C_{*k} - C_k)\ (k = 1, 2, \ldots, n_s) \quad (7a)$$

where C is the depth-averaged suspended-load concentration by volume (m^3/m^3), the concentration by mass is therefore $\rho_s C_k$ (kg/m^3) with ρ_s the sediment density (kg/m^3); C_{*k} is the suspended-load transport capacity; ω_{sk} is the settling velocity (m/s); α is the adaptation coefficient for the suspended load; n_s is the number of size classes; the correction factor

$$\beta_{sk} = (\int_{z_b+\delta}^{\eta} u_s C_k dz)/(U_s \int_{z_b+\delta}^{\eta} C_k dz) = \overline{U_s}/U \quad (7b)$$

represents the lag between the flow and the suspended-load transport (≤ 1) with δ the thickness of the bed-load zone near the bed, u_s the velocity of the suspended load (m/s), and U_s the velocity magnitude (m/s). The suspended-load diffusivity coefficient is $\varepsilon_s = \nu_t/\sigma_c$, with the Schmidt number $0.5 \leq \sigma_c \leq 1$ and the dispersion terms $D_{sx} = -\frac{1}{h}\int_{z_b}^{\eta}(u_{3d} - u_{2d})(c_{3d} - C_{2d})dz$ and $D_{sy} = -\frac{1}{h}\int_{z_b}^{\eta}(v_{3d} - v_{2d})(c_{3d} - C_{2d})dz$ account for the nonuniform distributions of flow velocity and sediment concentration over flow depth, with small u_{3d}, v_{3d}, and c_{3d} in 3D. In this study, the dispersion terms were ignored so that $D_{sx} = 0$ and $D_{sy} = 0$.

In Equation (7), there is one calibration parameter, the adaptation coefficient α for the suspended load, and three input variables, including the settling velocity ω_s, the suspended-load transport capacity C_*, and the correction factor β_s, to be determined by semiempirical formulas. As mentioned previously, there are so many formulas available with their own advantages, limitations, and application ranges, this study does not focus on the comparisons of those formulas for "best picks"; we selected appropriate formulas that provided reasonable results during the validation process.

For the settling velocity, Zhang's formula [18] for naturally-worn sediment particles was used, which assumes a combination of drag force in the laminar region and the turbulent region.

$$\omega_s = \sqrt{(13.95\frac{\nu}{d})^2 + 1.09(\frac{\rho_s}{\rho} - 1)gd} - 13.95\frac{\nu}{d} \quad (8)$$

where d is the particle diameter (m) and ν is the kinematical viscosity (m^2/s). Julien [37] and Cheng [38] also proposed similar formulas as Equation (8).

A variety of formulas have been developed for the suspended-load transport capacity. For example, Einstein [39] determined the suspended-load transport rate by integrating the local sediment concentration over the suspended-load zone; Zhang [18], based on the energy balance of sediment-laden flow, derived the relation between suspended-load transport capacity C_* and parameter $U^3/(gR\omega_s)$; and Bagnold's formula [40] was based on the stream power concept. The formula proposed by Wu et al. [41], based on the stream power concept, was selected:

$$q_{s*k} = 0.0000262[(\frac{\tau}{\tau_{ck}} - 1)\frac{U}{\omega_{sk}}]^{1.74} \cdot p_{bk}\sqrt{(\gamma_s/\gamma - 1)gd^3} \tag{9}$$

where q_{s*k} is the fractional suspended transport capacity (m^2/s); γ_s and γ are the specific weight (N) of sediment and water, respectively; p_{bk} is the fraction of kth size class; and τ is the shear stress of entire cross section (Pa); $\tau_{ck} = 0.03(\gamma_s - \gamma)d_k(p_{hk}/p_{ek})^{0.6}$ with the hiding probability $p_{hk} = \sum_j p_{bj}d_j/(d_j + d_k)$ and the exposure probability $p_{ek} = \sum_j p_{bj}d_k/(d_j + d_k)$.

The correction factor β_s, the lag between the flow and sediment velocities, can be ignored for fine sediments, though for coarse sediments it is nonnegligible. According to [42], based on the logarithmic velocity distribution and the Rouse distribution for suspended load, the correction factor relation to the Rouse number R and the Chezy's coefficient C_h is defined as:

$$\beta_s = \begin{cases} 0.0289R^3 - 0.0448R^2 - 0.2977R + 0.994\ (C_h = 40) \\ 0.0222R^3 - 0.0433R^2 - 0.1825R + 0.9924\ (C_h = 60) \\ 0.0189R^3 - 0.044R^2 - 0.122R + 0.9924\ (C_h = 80) \\ 0.0113R^3 - 0.0227R^2 - 0.1051R + 0.9918\ (C_h = 100) \end{cases} \tag{10}$$

where $R = \omega_s/(\kappa U_*)$ is the Rouse number with κ the van Karman constant (=0.41) and $U_* = \sqrt{\tau_0/\rho}$ is the shear velocity (m/s). Interpolation may be performed with Equation (10) for the other C_h values.

2.2.2. Bed-Load Transport

The governing equation for bed loads of kth size class [19] reads as follows:

$$\frac{\partial(q_{bk}/u_{bk})}{\partial t} + \frac{\partial(\alpha_{bx}q_{bk})}{\partial x} + \frac{\partial(\alpha_{by}q_{bk})}{\partial y} = \frac{1}{L_b}(q_{b*k} - q_{bk}) \tag{11}$$

where q_{bk} is the bed-load transport rate by volume per unit time and width (m^2/s); q_{b*k} is the transport capacity (m^2/s); L_b is the bed-load adaptation length (m); and the direction cosines of bed-load movement are $\alpha_{bx} = u/U$ and $\alpha_{by} = v/U$ with U the velocity magnitude (m/s) when the effect of the bed slope is ignored.

To solve Equation (11), the bed-load velocity formula established by van Rijn [43] and the bed-load transport capacity formula developed by Wu et al. [41] were adopted. For the bed-load velocity, van Rijn [43] proposed the concept of the transport stage number T, which was defined as the excess bed-shear stress $T = (U'_*/U_{*cr})^2 - 1$, with $U'_* = Ug^{0.5}/[18\log(4h/d_{90})]$ the effective shear velocity and U_{*cr} the critical shear velocity given by the Shields diagram. Then, the bed-load velocity is estimated by

$$u_b = 1.5T^{0.6}\sqrt{(\rho_s/\rho - 1)gd}. \tag{12}$$

The bed-load transport capacity is related to the flow conditions and sediment properties. The formulas for bed-load transport capacity can be categorized into stream-power based (or velocity based) (i.e., Ref. [40]), shear-stress based (i.e., Refs. [44–46]), and probability-theory based (i.e., Ref. [38]). Wu et al. [41] proposed the use of the nondimensional excess grain shear stress $T_k = (n'/n)^{3/2}\tau_b/\tau_{ck} - 1$ for the estimation

$$q_{b*k} = 0.0053[(\frac{n'}{n})^{3/2}\frac{\tau_b}{\tau_{ck}} - 1]^{2.2} \cdot p_{bk}\sqrt{(\gamma_s/\gamma - 1)gd_k^3} \tag{13}$$

where q_{b*k} is the fractional bed-load transport capacity (m^2/s); τ_{ck} is the critical shear stress (Pa); $n' = d_{50}^{1/6}/20$, and n is the Manning's roughness coefficient ($m^{-1/3}s$). Note that the hiding and exposure effect in nonuniform bed material is considered through τ_{ck}.

In Equations (7) and (11), the adaptation coefficient, α, for suspended load, and the adaptation length, L_b, for bed load, are critical to predicting nonequilibrium sediment transport. Physically, the adaptation length L_b represents the distance required for the sediment transport to move from nonequilibrium to equilibrium conditions, while the adaptation coefficient α originally is related to the near-bed suspended-load concentration C_b, with $C_b = \alpha C$. To be consistent, an adaptation length for the suspended load (m) can also be defined as $L_s = Uh/(\alpha\omega_s)$. In practice, α and L_b are case-dependent calibration parameters. The adaptation length for the bed load, L_b, is closely related to the bed-form characteristics and can be evaluated using the characteristic lengths, such as channel width, sand-dune length, alternate-bar length, water depth, and mesh-cell length. The adaptation coefficient α is in the range of [0, 1]. Based on the work of Galappatti and Vreugdenhil [47], Armanini and de Silvio [48] proposed the following function to evaluate α:

$$\alpha = 1/\{\frac{a}{h} + (1 - \frac{a}{h})\exp[-1.5(\frac{a}{h})^{-1/6}\frac{\omega_s}{U_*}]\} \tag{14}$$

where a is related to the zero-velocity distance in the logarithmic velocity distribution, defined as $a = 33h/\exp(1 + \kappa C_h/\sqrt{g})$.

The suspended-load transport Equation (7) and the bed-load transport Equation (11) can be combined into a total-load transport equation [20,26]. The single total-load equation is computationally more convenient since only one equation is solved, though it requires an additional fraction parameter to identify the suspended load and the bed load in the total load. In this study, however, the suspended-load equation and the bed-load equation will be solved separately, which is conceptually simple and clear.

2.2.3. Bed Changes and Sorting

The source terms on the right-hand side (RHS) of Equations (7a) and (11) represent the net exchange of sediment between the bed and the flowing water, including the suspended load in the water column and the bed load in the near-bed zone. Therefore, the bed changes can be calculated by

$$(1 - p'_m)\frac{\partial z_{bk}}{\partial t} = \alpha\omega_{sk}(C_k - C_{*k}) + \frac{1}{L_b}(q_{bk} - q_{b*k}) \tag{15}$$

where p'_m is the sediment porosity, and z_{bk} represents the bed elevation (m) contributed by kth size class.

In the vertical direction, the bed can be divided into at least three layers from top to bottom: the mixing layer (also called the active layer), the immediate subsurface layer, and the bottom subsurface layer. The bottom subsurface layer can be further divided into more sublayers. The exchanges of sediment between the water column and the bed leads to bed sorting with dynamic variations of sediment composition in each bed layer.

As is widely accepted, sediment exchange occurs at the mixing layer. The composition of the mixing layer can be found using a mass-balance approach (e.g., [19,25,26,49]). This study used the equation proposed by Wu [19] for the mixing layer.

$$\frac{\partial(A_m p_{bk})}{\partial t} = \frac{\partial z_{bk}}{\partial t} + p_{bk}^*(\frac{\partial A_m}{\partial t} - \frac{\partial z_b}{\partial t}) \tag{16}$$

where A_m is the thickness of the mixing layer and p_{bk} is the fraction of the kth size class in the mixing layer. When $\frac{\partial A_m}{\partial t} - \frac{\partial z_b}{\partial t} \geq 0$, bed changes remain in the current-mixing layer, so $p_{bk}^* = p_{bk}$; when $\frac{\partial A_m}{\partial t} - \frac{\partial z_b}{\partial t} < 0$, bed changes penetrate to the immediate subsurface layer, so $p_{bk}^* = p_{ik}$ (the fraction of kth size class in the immediate subsurface layer).

The thickness of the bed layers (the mixing layer, the immediate subsurface layer, and the bottom subsurface layer) are user-input variables for the simulation. For the mixing layer, a minimum thickness, $A_{\min}$, was set to maintain the active sediment exchanges in the mixing layer. This minimum thickness of the mixing layer, $A_{\min}$, can be either user-specified or determined by d_{50} and the sand dune height Δ_s [20].

$$A_{\min} = \max(2d_{50}, 0.5\Delta_s) \tag{17}$$

The mixing-layer thickness changes little during the computation, though the division of each layer may move upward with net deposition or downward with net erosion. In such cases, the composition changes of the mixing layer may induce corresponding changes in the other subsurface layers. For example, for the immediate subsurface layer, one can obtain [20]

$$\frac{\partial(A_i p_{ik})}{\partial t} = -p^*_{bk}(\frac{\partial A_m}{\partial t} - \frac{\partial z_b}{\partial t}) \tag{18}$$

where A_i is the thickness of the immediate subsurface layer.

For the nonerodible bed with only net depositions allowed, both erosions and the bed-layer thickness need to be limited.

3. Numerical Method

The governing equations of the flow (Equations (1)–(3) and sediment transport (Equations (7) and (11)) were discretized using the finite volume method (FVM) on a collocated unstructured hybrid-mesh system with mixed-triangle and quadrilateral cells. Compared to the structured-mesh system, the unstructured mesh is more suitable for geometrically complex domains with higher adaptivity.

3.1. FVM Discretization

Following the work of Zhang et al. [36], the discretization method is illustrated by the integral form of the suspended-load transport equation in the typical cells shown in Figure 1, where one triangle cell and one quadrilateral cell are neighbored with each other. In each cell, the centroid nodes (denoted by the superscript "c") and the vertex nodes (denoted by the superscript "v") are counterclockwise numbered. A similar discretization procedure can be carried out for the bed-load transport equation.

For the time integral, the first-order Euler forward scheme was used:

$$\int_{\Omega} \frac{\partial(hC_k/\beta_{sk})}{\partial t} d\Omega \approx \frac{(hC_k/\beta_{sk})^{n+1} - (hC_k/\beta_{sk})^n}{\Delta t} \cdot \Omega \tag{19}$$

where the superscript "n" denotes the time level.

For the convective fluxes,

$$\int_{\Omega} (\frac{\partial(uhC_k)}{\partial x} + \frac{\partial(vhC_k)}{\partial y}) d\Omega \approx \sum F_{0-m}(hC_k)_{0-m} \tag{20}$$

where $F_{0-m} = u_{0-m}\Delta y_{i-j} - v_{0-m}\Delta x_{i-j}$ is the flux; the subscripts "i" and "j" denote the two vertices of the edge "$0 - m$" ordered counterclockwise. Note that for coordinate calculations, the subscript "$i - j$" denotes the subtraction of coordinate "i" from "j"; that is, $\Delta x_{i-j} = x_j - x_i$ and $\Delta y_{i-j} = y_j - y_i$.

The second-order upwind scheme was used for the edge center "$0 - m$",

$$\phi_{0-m} \approx \begin{cases} \phi_0 + (\phi_x)_0 \Delta x_{0-F} + (\phi_y)_0 \Delta y_{0-F} & (F_{0-m} > 0) \\ \phi_m + (\phi_x)_m \Delta x_{m-F} + (\phi_y)_m \Delta y_{m-F} & (F_{0-m} < 0) \end{cases} \tag{21}$$

where $\phi_{0-m} = (hC_k)_{0-m}$ and, ϕ_x and ϕ_y represent the first-order derivatives with respect to x and y.

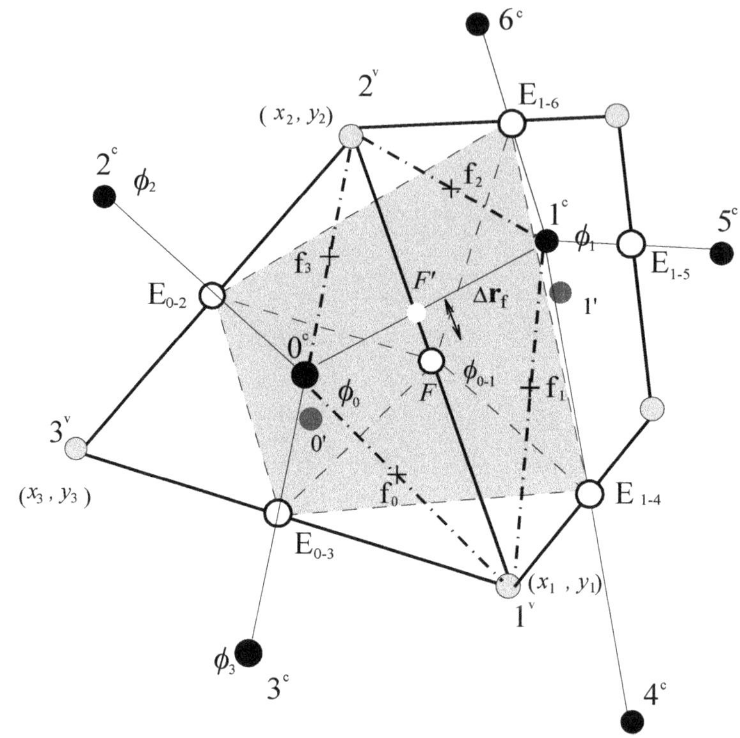

Figure 1. Typical triangle and quadrilateral cells: the black circle denotes the cell centers; the gray circle denotes the vertices; the big white circles denote the edge centers; the small white circle denotes the interception point of the edge and the line of two neighboring cell centers; the blank quadrilateral (0^c-1^v-1^c-2^v) with the dash-dot lines is formed by two vertices of the edge and two neighboring cell centers; and the quadrilateral filled with gray color is formed by four neighboring edge centers (E_{0-1}, E_{0-3}, E_{1-4}, and E_{1-6}) with E_{0-1} as centroid.

In Equation (21), the first term on the RHS will be treated implicitly, while the second term with gradients is treated as the source term.

$$S^{\phi} = \sum_m \{F^+_{0-m}\cdot[(\phi_x)_0\Delta x_{0-F} + (\phi_y)_0\Delta y_{0-F}] + F^-_{0-m}\cdot[(\phi_x)_m\Delta x_{m-F} + (\phi_y)_m\Delta y_{m-F})]\} \tag{22a}$$

$$F^{\pm}_{0-m} = \frac{1}{2}(F_{0-m} \pm |F_{0-m}|) \tag{22b}$$

For the diffusion fluxes, the Green Theorem was applied so that one can obtain:

$$\int_{\Omega_{0-m}} [\frac{\partial}{\partial x}[h\varepsilon_s\frac{\partial C_k}{\partial x}) + \frac{\partial}{\partial y}(h\varepsilon_s\frac{\partial C_k}{\partial y})]d\Omega \approx \sum_m [(h\varepsilon_s\frac{\partial C_k}{\partial x})_{0-m}\cdot\Delta y_{i-j} - (h\varepsilon_s\frac{\partial C_k}{\partial y})_{0-m}\cdot\Delta x_{i-j}] \tag{23}$$

where Ω_{0-m} is the area of the quadrilateral at the edge "$0 - m$" (refer to the quadrilateral 0^c-1^v-1^c-2^v in Figure 1).

For the source terms, one can obtain:

$$\int_{\Omega} [\alpha\omega_{sk}(C_{*k} - C_k) - S^{\phi}]d\Omega \approx [-\alpha\omega_{sk}C_k + \alpha\omega_{sk}C_{*k} - S^{\phi}]\cdot\Omega \tag{24}$$

where the first term with C_k at RHS will be treated implicitly.

3.2. Edge-Gradient Evaluation

On the RHS of Equation (23), the edge gradient was evaluated by the Green Theorem, and the values at the vertices are required, which are interpolated from the cell centers.

$$\left(\frac{\partial C}{\partial x}\right)_{0-m} = \frac{1}{\Omega_{0-m}}\oint Cdy \approx \frac{1}{2\Omega_{0-m}}[(C_m^c - C_0^c)\Delta y_{i-j} - (C_j^v - C_i^v)\Delta y_{0-m}] \tag{25a}$$

$$\left(\frac{\partial C}{\partial y}\right)_{0-m} = \frac{-1}{\Omega_{0-m}}\oint Cdx \approx \frac{-1}{2\Omega_{0-m}}[(C_m^c - C_0^c)\Delta x_{i-j} - (C_j^v - C_i^v)\Delta x_{0-m}] \tag{25b}$$

Note that the subscript "k" is omitted here for simplicity.

If the Taylor series expansion is used, no interpolation is needed, and one can obtain:

$$\left(\frac{\partial C}{\partial x}\right)_{0-m} = \frac{1}{\Omega_{0-m}}\oint Cdy \approx \frac{1}{\Omega_{0-m}}[(C_m^c - C_0^c + D_{0-m})\Delta y_{i-j} + (G_x)_{0-m}] \tag{26a}$$

$$\left(\frac{\partial C}{\partial y}\right)_{0-m} = \frac{-1}{\Omega_{0-m}}\oint Cdx \approx \frac{-1}{\Omega_{0-m}}[(C_m^c - C_0^c + D_{0-m})\Delta x_{i-j} + (G_y)_{0-m}] \tag{26b}$$

$$D_{0-m} = [(C_x)_m - (C_x)_0] \cdot \Delta x_{F-F'} + [(C_y)_m - (C_y)_0] \cdot \Delta y_{F-F'} \tag{26c}$$

$$\begin{aligned}(G_x)_{0-m} = {} & \tfrac{1}{2}[(C_x)_0\Delta x'_{0-i} + (C_y)_0\Delta y'_{0-i}] \cdot \Delta y'_{0-i} + \tfrac{1}{2}[(C_x)_0\Delta x'_{0-j} + (C_y)_0\Delta y'_{0-j}] \cdot \Delta y'_{j-0} + \\ & \tfrac{1}{2}[(C_x)_m\Delta x'_{m-i} + (C_y)_m\Delta y'_{m-i}] \cdot \Delta y'_{i-m} + \tfrac{1}{2}[(C_x)_m\Delta x'_{m-j} + (C_y)_m\Delta y'_{m-i}] \cdot \Delta y'_{j-m}\end{aligned} \tag{26d}$$

$$\begin{aligned}(G_y)_{0-m} = {} & \tfrac{1}{2}[(C_x)_0\Delta x'_{0-i} + (C_y)_0\Delta y'_{0-i}] \cdot \Delta x'_{0-i} + \tfrac{1}{2}[(C_x)_0\Delta x'_{0-j} + (C_y)_0\Delta y'_{0-j}] \cdot \Delta x'_{j-0} + \\ & \tfrac{1}{2}[(C_x)_m\Delta x'_{m-i} + (C_y)_m\Delta y'_{m-i}] \cdot \Delta x'_{i-m} + \tfrac{1}{2}[(C_x)_m\Delta x'_{m-j} + (C_y)_m\Delta y'_{m-i}] \cdot \Delta x'_{j-m}\end{aligned} \tag{26e}$$

where Ω_{0-m} denotes the areas of the quadrilateral at the edge $0-m$; C_x and C_y denote the first-order derivatives; D_{0-m} is the correction term for mesh irregularity (will automatically vanish in well-connected mesh), and all the coordinates with the superscript "$'$" count in the distance vector $\nabla \vec{r}_f$ as well. For example, $\Delta x'_{0-i} = x_i^v - (x_0^c + \Delta x_{F-F'})$ and $\Delta y'_{m-i} = y_i^v - (y_m^c + \Delta y_{F-F'})$, and so on.

This evaluation method for the edge gradient was also applied to the velocity gradient at edges in the momentum equations and the pressure gradient at edges when coupling the continuity equation and the momentum equations. Compared to the conventional method requiring vertex values, this method accounts for mesh irregularities and avoids interpolations for the vertex values.

3.3. Multipoint Momentum Interpolation Correction

After discretization, the momentum Equations (2) and (3) can be linearized as follows:

$$\frac{\boldsymbol{A}^{U*}}{r_u}\boldsymbol{U}^* + \sum_1^b A_b^U \boldsymbol{U}_b^* = \boldsymbol{U}^n - g\Delta t(\nabla\eta) + \frac{(1-r_u)}{r_u}\boldsymbol{A}^{U*}\boldsymbol{U}^m + S^U \tag{27}$$

where $\boldsymbol{U}^* = (u^*, v^*)$ is the provisional velocity (m/s); $\boldsymbol{U}^m$ is the velocity of the previous iteration (m/s) with the superscript "m" denoting the iteration step; r_u is the under-relaxation parameter for velocity; $\nabla\eta = (\partial\eta/\partial x, \partial\eta/\partial y)$ is the pressure gradient; $\boldsymbol{A}^{*U} = (A^{*u}, A^{*v})$ is the nondimensional matrix coefficient; $\boldsymbol{A}_b^U = (A_b^u, A_b^v)$ is the matrix coefficient for neighboring cells; and S^U is the source term.

For the collocated mesh system, the momentum interpolation (MI) method proposed by Rhie and Chow [50] is to interpolate the discretized momentum equation at the edge between two neighboring cells to evaluate the edge velocity and flux.

$$\boldsymbol{U}_{0-m}^* \approx \boldsymbol{IS}_{0-m} - \frac{g\Delta t}{\boldsymbol{A}_{0-m}^U}(\nabla\eta)_{0-m} \tag{28}$$

where $IS = U^* + \frac{g\Delta t}{A^U}(\nabla\eta)$ denotes the interpolation subject (m/s) at cell centers; IS_{0-m} denotes the cross-edge interpolation subject with $IS_{0-m} = s_{0-m}IS_0 + (1 - s_{0-m})IS_m$ with $s_{0-m}(\in [0,1])$ the linear interpolation coefficient; and the matrix coefficient at the edge is defined as $A^U_{0-m} = s_{0-m} \cdot A^U_0 + (1 - s_{0-m}) \cdot A^U_m$.

Zhang and Jia [51] extended the cross-edge MI to the surrounding edge centers by defining the multipoint interpolation subject as the averaged surrounding edge interpolation subject $IS^{MP}_{0-m} = \sum_1^4 w_n IS_{i-j}$ with w the weighting coefficient. Specifically, at edge "0–1" in Figure 1, the multipoint interpolation subject is $IS^{MP}_{0-1} = w_1 IS_{0-1} + w_2 IS_{0-2} + w_3 IS_{1-4} + w_4 IS_{1-6}$ with the superscript "MP" representing "multi-point". Therefore, the momentum interpolation with multipoint interpolation correction reads:

$$U^*_{0-m} \approx IS_{0-m} - \frac{g\Delta t}{A^U_{0-m}}(\nabla\eta)_{0-m} + (IS^{MP}_{0-m} - IS_{o-m}) \cdot r_I \tag{29}$$

where r_I is a relaxation factor in the range of [0, 1] to control the correction of the last term on the RHS of Equation (29). With the multipoint-momentum interpolation corrections, nonphysical oscillations can be removed, especially those possible oscillations induced by the wetting-and-drying process, which is common due to the morphological changes in the unsteady sediment transport.

3.4. Solution Procedure

All discretized equations are solved by the BiCGSTAB(l) (biconjugate-gradient stabilized method) solver [52]. At each time step, the flow and the sediment transport are coupled in the following way: (1) the flow field is calculated first based on the current bed conditions (bed elevation and bed material composition); (2) then, the sediment transport is simulated using the calculated flow field; and (3) finally, the bed changes and the bed sorting are calculated and bed elevations and sediment compositions are updated.

4. Examples and Application

In the following sections, the hybrid-sediment transport model will be demonstrated and validated by selected examples, including the degradation case [9] for the erosion process, the aggradation case [53] for the deposition process, and the long-term unsteady sediment transport in the East Fork River [54]. Finally, the model was applied to JiJi Reservoir in Taiwan [1]. All computational meshes were generated by CCHE-MESH [55], a quality mesh generator for both structured and unstructured meshes.

4.1. Bed Degradation

The first example is based on the benchmark experiments conducted by Ashida and Michiue [9] to demonstrate the model's capability of handling bed degradation and armoring processes under clear-water conditions. The experimental flume was 20 m long and 0.8 m wide with a slope of 1%. On the flume bed, there were 12 size classes of sediment particles (Table 1) ranging from sand to gravel with the median size of 1.5 mm and the initial bed material thickness of 0.113 m. A steady flow of 0.0314 m^3/s was imposed at the inlet, while the water level of 0.06 m was maintained at the outlet.

In this simulation, the bed-load transport dominated, and the bed-load transport model was run on a 5 × 100 rectangle mesh. Bed roughness was calibrated as 0.023 $m^{-1/3}s$, and the time step was 1 s. For this bed degradation case, Wu [20] investigated the effects of the mixing layer thickness (varying between d_{50} and $2d_{50}$) and the bed-load adaptation length (time dependent, water depth dependent, or constant) on the simulations using the one-dimensional model. In this study, the mixing-layer thickness was set as $2d_{50}$ and the bed-load adaptation length was set as the averaged sand dune length (≈ 7.3 times the water depth).

Table 1. Size classes for degradation experiment (Ashida and Michiue [3]).

Size (mm)	0.25	0.35	0.5	0.7	0.9	1.25	1.75	2.5	3.5	5	7	9
Lower Bound	0.2	0.3	0.4	0.6	0.8	1	1.5	2	3	4	6	8
Upper Bound	0.3	0.4	0.6	0.8	1	1.5	2	3	4	6	8	10
Fraction	0.075	0.125	0.165	0.035	0.035	0.065	0.04	0.09	0.1	0.195	0.05	0.025

With the clear water condition, bed erosion occurred due to bed-load sediment transport. During the scouring process, a layer of coarse bed materials (called an armoring layer) may be formed, which could protect the bed from scouring, resulting in slowing or even stopping erosion (called armoring effects). In this bed degradation case, the armoring effects were significant. Figure 2 compares the measured erosion depth development with time at x = 7 m, 10 m, and 13 m to the simulation results. As can be seen, the erosion developed rapidly in the first 100 min and then became much milder afterward due to the armoring effects. The simulated erosion development profiles at three locations agreed well with the measured ones.

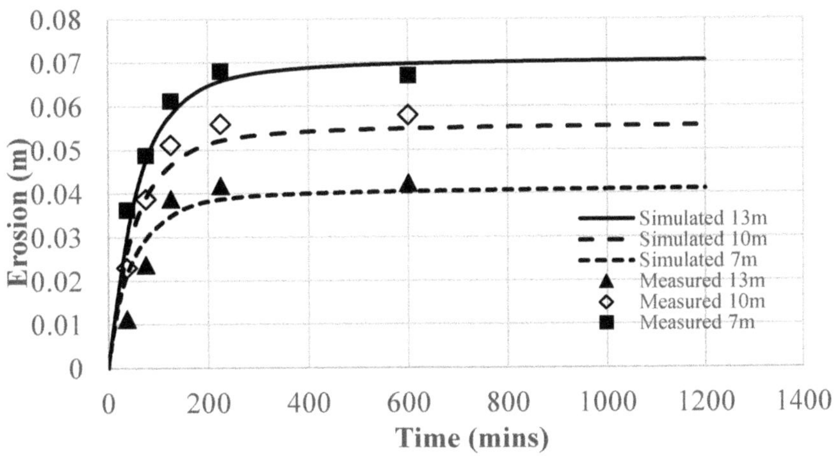

Figure 2. Erosion development profiles with time.

The armoring effect is closely related to the bed sorting calculation. Figure 3 compares the simulated bed composition with the measured one at x = 10 m. The model succeeded in simulating the armoring effects. The initial d_{50} was 1.5 mm and the measured final d_{50} was 5.4 mm, while the simulated final d_{50} was 5.6 mm. However, the simulated armoring layer was formed earlier, resulting in a slightly coarser bed-armoring layer. In general, the bed-sorting calculation was satisfactory for this case.

4.2. Bed Aggradation

The benchmark experiments conducted by Seal et al. [53] were selected to demonstrate the bed aggradation process. The experiment was designed to investigate longitudinal deposition formed by feeding poorly sorted sediment from upstream. The flume was 45 m long and 0.3 m wide with a slope of 2%. The input-sediment particle size varied from 0.125 mm (fine sand) to 64 mm (coarse gravel) with a median size of 6 mm. Figure 4 shows a sketch of this flume experiment.

In the experiment, there were three runs, and run two was selected in this study. For run two, at the inlet, a constant flow discharge of 0.049 m^3/s and a constant sediment discharge of 0.0942 kg/s were imposed, while at the outlet, the water level was 0.45 m. Table 2 lists the input sediment mixture consisting of 10 size classes. The sediment transport model ran for 32.4 h on a 5 × 200 rectangle mesh with the bed roughness of 0.025 $m^{-1/3}s$

and a time step of 1 s. The adaptation coefficient of the suspended load, α, was calibrated as 0.3, while the adaptation length of the bed load, L_b, was 0.3 m (the channel width).

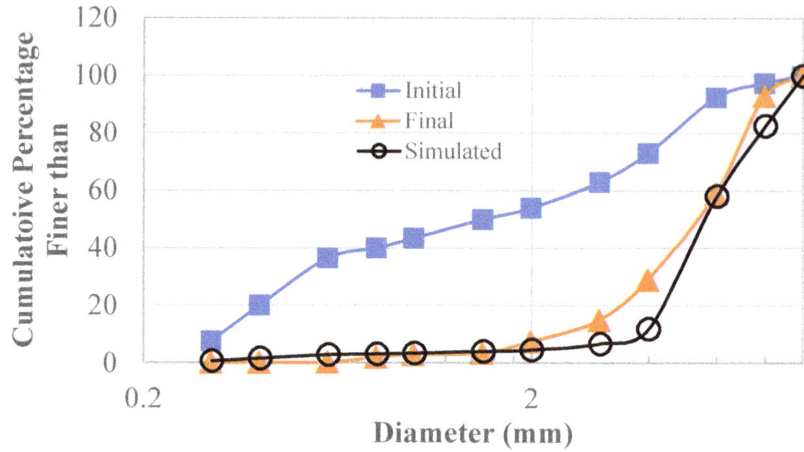

Figure 3. Bed compositions at $x = 10$ m.

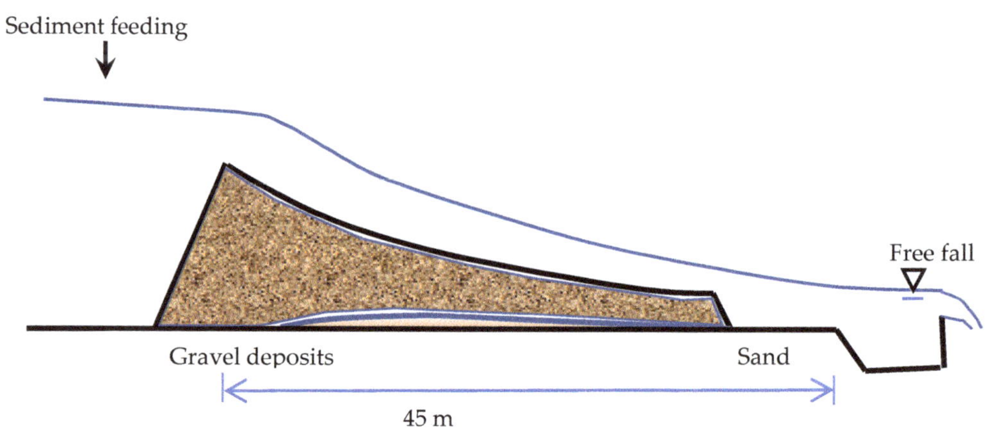

Figure 4. Sideview of the layout of the bed-aggradation experiment by Seal et al. [51].

Table 2. Input sediment mixture for the bed aggradation experiment (Seal et al. [32]).

Size (mm)	**0.125**	**0.25**	**0.5**	**1**	**2**	**4**	**8**	**16**	**32**	**64**
Fraction	0.022	0.104	0.102	0.07	0.04	0.172	0.12	0.168	0.124	0.078

Due to the constant feeding of coarse sediments, sediment deposition developed in this flume. Figure 5 compares the simulated bed profiles at different times to the measured ones. The shape of the bed profiles, including the ending slopes, was well captured by the model, though a small phase shift at 22 h was observed, resulting in an over-estimated deposition. However, the simulation results generally agreed well with the measurements.

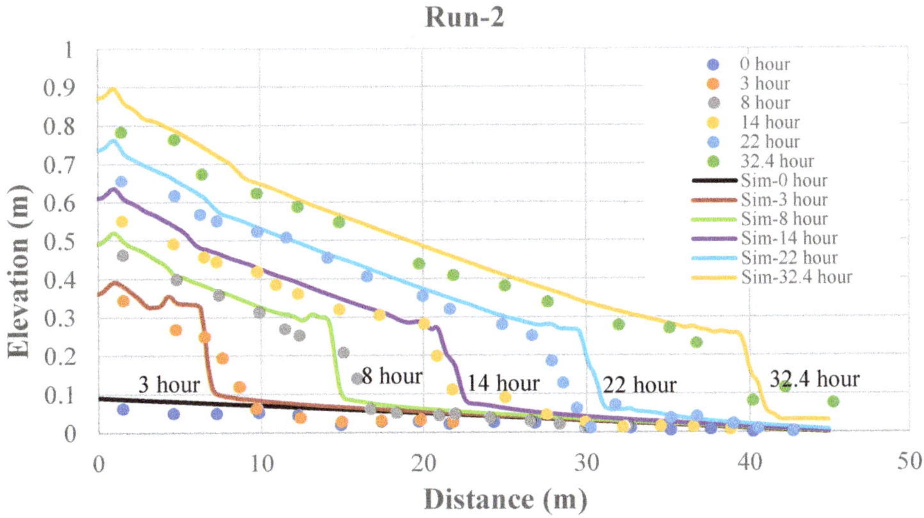

Figure 5. Longitudinal bed profiles at different times.

4.3. East Fork River

The East Fork River is a typical meandering river located in Wyoming State that was used for the study of bed-load transport in the 1970s. The selected reach was about 3.3 km long with the channel width varying from 16 m to 45 m and ending at a bed-load trap across the river (Figure 6). The flow in this river was influenced by spring runoff due to snowmelt, with rising flow rates in the morning, peak rates at midday, and declining flows in the afternoon. This field case was selected to demonstrate the capability of the model for long-term sediment transport under unsteady turbulent-flow conditions.

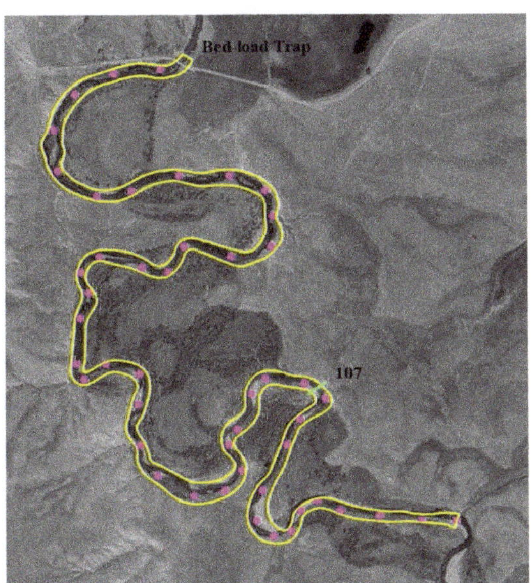

Figure 6. Selected reach of East Fork River. Solid circles represent the approximate locations of bed samples.

The simulation period was 2–19 June 1979. Figure 7 shows the hydrograph and the stage graph at the inlet and outlet. This model generated an unstructured quadrilateral mesh of this reach with minimum and maximum edge lengths equal to 0.46 m and 13.99 m (Figure 8). With the variation of flow, the wetting-and-drying process developed with time in this study reach, which may have induced nonphysical numerical oscillations in the model. The multipoint momentum interpolation correction technique [51] was used to remove the oscillations and maintain numerical stability and accuracy.

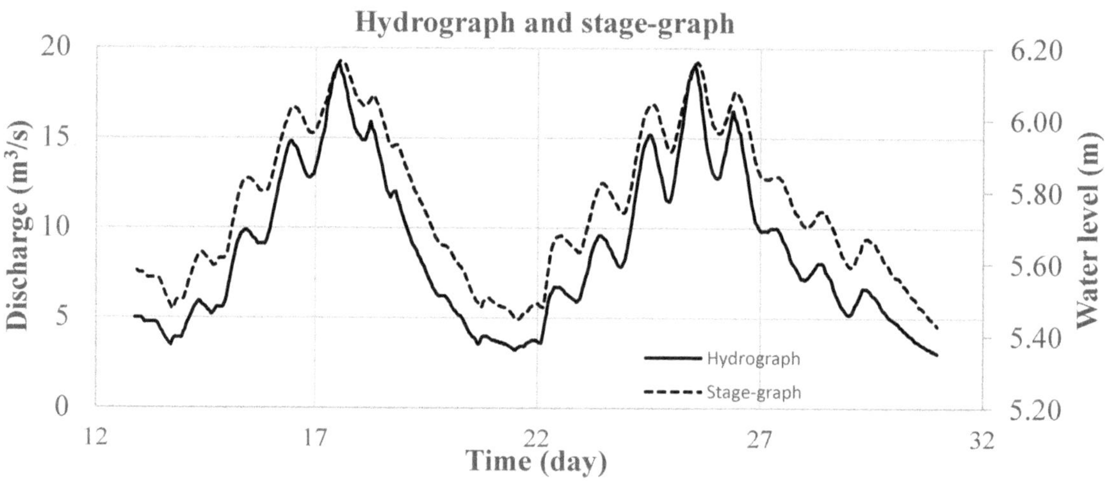

Figure 7. Hydrograph at the inlet and stage graph at the outlet.

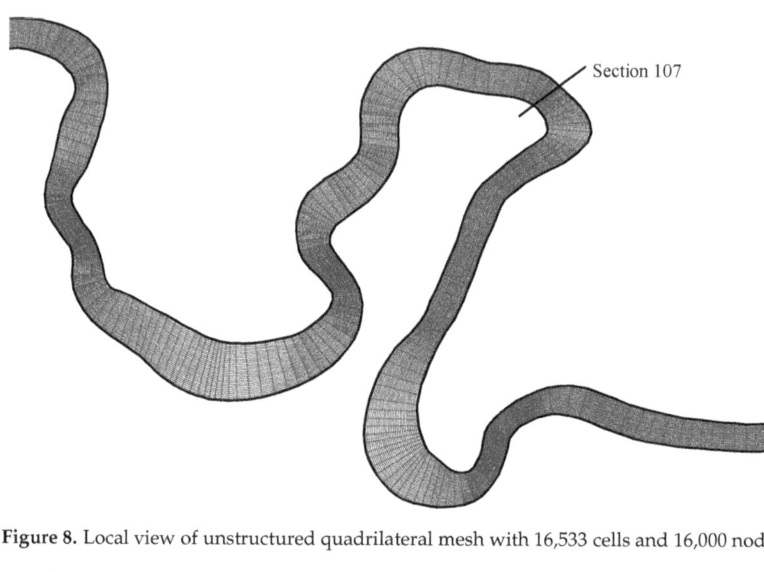

Figure 8. Local view of unstructured quadrilateral mesh with 16,533 cells and 16,000 nodes.

The measured water surface elevation at section 107 was used to calibrate the Manning coefficient, and Figure 9 compares the simulated and measured water surface profiles at section 107 using the calibrated Manning coefficient of 0.03 $m^{-1/3}s$ for the whole reach and the time step of 3 s. As can be seen, the simulated profile matched well with the measured one.

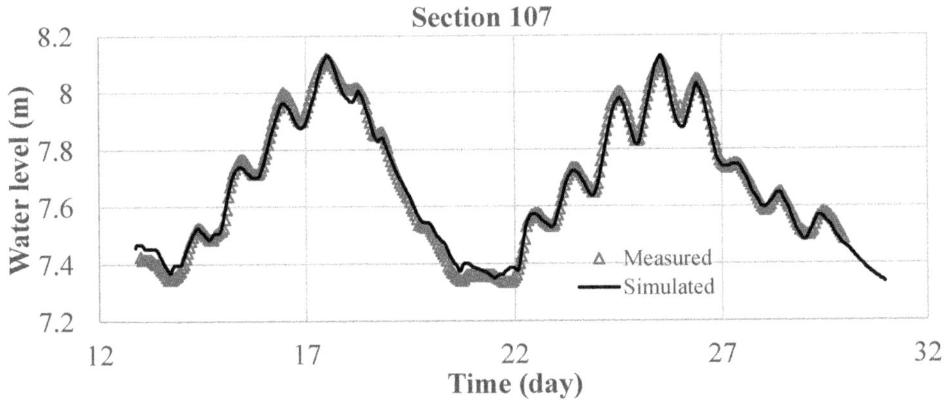

Figure 9. Water surface profiles with time at section 107.

Table 3 lists the initial nine size classes of sediment mixtures ranging from 0.088 mm (fine sand) to 32 mm (coarse gravel). Bed-sample information was obtained from the USGS technical report [52] for the selected reach. Although the bed load was dominant, both the suspended-load and bed-load transport were simulated. In this simulation, the adaptation coefficient of the suspended load, α, was calibrated as 0.001, while the adaptation length of the bedload, L_b, was 60 m, about two times the average channel width.

Table 3. Sediment size classes for the East Fork River (Meade et al. [52]).

Size (mm)	**0.088**	**0.177**	**0.354**	**0.707**	**1.41**	**2.83**	**5.66**	**11.3**	**32**
Fraction	0.044	0.00038	0.02	0.478	0.233	0.145	0.093	0.02	0.01

At the outlet, the measured sediment flux at the bed trap was available to compare with the simulated one (Figure 10). According to the comparisons, the sediment-transport model captured the general trend of the variation of the sediment discharge with time, except for the overestimation of the first peak and the underestimation of the second peak. The simulation results were also identical to the ones from Wu [19].

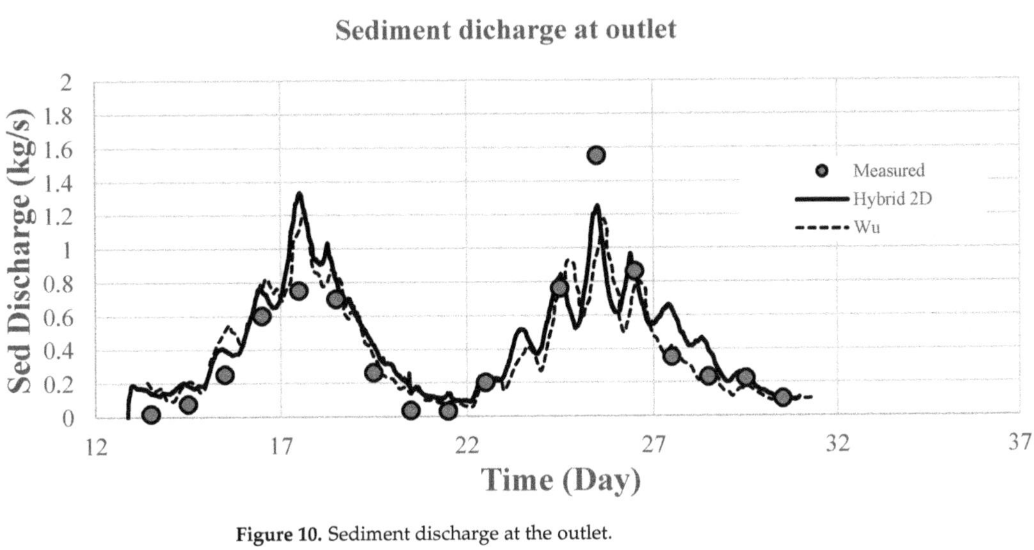

Figure 10. Sediment discharge at the outlet.

4.4. *JiJi Reservoir*

The sediment-transport model was used to simulate the deposition process in the JiJi Reservoir (Figure 11) in Taiwan, which was previously simulated by the CCHE2D sediment-transport model using structured meshes [1]. The JiJi Weir (Figure 11) was built across a mountain river, Chuoshui Creek, in Taiwan. The flow pattern of Chuoshui Creek was strongly affected by the precipitation pattern, with a very small discharge in dry seasons and extremely large flows in typhoon seasons. Due to the steep slopes and exposed soils, the flood water could transport high sediment loads and deposit them in the reservoir, resulting in rapid reductions in reservoir storage capacity. Once the reservoir was full, the sediment would be flushed through the spillway of the weir to the downstream channel.

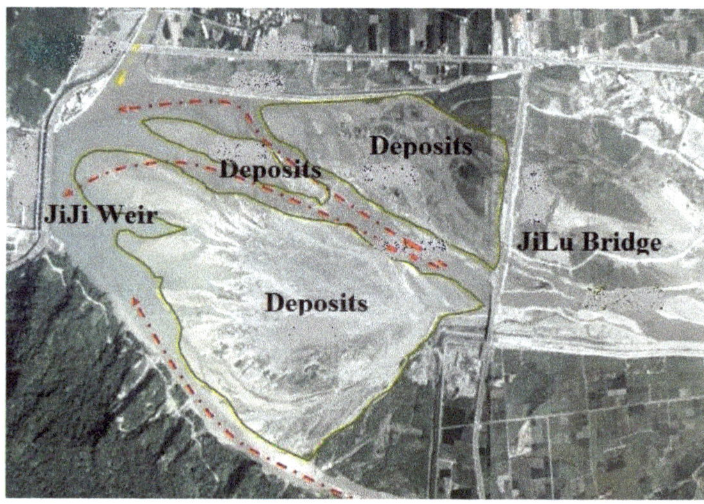

Figure 11. JiJi Reservoir.

Figure 11 shows the study domain from JiLu Bridge to JiJi Weir, which was about 2 km long. As can be seen, the reservoir suffered from serious sedimentation problems. In 2003, the total amount of sediment yield was about 14.77 million m^3, though only about 13% was transported downstream. According to the bed samples measured in July of 2004 (Figure 12), coarse sediments were deposited mainly upstream of the reservoir, close to the JiLu bridge, while fine sediments were deposited within about 1000 m of the weir.

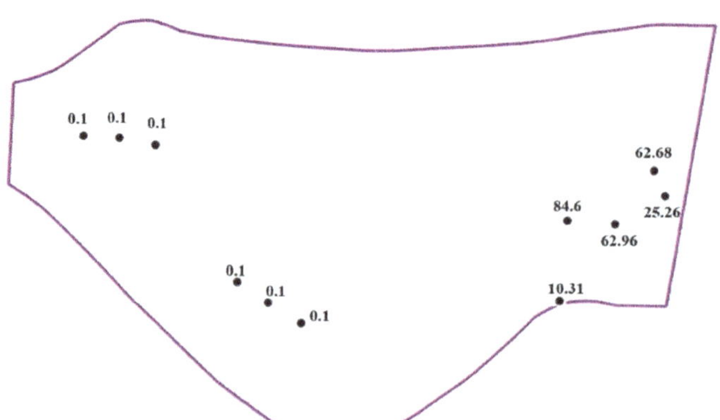

Figure 12. Bed sample locations with d_{50} in mm.

Table 4 lists six sediment-size classes used in the simulation, varying from medium silts to small boulders. The bed samples of 2004 (Figure 12) were used as the initial bed composition within the reservoir. Figure 13 shows the initial bed elevation and the quadrilateral mesh generated for this reservoir based on the measured topography data in 2004.

Table 4. Size classes in the JiJi Reservoir.

Size (mm)	0.0272	0.297	2.38	9.52	152	457
Lower Bound	0.001	0.074	0.59	4.76	19.1	305
Upper Bound	0.074	0.59	4.76	19.1	305	610

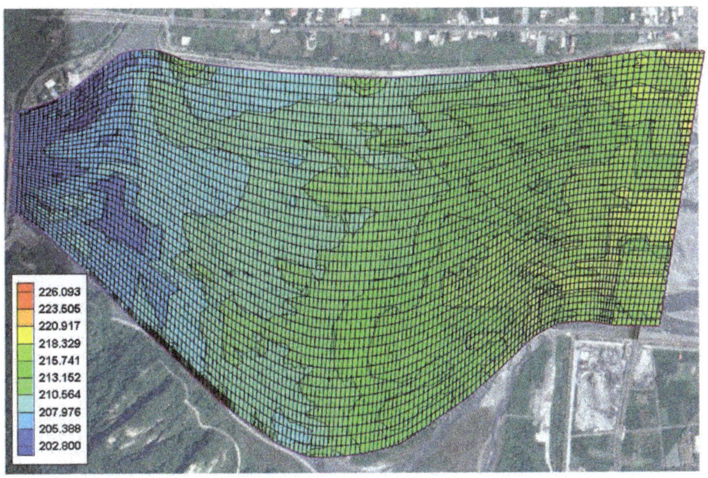

Figure 13. Initial bed elevations and computational mesh for the JiJi Reservoir with 6440 nodes and 6240 cells.

The simulation period was from July 2004 to October 2007. Due to the flow pattern of Chuoshui Creek, small discharges were assumed to have little effect on the sediment transport in the reservoir, which were therefore ignored in the simulation. The actual hydrograph and sedigraph consisted of all ten typhoon events during the study period with peak flow discharges varying from 1020 m^3/s to 8343 m^3/s, as shown in Figure 14a. Figure 14b shows the corresponding stage hydrograph at JiJi Weir. Such treatment shortened the simulation time from 3 years to 33 days.

Both suspended-load and bed-load transport models were applied to this reservoir with the adaptation coefficient of the suspended load, α, set as 0.01, and the adaptation length of bed load $L_b = 4000$ m.

Figure 15a shows the measured bed change pattern from 2004 to 2007. At the downstream end, sediment deposits approached the weir and formed a large hump across the reservoir, though a large scour hole also developed about 100 m from the weir. In the middle, sediment transport followed a general pattern of erosion on the left and deposition on the right, while erosion dominated upstream close to the JiLu Bridge. Note that the erosion on the left of the deposition delta (Figure 11) was partially caused by dredging activities with unknown amounts ([1,2]). Comparing to the measured values, the simulated bed changes (Figure 15b) captured the humps near the weir and the general deposition pattern on the right but missed the hole near the weir and predicted slight deposition on the left, as expected.

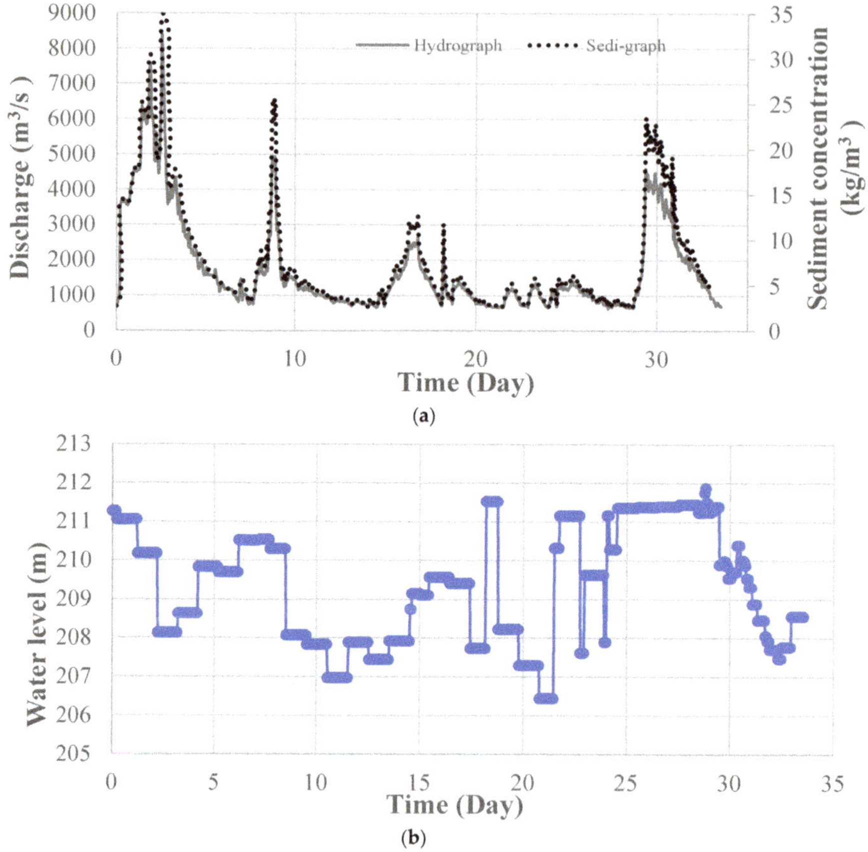

Figure 14. Boundary conditions (**a**) Hydrograph and sedigraph at JiLu Bridge. (**b**) Stage hydrograph at JiJi Weir.

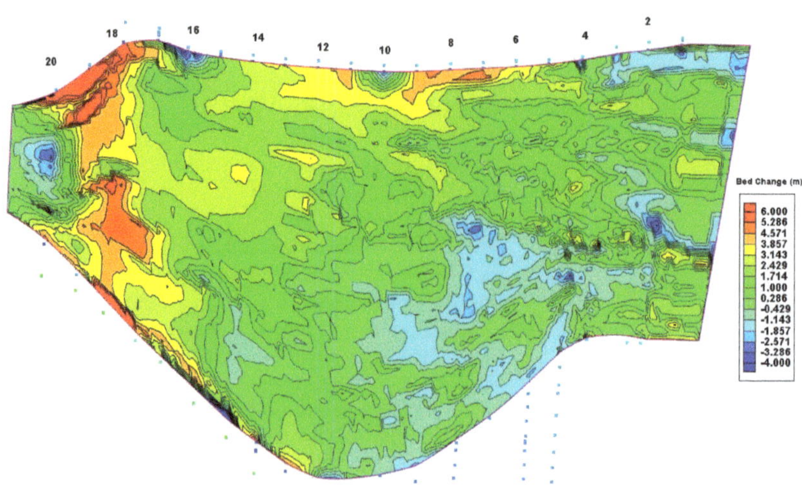

(**a**)

Figure 15. *Cont.*

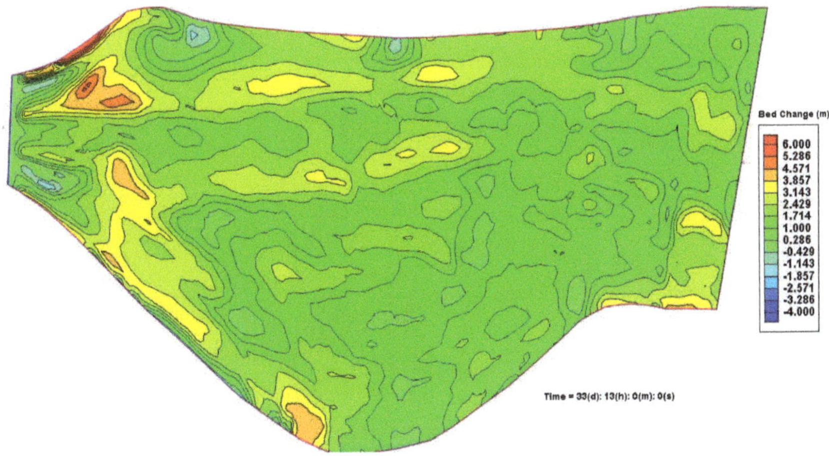

(**b**)

Figure 15. Bed changes from 2004 to 2007 in JiJi Reservoir. (**a**) Measured bed changes (**b**) Simulated bed changes.

Twenty measured cross sections were available within the JiJi Reservoir, and their locations are indicated in Figure 15a. Ten measured cross sections were compared with the simulated cross-section profiles, and the simulated bed-change patterns match them well (Figure 15b). As shown in Figure 16, from CS-2 to CS-12, the simulation predicted slight deposition on the left delta but captured the deposition on the right side. From CS-14 to CS-18, good agreement between the simulations and measurements was obtained. At CS-20, the simulation missed the scour hole in the middle of the cross section but predicted two smaller ones on two sides. Considering the uncertainties brought by the manual dredging activities on the left delta that were not considered in the simulation, the overall simulation results were reasonable.

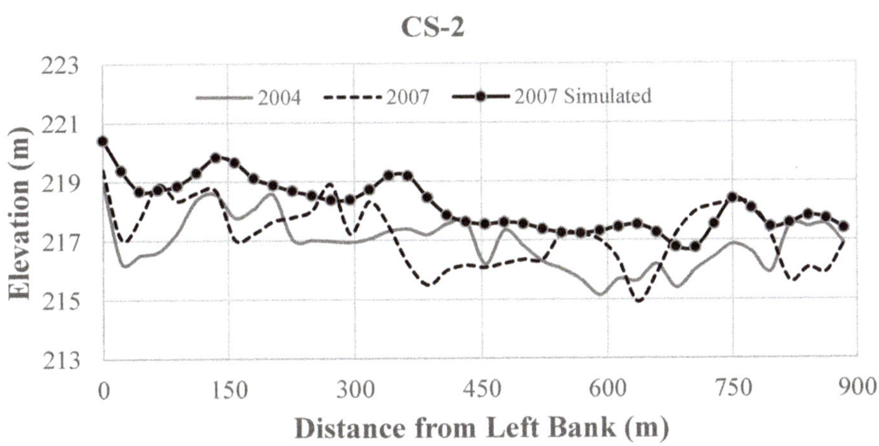

Figure 16. *Cont.*

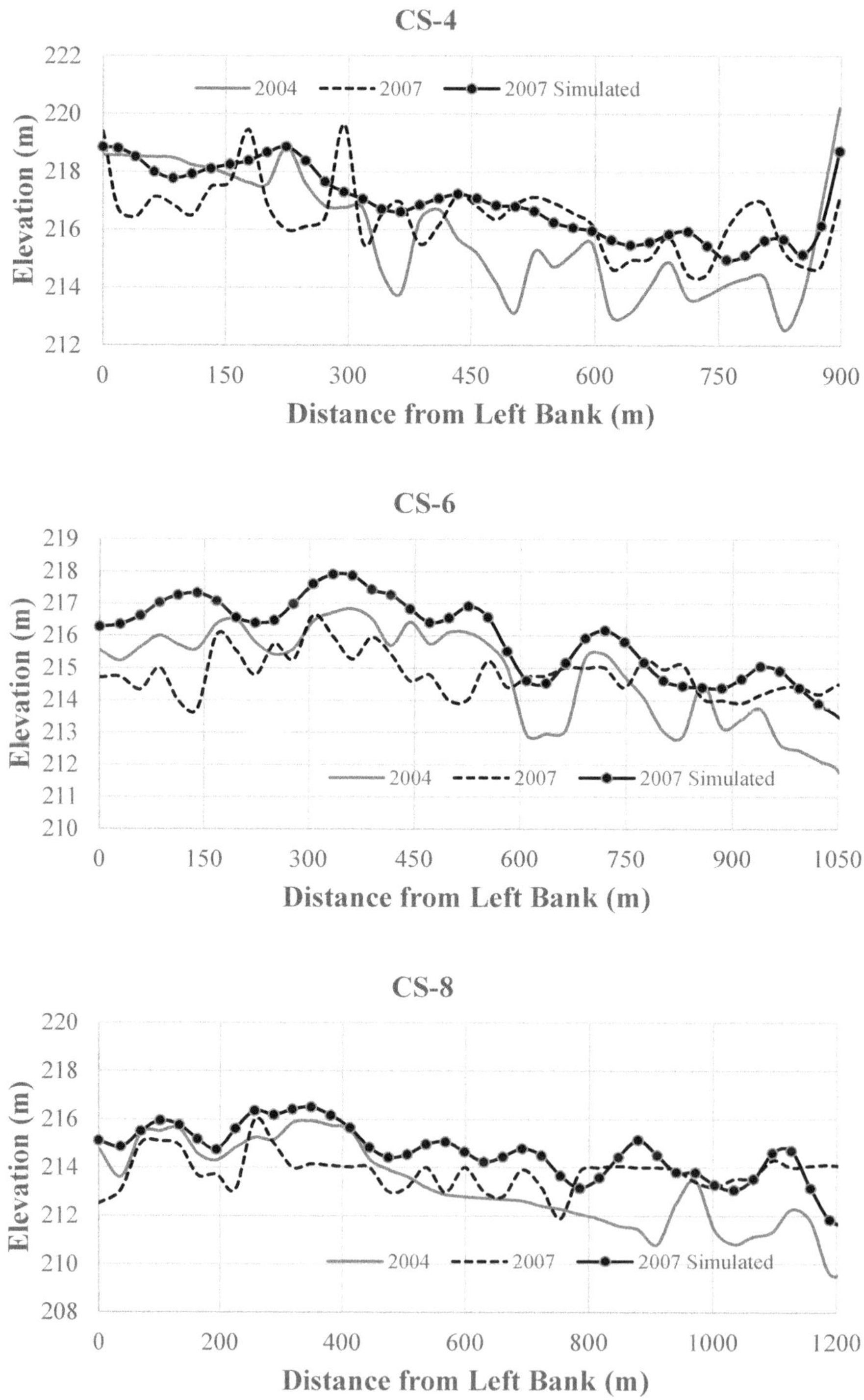

Figure 16. *Cont.*

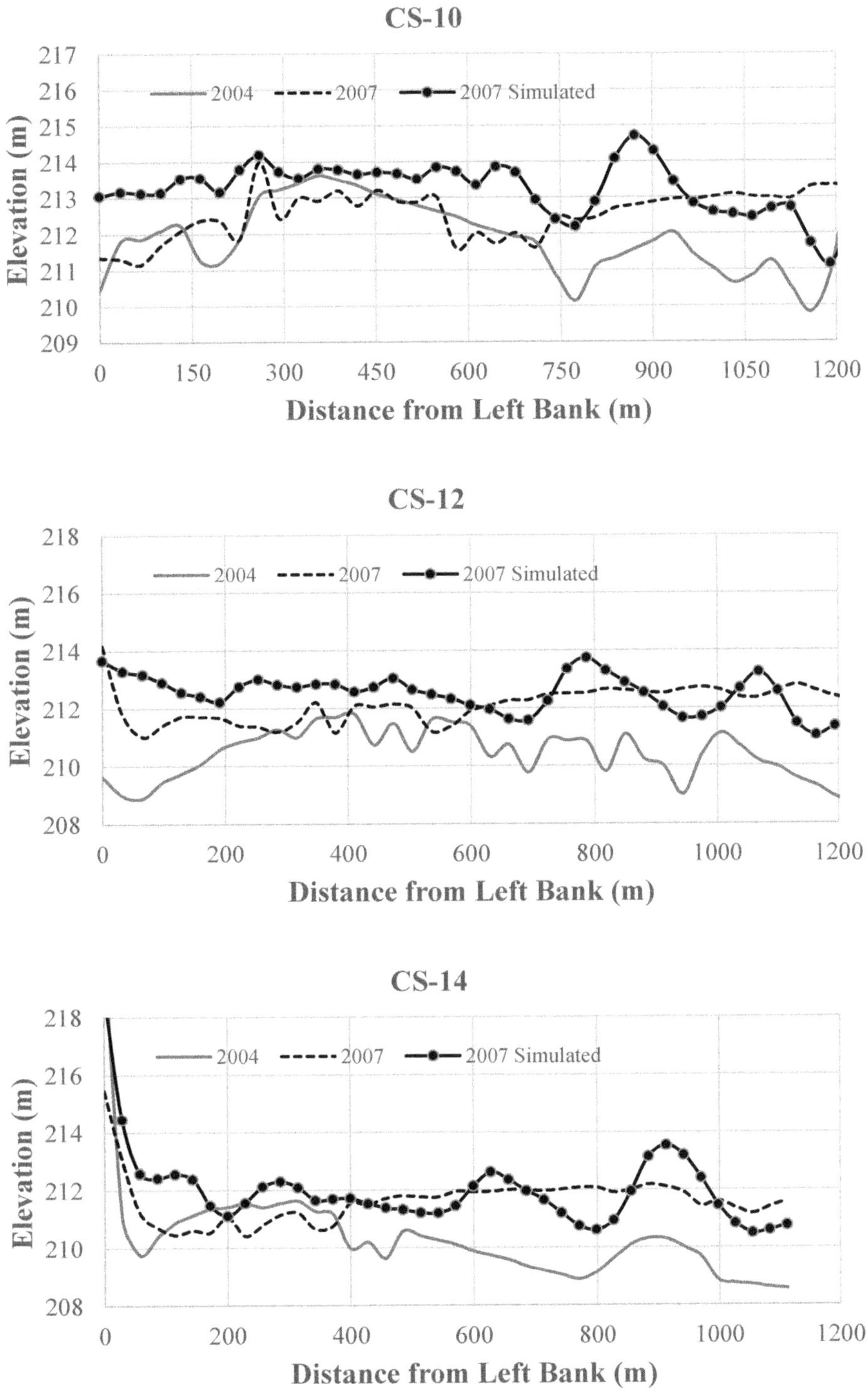

Figure 16. *Cont.*

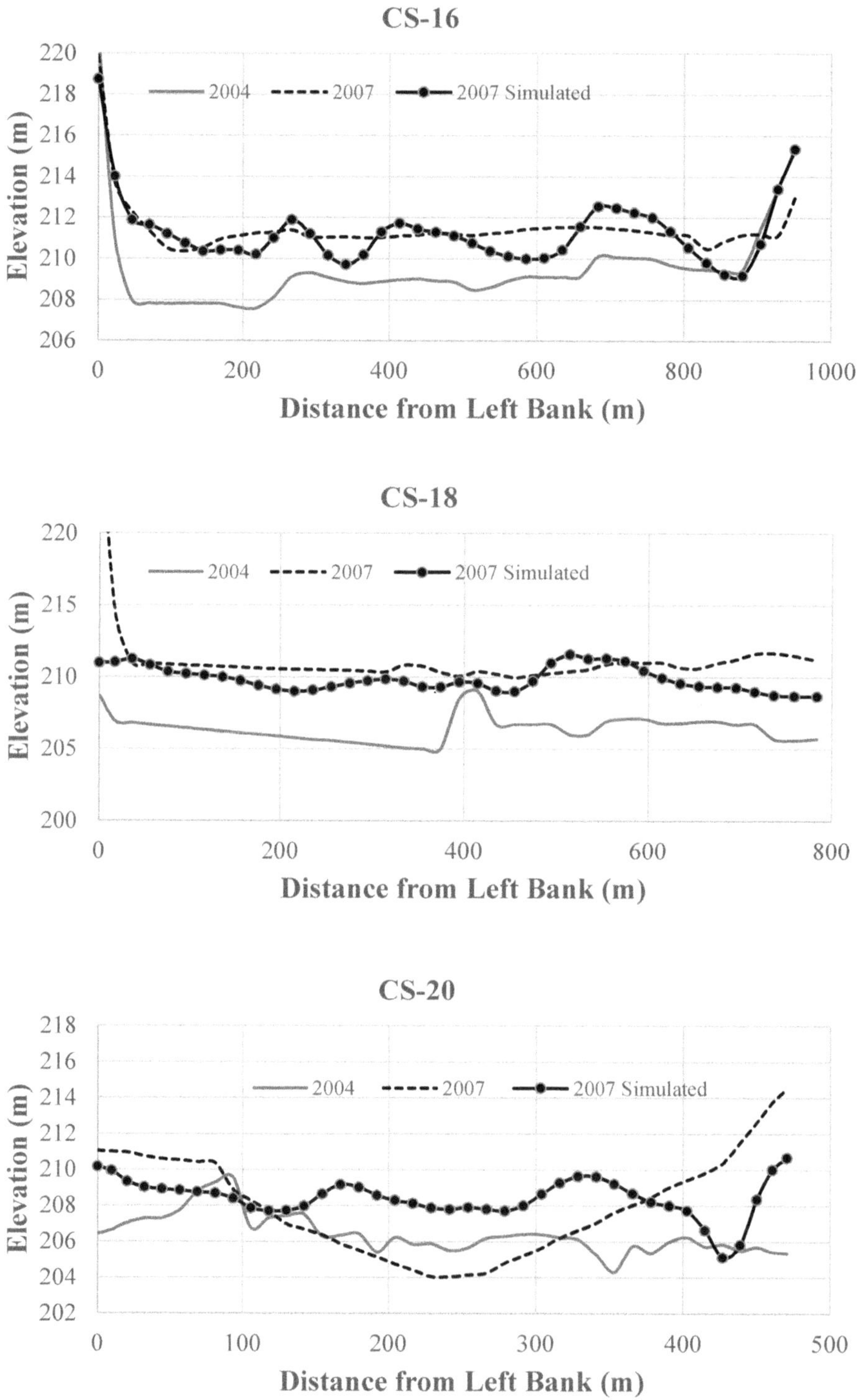

Figure 16. Comparisons of cross section profiles.

4.5. Discussion

Two laboratory-scaled cases and two field-scaled cases validated and demonstrated the developed sediment-transport model. For all cases, the fractional sediment transport played important roles, which was one of the model limitations identified by Papanicolaou et al. [34]. Multisized sediment particles may have interactions among each other in the transport or movement, such as collisions, flocculation, and disaggregation, which would affect the sediment-transport rates, concentrations, and settling behavior. Such interactions are better considered in two-phase flow models [21]. In this study, both the suspended-load and the bed-load transport-capacity formulas [41] considered the hiding-and-exposure effects in sediment mixtures, which partially and indirectly address the particle–particle interactions.

According to the comparisons with the measured data, the developed model obviously had better performances in laboratory cases than in field cases. One reason lies in the relatively simple and easy-to-control flow and sediment initial and boundary conditions in laboratory cases. In field cases, there are few well-monitored sediment data, and data scarcity has been a concern in sediment transport modeling. Specifically, in the case of the JiJi Reservoir, only limited bed samples were available, there was no inflow fractional sediment data, the measured cross section topography data were insufficient, and the dredging activities were not well recorded.

As stated previously, the flow model is the backbone of the sediment-transport model. Accurate flow calculations would provide solid conditions for qualitatively correct sediment-transport modeling. To ensure stable and accurate flow calculations, this study adopted second-order edge-gradient evaluation to consider corrections from mesh irregularities [36], and multipoint momentum-interpolation corrections for nonphysical oscillations in the wetting-and-drying process [51].

5. Conclusions

In this study, a 2D single-phase nonequilibrium sediment-transport model has been developed for nonuniform, noncohesive sediments in unsteady turbulent flows. The model was discretized using FVM on a hybrid unstructured-mesh system with high adaptivity for geometrically-complex domains. The conventional suspended-load transport equation and the bed-load transport equation were solved separately in the model. For the suspended-load gradient, the velocity gradient, and the pressure gradient at the edges, a second-order accurate method was used to take account of the mesh irregularities and avoid interpolations for the vertex values. For the possible nonphysical oscillations induced in the wetting-and-drying process, a multipoint momentum interpolation correction method was integrated to maintain both numerical stability and accuracy.

The sediment transport predictions were validated by a benchmark degradation case for the erosion process with armoring effects, a benchmark aggradation case for the deposition process, and a naturally meandering river was used for the validation of long-term unsteady sediment-transport processes. In general, the simulation results agreed well with the measured data, which indicates success in the development of this model. Finally, the model was successfully applied to simulate sediment transport in the JiJi Reservoir, which was significantly affected by typhoon events. In the future, this sediment-transport model will be improved and expanded in: (1) more applications for soil erosion and gully erosion in agricultural lands and (2) a cohesive sediment-transport process.

Author Contributions: Conceptualization, Y.Z.; methodology, Y.Z.; validation, Y.Z.; writing—original draft preparation, Y.Z.; writing—review and editing, D.W. and M.A.-H.; project administration, M.A.-H. All authors have read and agreed to the published version of the manuscript.

Funding: This work is supported by the U.S. Department of Agriculture, Agricultural Research Service, through the Cooperative Agreement No. 6060-13000-030-00D between the USDA-ARS National Sedimentation Laboratory (USDA-ARS-NSL) and the University of Mississippi/National Center for Computational Hydroscience and Engineering (NCCHE).

Informed Consent Statement: Not Applicable.

Data Availability Statement: All data in examples and application are available upon request.

Acknowledgments: The authors would like to thank Ron Bingner from National Sedimentation Laboratory (NSL), USDA Agricultural Research Service (ARS) for his suggestions and help in this study.

Conflicts of Interest: The authors declare that they have no known competing financial interests or personal relationships that could have appeared to influence the work reported in this paper.

References

1. Zhang, Y.; Jia, Y.; Yeh, K.C.; Liao, C.T.; Wang, S.S.Y. Numerical Simulation of Sediment Transport Simulation of JiJi Weir Reservoir. In Proceedings of the EWRI World Environment & Water Resources Congress 2009, Kansas City, MI, USA, 17–21 May 2009.
2. Yen, Y.H.; Chen, T.C.; Liu, G.X. Best Model for Operations of JiJi Weir. In Proceedings of the 3rd International Workshop on Sediment Bypass Tunnels, Taipei, Taiwan, 9–12 April 2019.
3. Turner, R.E.; Baustian, J.J.; Swenson, E.M.; Spicer, J.S. Wetland Sedimentation from Hurricanes Katrina and Rita. *Sci. (Am. Assoc. Adv. Sci.)* **2006**, *314*, 449–452. [CrossRef] [PubMed]
4. Liu, K.; Chen, Q.; Hu, K.L.; Xu, K.H.; Twilley, R.R. Modeling Hurricane-induced wetland-bay and bay-shelf sediment fluxes. *Coast. Eng.* **2018**, *135*, 77–90. [CrossRef]
5. Zhang, Y.; Jia, Y.; Yeh, K.C.; Liao, C.T.; Wang, S.S.Y. Sediment Transport Simulation of Chuoshui Creek. In Proceedings of the EWRI World Environment & Water Resources Congress 2010, Providence, RI, USA, 16–20 May 2010.
6. Jia, Y.; Zhang, Y.; Yeh, K.C.; Liao, C.T. *Modeling River Morphodynamic Process Using a Depth-Averaged Computational Model and an Application to a Mountain River*; IntechOpen: London, UK, 2019. [CrossRef]
7. Ding, Y.; Frey, A.E.; Kim, S.; Permenter, R.E. Probabilistic shoreline change modeling and risk estimation of erosion. *Costal Eng.* **2018**, *36*, 1. [CrossRef]
8. Dong, P.; Wu, X.Z. Application of a stochastic differential equation to the prediction of shoreline evolution. *Stoch. Environ. Res. Risk. Assess* **2013**, *27*, 1799–1814. [CrossRef]
9. Ashida, K.; Michiue, M. An investigation of river bed degradation downstream of a dam. In Proceedings of the 14th Congress of the IAHR, Paris, France, 3 September 1971.
10. Zhang, Y.; Jia, Y.; Yeh, K.C.; Liao, C.T. *Erosion Control at Downstream of Reservoir Using in-Stream Weirs*; IntechOpen: London, UK, 2022. [CrossRef]
11. Hoffmans, G.; Pilarczyk, K. Local scour downstream of hydraulic structures. *J. Hydraul. Eng.* **1995**, *121*, 326–340. [CrossRef]
12. Jia, Y.; Altinakar, M.; Guney, M.S. Three-dimensional Numerical Simulations Of Local Scouring Around Bridge Piers. *J. Hydraul. Res.* **2018**, *56*, 351–366. [CrossRef]
13. Vandaele, K.; Poesen, J.; Govers, G.; van Wesemael, B. Geomorphic threshold conditions for ephemeral gully incision. *Geomorphology* **1996**, *16*, 161–173. [CrossRef]
14. Roberts, M.E.; Burrows, R.M.; Thwaites, R.N.; Hamilton, D.P. Modelling classical gullies—A review. *Geomorphlogy* **2022**, *407*, 108216. [CrossRef]
15. Spasojevic, M.; Holly, F.M., Jr. 2-D bed evolution in natural watercourses—New simulation approach. *J. Waterw. Port Coast. Ocean. Eng.* **1990**, *116*, 425–443. [CrossRef]
16. Karki, S.; Nakagawa, H.; Zhang, Y. How Good Are 2D-Morphadynamic Models in Reproducing Bed Morphology Of Meandering Channels? Analyzing the Strengths And Limitations Of Three Popular 2D Models. In Proceedings of the 1st IAHR Young Professionals Congress, Virtual, 17–18 November 2020.
17. Chao, X.; Jia, Y.; Shields, F.D., Jr.; Wang, S.S.Y.; Cooper, C.M. Three-Dimensional Numerical Modelling of Water Quality and Sediment-Associated Processes with Application to a Mississippi Delta Lake. *J. Environ. Manag.* **2010**, *91*, 1456–1466. [CrossRef]
18. Zhang, R.J. *River Dynamics*; Industry Press: Beijing, China, 1961. (In Chinese)
19. Wu, W. Depth-averaged two-dimensional numerical modeling of unsteady flow and non-uniform sediment transport in open channels. *J. Hydraul. Eng.* **2004**, *130*, 1013–1024. [CrossRef]
20. Wu, W. *Computational River Dynamics*; Taylor & Francis Group: London, UK, 2007.
21. Chauchat, J.; Cheng, Z.; Nagel, T.; Bonamy, C.; Hsu, T.J. SedFoam-2.0: 2 3-D two-phase flow numerical model for sediment transport. *Geosci. Model Dev.* **2017**, *10*, 4372–4392. [CrossRef]
22. Dong, P.; Zhang, K. Two-phase flow modelling of sediment motions in oscillatory sheet flow. *Coast. Eng.* **1999**, *36*, 87–109. [CrossRef]
23. Cheng, Z.; Hsu, T.J.; Calantoni, J. SedFoam: A multi-dimensional Eulerian two-phase model for sediment transport and its application to momentary bed failure. *Coast. Eng.* **2017**, *119*, 32–50. [CrossRef]
24. Jia, Y.; Wang, S.S.Y. Numerical model for channel flow and morphological change studies. *J. Hydraul. Eng. ASCE* **1999**, *125*, 924–933. [CrossRef]
25. Horvat, M.; Horvat, Z. Sediment transport and bed evolution model for complex river systems. *Environ. Monit. Assess.* **2020**, *192*, 242. [CrossRef] [PubMed]
26. Lai, Y. A Two-Dimensional Depth-Averaged Sediment Transport Mobile-Bed Model with Polygonal Meshes. *Water* **2020**, *12*, 1032. [CrossRef]

27. Wu, W.; Vieira, D.A.; Wang, S.S.Y. A 1-D numerical model for nonuniform sediment transport under unsteady flows in channel networks. *J. Hydraulic Eng. ASCE* **2004**, *130*, 914–923. [CrossRef]
28. DHI. *MIKE 11 Users' Manual*; Danish Hydraulic Institute: Horsholm, Denmark, 2005.
29. Gibson, S.; Brunner, G.; Piper, S.; Jensen, M. Sediment Transport Computations in HEC-RAS. In Proceedings of the Eighth Federal Interagency Sedimentation Conference (8th FISC), Reno, NV, USA, 2–6 April 2006; pp. 57–64.
30. Huang, J.; Greimann, B.P. *GSTAR-1D, General Sediment Transport for Alluvial Rivers—One Dimension*; Technical Service Center, U.S. Bureau of Reclamation: Denver, CO, USA, 2007.
31. Jia, Y. *CCHE3D Technical Manual, National Center for Computational Hydroscience and Engineering Technical Report*; The University of Mississippi: Oxford, MS, USA, 2013.
32. Deltares. Delft3D-FLOW. In *Simulation of Multi-Dimensional Hydrodynamic Flow and Transport Phenomena, including Sediments–User Manual*; Version 3.04, rev. 11114; Deltares: Delft, The Netherlands, 2010.
33. Wu, W. *CCHE2D Sediment Transport Model*; Technical Report No. NCCHE-TR-2001-3, NCCHE; The University of Mississippi: Oxford, MS, USA, 2001.
34. Papanicolaou, A.T.N.; Elhakeem, M.; Krallis, G.; Prakash, S.; Edinger, J. Sediment Transport Modeling Review-Current and Future Development. *ASCE J. Hydraul. Eng.* **2008**, *134*, 1–14. [CrossRef]
35. Zhang, Y.; Jia, Y. Hybrid mesh generation using advancing reduction technique. *Int. J. Numer. Methods Fluids* **2015**, *79*, 514–535. [CrossRef]
36. Zhang, Y.; Jia, Y.; Zhu, T.T. Edge Gradients Evaluation in 2D hybrid FVM Model. *J. Hydraul. Res.* **2015**, *53*, 423–439. [CrossRef]
37. Julien, P.Y. *Erosion and Deposition*; Cambridge University Press: Cambridge, UK, 1995.
38. Cheng, N.S. Simplified settling velocity formula for sediment particle. *J. Hydraulic Eng. ASCE* **1997**, *123*, 149–152. [CrossRef]
39. Einstein, H.A. *The Bed-Load Function for Sediment Transportation in Open Channel Flows*; Technical Bulletin No. 1026; U.S. Department of Agriculture, Soil Conservation Service: Washington, DC, USA, 1950.
40. Bagnold, R.A. *An Approach to the Sediment Transport Problem from General Physics*; Professional Paper 422-J; USGS: Washington, DC, USA, 1966.
41. Wu, W.; Wang, S.S.Y.; Jia, Y. Nonuniform sediment transport in alluvial rivers. *J. Hydr. Res. IAHR* **2000**, *38*, 427–434. [CrossRef]
42. Wu, W.; Altinakar, M.; Wang, S.S.Y. Depth-average analysis of hysteresis between flow and sediment transport under unsteady conditions. *Int. J. Sediment Res.* **2006**, *21*, 101–112.
43. Van Rijn, L.C. Sediment transport, part I: Bed-load transport. *J. Hydraulic Eng. ASCE* **1984**, *110*, 1431–1456. [CrossRef]
44. Meyer-Peter, E.; Mueller, R. Formulas for bed-load transport. In *Report on Second Meeting of IAHR*; IAHR: Stockholm, Sweden, 1948; pp. 39–64.
45. Engelund, F.; Hansen, E. *A Monograph on Sediment Transport in Alluvial Streams*; Teknisk Vorlag: Copenhagen, Denmark, 1967.
46. Yalin, M.S. *Mechanics of Sediment Transport*; Pergamon Press: Oxford, UK, 1972.
47. Galappatti, G.; Vreugdenhil, C.B. A depth-integrated model for suspended sediment transport. *J. Hydr. Res. IAHR* **1985**, *23*, 359–377. [CrossRef]
48. Armanini, A.; di Silvio, G. Discussion on the paper 'A depth-integrated model for suspended sediment transport' by G. Galappatti and C.B. Vreugdenhil. *J. Hydr. Res.* **1986**, *24*, 437–441. [CrossRef]
49. Karim, F.; Kennedy, J.F. *IALLUVIAL: A Computer-Based Flow- and Sediment- Routing Model for Alluvial Streams and Its Application to the Missouri River*; Technical Rep. No. 250; IIHR, Univ. of Iowa: Iowa City, IA, USA, 1982.
50. Rhie, C.M.; Chow, W.L. Numerical study of the turbulent flow passed an airfoil with trailing edge separation. *AIAA J.* **1983**, *21*, 1525–1532. [CrossRef]
51. Zhang, Y.; Jia, Y. Multi-point momentum interpolation correction on collocated meshes. *J. Comput. Phys.* **2021**, *449*, 110783. [CrossRef]
52. Sleijpen, G.; Fokkema, D. BICGSTAB(L) For Linear Equations Involving Unsymmetric Matrices with Complex Spectrum. *Electron. Trans. Numer. Anal.* **1993**, *1*, 11–32.
53. Seal, R.; Paola, C.; Parker, G.; Southard, J.B.; Wilcok, P.R. Experiments on downstream fining of gravel: I. Narrow-channel Runs. *ASCE J. Hydraul. Eng.* **1997**, *123*, 874–884. [CrossRef]
54. Meade, R.H.; Myrick, R.M.; Emmett, W.W. Field data describing the movement and storage of sediment in the East Fork River. In *Wyoming, Part II Bed Elevations, 1979*; USGS Open-File Rep.; USGS: Denver, CO, USA, 1980; pp. 80–1190.
55. Zhang, Y. *CCHE-MESH 2D Structured and Unstructured Mesh Generator—Users' Manual version 5.x*; NCCHE-TR-2017-02; The University of Mississippi: Oxford, MS, USA, April 2017.

applied sciences

MDPI

Article

Numerical Study of the Flow and Blockage Ratio of Cylindrical Pier Local Scour

Mario Hurtado-Herrera [1,*], Wei Zhang [2], Abdelkader Hammouti [1,3], Damien Pham Van Bang [1,3] and Kim Dan Nguyen [4]

[1] Laboratory for Hydraulics and Environment, INRS-ETE, Québec City, QC G1K 9A9, Canada; abdelkader.hammouti@inrs.ca (A.H.)
[2] Department of Civil Engineering, Xi'an Jiaotong-Liverpool University, Suzhou 215123, China
[3] Department of Civil and Environmental Engineering, ETS, Université du Quebec, Montreal, QC H3C 1K3, Canada
[4] Laboratory for Hydraulics Saint Venant, Ecole des Ponts, EDF-CEREMA, 78400 Chatou, France
* Correspondence: mario_roberto.hurtado_herrera@inrs.ca

Abstract: A three-dimensional large eddy simulation model is used to simulate the turbulent flow dynamics around a circular pier in live-bed and clear-water scour conditions. The Navier–Stokes equations are transformed into a σ-coordinate system and solved using a second-order unstructured triangular finite-volume method. We simulate the bed evolution by solving the Exner-Polya equation assisted by a sand-slide model as a correction method. The bedload transport rate is based on the model of Engelund and Fredsœ. The model was validated for live-bed conditions in a wide channel and clear-water conditions in a narrow channel against the experimental data found in the literature. The in-house model NSMP3D can successfully produce both the live-bed and clear-water scouring throughout a stable long-term simulation. The flow model was used to study the effects of the blockage ratio in the flow near the pier in clear-water conditions, particularly the contraction effect at the zone where the scour hole starts to form. The scour depth in the clear water simulations is generally deeper than the live-bed simulations. In clear-water, the results show that the present model is able to qualitatively and quantitatively capture the hydrodynamic and morphodynamic processes near the bed. In comparison to the wide channel situation, the simulations indicate that the scour rate is faster in the narrow channel case.

Keywords: 3D flow simulation; clear-water flow; live-bed flow; horseshoe vortex; local scour; narrow channel

Citation: Hurtado-Herrera, M.; Zhang, W.; Hammouti, A.; Pham Van Bang, D.; Nguyen, K.D. Numerical Study of the Flow and Blockage Ratio of Cylindrical Pier Local Scour. *Appl. Sci.* **2023**, *13*, 11501. https://doi.org/10.3390/app132011501

Academic Editors: Manousos Valyrakis, Gordon Gilja, Thomas Pahtz, Panagiotis Michalis and Oral Yagci

Received: 25 July 2023
Revised: 9 October 2023
Accepted: 12 October 2023
Published: 20 October 2023

1. Introduction

Scour erosion refers to an increase in local sediment transport at the base of a submerged structure due to the vortices generated by fluid–structure interactions. According to Raudkivi and Ettema, 1983 [1], scour in rivers can be classified into three categories:

(i) General scour, which occurs irrespective of the presence of an obstacle.
(ii) Contraction scour, which occurs as a result of constriction due to lateral groins, spur dikes, or bridge abutments.
(iii) Local scour, which is only observed at the base of the obstacle and does not lead to sediment transport far from the pier.

Raudkivi and Ettema, 1983 [1], distinguish the first and last types as live-bed scour and clear-water scour, respectively. The shear stress exerted by the approaching flow determines the type of scour at the base of a structure. Clear-water scour occurs as a result of inflow shear stresses which are less than or equal to the threshold for sediment motion, whereas live-bed scour occurs for inflow shear stresses that are greater than this threshold (Melville et al., 1999 [2]). In the literature, this threshold between clear-water and live-bed conditions has been described in a number of ways. Ettema et al., 2017 [3], expressed it as the mean velocity needed for the entrainment of the bed, while Roulund et al., 2005 [4], expressed it as a critical value for the Shields number associated with the friction velocity, u_{τ}.

Examples of clear-water conditions can be found in nature, such as at the Tahrir and Imbaba bridges in Cairo, Egypt (Stevens et al., 1991 [5]), and also in laboratory studies (Williams et al., 2018 [6]). Scouring at the base of bridge piers has been found to be the third global cause of bridge collapse (Proske, 2018 [7]) after impact and flooding. However, the leading causes of bridge collapse vary significantly in dependence with the geographical situation, construction and inspection standards (Arneson et al., 2012 [8]). As a result, some regional studies have found that bridge scouring is responsible for between 50% and 64% of bridge failures [9–12]. This problem highlights the importance of studying the scouring process; hence, there is interest in numerical models capable of accurately modeling this case.

The characteristics of the flow around a cylindrical pier are well known, and the turbulent structures present have been extensively described in the past [4,13–15]. Upstream from the cylinder, the main structure responsible for scouring is a horseshoe vortex (HSV) system. This is also the case downstream in live-bed conditions, but in clear-water conditions, the HSV competes against wake vortices (WV) in the deposition and scouring processes (Lachaussée et al., 2018 [16]). However, despite the many existing studies on pier scouring, our current understanding of this process remains incomplete. A recent study by Lai et al., 2022 [17], pointed out that the gap between the numerical predictions of equilibrium scouring and experimental results can be traced to the limitations of the current sediment transport theory. As Lachaussée, 2018 [18], observed, there is a noteworthy disparity between studies of live-bed scouring [4,19–25] and clear-water scouring [26–29], the latter being much less studied.

There are many empirical and semi-empirical formulae for predicting bedload transport available in the literature (VanRijn, 1993 [30], but their predictions differ in certain regimes. Most sediment models need a sand-slide model to control the angles of scouring and deposition (Zhang, 2019 [31]). In clear-water conditions, the majority of models consider the bed load [27–29], although the suspended load has been considered for flow velocities in the upper spectrum (Olsen and Melaaen, 1993 [26]. The usual approaches for turbulence modeling include RANS [26,27] and URANS (Khosronejad et al., 2012 [28]) approaches. The scour equilibrium depth at the upstream face of the cylinder is frequently reported in studies, while the dune downstream and the kinematics of scour hole evolution are rarely described (Lai et al., 2022 [17]). Furthermore, while a few studies have been performed in computational domains that could be considered narrow channels [26,28], the effects of the channel width on sediment transport tend to be dismissed. This is because the traditional approach seeks to emulate an open channel by using virtual lateral walls boundaries.

In a fluid–structure problem tested between two lateral walls, the term "narrow channel" is used when the lateral walls affect the flow pattern in the vicinity of the structure and at its wake. A number of conditions will determine whether the wall effect is negligible or not. Chen et al., 1995 [32], observed a delay in the transition to vortex shedding, compared with the "free-stream" case, for blockage ratios (i.e., the ratio between the diameter D of the obstacle and the width W of the channel) $D/W \geq 0.1$. Furthermore, Singha and Sinhamahapatra, 2010 [33] describe three factors in which the presence of the side walls will alter the flow structure: (i) the incoming flow will be sheared, (ii) the flow through the gap directly above the obstacle will be higher than the "free-stream" case to maintain the mass continuity, and (iii) the overall vortex structure will be altered due to the interaction between the wall vortices and the vortices from the obstacle. Regarding the effect of the blockage ratio over the scour dynamics, Mignot et al., 2015 [34], state that the blockage ratio will not influence the scouring process when the obstacle is separated from the side walls by a fully developed flow with a supercritical Froude number. Lachaussée, 2018 [18], noted that the critical Shields number for the presence of horseshoe vortex-induced scour will also deviate from the "free-stream" case for blockage ratios $D/W \geq 0.1$.

In the present study, we propose a numerical investigation of the combined effects of the velocity and the blockage ratio over the scouring process around a vertical cylinder. The numerical model uses a second-order finite-volume method in σ-coordinates, with a Large Eddy Simulation (LES) method to model the turbulence near the sediment bed and the cylinder. This model is used alongside the Exner–Polya equation with the bedload function of Engelund and Fredsøe, 1976 [35], similar to the one proposed by Zhang et al., 2021 [19]. The narrow channel is emulated using a large blockage ratio and non-slip lateral boundaries, and the wide channel is emulated by a small blockage ratio alongside slip lateral boundaries. Table 1 highlights the position of the present study by presenting an overview of different configurations used to simulate the scour process in the literature. To the best of our knowledge, the present study presents a novel study of the effect of the blockage ratio and the flow velocity over the scour evolution at the base of a circular pier.

The present study is structured as follows. Section 2 covers the mathematical model, including the fluid and sediment models. Section 3 focuses on the numerical methods implemented for the solution of the mathematical model. Section 4 first reproduces the scour depth evolution of a live-bed experiment in a wide channel in Roulund et al., 2005 [4]. Then, the model is validated with a clear-water case in Lachaussée, 2018 [18]. Finally, simulations with the same flow conditions of the clear-water case in two channels with different blockage ratios are presented to assess the effect of the flow confinement in the scour evolution. Section 5 gives a discussion of the numerical results and the conclusions of the work, as well as some perspectives on the combined effects of the velocity, blockage, and relative coarseness, or the shape of the obstacle, over the scour depth.

Table 1. Comparison of numerical techniques used in clear-water scour simulation in the literature.

Reference	Numerical Method	Turbulence Model	Sediment Load Model	Re_D	D/W	Lateral Wall B. C.
Olsen et al., 1993 [26]	FVM	RANS ($k-\varepsilon$)	Suspended load (Van Rijn)	5.02×10^4	0.2	Slip
Zhao et al., 2010 [27]	FEM	RANS ($k-\omega$)	Bedload (Engelund and Fredsøe) and suspended load (convection-diffusion)	3.9×10^4	0.05	Slip
Khosronejad et al., 2012 [28]	UFVM	URANS	Bedload (Van Rijn)	2.9–6.3×10^4	0.14	Slip
Link et al., 2012 [29]	FVM	DES	Bedload (Lagrangian model)	3.25×10^4	0.07	Slip
Bento et al., 2023 [36]	FDM	RANS ($k-\varepsilon$)	Suspended load	3.5×10^4	0.027–0.055	Non-slip
Present study	UFVM	LES	Bedload (Engelund and Fredsøe)	1.9×10^3	0.1	Non-slip

2. Mathematical Model

To investigate the evolution processes of local scour around the cylinder in a rectangular channel, the flow model solves the Navier–Stokes (N-S) equations using a large eddy simulation (LES). In the present study, the free surface is neglected because a rigid roof is used to isolate the effects of the wake vortices on the scouring process (Lachaussée et al., 2018 [16]). In addition, the model computes near-bed transport separately using an Exner equation, in which the bedload transport is calculated with the formula given by Engelund and Fredsøe, 1976 [35]. The computational domain is illustrated by Figure 1.

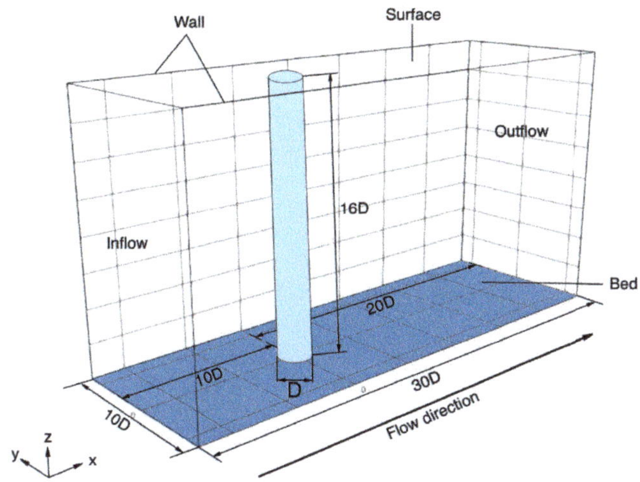

Figure 1. Sketch of the computational domain used for the numerical simulations (narrow channel case).

2.1. Hydrodynamic Model

For the flow, the governing equations employed for LES are obtained by filtering the Navier–Stokes equations (Equations (1) and (2)).

$$\frac{\partial \overline{u}_i}{\partial t} + \frac{\partial \overline{u}_i \overline{u}_j}{\partial x_j} = -\frac{1}{\rho}\frac{\partial \overline{p}}{\partial x_i} + \frac{\partial}{\partial x_j}\left(\nu \frac{\partial \overline{u}_i}{\partial x_j}\right) - \frac{\partial \tau_{ij}}{\partial x_j} + f_i, \quad (1)$$

$$\frac{\partial \overline{u}_i}{\partial x_i} = 0, \quad (2)$$

where $\overline{u}_i$ is the i-th filtered velocity component in Cartesian coordinates, $\overline{p}$ is the filtered pressure, ρ is the density, and ν is the kinematic viscosity. τ_{ij} is the sub-grid stress (SGS) tensor required for the LES formulation, defined in Equation (3).

$$\tau_{ij} = 2\nu_t \overline{S}_{ij} + \frac{1}{3}\tau_{jj}\delta_{ij}, \qquad \overline{S}_{ij} = \frac{1}{2}\left(\frac{\partial \bar{u}_i}{\partial x_j} + \frac{\partial \bar{u}_j}{\partial x_i}\right) \quad (3)$$

where ν_t is the turbulent viscosity at the subgrid scale, $\overline{S}_{ij}$ is the rate-of strain tensor at the resolved scale, and δ_{ij} is the Kronecker delta. The eddy viscosity is modeled using Equation (4) using the Smagorinsky model.

$$\nu_t = L_s^2 |\bar{S}|, \qquad |\bar{S}| = \sqrt{2\bar{S}_{ij}\bar{S}_{ij}}, \quad (4)$$

where L_s is the sub-grid length scale. This length scale cannot be described with a constant value when it is located in the near-wall region; instead, it decreases as it approaches the wall. This effect is captured by the Function (5).

$$\frac{1}{L_s^n} = \frac{1}{(C_s\Delta)^n} + \frac{1}{(\kappa L_w)^n}, \quad (5)$$

where $n = 2$ is the Mason wall matching power (Mason and Thomson, 1992 [37]), $C_s = 0.1$ is the Smagorinsky constant, $\Delta = V^{1/3}$ s the average spacing (where V is the volume of the cell,) $\kappa = 0.4$ is the Von Kármán constant, and L_w is the distance from the cell center of a control volume to the wall.

The present study employs the diameter of the pier (D) and the mean velocity (U) at the inflow boundary as reference scales for the length and velocity, respectively. Using these length and velocity scales, we then define the time scale $T = D/U$. By setting these reference scales, the governing equations become in terms of the pier Reynolds number, defined in Equation (6), which is based on the cylinder diameter, the mean flow velocity, and kinematic viscosity ($\nu = 1 \times 10^{-6}\,\mathrm{m}^2/\mathrm{s}$).

$$\mathrm{Re}_D = \frac{UD}{\nu}. \tag{6}$$

2.2. Morphodynamic Model

The present study analyzes sediment transport in the clear-water and live-bed regimes. Despite this, the Rouse number (W_s/u_τ, where W_s is the settling velocity) is not high enough for the suspended load to be significant; therefore, only the bedload formulation is considered. Thus, the morphological evolution of the sediment bed is described using Equation (7), also known as the Exner–Polya equation.

$$(1-\eta)\frac{\partial z_b}{\partial t} + \frac{\partial q_{b_i}}{\partial x_i} = 0, \tag{7}$$

where $\eta = 0.4$ is the porosity of the bed, z_b is the bed elevation, and q_{b_i} ($i = 1,2$) is the bedload transport rate given by Equation (8), proposed by Engelund and Fredøe, 1976 [35].

$$q_{b_i} = \frac{\pi d}{6} p_{EF} u_{bi}, \tag{8}$$

where d is the particle diameter, p_{EF} is the percentage of particles in motion modeled, and u_{b_i} ($i = 1,2$) are the mean velocity components of a sediment particle in movement. The value p_{EF} is given by Equation (9).

$$p_{EF} = \begin{cases} \left[1 + \left(\dfrac{\pi \mu_d}{6(\theta - \theta_c)}\right)^4\right]^{-\frac{1}{4}} & , \quad \theta > \theta_c \\ 0 & , \quad \theta \le \theta_c. \end{cases} \tag{9}$$

Here, μ_d is the dynamic friction coefficient, θ is the Shields number, and θ_c is its critical value for the initiation of motion. The mean velocity is given by the Equation (10).

$$u_{b_i} = a\left(1 - 0.7\sqrt{\frac{\theta_c}{\theta}}\right)u_{\tau_i}, \tag{10}$$

where a is an empirical constant and $\vec{u}_\tau = (u_{\tau_1}, u_{\tau_2})$ is the friction velocity. As implemented in Zhang et al., 2020 [20], θ_c is modelled using Equation (11).

$$\theta_c = \theta_{c_0}\left(\cos\beta\sqrt{1 - \frac{\sin^2\alpha\tan^2\beta}{\mu_s^2}} - \frac{\cos\alpha\sin\beta}{\mu_s}\right), \tag{11}$$

where θ_{c_0} is the critical Shields number for the initiation of motion in a bed with slope zero, μ_s is the static friction coefficient, β is the maximum angle of repose, and α is the difference between the angle of the flow near the bed and the maximum angle of repose β. The friction velocity is calculated by solving Equation (12), as proposed by Nikuradse, 1950 [38]:

$$\frac{u}{u_\tau} = \kappa^{-1}\ln\left(\frac{z}{z_0}\right), \tag{12}$$

where z_0 is governed by the friction Reynolds number $\mathrm{Re}_\tau = u_\tau k_s/\nu$, with k_s being the so called Nikuradse sand roughness. Based on the value of Re_τ, three hydraulic regimes are defined to compute z_0. The flow is hydraulically smooth for $\mathrm{Re}_\tau \leq 5$, and rough for $\mathrm{Re}_\tau \geq 70$. Soulsby, 1997 [39], developed a transitional formula in Equation (13) to link both regimes:

$$z_0 = \begin{cases} \dfrac{\nu}{9\,u_\tau} & ,\text{for } Re_\tau \leq 5 \\[2ex] \dfrac{\nu}{9\,u_\tau} + \dfrac{k_s}{30}\left[1 - \exp\left(-\dfrac{u_\tau k_s}{27\,\nu}\right)\right] & ,\text{for } 5 < Re_\tau < 70 \\[2ex] \dfrac{k_s}{30} & ,\text{for } Re_\tau \geq 70 \end{cases} \tag{13}$$

2.3. Sand-Slide Model for the Sediment Bed

In addition to the Exner–Polya equation, a sand-slide model is implemented to avoid the occurrence of bed slopes which are larger than the physical value of the angle of repose. The present study applies the mass-conservation-based algorithm of Khosronejad, as presented in Zhang et al., 2020 [20]. The bed slope is defined by the elevation gradient between a point and any neighboring horizontal cell center. Thus, if the slope exceeds the maximum angle of repose of the material, then a correction will be made to the bed elevation based on a material angle of repose ϕ. The bed corrections are obtained by mass conservation using the area projection of the cells. This area projection is an appropriate choice because there is no change in the cell area, as the mesh points move only in the vertical direction. Details about this algorithm in a triangular mesh can be found in Zhang et al., 2020 [20].

2.4. Narrow Channel Boundary Conditions

For the narrow channel case (i.e., $D/W \leq 0.1$), the shearing zone near the lateral walls cannot be neglected [32–34]. Therefore, the lateral wall boundaries adopt non-slip conditions, as well as the bottom boundary. The surface is a rigid boundary with slip conditions, and the outflow boundary adopts non-reflective conditions. For the inflow boundary, velocity is defined by a theoretical vertical profile fit for the desired friction velocity given by Equation (14) (Swamee, 1933 [40]).

$$u_+ = \left(z_+^{-1/p} + [\kappa^{-1}\ln(c + 9z_+)]^{-1/p}\right)^{-p}, \tag{14}$$

where $u = u_\tau u_+$, $z = \delta_\nu z_+$, and $\delta_\nu = \nu u_\tau^{-1}$. Here, we consider $c = 1$ and $p = 3/10$. Equivalences $\hat{u} = u_+ \frac{u_\tau}{U}$ and $\hat{z} = z_+ \frac{\delta_\nu}{D}$ are used in Equation (1) due to their dimensionless format.

3. Numerical Method

The numerical model used herein is an evolved version of the program initially developed by Uh Zapata et al., 2014 [41], to study hydrodynamics problems in irregular domains. Versions of the model have been used to study the morphodynamic evolution of sediment [20,42], including the scour around a cylindrical pier in live-bed conditions.

The mobile bed induces variable bathymetry and can lead to discretization problems. In order to avoid these problems, the σ-transformation (defined in the study of Phillips, 1957 [43]) is applied to the system. It consists of the simplification of the geometry of the domain in the vertical direction. This is achieved by transforming the system into the coordinate system defined in Equation (15).

$$t^* = t,\ x^* = x,\ y^* = y,\ \sigma = \frac{z + h}{h}; \tag{15}$$

where $h(x^*, y^*, t^*)$ and $z \in [-h, 0]$ and $\sigma \in [0, 1]$. An in - explanation of the implementation can be found in Uh Zapata et al., 2014 [41]. The velocity field and pressure are decoupled using the projection method initially proposed by Chorin, 1968 [44]. The resulting linear systems are solved by the Successive Over-Relaxation method (SOR). Finally, the whole program operates in parallel using the Message Passing Interface (MPI). Further details can be found in Uh Zapata et al., 2019 [45].

3.1. Configurations of the Computational Domain

As a result of the implementation of the σ-transformation, the research code can only model partially submerged prismatic obstacles. However, as stated in Lachaussée, 2018 [18], the influence of the height of the obstacle over the scouring process is negligible for H/D larger than 3 (H being the height of the pier), which is the case studied herein. Therefore, a test of height $H = 0.09$ m is equivalent to one with $H = 0.16$ m due to its diameter being $D = 0.01$ m.

Figure 1 shows a global view of the computational domain for the narrow channel case in the present study. The configuration and dimensions are the same as those used in the laboratory experiments performed by Lachaussée, 2018 [18]. The numerical domain is a $30D$ long, $10D$ wide, and $16D$ deep cuboid with a cylinder situated at the axes' origin, such that the domain is symmetrical in the y-direction with $10D$ upstream and $20D$ downstream of the cylinder.

Figure 2 shows a two-dimensional view of the mesh used in the simulations. The mesh consists of layers of stacked prismatic elements, which are extended from an unstructured triangular 2D mesh in the horizontal direction. To obtain the turbulent flow around the cylinder, the minimum size of the mesh (triangle length l) is set about $l = 0.015D$ and $l = 0.02D$ in the horizontal and vertical directions, respectively. The largest triangles are close to the walls with an average length of $l = 0.25D$ (zone 1). In the middle region (zone 2), the mesh reaches a maximum of $l = 0.1D$. Note that higher mesh density is used close to the cylinder (zone 3) to capture the vortex structure. The total number of vertex and cell-centered points in the horizontal are 20,123 and 39,721, respectively. For this mesh, we have 205 vertex points around the cylinder. In the vertical direction, the mesh is comprises 96 layers of elements and the height of the cells increases geometrically as they are further from the bed. The first cell center near the bed is set to be situated in the viscous sub-layer ($z_+ < 5$). In the clear-water case, the first cell center close to the bed is at $z_+ = 2.82$, $u_+ = 23.68$ at the cell center closest to the surface.

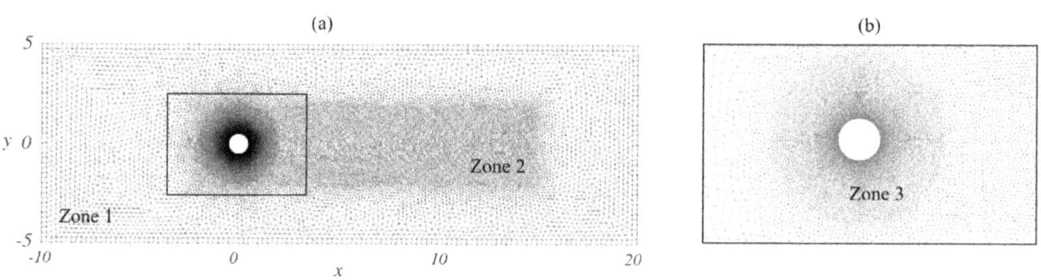

Figure 2. (**a**) Horizontal view of the mesh used for the numerical simulations. (**b**) Detail of the cells near the cylinder.

3.2. Initial and Boundary Conditions

To validate the numerical model in the clear-water case, we set the problem using the PIV measurements by Lachaussée, 2018 [18]. These measurements were performed close to the bed with a vertical resolution of $\Delta z_{PIV} = 8 \times 10^{-4}$ m. To avoid the influence of waves on the scouring process, a plexiglass sheet is put over the water surface to prevent a free surface. Due to the flow conditions, the Froude number $Fr = U/(gh)^2$ is small, so implementing a rigid roof as a surface boundary condition is appropriate. The experimental configuration produces a flow with a mean velocity $U \approx 0.16$ m/s and friction velocity $u_\tau \approx 8 \pm 1$ mm/s. As $\mathrm{Re}_\tau \approx 5.4$ in this case, according to Soulsby's formulation (Equation (13)), the flow approaching the obstacle is in the transitional regime, but very close to being hydraulically smooth. Naturally, as the flow approaches the obstacle and its friction velocity increases, the flow regime is consolidated as transitional.

For the clear-water simulations, we adjust the Swamee profile (14) to the PIV data while confirming that the friction velocity is conserved, and then we apply the profile to the inflow boundary condition. As a result, considering $D = 0.01$ m, the pier Reynolds number in the numerical model is $\mathrm{Re}_D \approx 2000$.

For the live-bed simulations, the Swamee profile (14) is adjusted to the flow velocity and pier size described in Roulund et al., 2005 [4]. The mean velocity in this case is $U = 0.46$ m/s, the pier diameter is $D = 0.1$ m, and, therefore, the pier Reynolds number is $\mathrm{Re}_D = 46{,}000$.

In the present study, the same parameters are used in all the simulations for sediment density, fluid sediment, grain size, sand roughness, bed porosity, critical Shields number for a horizontal slope, and static and dynamic friction coefficients. The parameters used in the tests are summarized in Table 2.

Table 2. Flow and sediment parameters used for all simulations.

Parameter	Symbol	Clear-Water Case	Live-Bed Case
Mean velocity	U	19.23 cm/s	46 cm/s
Reynolds number	Re_D	1923	46,000
Friction velocity	u_τ	0.9 cm/s	1.8 cm/s
Water depth	H	16 cm	40 cm
Cylinder diameter	D	1 cm	10 cm
Sediment density	ρ_s	2500 kg m^{-3}	2500 kg m^{-3}
Fluid density	ρ	1000 kg m^{-3}	1000 kg m^{-3}
Grain size	d	0.27 mm	0.27 mm
Sand roughness	k_s	0.67×10^{-3} m	0.67×10^{-3} m
Friction Reynolds number	Re_τ	6	12
Bed porosity	η	0.4	0.4
Critical Shields number	θ_{c0}	0.04	0.04
Dynamic friction coefficient	μ_d	0.62	0.62
Static friction coefficient	μ_s	0.63	0.63

From the value of θ_{c0}, it follows that the critical friction velocity for the incipient motion of a solid particle on a flat bed is given by Equation (16).

$$u_{\tau crit} = \left[\theta_{c0} g d(\rho_s/\rho - 1)\right]^{1/2} = 1.27 \times 10^{-2}\ \mathrm{m/s}. \tag{16}$$

Therefore, the ratio $u_\tau/u_{\tau crit}$ is equal to 0.72, i.e., clear-water conditions.

4. Numerical Results

4.1. Live-Bed and Wide Channel Case

As stated in Section 2, the in-house model was validated in Zhang et al., 2020 [20] by reproducing the experimental results in Roulund et al., 2005 [4]. The flow and scour conditions are detailed in the live-bed case in Table 2. In the present study, the Swamee profile defined in Equation (14) is implemented, which differs from the Poiseuille profile utilized in Zhang et al., 2020 [20]. Additionally, the blockage ratio for the wide channel in the present study is $D/W = 1/16$, while the blockage ratio in Zhang et al., 2020 [20] is $D/W = 1/30$. However, for slip walls, it is considered that the sensitivity of the flow near the obstacle to D/W is minimal when $D/W < 0.1$ (Chen et al., 1995 [32] and Lachaussée, 2018 [18]). Therefore, the difference in width can be considered negligible.

Figure 3 compares the experimental measurements from Roulund et al., 2005 [4] corresponding to the rough bed case with a cylinder Reynolds number $Re_D = 46{,}000$ with the numerical results of Roulund et al., 2005 [4], and the present study. These results show that the scour depth development over time can still be reproduced after the implementation of the current inflow profile. Furthermore, the current estimation is closer to the experimental data presented originally in Roulund et al., 2005 [4] in terms of scour depth evolution over time.

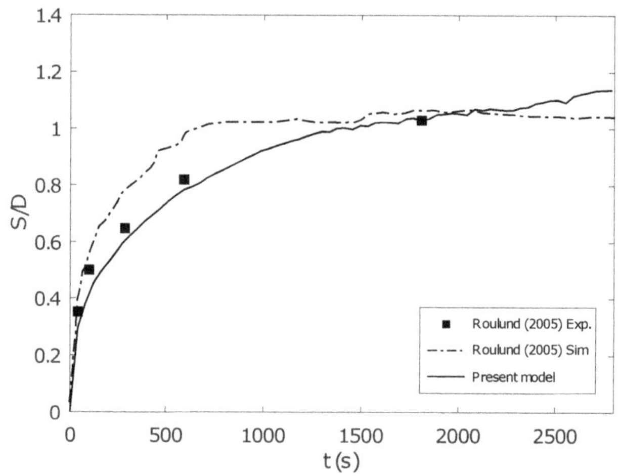

Figure 3. Scour depth evolution upstream of the cylinder for the live-bed case [4].

4.2. Clear-Water Case

This section shows the capability of the proposed numerical model for reproducing the hydrodynamics and morphodynamics behavior around the cylinder in clear-water. The flow results are presented in Figures 4–7, and detailed in the section "Hydrodynamic validation", and the morphological results are presented in Figures 8–14, and detailed in the subsection "Morphodynamic validation". The numerical simulations are performed in two stages to avoid inaccuracies produced by the initial state of the flow. We only compute the hydrodynamic model in the first stage until the flow stabilizes. In the second stage, we calculate both the flow and sediment evolution. In the beginning, the initial condition of the pressure is zero, and the inflow Swamee boundary condition is imposed throughout the domain.

4.2.1. Hydrodynamic Validation

In order to examine the inflow field, the vertical velocity component and the friction velocity are discussed and compared with experimental data in this section. The evolution of the flow is analyzed to find the time needed for the velocity to reach an equilibrium. The study of morphological evolution begins at the conclusion of this warm-up time.

Figure 4 presents the time evolution of the velocity at the first vertical cell and the friction velocity results. Note that both quantities stabilize as expected. Moreover, there are no significant changes after $t/T = 60$. This time agrees with the non-dimensional time necessary for the flow to travel twice the distance from the inlet to the outlet. It is also equivalent to dimensional 3.12 s using a time scale of $T = 0.052$ s. Figure 5 compares the vertical profile of the resolved approaching flow at $x/D = -5$ and $t/T = 100$ with the experimental measurements in Lachaussée, 2018 [18]. This velocity profile is close to the PIV measurements for $z/D < 2$, which is the critical region for scour erosion. In the present study, the warm-up time is set to $t/T = 100$.

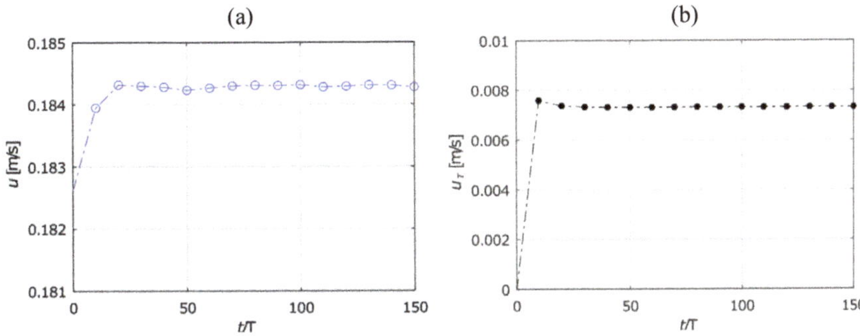

Figure 4. Time analysis of (**a**) the velocity at the first vertical cell above the bed, and (**b**) the friction velocity for a flow with $\mathrm{Re}_D \approx 2000$ at $(x, y) = (-5, 0)$.

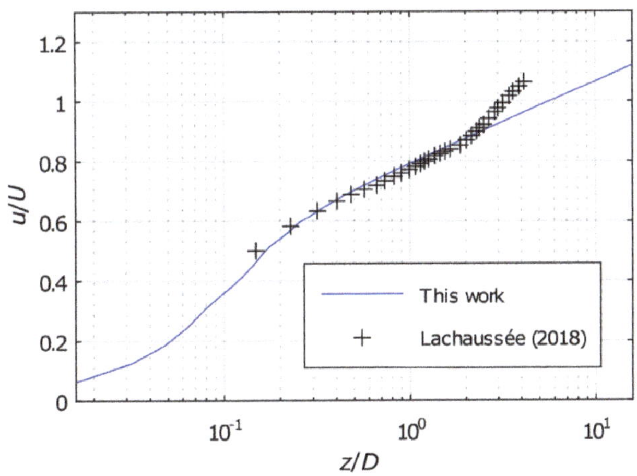

Figure 5. Comparison between the numerical vertical velocity profile at $(x, y) = (-5, 0)$ and the PIV measurements in Lachaussée, 2018 [18] for the flow with $\mathrm{Re}_D \approx 2000$ and $U = 0.1923$ m/s.

Figure 6 displays the three-dimensional instantaneous flow at the non-dimensional time $t/T = 100$, just before the beginning of the second stage. The cylinder is represented by the gray surface and the flow by the black streamlines. Note the complex patterns of the streamlines close to the cylinder. The adverse pressure gradients induced by the cylinder obstruction create downflows in front of the cylinder and generate a horseshoe vortex (HV) near the bottom and wake region behind the cylinder. Figure 6a shows the main coherent structure in an instantaneous flow associated with an HV using the Q criterion. We clearly observed a clockwise vortex close to the bottom in Figure 6b, which also presents 3D streamlines of the wake region behind the cylinder. Figure 6c presents 3D streamlines of the wake region behind the cylinder. We note fluid particles situated close to the bed could be first entrained into the core of this region, and then, from there, toward the surface by an upwelling anti-clockwise vortex. This counter-clockwise rotating vortex would be responsible for the scour mechanisms downstream of the cylinder.

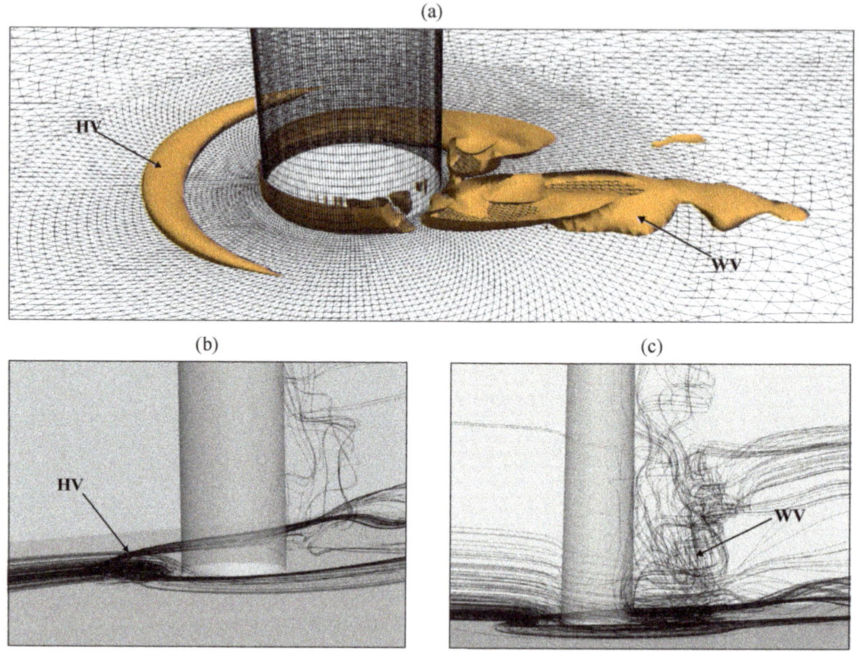

Figure 6. Three-dimensional instantaneous flow visualization at $t/T = 100$: (**a**) Detailed view of the vortex structures at the base of the cylinder. (**b**) Front and (**c**) side view depicting stream lines.

Finally, Figure 7 shows the horizontal profile of the friction velocity amplification over the line $y = 0$ at different time stages. Here, $u_{\tau crit} = 1.27 \times 10^{-2}$ m/s as defined in Equation (16). In this case, the sediment transport starts at $t/T = 100$. Before a dune is formed, the friction velocity slightly surpasses the critical value. It is worth noting that the distance between the inflow boundary and the cylinder is enough to stabilize the flow. Thus, the friction velocity reaches an equilibrium value of 8 ± 1 mm/s, remaining unperturbed as long as it is far enough from the cylinder. This value agrees with the measurements given by Lachaussée, 2018 [18].

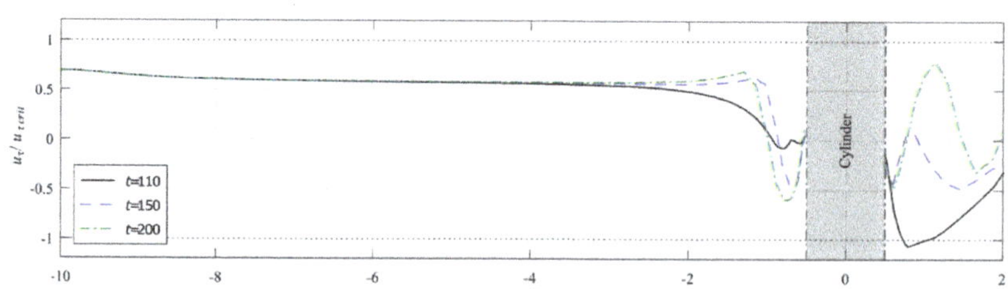

Figure 7. Friction velocity amplification along the x direction with $y = 0$ for $\mathrm{Re}_D \approx 2000$.

4.2.2. Morphodynamic Validation

This section is devoted to the analysis of the scour around the cylinder. The morphological model is tested for $\mathrm{Re}_D \approx 2000$, in the clear-water regime. Any friction velocity greater than 1.27×10^{-2} m/s will result in a Shields number satisfying the condition $\theta > \theta_{c0}$, and, thus, it will produce sediment transport. Figure 8 illustrates the early evolution of the obtained scour hole at non-dimensional morphological times $t/T*$ equal to 100, 200, 400, and 800. At these early stages, the development of the scour hole agrees qualitatively with the one presented for live-bed conditions in Zhang et al., 2021 [19]. $T*$ is the morphological time scale defined by Sumer, 1992 [46] as follows:

$$T* = \frac{D^2}{\sqrt{g(s-1)d^3}}, \tag{17}$$

where d is the grain size and $s = \rho_s/\rho$ is the relative density of the sediment. In this case, $T* = 5.87$.

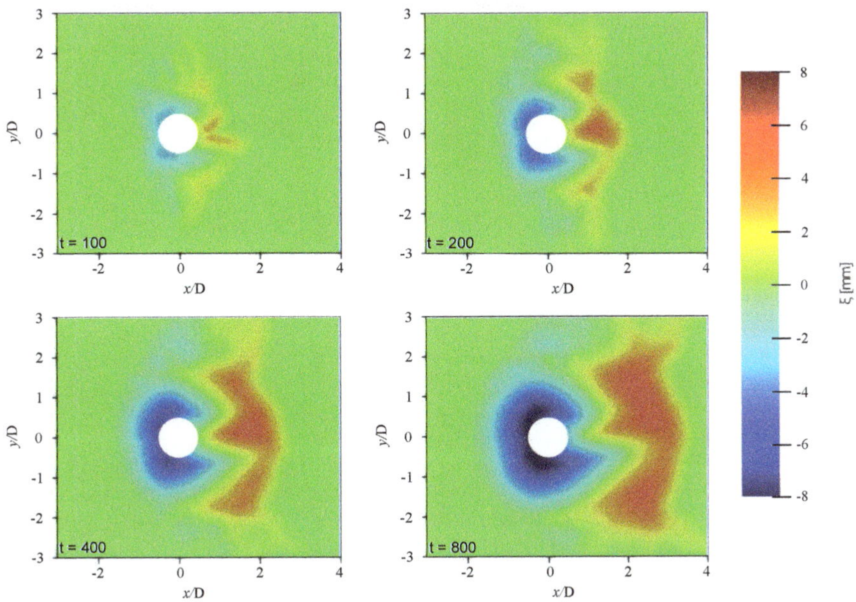

Figure 8. Bed elevation near the cylinder at different moments in time, demonstrating the evolution of the scour hole.

Figure 9 presents the horizontal profiles of the same results shown in Figure 8, offering a more quantitative insight into the scour hole and the dune. The scour hole upstream of the cylinder presents a consistent angle of approximately 32°, which is consistent with the prescribed value for the maximum angle of repose in Lachaussée et al., 2018 [16]. By contrast, the dune downstream of the cylinder does not have a consistent shape throughout its formation. While the lee-side angle of the dune is unchanged once it reaches the same slope as the scour hole, the stoss side of the dune presents more inconsistency. This last behavior might result from the wake vortices interacting directly on the stoss side of the dune.

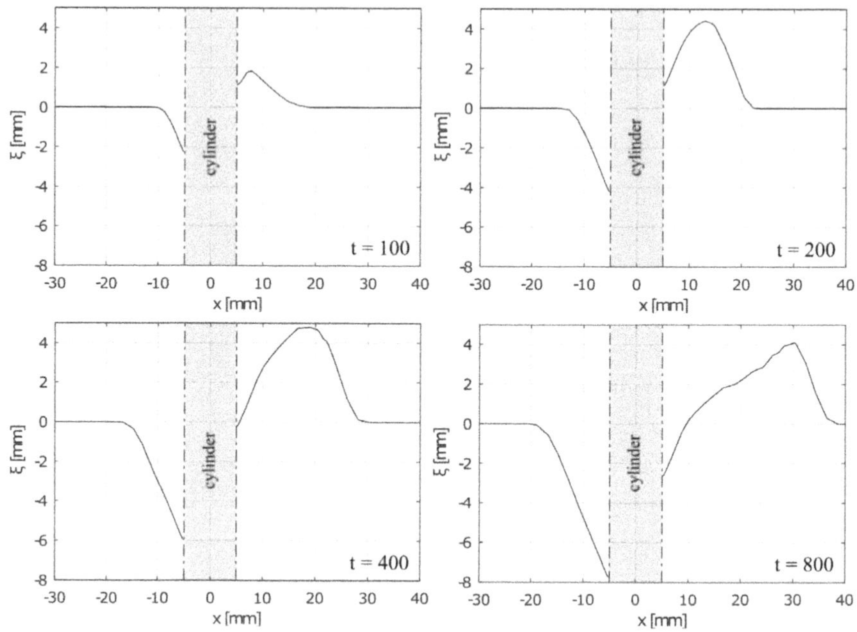

Figure 9. Horizontal profile of the bed elevation at $y = 0$, corresponding to the same states shown in Figure 8.

Additionally, the presence of a horseshoe vortex in the studied range of Re_D is in agreement with Baker, 1979 [47]. The presence of scour induced by the horseshoe vortex (herein referred to as "horseshoe scour") in clear-water regime agrees with Lachaussée et al., 2018 [16]. Three measures are compared with their data: the maximum depth ξ_m and the characteristic lengths L_x and L_y, illustrated in Figure 10. L_x is defined as the distance between the pier and the point of the bed where its depth is $\xi_m/2$. L_y is also defined as the distance between the pier and the point of the bed where its depth is $\xi_m/2$.

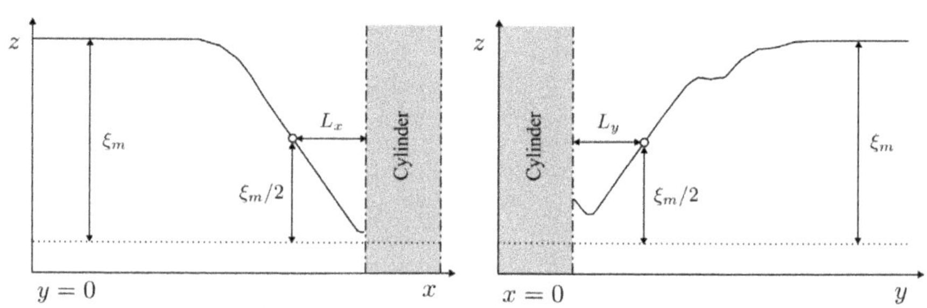

Figure 10. Characteristic lengths of the scour hole.

The time evolution of ξ_m, L_x, and L_y can be fit to the exponential law

$$f(t) = A\left(1 - B\exp\left(-\frac{t}{T_a}\right) - (1-B)\exp\left(-\frac{t}{10T_a}\right)\right), \tag{18}$$

where A, B, and T_a are empirical parameters. This is illustrated in Figure 11a, where the square, triangle, and circle markers are the computed maximum depth, and the upstream and lateral characteristic lengths, respectively. The solid and dashed lines are the adjusted Equation (18) to fit the evolution of ξ_m, L_x, and L_y.

Figure 11 presents a comparison between the numerical and experimental results with the model Equation (18). Figure 11a illustrates the fitting of the Equation (18) to the numerical data. The values for the empirical parameters are $A = -10.3$, $B = 0.6$, and $T_a = 1000$ for the time evolution of ξ_m; $A = 14.5$, $B = 0.42$, and $T_a = 1000$ for the time evolution of L_x, and $A = 15$, $B = 0.52$, and $T_a = 1000$ for the time evolution of L_y.

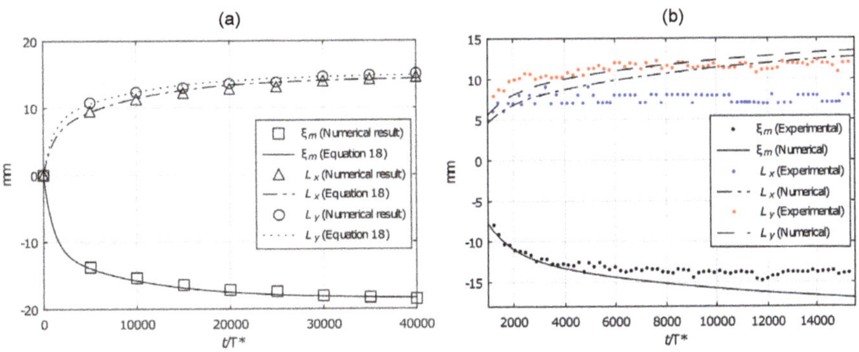

Figure 11. (**a**) Time evolution of ξ_m, L_x, and L_y. (**b**) Comparison between the numerical and experimental data.

As illustrated by Figure 11b, initially, the scour evolution in the numerical results fits well with the experimental results of Lachaussée, 2018 [18], but the model eventually over-predicts the maximum scour depth. With regard to the characteristic lengths, the numerical results present a better fit for L_y than L_x. This is because the numerical results present a rounder scour hole than the one in the experiment. As a comment, we note that the maximum depth of the predicted scour hole is often situated in a region that, in the experimental case, was obscured by the pier, and thus could not be measured.

Figure 12 shows a comparison between the numerical results of the present study and the experimental results with equivalent flow and sediment conditions reported in the thesis of Lachaussée, 2018 [18]. Both the scour hole and the dune have a similar magnitude and angle, with some discrepancy around the limits of the scour hole and downstream of the dune. The numerical and experimental data in the figure both correspond to an instant in which the scour hole has not yet reached an equilibrium.

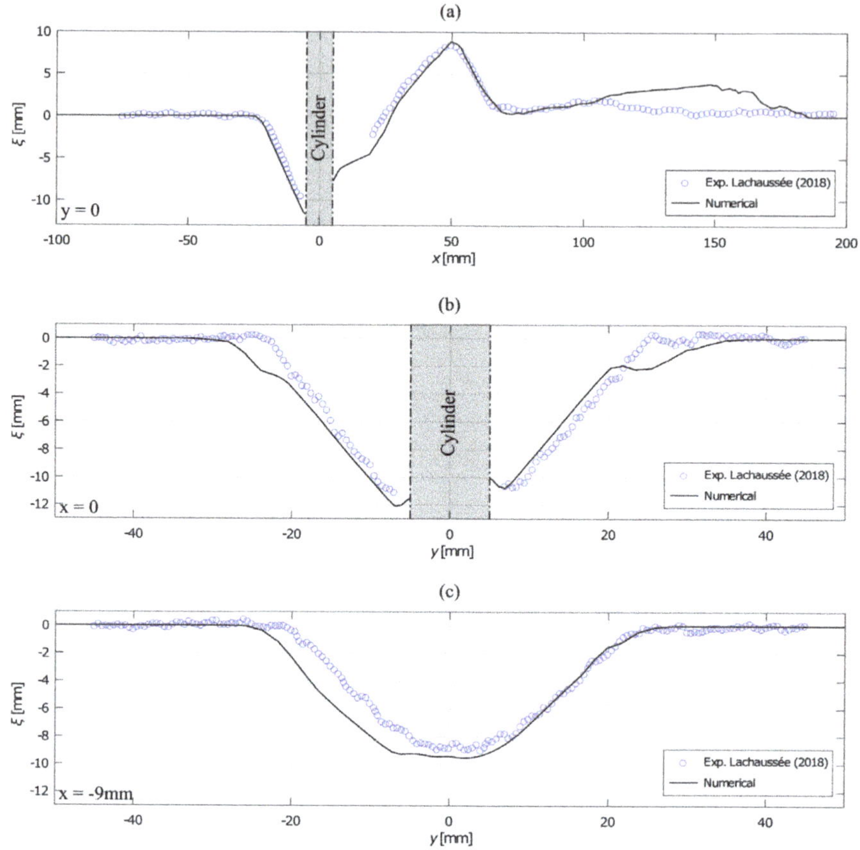

Figure 12. Comparison between the numerical results at a non-dimensional time $t/T = 3500$ and the experimental results reported in Lachaussée, 2018 [18] at time $t = 2.4$ h. The horizontal profiles were extracted along the lines $y = 0$ (**a**), $x = 0$ (**b**), and $x = -9$ mm (**c**).

Figure 13 shows a topographical comparison at the same time as Figure 12. The shape and magnitude of the predicted scour hole is very similar to that of the experimental results. However, while the position and magnitude of the dune in the numerical results agree with the experimental ones, the width is different. A shallow scour hole is present at the lee-side of the dune, which is in agreement with the experimental results, although the one predicted is larger.

This second scour hole seems to be a result of the non-slip conditions at the lateral walls, as these conditions model the physical walls of a narrow channel. This scour hole is an example of wake scour, as the wake of the cylinder is trapped inside the channel, producing scour at the lee-side of the dune. A 3D view is shown in Figure 14.

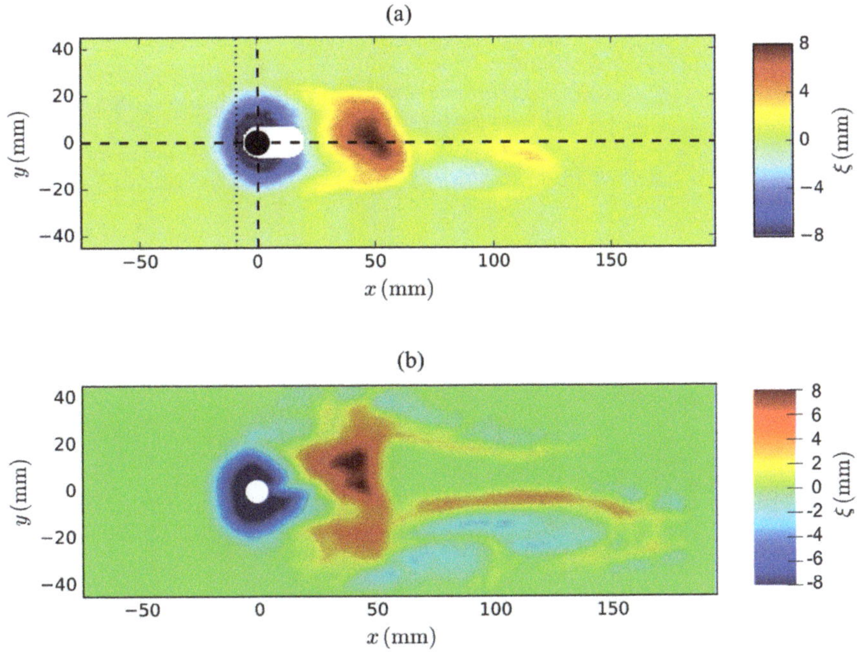

Figure 13. (**a**) Topographical map of the experimental results reported in Lachaussée, 2018 [18] at time $t = 2.4$ h. (**b**) Topographical map of the state of the sediment bed for the numerical results at non-dimensional time $t/T = 3500$.

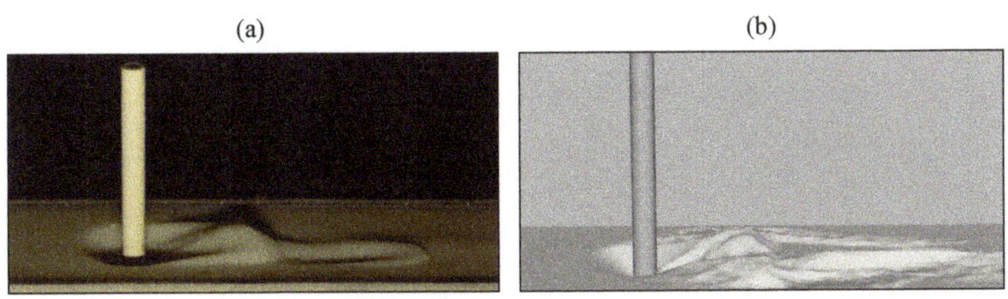

Figure 14. (**a**) Snapshot of the scour in development at time $t = 2.4$ h (Lachaussée, 2018 [18]). (**b**) Scour hole at non-dimensional time $t/T = 3500$.

4.3. Effects of the Channel Width on the Clear-Water Local Scour Pattern

In a free-stream, the streamlines lateral to an obstacle will contract, resulting in flow acceleration. If the flow acceleration results in a supercritical Shields number, the sediment particles will be set in motion, resulting in scouring. In a narrow channel, the flow is confined between two lateral walls, and this confinement will accentuate the contraction of the streamlines, further accelerating the flow past the obstacle. Under this contraction effect, the dune will be pushed downstream. In clear-water conditions, the region of threshold overshoot has a larger expansion than in live-bed conditions. In this section, the narrow channel case has a blockage ratio of $D/W = 0.1$, and the wide channel has a blockage ratio $D/W = 0.0625$.

Figure 15 shows the friction velocity amplification $u_\tau/u_{\tau crit}$, where $u_{\tau crit}$ is as defined in Equation (16). The figures show the amplification effect of the friction velocity at the sides of the cylinder. Roulund et al., 2005 [4], describes this amplification as being caused by the contraction of the flow streamlines and the horseshoe vortex. The isolines in Figure 15a show that the largest amplification is located at an angle within the interval $43^\circ \leq \varphi \leq 75^\circ$. This region is approximately 45 percent wider than the equivalent region in the wide channel case, which is located at $48^\circ \leq \varphi \leq 70^\circ$ (Figure 15b). In the narrow channel case, the confinement of the flow results in a larger contracted area relative to the wide channel case, which will result in a larger sediment transport rate at the start of the development.

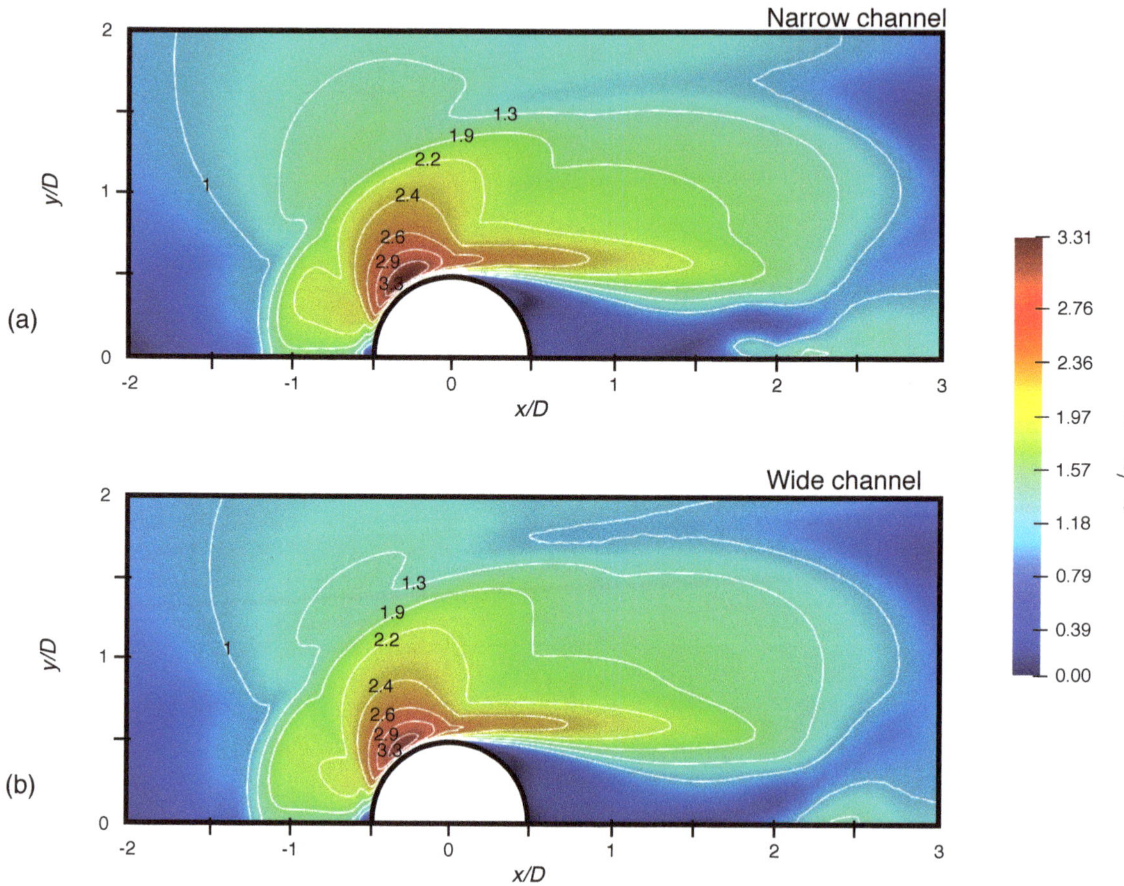

Figure 15. Friction velocity amplification distribution for $\mathrm{Re}_D = 2000$ in a rigid bed. (**a**) Narrow channel case ($D/W = 0.1$). (**b**) Wide channel case ($D/W = 6.25 \times 10^{-2}$).

Figure 16 compares the resulting bed topography of the clear-water simulations. It can be seen that both scour holes and dunes present similar characteristics. The horseshoe scour hole is circular without reaching the wake-side of the cylinder. The maximum scour depth is present in both cases at the lee-side of the cylinder, lateral to the stagnation point, at $|y/D| \approx 0.25D$. The scour hole in the narrow channel is deeper than the scour hole in the wide channel by approximately 15%.

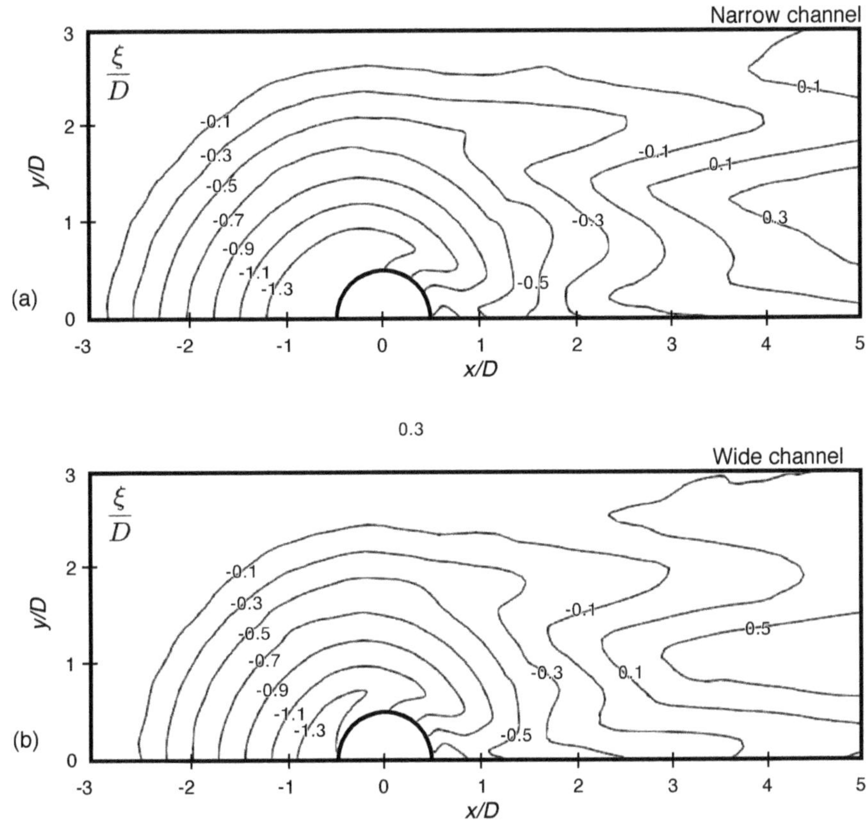

Figure 16. Comparison of the bed topography at equilibrium condition in clear-water conditions in the narrow channel (**a**) and the wide channel (**b**).

Figure 17 compares the time evolution of the normalized scour depth for the three scour simulations performed for the present study. Regarding clear-water, it can be observed that, in both cases, the scour depth evolution is largely monotonically increasing. This continuously growing scour behaviour, although not perfectly monotonic in lab studies, has been documented in experimental studies on clear-water scour, such as Lachaussée, 2018 [18], Kothyari et al., 2007 [48], and Melville et al., 1999 [2]. Moreover, the scour rate decreases over time in both cases. From the the beginning of the scour simulation, both cases presented a similar scour evolution, but the scour rate decreases more rapidly for the wide channel case. This initial similar behavior can be explained by the similar magnitude reached by the friction velocity in the rigid bed simulation (Figure 16). The area where the friction velocity overshoots its critical value is only slightly larger in the narrow channel case, so the faster scour rate does not become apparent in the early time of the simulation. As a result, the narrow channel effect results in a faster scour process and a deeper horseshoe scour hole in the clear-water test, as shown in Figure 16.

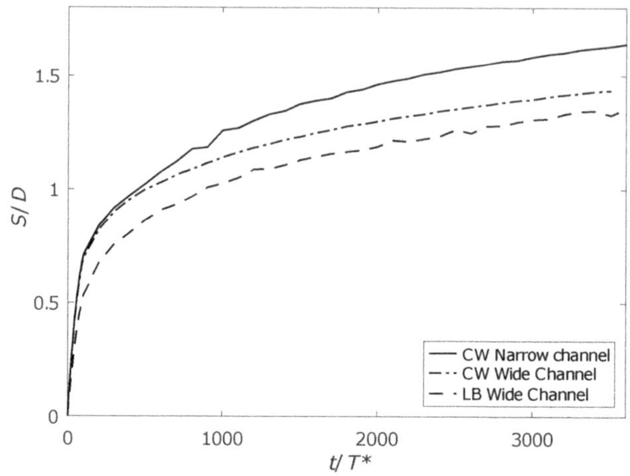

Figure 17. Scour depth evolution upstream of the cylinder for the clear-water (CW) and live-bed (LB) cases.

It is important to note that the live-bed tests were based on the configuration in Roulund et al., 2005 [4], while the clear-water conditions are based on the configuration in Lachaussée, 2018 [18]. A notable difference is the pier diameter, which affects the relative coarseness (D/d) of the model. In the Roulund configuration, $D/d \approx 370$, while in the Lachaussée case $D/d \approx 37$. Several studies have pointed out the influence of the relative coarseness on the effect of the blockage ratio over the scouring process. Tejada, 2014 [49], states that, for $D/d < 100$, the blockage ratio has minimal influence over the scour process. However, an experimental study by Williams et al., 2018 [6], was performed with a relative coarseness $D/d = 81$ and it was found that the larger blockage ratio produced a larger scour depth. This suggests that the relationship between the relative coarseness and the blockage ratio may be more complex than previously thought.

The effect of the roughness length z_0 over the scouring process is a similar consideration to the relative coarseness effect. The behavior of z_0, which directly influences the friction velocity, is directly influenced by the hydraulic regime of the flow (Equation (13)). It can be seen in Table 2 that the clear-water case is located in the transitional hydraulic regime, but very close to the smooth hydraulic regime at $\mathrm{Re}_D \approx 6$. Furthermore, considering the largest amplification factor seen in Figure 15, we obtain $\mathrm{Re}_D \approx 22.8$, which is within the transitional hydraulic regime. In contrast, the live-bed configuration (Table 2) results in a larger friction velocity amplification near the cylinder (Zhang et al., 2020 [20]), which locally overshoots the threshold for the rough hydraulic regime.

5. Conclusions

Despite important research efforts over the last decade, there are still difficulties in accurately predict the maximum value of S/D (Harasti et al., 2023 [50]). It is for this reason that a correct understanding of the kinematics of the scour at the base of an obstacle are vital for designing and constructing sustainable hydraulic structures. This study presents a numerical model using the LES approach. The code was initially developed by Zhang et al., 2020 [20] for a wide channel in live-bed conditions in the turbulent rough regime ($\mathrm{Re}_D \approx 46{,}000$). In the present study, we successfully studied the combined effect of the velocity and the blockage with a focus on an approaching flow near the hydraulically smooth regime. We consider two different test cases. The first case presents a good agreement with the experimental findings of Roulund et al., 2005 [4] ($\mathrm{Re}_D \approx 46{,}000$), with the objective of acting as a control case to ensure that the model could still accurately predict the scouring process in the live-bed regime after the extension of the applicability to the clear-water regime and the generalization of the flow conditions. A moderately low

Reynolds number ($Re_D \approx 2000$) is used for the second case to study the model's ability to predict clear-water scour. Validation is achieved by qualitatively and quantitatively comparing the predicted scour depth and time evolution with a laboratory experiment reported in Lachaussée, 2018 [18].

We obtain fair agreement with the velocity profile obtained using PIV measurements in the narrow channel with non-slip lateral wall boundary conditions. Although the numerical model still has discrepancies at the lee-side of the dune and downstream of it, the accuracy of the predicted angle, size, and depth of the scour hole are very good near the obstacle. This is despite the difference in the height of the cylinder and despite the fact that the experimental set-up uses a submerged cylinder while the cylinder in the numerical set-up is only partially submerged. This behavior is consistent with the conclusion regarding the height–diameter ratio of the obstacle in Lachaussée, 2018 [18]. The model can also accurately predict the depth evolution of the scour hole. However, in the long-term simulation, the model reaches its equilibrium at a greater depth. This behavior could be a consequence of the sediment bed hypothesis as a continuum, the pier "shadow effect" on the measurement technique, or a combination of both.

The amplification to the flow contraction caused by the confinement is well described by the model. In the clear-water case, the narrow channel simulation presented a higher scour depth than the wide channel simulation. These results show that a large blockage ratio ($D/W \geq 0.1$) can have an important impact over the scour process, despite a low relative coarseness ($D/d < 100$). However, because all tests in the present study were performed for low Froude number values, it would be interesting to perform a similar study for a higher value Froude number to account for the free surface effects on the scouring process. Future study will focus on testing similar flow conditions with different pier geometries.

Author Contributions: M.H.-H.: Methodology, Writing, Validation; W.Z.: Methodology, Conceptualization; A.H.: Methodology, Writing; D.P.V.B.: Project administration, Methodology, Conceptualization, Supervision; K.D.N.: Methodology, Conceptualization. All authors have read and agreed to the published version of the manuscript.

Funding: This research was partially funded by the Quebec Ministry of Foreign Affairs (MRIF, project AVOCAP-2021-2023) and the Mexican Council of Science and Technology (Investigadoras e Investigadores por Mexico, CONACYT). Additional support from the Jiangsu Science and Technology Programme (BK20220285), the NSERC-Discovery program (RGPIN-2018-0677), and Compute Canada (project SINAPSE, No. 3148) is also acknowledged.

Institutional Review Board Statement: Not applicable.

Informed Consent Statement: Not applicable.

Data Availability Statement: Not applicable.

Acknowledgments: The authors would like to thank Miguel Uh Zapata for his invaluable input and support throughout the research process. The authors also extend special thanks to Compute Canada (contract No. 3148) for providing access to the computing facility.

Conflicts of Interest: The authors declare no conflict of interests.

References

1. Raudkivi, A.; Ettema, R. Clear-water scour at cylindrical piers. *J. Hydraul. Eng.* **1983**, *109*, 338–350. [CrossRef]
2. Melville, B.W.; Chiew, Y.M. Time scale for local scour at bridge piers. *J. Hydraul. Eng.* **1999**, *125*, 59–65. .:1(59). [CrossRef]
3. Ettema, R.; Constantinescu, G.; Melville, B.W. Flow-field complexity and design estimation of pier-scour depth: Sixty years since Laursen and Toch. *J. Hydraul. Eng.* **2017**, *143*, 03117006. . [CrossRef]
4. Roulund, A.; Sumer, B.M.; Fredsøe, J.; Michelsen, J. Numerical and experimental investigation of flow and scour around a circular pile. *J. Fluid Mech.* **2005**, *534*, 351–401. [CrossRef]
5. Stevens, M.A.; Gasser, M.M.; Saad, M.B.A.M. Wake Vortex Scour at Bridge Piers. *J. Hydraul. Eng.* **1991**, *117*, 891–904. .:7(891). [CrossRef]

6. Williams, P.; Bolisetti, T.; Balachandar, R. Blockage correction for pier scour experiments. *Can. J. Civ. Eng.* **2018**, *45*, 413–417. [CrossRef]
7. Proske, D. *Bridge Collapse Frequencies versus Failure Probabilities*; Springer: Berlin, Germany, 2018.
8. Arneson, L.; Zevenbergen, L.; Lagasse, P.; Clopper, P. *Evaluating Scour at Bridges*; Technical Report; Federal Highway Administration: Fort Collins, CO, USA, 2012.
9. Huber, F. Update: Bridge scour. *ASCE* **1991**, *61*, 62–63.
10. Wardhana, K.; Hadipriono, F.C. Analysis of recent bridge failures in the United States. *J. Perform. Constr. Facil.* **2003**, *17*, 144–150. [CrossRef]
11. Lin, C.; Han, J.; Bennett, C.; Parsons, R.L. Case history analysis of bridge failures due to scour. In *Climatic Effects on Pavement and Geotechnical Infrastructure*; American Society of Civil Engineers: Reston, VA, USA, 2014; pp. 204–216. [CrossRef]
12. Taricska, M. An Analysis of Recent Bridge Failures (2000–2012). Master's Thesis, Ohio State University, Columbus, OH, USA, 2014.
13. Kirkil, G.; Constantinescu, G. Nature of flow and turbulence structure around an in-stream vertical plate in a shallow channel and the implications for sediment erosion. *Water Resour. Res.* **2009**, *45*, W06412. [CrossRef]
14. Kirkil, G.; Constantinescu, G. A numerical study of the laminar necklace vortex system and its effect on the wake for a circular cylinder. *Phys. Fluids* **2012**, *24*, 073602. [CrossRef]
15. Kirkil, G.; Constantinescu, G. Effects of cylinder Reynolds number on the turbulent horseshoe vortex system and near wake of a surface-mounted circular cylinder. *Phys. Fluids* **2015**, *27*, 075102. [CrossRef]
16. Lachaussée, F.; Bertho, Y.; Morize, C.; Sauret, A.; Gondret, P. Competitive dynamics of two erosion patterns around a cylinder. *Phys. Rev. Fluids* **2018**, *3*, 012302. [CrossRef]
17. Lai, Y.G.; Liu, X.; Bombardelli, F.A.; Song, Y. Three-Dimensional Numerical Modeling of Local Scour: A State-of-the-Art Review and Perspective. *J. Hydraul. Eng.* **2022**, *148*, 03122002. [CrossRef]
18. Lachaussée, F. Érosion et Transport de Particules au Voisinage d'un Obstacle (In French). Ph.D. Thesis, Université Paris-Saclay Paris, France, 2018. Available online: https://www.theses.fr/2018SACLS377 (accessed on 24 July 2023).
19. Zhang, W.; Uh Zapata, M.; Pham Van Bang, D.; Nguyen, K.D. Three-Dimensional Hydrostatic Curved Channel Flow Simulations Using Non-Staggered Triangular Grids. *Water* **2022**, *14*, 174. [CrossRef]
20. Zhang, W.; Uh Zapata, M.; Bai, X.; Pham Van Bang, D.; Nguyen, K.D. Three-dimensional simulation of horseshoe vortex and local scour around a vertical cylinder using an unstructured finite-volume technique. *Int. J. Sediment Res.* **2020**, *35*, 295–306. [CrossRef]
21. Yu, P.; Zhu, L. Numerical simulation of local scour around bridge piers using novel inlet turbulent boundary conditions. *Ocean Eng.* **2020**, *218*, 108166. [CrossRef]
22. Jia, Y.; Altinakar, M.; Guney, S. Three-dimensional numerical simulations of local scouring around bridge piers. *J. Hydraul. Res.* **2017**, *56*, 351–366. [CrossRef]
23. Xiong, W.; Tang, P.; Kong, B. Computational Simulation of Live-Bed Bridge Scour Considering Suspended Sediment Loads. *J. Comput. Civ. Eng.* **2017**, *31*, 04017040. [CrossRef]
24. Baykal, C.; Sumer, B.M.; Fuhrman, D.R.; Jacobsen, N.G.; Fredsøe, J. Numerical investigation of flow and scour around a vertical circular cylinder. *Phil. Trans. R. Soc. A* **2015**, *373*, 104. [CrossRef]
25. Baykal, C.; Sumer, B.M.; Fuhrman, D.R.; Jacobsen, N.G.; Fredsøe, J. Numerical simulation of scour and backfilling processes around a circular pile in waves. *Coast. Eng.* **2017**, *122*, 87–107. [CrossRef]
26. Olsen, N.; Melaaen, C. Three-dimensional calculation of scour around cylinders. *J. Hydraul. Mech.* **1993**, *119*, 1048–1054. [CrossRef]
27. Zhao, M.; Cheng, L.; Zang, Z. Experimental and numerical investigation of local scour around a submerged vertical circular cylinder in steady currents. *Coast. Eng.* **2010**, *57*, 709–721. [CrossRef]
28. Khosronejad, A.; Flora, K.; Sotiropoulos, F. Experimental and computational investigation of local scour around bridge piers. *Water Resour.* **2012**, *37*, 73–85. [CrossRef]
29. Link, O.; González, C.; Maldonado, M.; Escauriaza, C. Coherent structure dynamics and sediment particle motion around a cylindrical pier in developing scour hole. *Acta Geophys.* **2012**, *60*, 1689–1719. [CrossRef]
30. Van Rijn, L.C. *Principles of Sediment Transport in Rivers, Estuaries And Coastal Seas*; Aqua Publications: Amsterdam, The Netherlands, 1993.
31. Zhang, W. 3D Numerical Simulation of Scour Erosion Around an Obstacle. Ph.D. Thesis, Université Paris-Est, Paris, France, 2019.
32. Chen, J.H.; Pritchard, W.G.; Tavener, S.J. Bifurcation for flow past a cylinder between parallel planes. *J. Fluid Mech.* **1995**, *284*, 23–41. [CrossRef]
33. Singha, S.; Sinhamahapatra, K.P. Flow past a circular cylinder between parallel walls at low Reynolds numbers. *Ocean Eng.* **2010**, *37*, 757–769. [CrossRef]
34. Mignot, E.; Moyne, T.; Doppler, D.; Riviere, N. Clear-Water Scouring Process in a Flow in Supercritical Regime. *J. Hydraul. Eng.* **2015**, *142*, 04015063. [CrossRef]
35. Engelund, F.; Fredsøe, J. A sediment transport model for straight alluvial channels. *Hydrol. Res.* **1976**, *7*, 293–306. [CrossRef]
36. Bento, A.M.; Pêgo, J.P.; Viseu, T.; Couto, L. Scour Development Around an Oblong Bridge Pier: A Numerical and Experimental Study. *Water* **2023**, *15*, 2867. [CrossRef]
37. Mason, P.J.; Thomson, D.J. Stochastic backscatter in large-eddy simulations of boundary layers. *J. Fluid Mech.* **1992**, *242*, 51. [CrossRef]

38. Nikuradse, J. *Laws of Flow in Rough Pipes;* Technical Memorandum 1292; National Advisory Committee for Aeronautics: Washington, DC, USA, 1950.
39. Roulsby, R. *Dynamics of Marine Sands: A Manual for Practical Applications;* Thomas Telford: London, UK, 1997.
40. Swamee, P.K. Generalized inner region velocity distribution equation. *J. Hydraul. Eng.* **1993**, *119*, 651–656. [CrossRef]
41. Uh Zapata, M.; Pham Van Bang, D.; Nguyen, K.D. An unstructured finite volume technique for the 3D Poisson equation on arbitrary geometry using a sigma-coordinate system. *Int. J. Numer. Methods Fluids* **2014**, *76*, 611–631. [CrossRef]
42. Uh Zapata, M.; Pham Van Bang, D.; Nguyen, K.D. Unstructured Finite-Volume Model of Sediment Scouring Due to Wave Impact on Vertical Seawalls. *J. Mar. Sci. Eng.* **2021**, *9*, 1440. [CrossRef]
43. Phillips, N.A. A coordinate system having some special advantages for numerical forecasting. *J. Meteor.* **1957**, *14*, 184–185. [CrossRef]
44. Chorin, A.J. Numerical solution of the Navier-Stokes equations. *Math. Comput.* **1968**, *22*, 745–762. [CrossRef]
45. Uh Zapata, M.; Zhang, W.; Pham Van Bang, D.; Nguyen, K.D. A parallel second-order unstructured finite volume method for 3D free-surface flows using a σ coordinate. *Comput. Fluids* **2019**, *190*, 15–29. [CrossRef]
46. Sumer, B.; Fredsøe, J.; Christiansen, N. Scour around vertical pile in waves. *J. Waterw. Port Coast. Ocean Eng.* **1992**, *118*, 15–31. [CrossRef]
47. Baker, C.J. The laminar horseshoe vortex. *J. Fluid Mech.* **1979**, *95*, 347–367. [CrossRef]
48. Kothyari, U.; HAger, W.H.; Oliveto, G. Generalized approach for clear-water scour at bridge foundation elements. *J. Hydraul. Eng.* **2007**, *133*, 1229–1240. [CrossRef]
49. Tejada, S. Effects of Blockage and Relative Coarseness on Clear Water Bridge Pier Scour. Master's Thesis, University of Windsor, Windsor, ON, Canada, 2014. Available online: https://scholar.uwindsor.ca/etd/5055/ (accessed on 24 July 2023).
50. Harasti, A.; Gilja, G.; Adžaga, N.; Žic, M. Analysis of Variables Influencing Scour on Large Sand-Bed Rivers Conducted Using Field Data. *Appl. Sci.* **2023**, *13*, 5365. [CrossRef]

applied sciences

MDPI

Article

Analysis of Variables Influencing Scour on Large Sand-Bed Rivers Conducted Using Field Data

Antonija Harasti [1], Gordon Gilja [1,*], Nikola Adžaga [2] and Mark Žic [3]

1 Department of Hydroscience and Engineering, Faculty of Civil Engineering, University of Zagreb, Fra Andrije Kacica Miosica 26, 10000 Zagreb, Croatia; antonija.harasti@grad.unizg.hr

2 Department of Mathematics, Faculty of Civil Engineering, University of Zagreb, Fra Andrije Kacica Miosica 26, 10000 Zagreb, Croatia; nikola.adzaga@grad.unizg.hr

3 Division of Materials Physics, Ruder Boskovic Institute, Bijenicka Cesta 54, 10000 Zagreb, Croatia; mark.zic@irb.hr

* Correspondence: gordon.gilja@grad.unizg.hr

Abstract: Throughout the lifespan of a bridge, morphological changes in the riverbed affect the variable action-imposed loads on the structure. This emphasizes the need for accurate and reliable data that can be used in model-based projections targeted for the identification of risk associated with bridge failure induced by scour. The aim of this paper is to provide an analysis of scour depth estimation on large sand-bed rivers under the clear water regime, detect the most influential (i.e., explanatory) variables, and examine the relationship between them and scour depth as a response variable. A dataset used for the analysis was obtained from the United States Geological Survey's extensive field database of local scour at bridge piers, i.e., the Pier-Scour Database (PSDB-2014). The original database was filtered to exclude the data that did not reflect large sand-bed rivers, and several influential variables were omitted by using the principal component analysis. This reduction process resulted in 10 influential variables that were used in multiple non-linear regression scour modeling (MNLR). Two MNLR models (i.e., non-dimensional and dimensional models) were prepared for scour estimation; however, the dimensional model slightly overperformed the other one. According to the Pearson correlation coefficients (r), the most influential variables for estimating scour depth were as follows: Effective pier width ($r = 0.625$), flow depth ($r = 0.492$), and critical and local velocity ($r = 0.474$ and $r = 0.436$), respectively. In the compounded hydraulic-sediment category, critical velocity had the greatest impact (i.e., the highest correlation coefficient) on scour depth in comparison to densimetric Froude and critical Froude numbers that were characterized by correlation coefficients of $r = 0.427$ and $r = 0.323$, respectively. The remaining four variables (local and critical bed shear stress, Froude number, and particle Reynolds number) exhibited a very weak correlation with scour depth, with $r < 0.3$.

Keywords: bridge scour; sand-bed; principal component analysis; multiple non-linear regression; PSDB-2014

Citation: Harasti, A.; Gilja, G.; Adžaga, N.; Žic, M. Analysis of Variables Influencing Scour on Large Sand-Bed Rivers Conducted Using Field Data. *Appl. Sci.* **2023**, *13*, 5365. https://doi.org/10.3390/app13095365

Academic Editor: Cheng-Yu Ku

Received: 15 March 2023
Revised: 20 April 2023
Accepted: 21 April 2023
Published: 25 April 2023

1. Introduction

The courses of large European rivers have been altered over the centuries of urbanization to provide support for development, such as flood protection, fairway routes, energy production, and water use for domestic and industrial consumption, agriculture, or recreation [1]. The proximity of rivers and large cities inevitably means that river flow interacts with the built environment, most often with hard and critical infrastructure [2]. Bridges stand out from other structures in the infrastructure since their elements are typically placed in the main channel, and thus, they are continuously exposed to river flow. In this context, bridges are subjected to multiple hazards of various severity that can occur simultaneously, such as flooding, scouring, and seismic loading, resulting in bridge failure.

Throughout the lifespan of the bridge, the riverbed morphology changes, which in turn affects the variable action-imposed loads on the structure [3] and increases the risk of failure [4]. Numerous researchers have identified scour as the most common cause of failure among all other hydraulic-related causes [5–7]. Climate change has a significant impact on bridge safety since their long life span does not allow a reliable estimation of the impact of climate change [8,9]. The annual costs of adaptation to reduce the risk of bridge scour under projected future climate change scenarios are estimated at €541 million in Europe for the period 2040–2070 [10]. Therefore, potential threats to bridge safety have been identified during maintenance, which ensures their resilience [11,12] and reduces potential direct and indirect losses [13]. Obstacles to reliable estimation of potential scour are numerous: the unknown accuracy of field data, the uncertainty of laboratory data upscaled to the prototype scale, and an overall lack of understanding of the interaction between the turbulent flow and the erodible riverbed. Scour is initiated when the flow velocity exceeds a threshold for entrainment of sediment, and it lasts until after the flood peak [14]. Thereafter, as the flood stage recedes, the scour hole refills and its depth decreases [15,16]. As scour is the most common cause of bridge failure, there is an apparent need for accurate and reliable data that can be used in model-based projections of the risks associated with bridge scour.

Most of the existing scour prediction models available in the literature are derived from laboratory data [17]. Using a pump and a tail gate in the laboratory provides controlled hydraulic conditions for experimental data collection. Establishing a specific flow environment in a laboratory limits the range of laboratory conclusions [18]. Furthermore, upscaling to field prototype bridge piers is a daunting task that can lead to unreliable extrapolated data. It is fair to say that laboratory-based scour models are adequate only for the limited range of variables for which they were developed. With the aim of avoiding extrapolation issues, the accuracy of these models should be additionally confirmed with field measurements. Therefore, it is necessary to gain a perspective on scour potential under field conditions [19]. Regardless of the data collection method, most of the scour prediction equations are based on conventional regression methods to provide explicit expressions that are easy-to-use and readily applicable to practice. However, conventional methods are limited to conditions similar to those under which the data were collected. This leads to biased results that often overestimate scour depth, i.e., a so-called conservative approach [20,21]. To describe the scour process, the scour database should be extensive and structured because many local variables are required to capture the complex flow environment in the bridge opening.

Recently, artificial intelligence (AI) has become a prominent tool that has the capability to capture complex scour processes governed by a large number of influencing variables. Recent studies have used AI techniques for scour prediction. Dong et al. [22] developed four machine-learning models: a back propagation neural network (BP NN), a genetic algorithm neural network (GA NN), a convolutional neural network (CNN), and a deep belief network (DBN NN) to predict local scour based on a combination of laboratory tests and field observations. Pandey et al. [23] developed scour equations using a genetic algorithm (GA) on a large experimental dataset. Rady [24] applied the adaptive neural fuzzy inference system (ANFIS) and genetic programming (GP) methods for pier scour assessment. Muzzammil et al. [25] implemented gene expression programming (GEP) to derive a new equation for pier scour estimation in a bed of a mixture of clay and sand. To evaluate the performances of developed AI models, the abovementioned authors conducted comparisons with traditional methods (regression technique, dimensional analysis, and functional relations) and concluded that AI methods showed better scour depth prediction. Although AI-based models have proven to be more accurate than conventional models, they have a tendency to overfit data. Generally, AI models learn to fit the selected data set as closely as possible, but learning the model for a particular data set might make it inappropriate for unseen data. Moreover, in the case of insufficient scouring of input data, it might occur that the AI model has a biased outcome due to overfitting issues.

The United States Geological Survey (USGS) has published an extensive field database of local scour at bridge piers, "A Pier-Scour Database" [26], collected for 270 bridges spanning over 433 rivers in 29 countries, resulting in a total of 1858 measurements of flow, bridge geometry, and river morphology. The database is structured as a digital spreadsheet with measurements organized as rows and values of variables under each measurement organized in columns, e.g., scour depth around bridge piers (d_s), pier width (b), pier length (l), angle of flow attack (θ), effective pier width (b_{ef}), pier type and nose shape, approach flow depth (y), approach flow velocity (v), bed material type, median sediment size (d_{50}), geometric standard deviation of sediment particles (σ_g), approximate recurrence interval for measured flow rate (RI), stream slope (S), date, type, time, and location of measurements, etc.

As this extensive PSDB-2014 database has been available since 2014, it has been used in several studies focusing on pier scour. Benedict and Caldwell [18] aimed to define the maximum scour depth in South Carolina by developing an envelope curve. They evaluate the upper bound of the flow-scour relationship by combining pier scour data from other sources for effective pier widths up to 9 m. Benedict and Knight [27] evaluated the HEC-18 model by comparing the PSDB-2014 experimental database since the HEC-18 model was developed on a scarce laboratory dataset. Rathod and Manekar [28] designed a new universal scour model utilizing the GEP method on the PSDB-2014 database. They compared the new GEP model with five existing conventional scour equations and found a good correlation with Jain and Fischer model Equation (14). Pandey et al. [29] used the PSDB-2014 database to modify laboratory-based Melville and Coleman's equation and make it applicable to field conditions. The authors filtered the PSDB-2014 database to retain only non-uniform gravel beds in clear-water conditions and introduced new K-factors. Ali and Günal [30] collected a new laboratory dataset and complemented it with the corresponding PSDB-2014 laboratory subset to train the ANN model with the Levenberg-Marquardt algorithm. Shahriar et al. [31] used the PSDB-2014 database to compare four pier scour models and assess their performance through error statistics and the probabilistic distribution of predictions.

This paper presents a preliminary analysis of the variables influencing the pier scour on large sand-bed rivers under the clear-water regime. Due to the lack of scour equations based on field measurements and the low accuracy of the laboratory-derived equations deployed for field data, this study uses a comprehensive PSDB-2014 database. The objective of this paper is to identify the most influential variables affecting scour on large sand-bed rivers using the structure as presented by the flow chart (Figure 1). First, the collection of a large database is required, i.e., in our case, obtaining the PSDB-2014 database. Second, filling the original database with more potentially important variables that were not originally included. Herein, these variables are designed as additionally calculated variables, such as Froude number, shear stress, and other similar variables that cannot be directly measured. Third, filter measurements from the collected database that are not related to specific environmental conditions, i.e., eliminate measurements to obtain a relevant dataset. Fourth, reduce unnecessary variables to the filtered subset with only influencing variables using principal component analysis (PCA). Fifth, perform sensitivity analysis to examine the strength of correlation between the influencing variables and the scour depth using the Pearson correlation matrix. At this point, a reliable scour prediction model can be designed and tested. In this paper, a dimensional regression-based model is proposed for scour depth estimation in large sand-bed rivers in clear water regimes. Several already existing models applicable to the aforementioned conditions were selected and compared with the model developed in this study to put this investigation in a relative perspective.

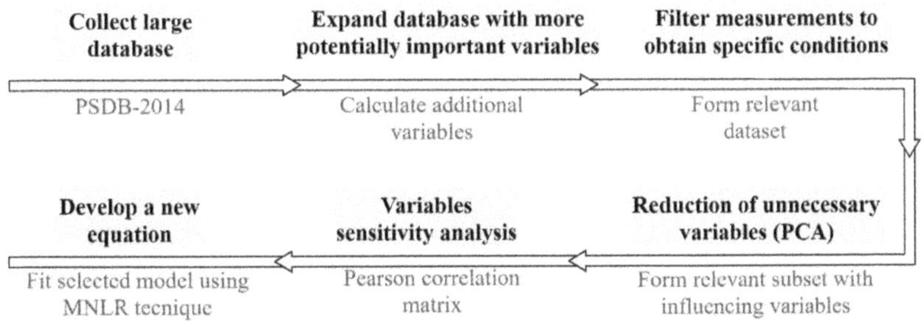

Figure 1. Flowchart of the proposed methodology to identify the most influential variables before developing a new model.

2. Methodology

2.1. Influencing Variables

The original PSDB-2014 database covers a wide range of directly measured flow, sediment, and geometry variables. From all measured variables in the original PSDB-2014 database, the following 13 variables were selected for the purpose of this study: d_s, b, l, θ, b_{ef}, pier type, nose shape, y, v, d_{50}, σ_g, RI, and S. Since this dataset contains only directly measured variables, it is of keen interest to include several other variables commonly used in scour prediction equations. Thus, this study introduces 8 additional variables calculated using the measured data from the PSDB-2014 database: channel width (B), Froude number (Fr), local bed shear stress (τ), critical bed shear stress (τ_c), critical velocity (v_c), critical Froude number for incipient motion (Fr_c), densimetric Froude number (Fr_d), and particle Reynolds number (Re_p) (Table 1). By including the aforementioned additional variables, information about the flow regime, clear-water or live-bed scour, and particle entrainment threshold conditions are taken into account. These additional variables are not given in the original PSDB-2014 database, but they are often used in scour research (e.g., [32,33]). With this addition, a comprehensive dataset of 21 independent variables was created with essential information on flow, sediment, and morphology at each location.

Table 1. Additional variables included in the analysis.

Additional Variable	Equation	
Froude number	$Fr = \frac{v}{\sqrt{g\cdot y}}$	(1)
local shear stress	$\tau = \gamma_w\cdot y\cdot S$	(2)
critical shear stress [34]	$\tau_c = 0.25\cdot\left(d_{50}\cdot\left[\frac{\left(\frac{\gamma_s}{\gamma_w}-1\right)\cdot g}{\nu^{2*}}\right]^{\frac{1}{3}}\right)^{-0.6}\cdot g\cdot(\rho_s-\rho_w)\cdot d_{50}\cdot tan(\Phi^*)$	(3)
critical velocity [35]	$v_c = \sqrt{\rho_{rel}{}^*\cdot g\cdot d_{50}}\cdot\left(0.0024\cdot\left(\frac{y}{d_{50}}\right)+2.34\right)$	(4)
critical Froude number	$Fr_c = \frac{v_c}{\sqrt{g\cdot y}}$	(5)
densimetric Froude number	$Fr_d = \frac{v}{\sqrt{(\rho_{rel}{}^*-1)\cdot g\cdot d_{50}}}$	(6)
particle Reynolds number	$Re_p = \frac{d_{50}\cdot\sqrt{\rho_{rel}{}^*\cdot g\cdot d_{50}}}{\nu^*}$	(7)

* ν is kinematic viscosity of fluid. Φ is the angle of repose of the particle [36]. ρ_{rel} is the submerged relative mass density of sediment particles.

2.2. Data Filtering

After the expansion of the subset (containing 13 initial variables) with the additional computed 8 variables, data filtering was conducted to retain only data relevant for the aim of the paper, i.e., to study measurements referring to large sand-bed rivers. To obtain this data, several sets of filtering criteria were defined that correspond to the flow regime and riverbed morphology of the large rivers.

The first set of filtering criteria is related to the flow regime. The flow of large rivers is subcritical due to the nature of the flow in lowland areas; hence, the Froude number can be used to distinguish subcritical and supercritical flow regimes. In open-channel flow, the critical flow regime is lower than its theoretical value ($Fr = 1$) since the flow transition from wide channel to contraction is gradual. Different studies identified the transition between subcritical and supercritical flow regimes to be within the interval $0.7 < Fr < 0.85$ for field conditions. For example, in the study by Azamathulla et al. [37], Fr values varied from 0.7 to 0.8, as in several other reports [18,29,38]. In this study, measurements with $Fr > 0.75$ were filtered out, i.e., applying Rady's approach [24]. Additionally, erroneous data related to Froude number were also filtered out—measurements of approach flow velocity (v) with missing or zero values were removed.

The second set of filtering criteria was applied to preserve the data corresponding to large sand-bed rivers. To restrain the dataset to sand-bed rivers, the following criteria were used: sediment particle median size (d_{50}) ranging between 0.0625 mm and 2 mm. The geometrical standard deviation of the sediment (σ_g) presents a measure of the non-uniformity of the sediment [28]. Uniform sediment gradation associated with uniform sand is considered to be less than 1.3 [27] because larger σ_g values would indicate the presence of coarser gravel particles. After applied filtration, the average value of σ_g is 3.0, which shows slightly non-uniform sediments, which indicates potential bed armoring and consequently reduces scour depth [39]. To restrain the dataset to large rivers, measurements of flow depth (y) lower than 1.5 m were excluded from further analysis. The original PSDB-2014 database [26] contains no information regarding the river widths. Benedict and Knight [40] have applied basin data to correlate known channel widths against drainage area, flow depth, and stream slopes. To validate filtering criteria for large rivers, correlation by Benedict and Knight [40] was used herein to calculate channel widths, of which 69% are greater than 30 m.

Finally, the filter that takes into account the scour depth error margin range has been applied. If only large rivers are considered, the accuracy of riverbed morphology measurement has to be taken into account, especially when equipment limitations do not allow detailed riverbed mapping. Therefore, from the original PSDB-2014 database, an additional 140 measurements were excluded where scour depth was in the error margin range of less than 0.2 m. Moreover, the original PSDB-2014 database contains multiple measurements collected at the same location under different hydrologic events, including multiple scour depth values. To prepare the database for the analysis of cause-and-effect relationships, it is important to maintain the independence of the measurements, i.e., only one measurement should be retained for each bridge. Therefore, for the second set of filtering criteria, only the measurements with the largest scour depth were considered (568 exclusions). The association of scour depth with hydraulic conditions in the PSDB-2014 database is not extensively documented. For some multiple measurements, scour depth and flow depth are inversely proportional, which may be a consequence of an undocumented hydraulic event, e.g., a recent flood. In addition to the second set of filtering criteria, only the most recent measurements have been retained (291 exclusions). Some measurements date from the early 1900s, so their reliability is low due to technical limitations such as the development of guidelines and instruments for bathymetric and hydraulic surveys. Overall, a total of 859 measurements were excluded.

After filtering the original PSDB-2014 database, a total of 348 independent measurements have remained relevant for estimating maximum scour depth on large, alluvial, and sand-bed rivers. Since scour depth is 10% greater in clear water than in live-bed conditions [41], the clear-water regime is the desired condition for estimating scour depth in this study. To distinguish whether flow conditions are clear-water or live-bed scour, the expression $Fr - Fr_c > 0.2$ by Jain and Fischer is used for validation [38]. Interestingly, after filtering by using the aforesaid expression, 98% of the 348 remaining measurements fall within the clear-water scour conditions, even though the clear-water criterion was not applied in the filter, which validates the reliability of the applied data filters. The range of original and additional variables before and after filtration is shown in Table 2.

Table 2. Range of original and additional variables with their symbols, measurement units, ranges, average values, and standard deviations before and after filtering.

	Variable	Before Filtering			After Filtering		
		Range	Average	Standard Deviation	Range	Average	Standard Deviation
Original	d_s	0–10.4	1.1	1.3	0.21–7.8	1.5	1.3
	b_{ef}	0.21–28.7	2.3	2.5	0.24–11.6	2.0	1.9
	b	0.21–19.5	1.6	1.6	0.24–11.6	1.3	1.4
	l	0.21–39.6	6.3	5.3	0.37–25.3	6.3	4.7
	θ	0–85.0	6.1	10.3	0–600	6.6	9.6
	d_{50}	0.001–228.6	14.7	25.0	0.06–1.82	0.59	0.41
	y	0–22.5	3.9	3.2	1.5–22.5	5.6	3.4
	v	0–5.4	1.4	0.8	0.20–3.9	1.4	0.71
	RI	1–500	53.6	50.9	1–500	63.1	69.4
	σ_g	1.2–20.3	3.3	2.8	1.4–20.3	3.0	1.2
	S	0.00007–0.02	0.00086	0.00152	0.00007–0.0036	0.00052	0.00044
Additional	B	5.4–692.5	71.5	67.1	10.5–692.5	75.3	80.7
	v_c	0.15–55.7	2.4	3.2	0.76–11.6	2.8	1.7
	τ	0–180	21.3	21.8	0.015–1.7	0.25	0.21
	τ_c	0.025–3.9	0.87	0.77	0.13–0.56	0.32	0.10
	Fr	0–1.98	0.28	0.21	0.039–0.55	0.20	0.10
	Fr_c	0.19–5.77	0.40	0.34	0.19–1.0	0.38	0.13
	Fr_d	0–629	11.1	23.4	1.6–88.9	16.0	9.9
	Re_p	0.0025–274,834	8222	19,074	1.2–195.2	42.1	43.7
Original	Pier type	Single and group			Not affected		
	Pier nose shape (drag coefficient) [42]	Cylindrical (1.2), Round (1.33), Square (2.0), Sharp (1.0), Triangular (1.72)					

2.3. Variable Reduction

Estimation of scour depth requires the detection of the most significant input variables relevant to the complex scour process in large sand-bed rivers. This can be achieved using principal component analysis (PCA), which has already been proven to be an effective decision-making tool for dimensionality reduction when many variables are involved [43]. PCA determines which variable influences scour depth the most by analyzing how much each variable contributes to the variance. Therefore, PCA was applied to the filtered dataset of 348 measurements and a total of 21 variables (including the observed scour depth d_s). PCA interprets data with principal components—linear combinations of the input variables that are orthogonal to each other. In this study, there are 21 principal components as well as 21 input variables. The first principal component explains the largest amount of variability in the dataset; the second principal component accounts for the next largest variance; etc. The principal components are lines in the coordinate system with a corresponding eigenvalue—the sum of the squared distances between the orthogonally projected observed data onto the line and the origin of the coordinate system.

Variable reduction in PCA analysis is based on the number of retained principal components without diminishing the total variance of the scour dataset. A commonly used criterion for reducing the number of variables is the Kaiser-Guttman criterion, which considers only those principal components whose eigenvalues are greater than 1. Applying the Kaiser-Guttman criterion for variable reduction would keep the first 7 principal components and thus eliminate only 2 variables. Since the Kaiser-Guttman criterion was previously proven to be inaccurate because it overestimates the number of principal components [44], it was decided to use a more stringent criterion. If we discard the principal components whose eigenvalues are less than 2, then the first three principal components remain. The first three principal components explain 56% of the variance of the entire dataset. The variables that contribute the most to the first three principal components can be identified as influential (b_{ef}, y, d_{50}, v, v_{cr}, τ, τ_{cr}, Fr, Fr_c, Fr_d, Re_p, B, b, l, and S), while other variables (σ_g, θ, RI, pier type, and nose shape) can be excluded from further analysis. The results of the PCA analysis are presented in the form of a loading plot, in which the variable vectors are positioned with respect to the first two principal components (Figure 2a). Variables located near the center of the coordinate system have the smallest impact on the variance of the dataset (within the inner dashed circle) and are therefore eliminated from further analysis. When two variable vectors are perpendicular to each other, they are not correlated at all, while variables whose vectors are close to each other are highly correlated. The contribution of each variable to the first three principal components can be evaluated by the squared cosine values (Figure 2b). Since the principal components are axes of the rotated coordinate system, the squared cosine represents the quality of the variables after rotation. The values in bold are the maximum squared cosine values for each variable. Therefore, only the variables that have the maximum squared cosine for one of the first three principal components are retained for further analysis.

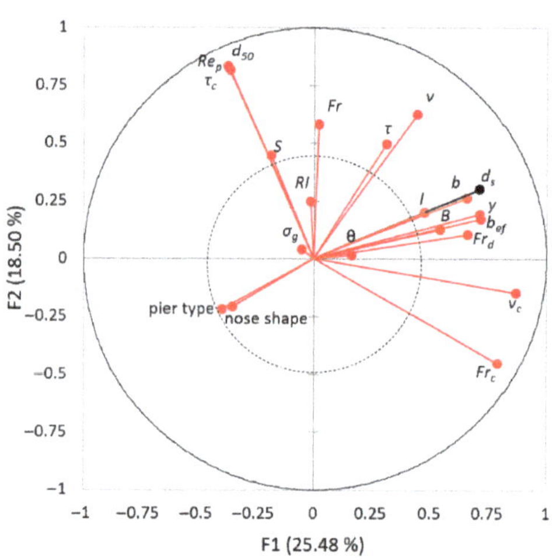

(a)

	F1	F2	F3
d_s	**0.513**	0.091	0.000
b_{ef}	**0.514**	0.030	0.009
y	**0.508**	0.038	0.120
d_{50}	0.140	**0.691**	0.123
v	0.196	**0.389**	0.154
v_{cr}	**0.754**	0.022	0.013
τ	0.097	**0.248**	0.126
τ_{cr}	0.139	**0.680**	0.126
Fr	0.000	0.338	**0.387**
Fr_c	**0.623**	0.203	0.005
Fr_d	**0.436**	0.011	0.337
Re_p	0.133	**0.665**	0.115
B	**0.294**	0.016	0.128
b	**0.430**	0.070	0.000
l	**0.224**	0.039	0.174
S	0.035	0.200	**0.297**
σ_g	0.003	0.002	0.001
θ	0.026	0.000	0.003
RI	0.000	0.062	0.220
Pier	0.160	0.048	0.119
Nose	0.125	0.044	0.037

(b)

Figure 2. Results of PCA analysis: (**a**) Loading plot that presents a total of 21 variables in the coordinate system with the first two principal components on the axes; (**b**) Contribution of each variable to the retained first three principal components throughout the squared cosines of the variables. Values in bold correspond for each variable to the principal component for which the squared cosine is the largest.

Based on the PCA results, 5 variables were discarded from further analyses, and all remaining variables can be classified into the following categories: pier geometry (b, l, and b_{ef}), hydraulic (y, v, τ, and Fr), compounded hydraulic-sediment variables (v_c, Fr_c, and Fr_d), sediment properties (d_{50}, τ_c, and Re_p), and channel properties (B and S). Previous studies have concluded that d_{50} can be eliminated from the development of the scour equation since its range of values is small when compared to the other variables, which consequently leads to biased results [19]. Since only field data are used in this study, d_{50} is expected to be negligible due to reaching large values of the ratio b/d_{50} [45,46]. In order to obtain information on sediment properties, only τc and Rep were retained for developing the equation, while d_{50} was discarded.

Similarly, the variables describing pier geometry (θ, b, l and b_{ef}) are also close together. The angle of flow attack (θ) is not a reliable variable for scour prediction because flow direction depends on the water level and changes over time. This makes the measurement of θ arbitrary and should be used only when it can be reliably estimated [24]. The former debate was supported by the PCA results, which suggested that θ and l should not be used in scour prediction, leaving b and b_{ef} as the only descriptors of pier geometry. Taking into consideration the difficult approximation of complex pier geometry (in the case of a group of piers, pile caps, etc.), this study uses b_{ef} as the pier geometry variable, which is reduced to b when no other data is known but can be adjusted for pier alignment when necessary.

Both variables from the channel properties category (B and S) can be excluded. Although river width was an important variable for validating the filter applied to obtain a database corresponding to the large rivers, it is not necessary to include it in the development of the equation since this paper focuses on local scour. Stream slope is a local variable, originally measured in PSDB-2014 near the bridge. Since S was only used for estimating shear stress, a variable that estimates the drag force of flowing water, it can be eliminated from scour equation development.

The remaining variables are a combination of commonly used variables in scour prediction as well as variables that are rarely used in scour research on field conditions (τ, τ_c, Re_p). Although τ has been previously recognized as an influential variable for scour prediction [47], it is usually omitted from conventional scour equations due to difficulties in obtaining direct measurements in a complex flow environment [19]. However, τ and v are both variables that reflect drag forces in front of the pier. Since it is challenging to measure v in situ during the flood, perhaps it can be replaced by τ. Locally measured variables (y and S) in the original PSDB-2014 database allow τ to be estimated for each pier. Since it is not sufficiently investigated how much local value of τ contributes to d_s compared to v, its importance for the local scour process will be evaluated in this study. The incipient motion of sediment starts when τ exceeds the critical shear stress (τ_c). Instead, in equations for estimating scour depth, v_c is mainly used because it is easier to measure and is proportional to τ_c. In this study, it is decided to retain both critical variables τ_c and v_c, in order to evaluate their contribution to d_s estimation. Particle Reynolds number (Re_p), which is a measure of eddy currents around particles, is an important variable for estimating the transport of sediment mixtures [48]. Previous research by Vonkeman and Basson [49] demonstrates that replacing v_c with Re_p in the HEC-18 scour equation can improve scour depth prediction.

After the variable reduction analysis performed on the filtered dataset, 10 influential variables remained (Table 3), forming a filtered subset relevant to the development of functional dependencies. Since the focus of this study is on the functional dependencies between scour depth and selected variables, dependent variables were retained with the aim of testing which of them has a greater impact on scour depth. Full variable names can be seen at the end of a paper in the list of symbols.

Table 3. Influential variables selected for developing new regression-based models.

Influential variables	b_{ef}	y	v	v_c	τ	τ_c	Fr	Fr_c	Fr_d	Re_p

3. Results

3.1. MNLR—Multiple Nonlinear Regression

Multiple nonlinear regression (MNLR) is a mathematical method that establishes a functional dependence between variables to predict the target variable. The optimization process finds the best fit for the model based on the principle of least squares. In Section 2.3, 10 influential variables were chosen as suitable for scour analysis on large sand-bed rivers. In order to model the data between the response d_s and the explanatory influential variables, the following dimensional and non-dimensional models were proposed:

$$d_s = a \cdot \left(b_{ef}\right)^b \cdot (y)^c \cdot (v)^d \cdot (v_c)^e \cdot (\tau)^f \cdot (\tau_c)^g \cdot (Fr)^h \cdot (Fr_c)^i \cdot (Fr_d)^j \cdot (Re_p)^k, \quad (8)$$

and non-dimensional model:

$$\frac{d_s}{y} = a \cdot \left(\frac{b_{ef}}{y}\right)^b \cdot \left(\frac{v}{v_c}\right)^d \cdot \left(\frac{\tau}{\tau_c}\right)^f \cdot (Fr)^h \cdot (Fr_c)^i \cdot (Fr_d)^j \cdot (Re_p)^k. \quad (9)$$

Both models were fitted to the filtered dataset by using MNLR, and the extracted parameters (*a*, *b*, *c*, *d*, *e*, *f*, *g*, *h*, *i*, *j*, and *k*) were used to form two new regression-based models presented in Table 4.

Table 4. Parameters obtained applying MNLR technique on both dimensional and non-dimensional models.

	a	*b*	*c*	*d*	*e*	*f*	*g*	*h*	*i*	*j*	*k*
dimensional	3.29	0.49	1.19	−0.91	−0.99	−0.011	0.019	0.38	0.26	0.97	−0.44
non-dimensional	0.002	0.48		−0.90		−0.047		−0.33	−2.50	1.16	−0.094

The performance of MNLR dimensional and non-dimensional models was validated with a commonly used statistical measure, i.e., the coefficient of determination (R^2). R^2 shows how closely the scour depth, predicted by the MNLR dimensional and nondimensional equations, resembles the observed scour depth with values of 0.51 and 0.45, respectively. The same observation was made by Ali and Günal [30], who noticed that dimensional scour data were more accurate than those based on dimensionless data. Based on R^2, the dimensional MNLR model was adopted for further analysis. Too many influencing variables reduce the number of degrees of freedom and also increase R^2. Hereby, the adjusted R^2 is useful because it corrects the original R^2 in such a way that if the number of variables increases, the number of degrees of freedom decreases, as does the adjusted R^2. For the developed model, the adjusted $R^2 = 0.50$ does not change significantly in comparison with the original R^2, which confirms that the number of selected variables is sufficient and not overused.

The dependency between measured scour depth values (d_s) and those predicted by the MNLR dimensional model ($d_{s,pred}$) is presented in Figure 3. The gray continuous line shows ideal agreement between measured and predicted values, i.e., $d_{s,pred} = d_s$. Data points that lie below the line of agreement indicate overprediction (54% of predicted values are greater than actual measured values), while data points that lie above the abscissa represent underprediction (46% of predicted values are lower than measured values).

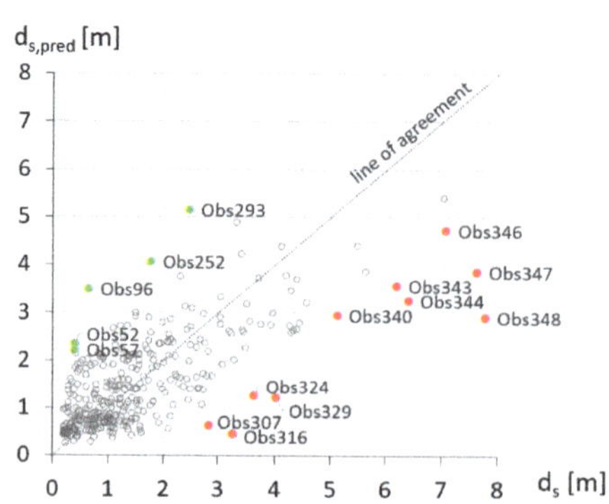

Figure 3. Scattered plot of measured (d_s) and scour depth predicted ($d_{s,pred}$) by MNLR dimensional model. Isolated points are outliers where red dots present underprediction and green dots present overprediction.

A prediction interval of 95% was calculated using the following equation:

$$= d_s \mp t_{95} \cdot SE \cdot \sqrt{1 + \frac{1}{n} + \frac{\left(d_s - \overline{d_s}\right)}{SS_{xx}}} \tag{10}$$

where t_{95} is the two-tailed Student's T-Distribution for 95% fit, SE is the standard error, which considers the squared ratio of the residual sum of squares by degrees of freedom, n is the number of entries, $(\overline{d_s})$ is the average of all measured scour depths, and finally SS_{xx} is the sum of squares of the deviations of measured scour depths from their mean value.

The individual dependance of each variable on the residual (the difference between measured and predicted scour depth) is presented as a series of graphs merged into Figure 4. Upper and lower limits represent the 95% prediction intervals in Equation (10) illustrated by the red dashed lines for the purpose of detecting potential outliers. This means that 95% of all measurements would deviate from predicted values within the region between the dashed red lines. If the data falls outside the 95% prediction interval region, it can be marked as a potential outlier. The red dots show the data with the best prediction—residuals within 5% of the total deviation. A common feature for all the graphs is the homoscedasticity of residuals over the entire range of explanatory variables, indicating that there is no bias in the model in any part of the data range.

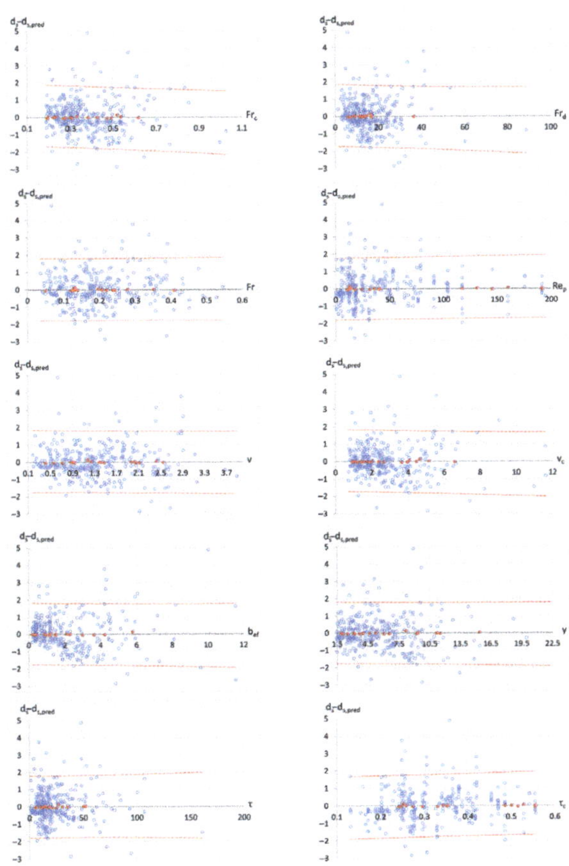

Figure 4. Dependence of influential variables on the scour depth residual. Upper and lower limits illustrated by the red dashed lines represent the 95% prediction intervals.

Since all the scour data only contains measured variables, there is no additional metadata that could be used to expand the filtered dataset, and therefore only collected variables were analyzed in order to detect outliers. All points lying outside of the prediction interval were considered potential outliers. There are 5 outliers detected based on model overprediction, whose residuals have a negative value, and 10 outliers based on model underprediction, whose residuals have a positive value. Outliers are detected based on prediction intervals in Figure 4 and presented as isolated points with associated IDs in Figure 3, where green dots illustrate overprediction and red dots underprediction. Potential outliers and their corresponding influential variable values are presented in Table 5, sorted from minimal to maximum residual. The lack of data does not allow the detection of true outliers, and therefore, outliers can only be detected based on their common features. The only identifiable common feature of all potential outliers is the Froude number value. One of the data points (#Obs348) has a very low Froude number value (Fr = 0.07), making it unreliable, while the other one (#Obs293) is at the boundary conditions for clear-water scour ($Fr - Fr_c$ = 0.06). Therefore, of the initial 15 potential outliers, two were detected as true outliers and removed from further analyses. When outliers were removed from the model, the performance of the dimensional model increased from R^2 = 0.50 to R^2 = 0.54.

Table 5. List of potential outliers with corresponding measured and calculated influential variable values.

ID	b_{ef}	y	v	v_c	τ	τ_c	Fr	Fr_c	Fr_d	Re_p	d_s	$d_{s,pred}$	**Residuals**
Obs96	4.88	9.81	2.13	5.11	38.51	0.27	0.22	0.52	27.57	17.90	0.67	3.50	−2.83
Obs293	11.58	4.85	2.68	2.30	95.08	0.31	0.39	0.33	29.82	28.11	2.50	5.15	−2.65
Obs252	5.54	13.47	2.35	7.44	72.45	0.26	0.20	0.65	32.61	14.39	1.80	4.08	−2.28
Obs52	9.63	3.60	0.79	1.30	7.76	0.49	0.13	0.22	5.46	117.86	0.43	2.36	−1.93
Obs57	2.50	9.33	1.55	4.09	17.38	0.33	0.16	0.43	16.63	31.55	0.43	2.22	−1.79
Obs340	5.97	8.35	1.22	3.28	11.47	0.37	0.13	0.36	11.37	47.57	5.15	2.95	2.20
Obs307	0.61	2.77	1.13	1.14	5.17	0.42	0.22	0.22	8.86	79.52	2.87	0.65	2.22
Obs346	6.10	22.52	2.43	9.11	66.29	0.34	0.16	0.61	24.65	36.96	7.10	4.73	2.37
Obs324	1.07	4.54	1.77	2.17	53.46	0.31	0.26	0.33	19.65	28.11	3.66	1.27	2.38
Obs343	4.33	5.67	2.93	3.32	32.81	0.25	0.39	0.45	41.99	13.07	6.22	3.57	2.65
Obs329	1.07	4.15	1.71	2.00	48.80	0.31	0.27	0.31	18.97	28.11	4.05	1.23	2.82
Obs316	0.61	2.19	0.67	0.97	4.52	0.42	0.14	0.21	5.27	79.52	3.29	0.47	2.82
Obs344	4.37	5.21	2.44	2.98	16.62	0.26	0.34	0.42	33.88	14.39	6.43	3.26	3.17
Obs347	4.27	9.78	2.90	5.62	9.60	0.25	0.30	0.57	41.55	13.07	7.65	3.86	3.79
Obs348	10.09	9.00	0.65	3.60	48.40	0.36	0.07	0.38	6.24	43.61	7.80	2.90	4.90

3.2. Comparison with Different Scour Models

To evaluate the relative performance of the dimensional model developed on specifically filtered data in the present study, a comparison with already developed empirical scour models was conducted (Figure 5). The link between selected models (Table 6) is their applicability for scour depth prediction in sand-bed rivers during the clear-water regime. However, there are some differences between them based on: variables used for scour prediction; source of large-scale data (field, combination of laboratory and field, and numerical data); and equation development technique (1 conventional MNLR equation, 2 recent MNLR equations, and 2 GEP methods). The conventional MNLR technique is an approach to developing functional dependencies using dimensional analysis and the Buckingham PI theorem to attain dimensional consistency. GEP is one of the evolutionary algorithms that can create simple and explicit equations and is therefore often used for developing scour depth expressions. Two governing variables that are included in all scour equations are water depth (y) and pier width (b), as presented in Table 6.

The comparisons between previously developed empirical models (Table 6) and the dimensional model developed in this study, Equation (17), are presented in Figure 5, where the line of agreement indicates the perfect match between observed and predicted data, i.e., $d_{s,pred} = d_s$. The circle points, obtained by using our dimensional model and the filtered dataset, are the same as in Figure 3. However, Figure 5 also displays the lines that were constructed by using predicted $d_{s,pred}$ data obtained by utilizing different models from Table 6. The following discussions will explain the performance of all models from Table 6.

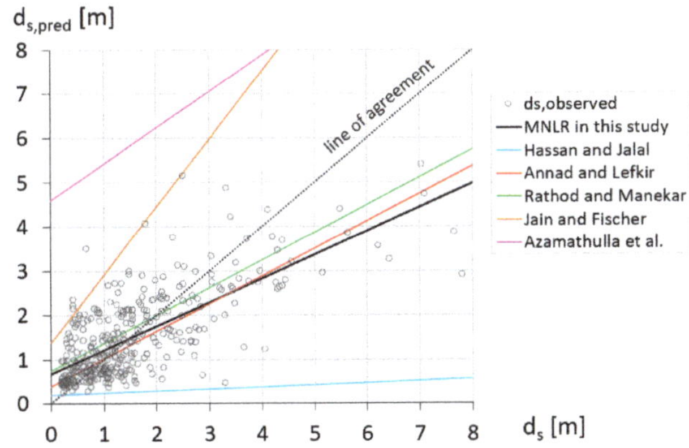

Figure 5. Comparison of dimensional model developed in this study with various previously developed scour equations (Table 6). The displayed data (i.e., circles) are obtained by using our non-dimensional model and the filtered data. The lines present data computed by different models when using the filtered data.

Table 6. Summary of selected previously developed scour models and two models developed in this paper.

Author	Dataset	Method	Equation	
Annad and Lefkir, 2022 [19]	field (PSDB-2014)	MNLR	$d_s = 0.318 \cdot K_1{}^* \cdot K_2{}^* \cdot b^{0.76} \cdot y^{0.515}$	(11)
Rathod and Manekar, 2022 [28]	field and laboratory (PSDB-2014)	GEP	$d_s = y \cdot \left[\left(\frac{b}{b+1.6 \cdot y} \right) + \left(\frac{b \cdot v}{y \cdot (b+v)} \right) + \left(\frac{\frac{b}{y}}{\frac{34.12 \cdot y}{Fr_d} + 2\sigma_g} \right) \right]$	(12)
Hassan and Jalal, 2021 [50]	numerical on a large scale	GEP	$d_s = b \cdot \left[7.24 \cdot \frac{v}{v_c} \cdot \frac{b}{B} - \left(\frac{v}{v_c} \right)^2 \cdot \frac{b}{B} - \frac{v}{v_c} \cdot \left(\frac{b}{B} \right)^2 + \frac{\frac{b}{B}}{\frac{b}{B} \cdot \frac{v}{v_c} - \frac{y}{b} - \frac{b}{B}} + \left\{ \frac{y}{b} \cdot \frac{b}{B} \cdot Fr_{pier}{}^* + \left(\frac{b}{B} \right)^2 \cdot Fr_{pier}{}^* \right\} \cdot K_2{}^{2*} \cdot \frac{v}{v_c} \right]$	(13)
Jain and Fischer, 1979 [38]	field and laboratory	conventional MNLR	$d_s = 1.84 \cdot b \cdot \left(\frac{y}{b} \right)^{0.3} \cdot (Fr_c)^{0.25}$	(14)
Azamathulla et al., 2010 [37]	field	MNLR	$d_s = 1.82 \cdot y \cdot \left(\frac{d_{50}}{y} \right)^{0.042} \cdot \left(\frac{b}{y} \right)^{-0.28} \cdot \left(\frac{L}{y} \right)^{-0.37} \cdot Fr^{0.42} \cdot \sigma_g{}^{-0.031}$	(15)
Our non-dimensional model	field (PSDB-2014)	MNLR	$\frac{d_s}{y} = 0.002 \cdot \left(\frac{b_{ef}}{y} \right)^{0.48} \cdot \left(\frac{v}{v_c} \right)^{-0.90} \cdot \left(\frac{\tau}{\tau_c} \right)^{-0.047} \cdot (Fr)^{0.33} \cdot (Fr_c)^{-2.5} \cdot (Fr_d)^{1.16} \cdot (Re_p)^{-0.094}.$	(16)
Our dimensional model	field (PSDB-2014)	MNLR	$d_s = 3.29 \cdot (b_{ef})^{0.49} \cdot (y)^{1.19} \cdot (v)^{-0.91} \cdot (v_c)^{-0.99} \cdot (\tau)^{-0.011} \cdot (\tau_c)^{0.019} \cdot (Fr)^{0.38} \cdot (Fr_c)^{0.26} \cdot (Fr_d)^{0.97} \cdot (Re_p)^{-0.44}$	(17)

* K_1 is 1.2 for live-bed scour and 1 for clear-water scour. K_2 is 1.1 for round, 1 for square, and 0.9 for a sharply pierced nose. Fr_{pier} is the pier Froude number, calculated as follows: $Fr_{pier} = \frac{v}{\sqrt{g \cdot b_{ef}}}$.

Annad and Lefkir [19] (AL abbreviated) proposed a new scour equation obtained by the same dimensional MNLR technique and the identical PSDB-2014 database as used in this study. The authors have retained almost all observed measurements (1249 of 1858), except for those that lack information regarding bridge pier shape. However, they have included only 4 variables: b, y, pier shape nose correction factor (K_2), and live-bed vs. clear water correction factor (K_1). Considering that their approach was similar to ours (the same database and the same modeling technique), their equation almost matches the dimensional model developed in this study (Equation (17)). According to Figure 5, there are similarities between the performance of our model and that of the AL model that can be assigned to the fact that in the AL model there was no restrictive data filtering, i.e., almost the whole span of measurements was considered.

Rathod and Manekar [28] (RM abbreviated) developed the GEP model for scour depth estimation using the PSDB-2014 database. In order to develop a unique and universal scour equation, the authors integrated laboratory and field datasets to obtain larger variable ranges. Even though the RM model was developed using the GEP method and on the combined nature of the data (lab and field), the model has a similar performance in comparison to the model developed in this study, probably due to the application of the PSDB-2014 database. Assuming that the type of technique selected for model development does not affect scour prediction performance as much as the filtering of the dataset, the performances of the RM and AL models are expected to be very close, as shown in Figure 5.

Hassan and Jalal [50] (HJ abbreviated) numerically simulated scour around the bridge pier on a real scale to evaluate the performance of predicting scour depth with the GEP model. They used a numerical dataset consisting of 243 observations collected in a clear-water regime and calibrated with the results from the Melville laboratory model [51]. The authors carried out a sensitivity analysis of influential variables and showed that y has the greatest influence on the predicted d_s, followed by the ratio of velocities v/v_c, the ratio of pier width B/b, the pier Froude number Fr_{pier}, and finally the pier shape factor K_s with the weakest influence. According to Figure 5, the HJ model tends to underpredict scour depth. Unfortunately, only ranges of nondimensional variables are given in their study, so the real span of variable values remains unknown. However, it can be assumed that Hassan and Jalal [50] collected numerical data in terms of lower water depths since flow depth turned out to be the most significant variable. This assumption is supported by the work of Melville and Sutherland [52], as they claim that flow depth does not play an important role in scour when the y/b_{ef} ratio is above 2.6. The final difference is that they have taken into account the ratio of pier width B/b, so it can be assumed that contraction scour has a more important role in their dataset.

Jain and Fischer [38] (JF abbreviated) is one of the many existing regression-based scour equations. The JF model was selected for this comparison process among other regression-based models because previous research [28], which applied the PSDB-2014 database, claims that the JF model is superior to the other conventional models. The JF model is developed in conditions of higher Fr values and sand-bed sediment properties (0.25–2.5 mm with a median of about 1.5 mm). Even though Jain and Fischer collected data in similar conditions that were applied for data filtration in this study, it is still an overly predictive and conservative method. The first reason for overprediction in JF model performance is related to including laboratory-based data, and the second reason is the traditional methodology of developing equations where coefficients were generated to form an envelope for all collected data.

Azamathulla et al.'s [37] (Az abbreviated) model presented the lowest performance amongst all equations selected for comparison in this study. Data in [37] were collected from different studies available in the literature to develop their MNLR model based on 398 field measurements collected over a non-cohesive and uniform riverbed. A possible reason for such low performance could be the larger range of sediments, with d_{50} values ranging from 0.12 to 108 mm and a median of 54 mm. Although information regarding the angle of flow attack is not available in their study, the authors included pier length in their equation. However, it can be assumed that piers were skewed, as otherwise, it would have no effect on scour depth [53]. Azamathulla et al. also provided an explicit equation for the GEP model, but when the PSDB-2014 measurements were input into the equation, the computed results were unreasonable, such as negative values of scour depth. In addition, parts of an equation of the developed MNLR model are vague, such as the exponent of the variable σ_g.

3.3. Variable Sensitivity Analysis

The Pearson correlation coefficient (r) (Figure 6) interprets the strength of the correlation between selected variables in the filtered subset and measured scour depth (d_s). The Pearson correlation coefficient varies in range from -1 to 1. A value of -1 indicates a perfect negative correlation, a value of 1 a perfect positive correlation, and a value of 0 indicates no correlation at all. If variables are classified into four categories: pier geometry (b_{ef}), hydraulic (y, v, τ, and Fr), compounded hydraulic-sediment variables (v_c, Fr_c, and Fr_d), and sediment (τ_c, and Re_p), then the pier geometric variables are the most influential, followed by compounded hydraulic-sediment, hydraulic, and finally the sediment variables. Based on Pearson correlation coefficients, b_{ef} proved to be the most influential variable for estimating d_s (r = 0.625), which is the same observation that was already stated in previous investigations [28,30]. The second most influential variable is y (r = 0.492), followed closely by v_c and v, r = 0.474 and r = 0.436, respectively. In the compounded hydraulic-sediment category, v_c has the greatest impact on scour depth in comparison to Fr_d and Fr_c, which are defined by r = 0.427 and r = 0.323, respectively. The remaining four variables (τ, τ_c, Fr, and Re_p) show a very weak correlation with scour depth, with $r < 0.3$.

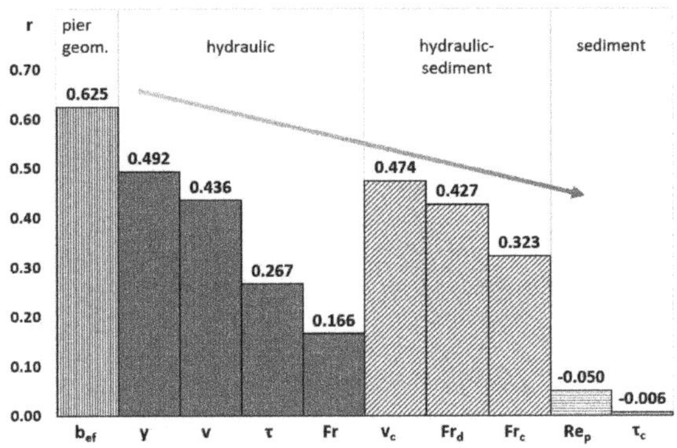

Figure 6. Pearson correlation coefficients (r) for selected influential variables that are classified into four variable categories. The selected variables are taken from the filtered dataset.

4. Discussion

In the present study, large, alluvial, and sand-bed rivers were taken into account with the intention of evaluating the significance of several influential variables in estimating maximum scour depth. Comparing four different categories, i.e., pier geometry (b_{ef}), hydraulic (y, v, τ, and Fr), sediment (τ_c, and Re_p), and compounded hydraulic-sediment variables (v_c, Frc, and Fr_d), it was elucidated that pier geometry variables are the most influential ones. Rating pier geometry variables above hydraulic variables was expected because more than 70% of the data used in this study exceeds the threshold value of y/b_{ef} = 2.6, over which the flow depth is no longer significant for the pier scour process [52].

The approach velocity and the flow depth have the next largest impact on scour depth, while local shear stress and Froude number remain the least important variables. The latter confirms the traditional approach assumptions that the equilibrium state of scour depth primarily depends on both the pier width and approach flow velocity, while flow depth is just an indirect effect of the downflow magnitude [38]. The introduction of new hydraulic variables such as τ and Fr is supposed not to significantly contribute to the variance of scour depth due to their unreliable estimation and small ranges of values. A similar conclusion can be drawn regarding the sediment variables. In this study, d_{50} was excluded from analysis, suspecting that it has a negligible effect, so two new sediment variables (τ_c, Re_p) were included in the scour depth equation. Even though it was expected that Re_p would play a more significant role owing to a larger standard deviation than d_{50}, both sediment variables τ_c and Re_p yielded the lowest Pearson correlation coefficients, which shed light on sediment properties in general (d_{50}, τ_c, and Re_p) as being insignificant on scour depth variance. In order to consider the effect of sediment properties, it was recommended to use compound hydraulic-sediment variables such as v_c, Fr_c, and Fr_d that consist of both the acting forces of the flow and sediment properties. In this study, the compounded hydraulic-sediment variables and hydraulic variables had a similar impact on the scour prediction because the average values of their Pearson correlation coefficients were 0.408 and 0.340, respectively. A comparable observation was made by Török et al. [54], who deem that shear Reynolds number, a function of grain size and shear velocity, is a more adequate variable for evaluating sediment transport than Re_p, d_{50}, or other variables whose evaluation is based only on sediment properties. Furthermore, at the beginning of this paper, there was a doubt concerning which variable should be used to assess the incipient motion of sediment, v_c or τ_c. The doubt originates from the fact that v_c is a function of y and d_{50}, while τ_c is a function of d_{50}. The assumption that the compound hydraulic-sediment variable would be more significant was verified by Pearson correlation coefficients of v_c and τ_c with values of 0.474 and -0.006, respectively. After v_c, which proved to be the most influential variable in the category of compounded hydraulic-sediment variables, Fr_d and Fr_c are next in the sequence. It was expected that Fr_d would show a stronger association with scour depth since it has been previously determined as the most influencing variable in the group of variables that describe sediment properties [55–57].

The introduction of sediment variables into the scour depth equation showed negligible effects. However, this observation does not mean that they can be completely excluded from the scour analysis, as sediment properties have a significant effect on the scour process. Equations developed for different compositions of riverbed sediments behave differently, as evidenced by the comparison process performed in this work. The Az model showed the lowest performance and was based on sediment ranges that tend toward gravel grain sizes. The fact that sediment variables play an important role indicates that it is necessary to take sediment size into account when filtering data sets to achieve certain environmental conditions. However, if sediment properties are selected as one of the scour variables in the equation, it is better to include them through compounded hydraulic-sediment variables than to take them directly because of their small contribution to the variance of the data set.

Recently, regression-based AI models have become widely used in scour studies because of their simple structure and their ability to find functional dependencies among a large number of scour-related variables. However, AI-based models tend to overfit the training dataset, and they are only appropriate for the range of variable values taken for the training. To avoid the aforesaid issues, it is important to appropriately choose specific environmental variables (flow regime, complex pier geometry, pier alignment, sediment uniformity and grain sizes, bridge in the bend, etc.) and their value ranges in order to filter the dataset adequately for AI-model training. If the data are too dispersed, the development of a unique and universal scour equation remains an option. For many years, researchers have been struggling to find a universal equation, which means finding an envelope for a dispersed cloud of data that will predict scour oversafely, i.e., the so-called conservative method.

To avoid overprediction, a best-fitting curve that passes through a cloud of datasets while minimizing the sum of the squared distances should be created. However, it is unlikely to develop such a curve with a machine learning algorithm trained to match the data as closely as possible to a dispersed and scattered dataset. For example, Rathod and Manekar [28] developed two GEP models: one based on a combination of field and laboratory data and one based on laboratory data only. The GEP model, whose development was based only on laboratory data, showed much better performance owing to smaller variable ranges and, consequently, a less scattered dataset. Furthermore, comparisons performed in this work (Figure 5) showed that the AL and RM models are superior to all other scour models. Although the models utilize different techniques such as MNLR and GEP, the deviation between them is quite small, and they have a high similarity in scour prediction. This possibly stems from the same dataset used (PSDB-2014), indicating that any prediction model's performance highly depends on the data and less on the method used for its development. The same can be said for the selection of influencing variables. In the AL and RM models, only 4 and 5 variables, respectively, were introduced, while in this study, 10 variables were considered for the development of the equation without improving the performance of the model. Without the detection of environmental conditions and the application of a specific filter, the data set will be too dispersed or too gathered, and neither the complex curve of the AI technique nor the trivialized envelope curve of the regression model will provide a more accurate prediction.

However, it must be emphasized that the dimensional MNLR model Equation (17) provided a low R^2 value (R^2 = 0.5) due to the presence of scatter in the dataset. Previous equations that used the same PSDB-2014 dataset (AL and RM) performed better, as indicated by higher R^2 or lower RMSE values. The reason could be that they used almost all field measurements without excluding multiple measurements for the same bridge. Such a collection of data with similar measurements could lead to overfitting. However, in this study, the independence of the measurements was maintained in the final set of filtration criteria by retaining only one measurement for each bridge, which eventually led to the exclusion of 46% from the original dataset. Another reason for the scatter is that the filtered dataset contains only field data where the scour regime is unknown; it is difficult to determine whether some measurements have reached an equilibrium or maximum state.

5. Conclusions

Most of the existing scour prediction equations available in the literature are derived from laboratory data since flume experiments provide controlled hydraulic conditions and straightforward data collection. On the other hand, the limited range of flow conditions in the flume limits the wider application of derived scour equations, i.e., their application to the prototype conditions. The USGS dataset PSDB-2014 provides a wide range of on-site measured flow, sediment, and geometry variables related to scour; thus, this dataset was the most representative source of scour data for large sand-bed rivers. The measured variables from the dataset were expanded to include eight additional calculated variables commonly used in scour prediction equations and consequently filtered to retain measurements taken only for large sand-bed rivers. After filtering the original dataset, 98% of the remaining measurements were within the clear-water scour conditions, even though the clear-water criterion was not applied in the filter, which validates the reliability of the applied data filter.

Several influential variables were removed by using the PCA, as well as two commonly used variables, θ and d_{50}, as they showed low impact on the scour depth. Since the measurement of θ is arbitrary and changes with flow severity, it should be used only when it can be reliably estimated, which is not often the case. Since the pier length was also eliminated by the PCA and taking into consideration the difficult approximation of complex pier geometry with a single variable, this study selected b_{ef} as the only pier geometry variable, combining the information of pier geometry and alignment with the flow, if available. PCA has eliminated the characteristic sediment size as well, which can be explained by focusing on sand-bed rivers where sand is uniform and therefore has a significantly smaller range than other variables. To take into account riverbed composition, compounded hydraulic-sediment variables (v_c, Fr_c, and Fr_d) were retained, and consequently, 10 influential variables were classified into four categories: pier geometry (b_{ef}), hydraulic (y, v, τ, and Fr), sediment (τ_c, and Re_p), hydraulic-sediment variables (v_c, Fr_c, and Fr_d), and compounded. Afterward, the variables from these categories were used to determine the scour model variables.

The proposed dimensional MNLR model designed for scour estimation on the selected data subset has a firm similarity with the two other models (AL and RM) developed using the same PSDB-2014 dataset but with different filters or methods. The comparison performed in this work indicates that the filtering method has a greater influence on the model's performance than the model type (MNLR or GEP). However, the MNLR model developed in this study, Equation (17), yielded a low R^2 value of 0.5 due to the scatter of the dataset, which could be a consequence of excluding multiple measurements for the same bridge locations. Since, at the time of the field measurement, it is not known whether the equilibrium or maximum state has been reached, dispersion in the results is expected.

Furthermore, the selection of influential variables that can be reliably measured or estimated is crucial when creating a database for scour estimation. Based on the Pearson correlation coefficient, bef proved to be the most influential variable for estimating ds in this study, and the second most influential variable is y, followed closely by v_c, v, Fr_d, and Fr_c. The remaining four variables (τ, τ_c, Fr, and Re_p) exhibit a very weak correlation with scour depth, probably resulting from errors in the measurement of variables used for their calculation. The fact that compounded hydraulic-sediment variables were highly influential for our dimensional model indicates that it is necessary to use sediment size when filtering data to reduce the uncertainty associated with the acquisition of a representative bed sample.

Author Contributions: Conceptualization, A.H. and G.G.; methodology, A.H., M.Ž. and N.A.; validation, A.H. and G.G.; formal analysis, A.H. and N.A.; investigation, G.G.; data curation, A.H.; writing—original draft preparation, A.H. and G.G.; writing—review and editing, G.G., N.A. and M.Ž.; visualization, A.H.; supervision, G.G.; project administration, G.G. All authors have read and agreed to the published version of the manuscript.

Funding: This work has been funded in part by the Croatian Science Foundation under the project R3PEAT (UIP-2019-04-4046).

Institutional Review Board Statement: Not applicable.

Informed Consent Statement: Not applicable.

Data Availability Statement: Not applicable.

Conflicts of Interest: The authors declare no conflict of interest.

Glossary

Symbol	Unit	Description
d_s	[m]	scour depth
v	[m/s]	local approach flow velocity (upstream of the pier)
v_c	[m/s]	critical velocity
y	[m]	approach water depth (upstream of the pier)
b	[m]	nominal pier width
l	[m]	pier length
b_{ef}	[m]	effective pier width normal to the flow
θ	[°]	angle of attack
B	[m]	the channel width
S	[1]	stream slope
τ	[Pa]	local bed shear stress
τ_c	[Pa]	critical bed shear stress
Fr	[1]	Froude number
Fr_c	[1]	critical Froude number for incipient motion
Fr_d	[1]	densimetric Froude number
Re_p	[1]	particle Reynolds number
RI	[years]	recurrence interval for measured flow rate
g	[m/s^2]	gravitational acceleration
d_{50}	[mm]	sediment median grain size
d_{95}	[mm]	the size at which 95% of the sediment particles are smaller
σ_g	[1]	geometrical standard deviation of sediment (measure of non-uniformity)
ρ_{rel}	[1]	submerged relative mass density of sediment particles ($\rho_{rel} = [(\rho_s - \rho)/\rho] - 1 = 1.65$)
ρ_s	[kg/m^3]	mass density of sediment particles (equal to 2650 kg/m^3)
ρ_w	[kg/m^3]	mass density of water (equal to 1000 kg/m^3)
γ_s	[N/m^3]	specific gravity of sediment (equal to 25,996.5 N/m^3)
γ_w	[N/m^3]	specific gravity of water (equal to 9810 N/m^3)
K_1	[1]	the live-bed vs. clear-water correction factor
K_2	[1]	the pier shape correction factor
ν	[m^2/s]	kinematic viscosity of fluid ($\nu = 1.6 \times 10^{-6}$ m^2/s)
Φ	[°]	angle of repose for sediments
r	[1]	Pearson correlation coefficient

References

1. Grizzetti, B.; Pistocchi, A.; Liquete, C.; Udias, A.; Bouraoui, F.; Van de Bund, W. Human pressures and ecological status of European rivers. *Sci. Rep.* **2017**, 7, 205. [CrossRef]
2. Sholtes, J.S.; Ubing, C.; Randle, T.J.; Fripp, J.; Cenderelli, D.; Baird, D.C. *Managing Infrastructure in the Stream Environment*; Advisory Committee on Water Information Subcommittee on Sedimentation: Austin, TX, USA, 2017; p. 65.
3. Lee, M.; Yoo, M.; Jung, H.-S.; Kim, K.H.; Lee, I.-W. Study on Dynamic Behavior of Bridge Pier by Impact Load Test Considering Scour. *Appl. Sci.* **2020**, 10, 6741. [CrossRef]

4. Kallias, A.N.; Imam, B. Probabilistic assessment of local scour in bridge piers under changing environmental conditions. *Struct. Infrastruct. Eng.* **2016**, *12*, 1228–1241. [CrossRef]
5. Imhof, D. Risk Assessment of Existing Bridge Structures. Ph.D. Thesis, University of Cambridge, Cambridge, UK, 2004.
6. Schaap, H.S.; Caner, A. Bridge collapses in Turkey: Causes and remedies. *Struct. Infrastruct. Eng.* **2022**, *18*, 694–709. [CrossRef]
7. Yao, C.; Briaud, J.-L.; Gardoni, P. Risk Analysis on Bridge Scour Failure. In Proceedings of the International Foundations Congress and Equipment Expo, San Antonio, TX, USA, 17–21 March 2015; pp. 1936–1945.
8. Nasr, A.; Björnsson, I.; Honfi, D.; Larsson Ivanov, O.; Johansson, J.; Kjellström, E. A review of the potential impacts of climate change on the safety and performance of bridges. *Sustain. Resilient Infrastruct.* **2021**, *6*, 192–212. [CrossRef]
9. Kundzewicz, Z.W.; Pińskwar, I. Are Pluvial and Fluvial Floods on the Rise? *Water* **2022**, *14*, 2612. [CrossRef]
10. Nemry, F.; Demirel, H. *Impacts of Climate Change on Transport: A Focus on Road and Rail Transport Infrastructures*; European Commission Joint Research Centre: Luxembourg, 2012.
11. Badroddin, M.; Chen, Z. Lifetime Resilience Measurement of River-Crossing Bridges with Scour Countermeasures under Multiple Hazards. *J. Eng. Mech.* **2021**, *147*, 04021058. [CrossRef]
12. Tubaldi, E.; White, C.J.; Patelli, E.; Mitoulis, S.A.; de Almeida, G.; Brown, J.; Cranston, M.; Hardman, M.; Koursari, E.; Lamb, R.; et al. Invited perspectives: Challenges and future directions in improving bridge flood resilience. *Nat. Hazards Earth Syst. Sci.* **2022**, *22*, 795–812. [CrossRef]
13. Imam, B.M.; Chryssanthopoulos, M.K. Causes and Consequences of Metallic Bridge Failures. *Struct. Eng. Int.* **2012**, *22*, 93–98. [CrossRef]
14. Borghei, S.M.; Kabiri-Samani, A.; Banihashem, S.A. Influence of unsteady flow hydrograph shape on local scouring around bridge pier. *Proc. Inst. Civ. Eng. -Water Manag.* **2012**, *165*, 473–480. [CrossRef]
15. Hung, C.-C.; Yau, W.-G. Behavior of scoured bridge piers subjected to flood-induced loads. *Eng. Struct.* **2014**, *80*, 241–250. [CrossRef]
16. Lu, J.-Y.; Hong, J.-H.; Su, C.-C.; Wang, C.-Y.; Lai, J.-S. Field Measurements and Simulation of Bridge Scour Depth Variations during Floods. *J. Hydraul. Eng.* **2008**, *134*, 810–821. [CrossRef]
17. Harasti, A.; Gilja, G.; Potočki, K.; Lacko, M. Scour at Bridge Piers Protected by the Riprap Sloping Structure: A Review. *Water* **2021**, *13*, 3606. [CrossRef]
18. Benedict, S.T.; Caldwell, A.W. Upper Bound of Pier Scour in Laboratory and Field Data. *Transp. Res. Rec.* **2016**, *2588*, 145–153. [CrossRef]
19. Annad, M.; Lefkir, A. New Formula for Calculating Local Scour around Bridge Piers. *Adv. Eng. Forum* **2022**, *45*, 57–64. [CrossRef]
20. Gaudio, R.; Grimaldi, C.; Tafarojnoruz, A.; Calomino, F. Comparison of formulae for the prediction of scour depth at piers. In Proceedings of the First European IAHR Congress, Edinburgh, UK, 4–6 May 2010; pp. 6–12.
21. Zhang, G.; Hsu, S.A.; Guo, T.; Zhao, X.; Augustine, A.D.; Zhang, L. *Evaluation of Design Methods to Determine Scour Depths for Bridge Structures*; FHWA/LA.11/491; Louisiana State University: Baton Rouge, LA, USA; Federal Highway Administration: Washington, DC, USA, 2013.
22. Dong, H.; Sun, Z.; Li, Z.; Chong, L.; Zhou, H. Artificial Intelligence for Predicting Local Scour Depth around Piers Based on Dimensional Analysis. *J. Coast. Res.* **2020**, *111*, 21–25. [CrossRef]
23. Pandey, M.; Zakwan, M.; Sharma, P.K.; Ahmad, Z. Multiple linear regression and genetic algorithm approaches to predict temporal scour depth near circular pier in non-cohesive sediment. *ISH J. Hydraul. Eng.* **2018**, *26*, 96–103. [CrossRef]
24. Rady, R.M.A.E.-H. Prediction of local scour around bridge piers: Artificial-intelligence-based modeling versus conventional regression methods. *Appl. Water Sci.* **2020**, *10*, 57. [CrossRef]
25. Muzzammil, M.; Alama, J.; Danish, M. Scour Prediction at Bridge Piers in Cohesive Bed Using Gene Expression Programming. *Aquat. Procedia* **2015**, *4*, 789–796. [CrossRef]
26. Benedict, S.T.; Caldwell, A.W. *A Pier-Scour Database: 2427 Field and Laboratory Measurements of Pier Scour*; Data Series 845; U.S. Geological Survey: Reston, VA, USA, 2014; p. 32.
27. Benedict, S.T.; Knight, T.P. Use of Laboratory and Field Data to Evaluate the Pier Scour Equation from Hydraulic Engineering Circular 18. *Transp. Res. Rec.* **2017**, *2638*, 113–121. [CrossRef]
28. Rathod, P.; Manekar, V.L. Gene expression programming to predict local scour using laboratory and field data. *ISH J. Hydraul. Eng.* **2022**, *28*, 143–151. [CrossRef]
29. Pandey, M.; Oliveto, G.; Pu, J.H.; Sharma, P.K.; Ojha, C.S.P. Pier Scour Prediction in Non-Uniform Gravel Beds. *Water* **2020**, *12*, 1696. [CrossRef]
30. Ali, A.S.A.; Günal, M. Artificial Neural Network for Estimation of Local Scour Depth Around Bridge Piers. *Arch. Hydro-Eng. Environ. Mech.* **2021**, *68*, 87–101. [CrossRef]
31. Shahriar, A.R.; Ortiz, A.C.; Montoya, B.M.; Gabr, M.A. Bridge Pier Scour: An overview of factors affecting the phenomenon and comparative evaluation of selected models. *Transp. Geotech.* **2021**, *28*, 100549. [CrossRef]
32. Qi, M.; Li, J.; Chen, Q. Comparison of existing equations for local scour at bridge piers: Parameter influence and validation. *Nat. Hazards* **2016**, *82*, 2089–2105. [CrossRef]
33. Guo, J.; Suaznabar, O.; Shan, H.; Shen, J. *Pier Scour in Clear-Water Conditions with Non-Uniform Bed Materials*; FHWA-HRT-12-022; Federal Highway Administration: Washington, DC, USA, 2012; p. 62.

34. Shields, A. *Application of Similarity Principles and Turbulence Research to Bed-Load Movement*; Hydrodynamics Laboratory: Washington, DC, USA, 1936; p. 47.
35. Shahmohammadi, R.; Afzalimehr, H.; Sui, J. Assessment of Critical Shear Stress and Threshold Velocity in Shallow Flow with Sand Particles. *Water* **2021**, *13*, 994. [CrossRef]
36. Julien, P.Y. *Erosion and Sedimentation*; Cambridge University Press: Cambridge, UK, 1995.
37. Azamathulla, H.M.; Ghani Aminuddin, A.; Zakaria Nor, A.; Guven, A. Genetic Programming to Predict Bridge Pier Scour. *J. Hydraul. Eng.* **2010**, *136*, 165–169. [CrossRef]
38. Jain, S.C.; Fischer, E.E. *Scour around Circular Bridge Piers at High Froude Numbers*; FHWA-RD-79-104; Federal Highway Administration: Washington, DC, USA, 1979; p. 70.
39. Dey, S.; Raikar, R.V. Clear-Water Scour at Piers in Sand Beds with an Armor Layer of Gravels. *J. Hydraul. Eng.* **2007**, *133*, 703–711. [CrossRef]
40. Benedict, S.T.; Knight, T.P. Benefits of Compiling and Analyzing Hydraulic-Design Data for Bridges. *Transp. Res. Rec.* **2021**, *2675*, 1073–1081. [CrossRef]
41. Garde, R.C.J.; Kothyari, U.C. Scour around bridge piers. *PINSA 64* **1998**, *4*, 569–580.
42. Subedi, A.S.; Sharma, S.; Islam, A.; Lamichhane, N. Quantification of the Effect of Bridge Pier Encasement on Headwater Elevation Using HEC-RAS. *Hydrology* **2019**, *6*, 25. [CrossRef]
43. Harasti, A.; Gilja, G.; Adžaga, N.; Škreb, K.A. Principal Component Analysis in development of empirical scour formulae. In Proceedings of the 7th IAHR Europe Congress, Athens, Greece, 7–9 September 2022; pp. 271–272.
44. Zwick, W.R.; Velicer, W.F. Comparison of five rules for determining the number of components to retain. *Psychol. Bull.* **1986**, *99*, 432–442. [CrossRef]
45. Breusers, H.N.C.; Raudkivi, A.J. *Scouring*, 1st ed.; Taylor and Francis Group: London, UK, 1991; p. 152.
46. Laursen, E.M. Scour at Bridge Crossings. *J. Hydraul. Div.* **1960**, *86*, 39–54. [CrossRef]
47. Kiraga, M.; Popek, Z. Bed Shear Stress Influence on Local Scour Geometry Properties in Various Flume Development Conditions. *Water* **2019**, *11*, 2346. [CrossRef]
48. Parker, G. Transport of Gravel and Sediment Mixtures. In *Sedimentation Engineering*; Garcia, M., Ed.; American Society of Civil Engineers: Reston, VA, USA, 2008; pp. 165–251.
49. Vonkeman, J.K.; Basson, G.R. Evaluation of empirical equations to predict bridge pier scour in a non-cohesive bed under clear-water conditions. *J. S. Afr. Inst. Civ. Eng.* **2019**, *61*, 2–20. [CrossRef]
50. Hassan, W.H.; Jalal, H.K. Prediction of the depth of local scouring at a bridge pier using a gene expression programming method. *SN Appl. Sci.* **2021**, *3*, 159. [CrossRef]
51. Melville, B.W. *Local Scour at Bridge Sites*; University of Auckland: Auckland, New Zealand, 1975; p. 227.
52. Melville, B.W.; Sutherland, A.J. Design Method for Local Scour at Bridge Piers. *J. Hydraul. Eng.* **1988**, *114*, 1210–1226. [CrossRef]
53. Richardson, E.V.; Davis, S.R. *Evaluating Scour at Bridges*, 4th ed.; FHWA NHI 01-001; Hydraulic Engineering Circular No. 18; Federal Highway Administration: Washington, DC, USA, 2001; p. 378.
54. Török, G.T.; Józsa, J.; Baranya, S. A Shear Reynolds Number-Based Classification Method of the Nonuniform Bed Load Transport. *Water* **2019**, *11*, 73. [CrossRef]
55. Oliveto, G.; Hager, W.H. Temporal Evolution of Clear-Water Pier and Abutment Scour. *J. Hydraul. Eng.* **2002**, *128*, 811–820. [CrossRef]
56. Oliveto, G.; Hager, W.H. Further Results to Time-Dependent Local Scour at Bridge Elements. *J. Hydraul. Eng.* **2005**, *131*, 97–105. [CrossRef]
57. Tan, S.M.; Lim, S.-Y.; Wei, M.; Cheng, N.-S. Application of Particle Densimetric Froude Number for Evaluating the Maximum Culvert Scour Depth. *J. Irrig. Drain. Eng.* **2020**, *146*, 04020020. [CrossRef]

applied sciences

MDPI

Article

Coherent Flow Structures Linked to the Impulse Criterion for Incipient Motion of Coarse Sediment

Khaldoon AlObaidi and Manousos Valyrakis *,†

Infrastructure and Environment Research Division, School of Engineering, University of Glasgow, Glasgow G12 8TA, UK

* Correspondence: mvalyrak@gmail.com

† Current address: Department of Civil Engineering, Aristotle University of Thessaloniki, 54124 Thessaloniki, Greece.

Abstract: Incipient motion has been a topic of investigation by researchers, engineers and scientists for more than a century. The main approach for studying sediment entrainment has been the static approach that uses temporal and spatial averaged flow parameters like bed shear stress and stream power to link them indirectly to sediment entrainment. Recent research outputs have shed light on the important role of turbulent fluctuations in the sediment transport process. It is suggested that the approach of using temporal and spatial averaged parameters fails to account for the dynamic and probabilistic nature of the entrainment process, as inherited by flow turbulence. This has led to the introduction of the only dynamic criteria in the literature for studying sediment entrainment, namely the impulse and energy criteria. These criteria take into account both the magnitude and duration of the turbulent flow event used for assessing the conditions that can result in sediment entrainment. In light of this, this work aims to assess whether there is a trend in terms of the type of flow structures that occur in sequence before and after the occurrences of the flow impulses that have resulted in the coarse particle's entrainment. To achieve this, we conducted a well-controlled laboratory experiment to investigate the incipient motion of a 7 cm diameter instrumented particle. Five runs of the experiment were performed at flowrates close to the threshold of motion. The instrumented particle was equipped with micro-electro-mechanical sensors (MEMS) to accurately measure its inertial dynamics and detect motion. The sensors recorded entrainment events, and these events were stochastically linked to the impulses occurring for the tested flow conditions. Quadrant analysis was used to investigate the type of flow structures that occurred before, during and after the occurrence of quadrant events with an impulse above the critical impulse. The findings herein associate coarse particle entrainments with energetic impulses linked primarily to sweep events (Q_4) and secondarily, sequence of sweeps (Q_4) and ejections (Q_1).

Keywords: instrumented particles; MEMS sensors; frequency of entrainment; incipient motion; turbulent flows; flow structures; quadrant analysis

Citation: AlObaidi, K.; Valyrakis, M. Coherent Flow Structures Linked to the Impulse Criterion for Incipient Motion of Coarse Sediment. *Appl. Sci.* **2023**, *13*, 10656. https://doi.org/10.3390/app131910656

Academic Editor: Kelin Hu

Received: 21 August 2023
Revised: 13 September 2023
Accepted: 15 September 2023
Published: 25 September 2023

1. Introduction

The topic of sediment transport has been studied by scientists, engineers and researchers for more than a century due to its application in the fields of environmental monitoring, infrastructure monitoring, geosciences and marine biology. Different criteria and correlations have been developed to assess the conditions that result in the entrainment of a sediment particle and to estimate the bed-load transport rate. These criteria and correlations do not represent the full complexity of the process, which depends on multiple parameters, since they are mostly based on simple physics or simplified laboratory experiments [1,2]. These correlations and criteria have shown there to be large discrepancies between each other and between the actual data [3]. Historically, there are two approaches that have been followed to study sediment entrainment: the static approach and the dynamic approach. The static approach uses time-averaged parameters like bed shear stress [4] or critical stream power [5,6] to define the conditions that result in sediment entrainment. Since turbulence by definition is chaotic and random around the averaged values of pressure or velocity [7], multiple researchers [8–14] have tried to emphasise the important role of turbulent fluctuations in the sediment transport process in an attempt to overcome the limitations of the time-averaged parameters approach or the static approach. Some researchers have reported that increasing the turbulence level in the boundary layer while maintaining the same time-averaged shear stress in the boundary layer has resulted in an increase in bed-load transport rates [15,16]. Other researchers [17,18] have suggested that instantaneous normal Reynolds shear stresses are more correlated with sediment movement compared to average or instantaneous shear stresses. Thereafter, [19] suggested that both the force magnitude and duration of the energy near to the bed turbulent events are relevant when predicting sediment entrainment. He also introduced the impulse criterion, which is the first dynamic criterion used for studying sediment entrainment reported in the literature. Valyrakis et al. [20] and Celik et al. [21] generalised the applicability of the concept and Valyrakis et al. [22] extended it to the energy criterion which is the second dynamic criterion that exists in the literature for assessing the conditions that can result in sediment entrainment. The energy criterion formulates that the complete removal of sediment from the bed matrix occurs only if the impinging flow event(s) supply enough mechanical energy.

As part of a further investigation into the relationship between turbulence and sediment entrainment, researchers have started to investigate whether there are specific flow events that are associated with the entrainment of sediment. Specifically, the observation of Kline et al. [23] showed that among the random and complex turbulent events, there are elementary organised structures with spatial and temporal coherence. This has generated interest among researchers in relation to further investigating the flow structures that exist in boundary layers, linking the observation of Kline et al. [23] to the sediment entrainment problem. At the time of this work, there has been an extensive research output on the investigation of the coherent flow structures in the boundary layers for rough and smooth wall flows [24–26]. As for the study of sediment transport, some researchers [12,14,27] have reported that although there are chaotic turbulent fluctuations, there are specific events or flow structures that are associated with sediment transport episodes. These events have both temporal and spatial coherence. The method of quadrant analysis has been used widely to report the type of events occurring during sediment entertainment and sediment transport. The method was developed by Wallace et al. [28] to quantify the visual observations of Corino and Brodkey [29] by dividing the turbulent velocity fluctuations based on their signs into Q_1 ($+u'$, $+w'$), Q_2 ($-u'$, $+w'$), Q_3 ($-u'$, $-w'$) and Q_4 ($+u'$, $-w'$). Ejection (Q_2) and sweep (Q_4) events contribute positively to Reynolds shear stresses. These events were named by Corino and Brodkey [29] based on their experimental observations. Outward (Q_1) and inward interactions (Q_3) contribute negatively to Reynolds shear stresses and were named by Wallace et al. [28] for a lack of a better descriptive terminology. Heathershaw and Thorne (1985) demonstrated that most of the sediment entrainment has been caused by sweep events (Q_4) and to a lesser extent by outward interactions (Q_1). Multiple researchers [30–40] have suggested that entrainment is associated with high Reynolds shear stress structures, Q_4 or sweep, while ejections are responsible for the particles' suspended motion. Additionally, there is a tendency for there to be an existence of a group of increasingly stronger Q_2 (ejection) events followed immediately by Q_4 (sweep) events during the burst–sweep cycle which have resulted in sediment entrainment. These two quadrants have been reported to contribute to Reynolds stresses and momentum transfer, which is much stronger near the wall region [41,42]. Nelson et al. [43], on the other hand, suggested that the sweep component of the burst–sweep cycle resulted in the majority of the sediment transport due to its common occurrence. The events of outward interactions of the same duration and magnitude can result in the same transportation as sweep events do.

In another stream of investigation into sediment transport, some researchers have tried to link sediment entrainment to turbulent flow structures and to fluctuating drag and lift using laser Doppler velocimetry (LDV) and particle image velocimetry (PIV) measurements in addition to pressure measurements [21,44–50]. The conclusion of all their studies is that there are Q_4 events associated with high positive drag force fluctuations at the time of sediment particle entrainment. As for the lift fluctuations, their role is less clear in sediment entrainment. Dwivedi et al. [46] reported positive lift and drag during sweeps, Paiement-Paradis et al. [51] suggested that sweep may induce negative lift and finally, Smart and Habersack [50] have reported there to be advected pressure fluctuations due to passing vortices which lead to a positive lift just before Q_4 events. These are unrelated to form drag or form lift. In general, there are not many studies that have investigated the relationship between coherent flow structures and pressure fluctuations.

As of the time of this work, there is no precise or universal criterion for determining the conditions that result in the initiation of sediment movement as an incipient motion. However, the impulse and energy criteria seem to inherit fundamental physical principles and account for all of the parameters that can affect the initiation of a particle's dislodgement. The literature is absent on establishing a link between the impulse criterion and the flow structures that have caused the entrainment of a sediment particle. The main aim of this work was to assess whether there was any manner of trend in terms of the type of flow structures that occur in sequence before and after the occurrence of the flow impulses that have resulted in the particle's entrainment. The analysis in this work was performed with the assumption that the highest flow impulses have resulted in a higher particle mobility. In other words, there is no dependence on the energy transferred to the particle due to the type of flow structure.

In this study Acoustic Doppler Velocimetry (ADV) was employed to conduct laboratory flume experiments near motion thresholds. A (7 cm diameter) instrumented particle with MEMS sensors [52–54] was also used to record its dynamics of entrainment (acceleration and rotational velocity) under these near bed surface turbulent flow conditions. The coarse particle's inertial dynamics were quantified, and the frequency of entrainment was estimated using normalized rolling entrainment results. The estimated frequency of flow impulses above the critical impulse was linked stochastically to the coarse particle's entrainment frequency. The study also used flow velocity measurements and quadrant analysis to assess the quadrant readings during flow impulses above the critical impulse before and after the instance of their occurrence.

2. Methodology

Before investigating the flow structures that have occurred during, before and after the occurrence of impulses above the critical impulse using a well-controlled flume experiment, the sensors of the instrumented particle were calibrated separately using simple and easy to validate theoretically physical motions (see [52] and Appendix 1 in [54]) to quantify the uncertainties in their readings. The casing of the instrumented particle is designed to accommodate different quantities of weighted lead strips, thus offering an adjustable density, which ranges from 1200 to 2900 kg/m^3, when the embedded sensors are also considered. A 2023 kg/m^3 solid particle density is used for this study, which is not far from the natural density of quartz pebbles, but it allows rendering the flow conditions used in the flume, close to the incipient motion threshold. The instrumented particle's calibration process consisted of calibrating the accelerometer and gyroscope separately. For the accelerometer's calibration, a free fall from a known height experiment was performed. The free fall height estimated using the accelerometer's reading was compared to the theoretical height estimated using Newton's Second Law of Motion and the uncertainty in the accelerometer's reading was quantified. As for the gyroscope's calibration, the rolling down an inclined plane experiment was performed. The total angular velocity down the inclined plane estimated using the gyroscope's readings was compared to the theoretical total angular velocity estimated using the law of conservation of energy. The uncertainty in the gyroscope's reading was quantified. The error values provided the inertial sensor fusion filter's input to achieve the uncertainty reduction. The final error in the sensor's readings only reached 1.1%, which is considered to be a very small value. The sensors were then inserted into the instrumented particle, which was used to conduct the flume experiment.

The experiments in this work took place in a tilting and water recirculating flume in the Water Engineering Laboratory at the University of Glasgow. The flume was 8 m in length and 0.9 m in width with a water capacity of 0.2 m^3/s and a carry flow up to 0.4 m deep was used (Figure 1). The bed was covered with water-worked uniform-sized gravel (d_{50} = 25 mm) to achieve an adequate hydraulic roughness. The 1 m test section was located approximately 5.25 m downstream from the flume inlet and 1.75 m upstream from the flume outlet to ensure that the hydraulically rough turbulent flow was fully developed, as shown in Figure 1.

Each run of the experiment consisted of performing the sensor measurements and flow velocity measurements. In general, there are two main approaches in assessing the relationship between the instantaneous flow hydrodynamic forcing (cause) and resulting particle entrainments (effect). This can be done deterministically, if the flow and particle instrumentation are synchronously taking records, or stochastically, by means of cross-examining the trends between these records, especially useful if these records have been taken separately but refer to the same particle and flow conditions. The use of the former deterministic in-synch records obtaining approach is only possible via the employment of non-intrusive velocimetry methods (such as laser Doppler velocimetry or particle image velocimetry). Given that the experiments carried herein use ADV, which can locally interfere with the particle, the latter stochastic assessment approach is used herein. This involves using the particle movement records towards indirectly inferring the flow impulses that are effective in entraining the instrumented particle, which are then stochastically compared to the directly measured flow hydrodynamics of the same flow. The particle entrainment and flow velocimetry records are obtained separately, having only one piece of instrumentation at a time to avoid either interfering with the other. In this manner, it is secured that there is no interference between the used instrumentation. The ADV is positioned 1 cm upstream of the face of the instrumented particle, and it is configured appropriately to record instantaneously the three components of the local flow field [55]. Flow velocity profiles are then taken after the instrumented particle has been removed, with a vertical resolution of about 7 mm. Likewise, the entrainment record of the instrumented particle is performed after the flow measurement instrumentation has been removed and can be obtained live (via the WiFi protocol) or otherwise saved in the local memory for retrieval after the conclusion of the flow experiment.

Five runs were performed, and the flow velocity measurements and the sensor measurements were used to investigate the flow structures that resulted in entraining the instrumented particle stochastically. For the stochastic analysis of particle entrainment, both the flow and particle entrainment time series are required to be long enough so that they have statistics that are representative and independent of the duration of the records. For the typical flow conditions assessed herein, such durations have been experimentally established to be 8 min for the ADV flow hydrodynamic records (12,000 samples at 25 Hz recording rate) and above half an hour for the instrumented particle entrainments records (more than 180,000 samples when recording at 100 Hz). These five runs were used for the analysis of the flow structures that occurred during, before and after the occurrence of impulses above the critical impulse. The acquired instantaneous flow velocity data were de-spiked using the methodology presented by Goring and Nikora [56], modified by Wahl [57] and Mori et al. [58]. The post-processed velocity measurements were then used to estimate the shear velocity (u_*) using the Grass method [10] as discussed in [54]. The shear velocity was used to estimate the particle Reynolds number and the Shields parameter for the five runs of the experiment. The results are shown in Section 3, Table 1.

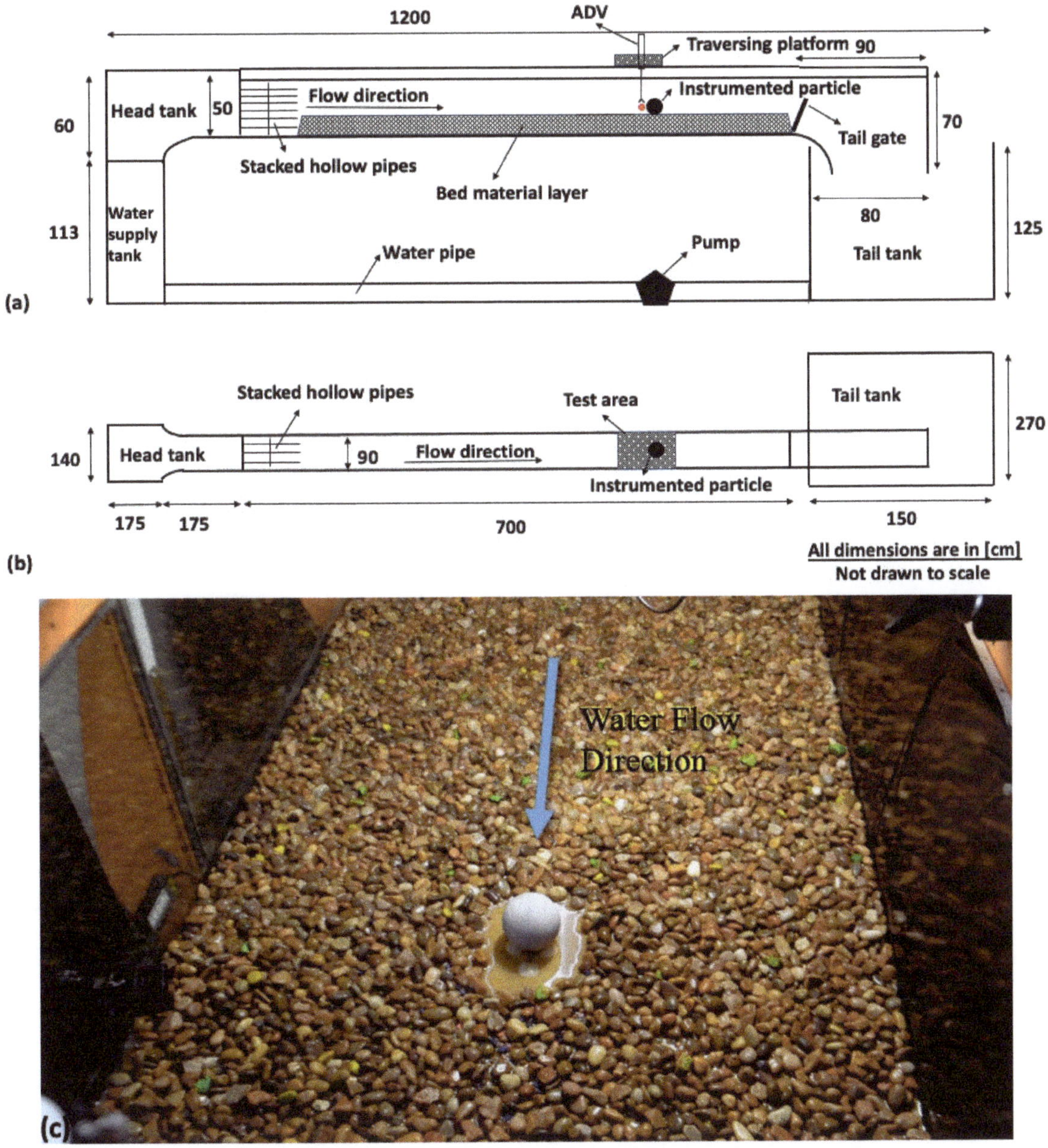

Figure 1. Illustration of the experimental setup: (**a**) sketch of the side view of the flume, also showing the test section and the relevant instrumentation (acoustic Doppler velocimetry and the instrumented particle), (**b**) the top view and (**c**) photo from the experimental setup, illustrating the use of the instrumented particle sitting on top of a dedicated 3D-printed micro-topography, transferring entrainment data in real time via the WiFi protocol (effective for submergence depths below 13.5 cm—see the WiFi antenna attached on the frame above the target particle), the high-speed video camera (used for visually assessing particle entrainments from the side) and the instrumented particle. The flow direction is also shown.

Table 1. The flow characteristics (mean flow velocity, mean flow depth, Froude number, flow and particle Reynolds number (*Re* and Re_*, respectively) and average bed surface Shields shear stresses) and resulting entrainment particle frequency for the 5 runs of the study.

Run	Flow Vel. [m/s]	Flow Depth [cm]	Fr	Re	$Re_* = \frac{\rho u_* D}{\mu}$	$\tau_* = \frac{\tau_o}{(\rho_s - \rho_w) g D}$	f_e [entrainment/min]
Run1	0.345	11.4	0.32	39,416	4528.20	0.0047	0.42
Run2	0.346	11.5	0.32	39,768	4774.57	0.0052	0.97
Run3	0.357	11.7	0.33	41,880	5067.38	0.0059	1.21
Run4	0.358	11.8	0.33	42,231	5451.77	0.0068	4.91
Run5	0.359	11.9	0.33	42,760	5862.61	0.0079	6.12

The instantaneous flow velocity measurements 1 cm upstream of the instrumented particle's face, in the longitudinal direction, and in the position in the middle of the top half of the particle's diameter, in the bed-normal direction, were used as the input to the theoretical equations presented in [21,22], to estimate the flow impulses (as is also shortly illustrated, in Figure 2). This corresponds to point velocimetry taken at a fixed bed-normal distance from the bed surface, which ranges from about 1/3 to 2/5 of the flow depth. This position has been referred to as the position of investigation in this work. The quadrant analysis method was performed using the instantaneous flow velocity measurements at the position of the investigation. As presented in the introduction section, in the quadrant analysis method, the turbulent velocity fluctuations were divided into 4 categories, or quadrants, based on their signs: Q_1 $(+u', +w')$, Q_2 $(-u', +w')$, Q_3 $(-u', -w')$ and Q_4 $(+u', -w')$. The turbulent velocity fluctuations were estimated based on Equations (1) and (2) below:

$$u'_i = u_i - \overline{u} \tag{1}$$

$$w'_i = w_i - \overline{w} \tag{2}$$

where u and w are the instantaneous flow velocity readings (stream-wise and normal to the bed surface components), u' and w' are the turbulent velocity fluctuations (stream-wise and normal to the bed surface components), $\overline{u}$ and $\overline{w}$ are the time-averaged velocity (stream-wise and normal components) and i is the ADV instantaneous reading. Using each instantaneous velocity reading, or basic ADV reading, the quadrant that corresponds to the reading was defined. An illustration of the analysis that was performed using the flow velocity measurements to estimate the flow impulses and determine the quadrants is shown in Figure 2 below.

As discussed in [54], the frequency of entrainment of the instrumented particle was estimated using the normalised roll results by the inertial sensor fusion of the instrumented particle's logged readings during the experiment. As for the flow impulses, which were estimated based on the illustration shown in Figure 2, they were used to estimate the frequency of flow impulses by dividing the total number of flow impulses by the total ADV recording time in each position, including the position of investigation. Given that the contribution of the Q_2 and Q_4 quadrants is more significant near the bed surface, it is expected that for smaller particle-size-to-flow-depth ratios, these quadrants will play a greater role in particle entrainment. Thus, it is expected that smaller particle sizes will have an even more pronounced contribution of the Q_2 and Q_4 events than found herein.

In the position of investigation, different theoretical critical impulse values (I_{cr}) were tested to determine the value that results in a good agreement between the frequency of entrainment of the instrumented particle and the frequency of flow impulses above that value, or above the critical impulse (f_I). Figure 3 below shows the selected critical value in addition to slightly higher and lower values for the critical impulse, for comparison purposes.

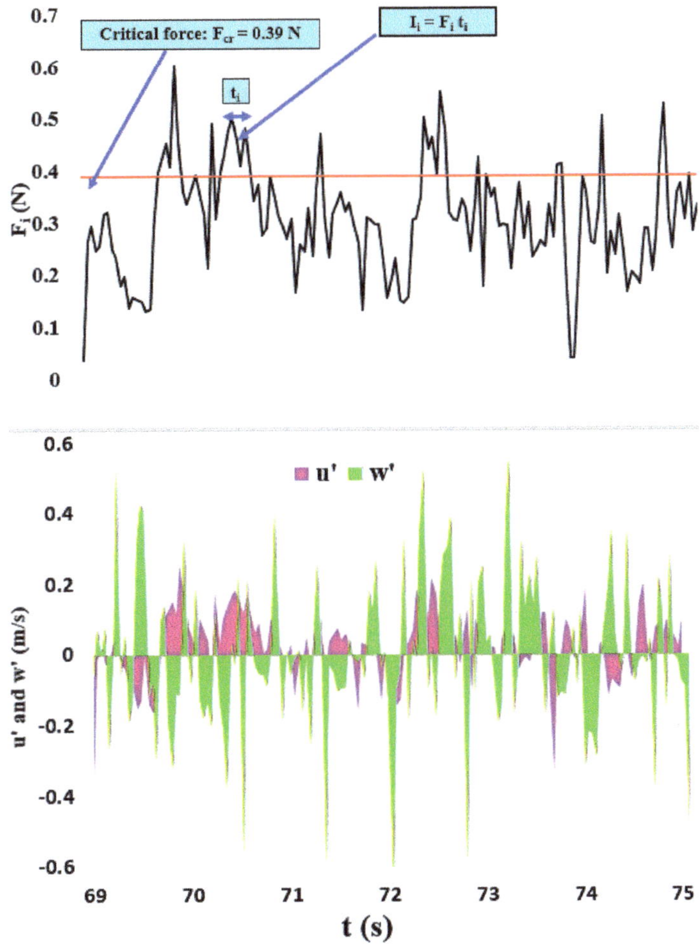

Figure 2. Illustration of force versus time showing the critical force (F_{cr} = 0.39 N, shown as a horizontal red line), the temporal fluctuations of the stream-wise velocity fluctuating component at a certain location, and the temporal fluctuations of the bed-normal component of the flow velocity at a certain location (u' and w') versus time.

It is clear from Figure 3 that the critical impulse is very close to 0.0455 m^2/s, i.e., the line of perfect agreement between the frequencies of flow impulses above the critical impulse and the frequency of entrainment of the instrumented particle. In a similar fashion, the theoretical equations presented in [21,22] can be used to derive values of critical impulses, by offering different input about the flow, grain and bed surface properties. and/or utilising appropriate assumptions (e.g., flow uniformity, grain shape, uniformity and packing of the bed surface grain sizes, local bed surface and grain arrangement slope). Due to dependence on all the above, there is no single reference value of the critical impulse that may set a grain into motion, but instead there is a range of flow conditions that may produce such results, depending on how frequently the generated flow impulses are in excess of the critical impulse level.

Based on the identified critical impulse value (0.0455 m^2/s) and a chosen value (0.0063 m^2/s) below the identified lower impulse bound (0.025 m^2/s), the flow impulses that have resulted in entraining the instrumented particle are being examined stochastically.

The reason a lower impulse value is also examined, is because a smaller critical value will produce the same but bigger magnitude impulses compared to the higher threshold critical impulse, for the same flow and bed surface hydrodynamics, but it will also generate a lot smaller magnitude and much more frequently occurring impulses, as shown in [22]. Thus, assessing the hydrodynamic interactions for a lower bound will allow offering a more conservative estimate for the effect of those flow structures on sediment transport. These are also expected to be of interest for the transport not only of sediment grains but also of smaller and/or more lightweight material (e.g., micro-plastics being caried along the fluvial vector) also exposed in the bed surface. Critically, assessing the occurrence of the different quadrant events and their sequence leading to entrainment, as validated via these bespoke flume experimental observations, is the overarching goal of this study. In the following section, the quadrants that occurred during, before and after the occurrence of flow impulses above the critical impulse were investigated in different windows of the ADV readings (1–5 readings).

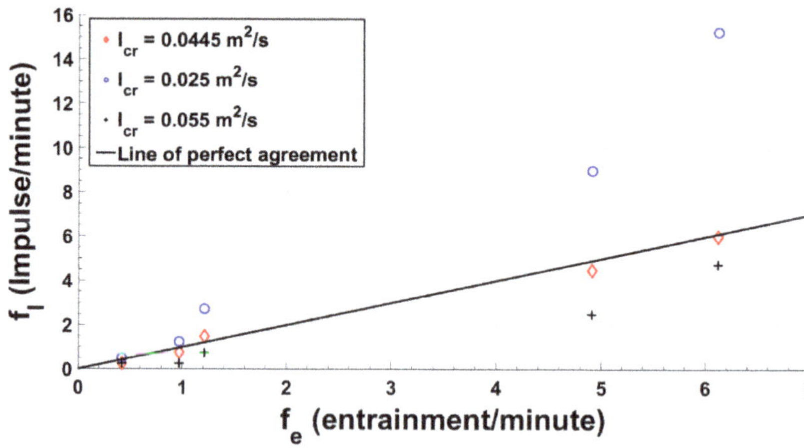

Figure 3. The frequency of entrainment of the instrumented particle, estimated using its reading during the experiment, versus the frequency of flow impulses above the three different values of critical impulse. The continuous straight line shows the line of perfect agreement between the observed and predicted values (please note the vertical and horizontal axis have different scaling).

3. Results

The flow depth, frequency of entrainment, particle Reynolds number and Shields parameter for the five runs of this study are shown in Table 1. It can easily be seen that while the flow depth is only slightly (e.g., less than 5%) changing with the increase of the flow rate, and the average shear stress eventually is less than doubling, the frequency of entrainments changes over an order of magnitude. This is to be expected for near threshold flow conditions and it also shows that the average flow metrics, such as mean bed surface shear stresses, are not sensitive in capturing the change in the probability of particle entrainment characterising the threshold flow conditions. Likewise, there is virtually no change for the mean flow velocities and Froude numbers, which illustrates that using equations which are indirectly employing mean flow parameters for assessing bed surface destabilisation or risk-to-scour around obstacles such as hydraulic infrastructure or instream vegetation is insufficient, while it could be directly done with instrumented particles [17].

The instrumented particle is only partially exposed to the flow, when the positioning of its local 3D-printed resting pocket relative to the top layers of bed surface particles is considered. The protrusion of the instrumented particle, defined as the ratio of the particle's size seen by the flow over the flow depth, is assessed to be fixed at about 0.4.

There have been many interesting studies assessing the role of particle protrusion into the flow e.g., [49], but here this is not further assessed given the infinitesimal changes of the flow depth relative to the particle's protrusion ratio. This is currently being explored using a novel setup that allows the instrumented particle to remain mobile while changing its protrusion to the bed surface, as opposed to earlier studies e.g., [49], and will be reported in a separate study.

The first step that was performed in the analysis was checking the quadrant readings that have occurred in the flow for the period of the ADV recording in the position of investigation. This was in the middle of the top half of the instrumented particle. The results of the quadrant readings that occurred in the flow for the period of the ADV recording in the position of investigation are shown in Figure 4.

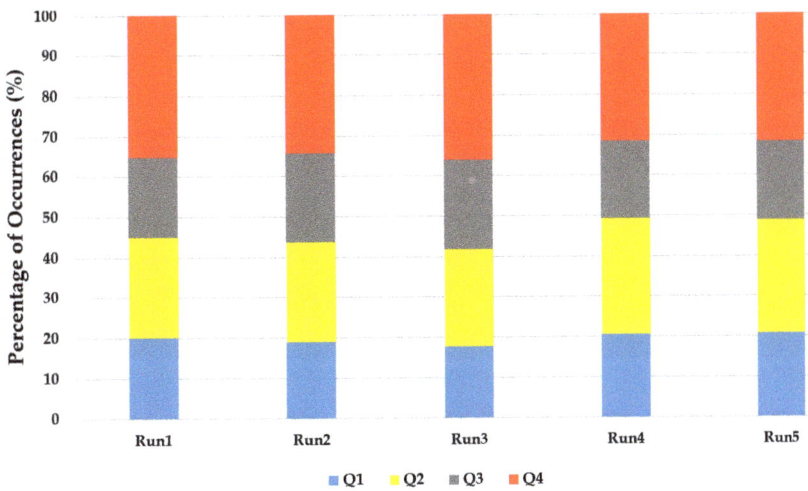

Figure 4. The percentages of the quadrant readings that have occurred in the flow during the periods of the ADV recordings in the position of investigation for the five runs of the experiment.

The percentages of the quadrant readings that have occurred in the flow during the periods of ADV recordings in the position of investigation for the five runs of the experiment were almost the same with a slight dominance found in Q_2 and Q_4 (i.e., sweep and ejection events).

The next step of the analysis was investigating the quadrant readings that have occurred during the turbulent flow events with an impulse above the critical impulse. The results are shown in Figure 5 below.

The impulse criterion only takes into account the turbulent flow events with a positive stream-wise velocity fluctuating component. The quadrant readings that have occurred during the events with an impulse above the critical impulse were Q_1 and Q_4 only, as is clear from Figure 5. The appearance of Q_4 within the impulse above the critical impulse is more dominant than Q_1 in the case of the three critical impulses of 0.0445 m^2/s as is clear from Figure 5.

In the next step, the quadrants that have occurred before and after the occurrence of flow impulses above the critical impulse were investigated for the different windows of the readings (1–5 readings). The results are shown in Figures 6–15.

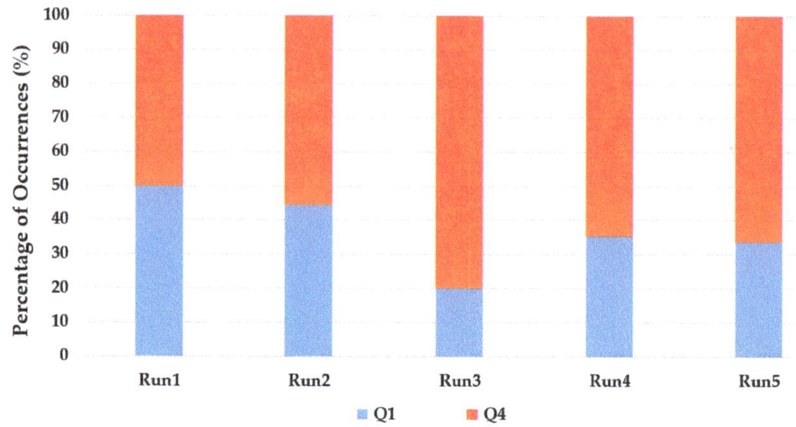

Figure 5. The percentages of the quadrant readings that have occurred during the events with an impulse above the critical impulse (I_{cr} = 0.0445 m^2/s).

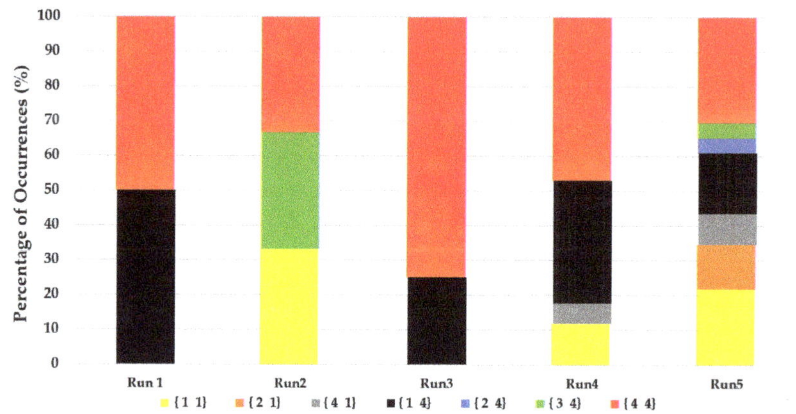

Figure 6. The percentages of the transitions between 1 quadrant reading before the impulse above the critical impulse and the first quadrant reading during the impulse above the critical impulse (for I_{cr} = 0.0445 m^2/s).

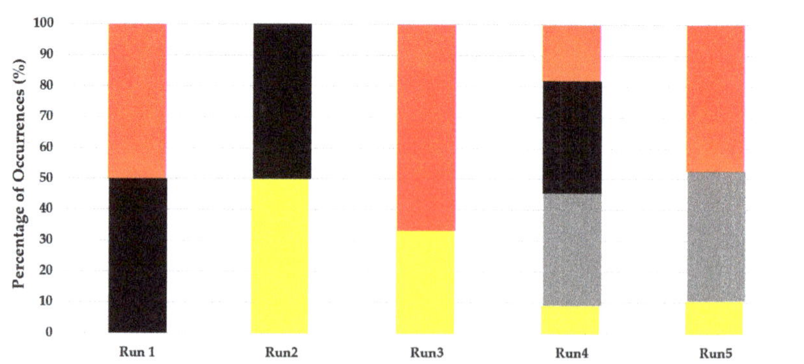

Figure 7. The percentages of the transitions between the last quadrant reading during the impulse above the critical impulse and 1 quadrant reading after that (for I_{cr} = 0.0445 m^2/s).

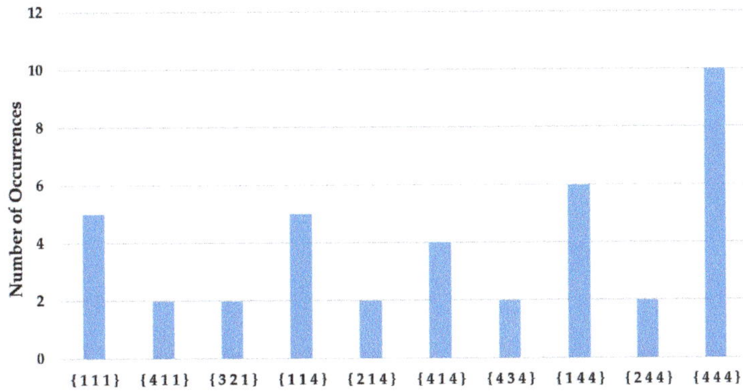

Figure 8. The number of occurrences of the most dominant transitions between two quadrant readings before the impulse above the critical impulse and the first quadrant reading during the impulse above the critical impulse (for I_{cr} = 0.0445 m^2/s).

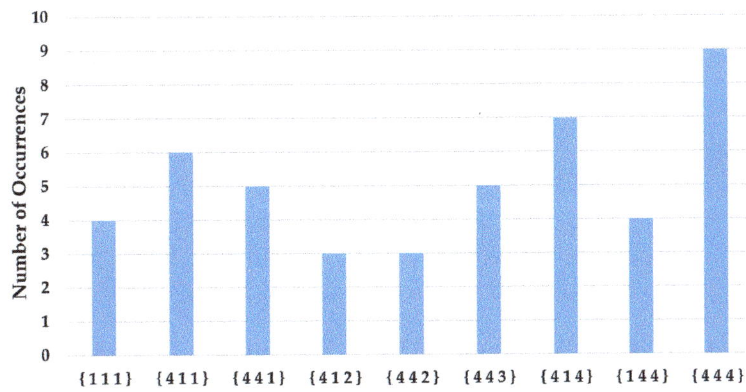

Figure 9. The number of occurrences of the most dominant transitions between the last quadrant reading during the impulse above the critical impulse and two quadrant readings after that (for I_{cr} = 0.0445 m^2/s).

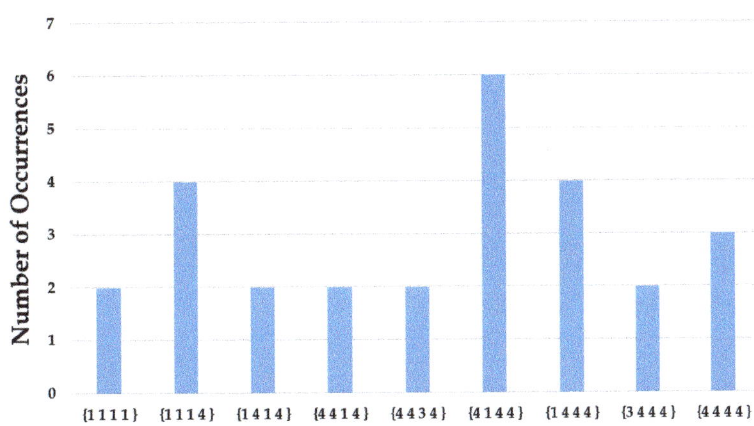

Figure 10. The number of occurrences of the most dominant transitions between three quadrant readings before the impulse above the critical impulse and the first quadrant reading during the impulse above the critical impulse (for I_{cr} = 0.0445 m^2/s).

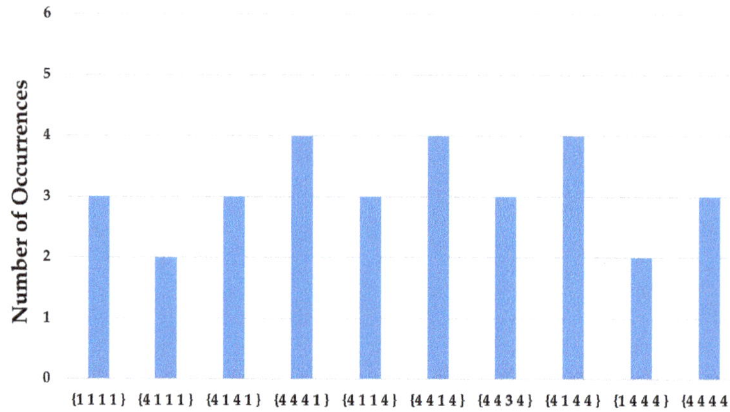

Figure 11. The number of occurrences of the most dominant transitions between the last quadrant reading during the impulse above the critical impulse and three quadrant readings after that (for I_{cr} = 0.0445 m^2/s).

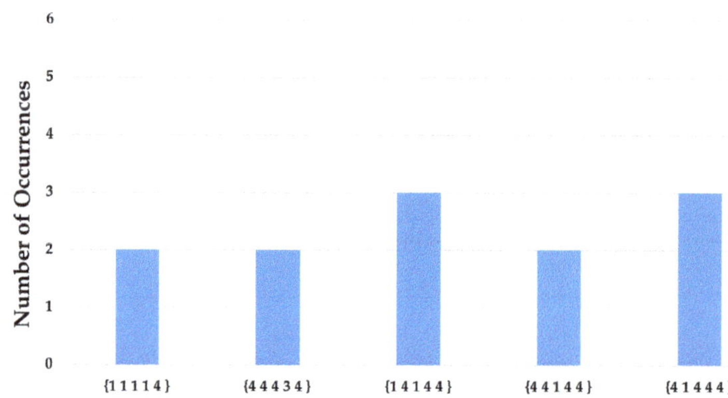

Figure 12. The number of occurrences of the most dominant transitions between four quadrant readings before the impulse above the critical impulse and the first quadrant reading during the impulse above the critical impulse (for I_{cr} = 0.0445 m^2/s).

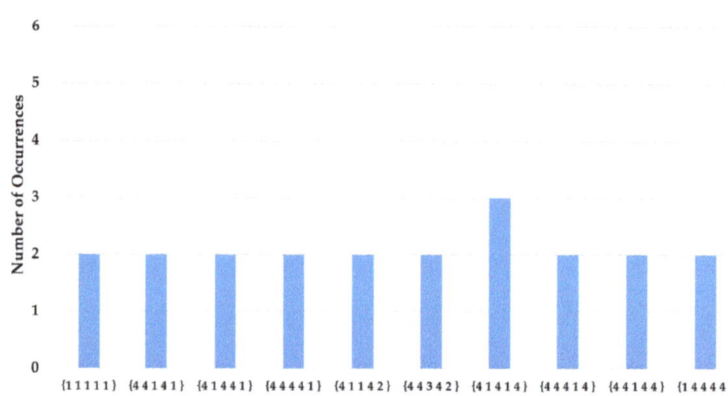

Figure 13. The number of occurrences of the most dominant transitions between the last quadrant reading during the impulse above the critical impulse and four quadrant readings after that (for I_{cr} = 0.0445 m^2/s).

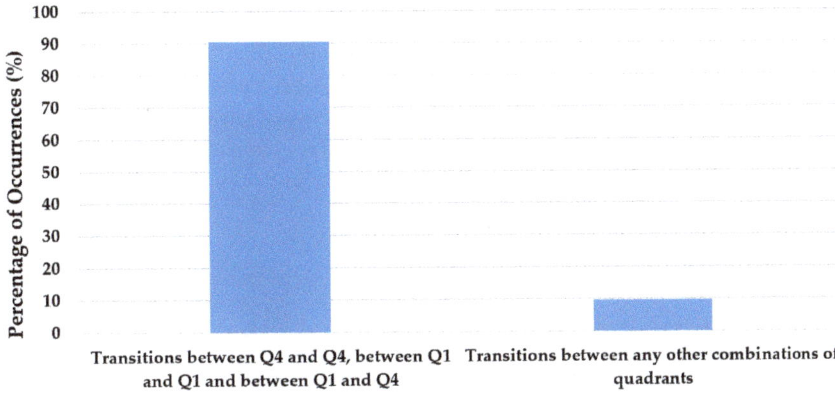

Figure 14. The percentages of the transitions between five quadrant readings before the impulse above the critical impulse and the first quadrant reading during the impulse above the critical impulse (for I_{cr} = 0.0445 m^2/s).

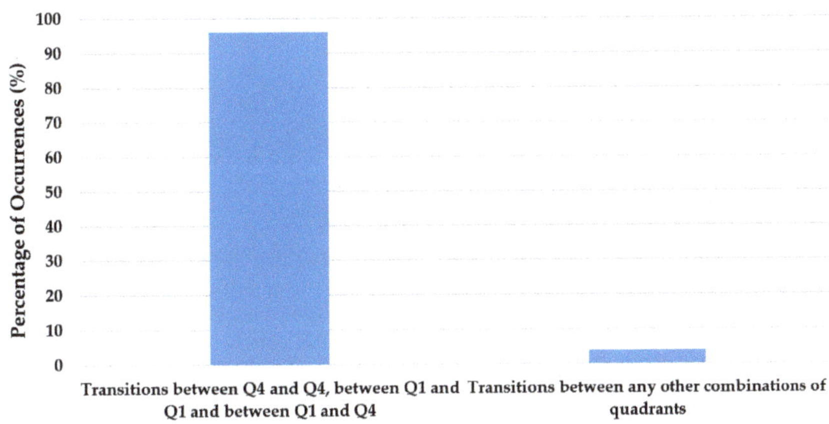

Figure 15. The percentages of the transitions between the last quadrant reading during the impulse above the critical impulse and five quadrant readings after that (for I_{cr} = 0.0445 m^2/s).

It is clear from Figure 6 that in the case of the critical impulse of 0.0445 m^2/s, most of the transitions between one quadrant reading before the occurrence of an impulse above the critical impulse and the first quadrant reading during the occurrence of an impulse above the critical impulse are between Q_4 and Q_4, Q_1 and Q_4 and Q_1 and Q_1. In the case of the transitions between the last quadrant reading during the occurrence of an impulse above the critical impulse and one quadrant reading after that, the result is between Q_4 and Q_4, Q_1 and Q_4 and Q_1 and Q_1, as it is clear from Figure 7. The same was observed for the case of transitions between two quadrant readings before and after the first and the last quadrant readings during the occurrences of impulses above the critical impulse. In other words, the transitions were between Q_4 and Q_4, Q_1 and Q_4 and Q_1 and Q_1, as shown in Figures 8 and 9.

As Figures 10–15 show, it is clear that most of the transitions between the first and the last quadrant readings during the occurrence of impulses above the critical impulse and three, four and five quadrants before and after these are again transitions between Q_4 and Q_4, Q_1 and Q_4 and Q_1 and Q_1.

The next step of the analysis that was performed was reducing the critical impulse to a value of 0.0063 m^2/s similar to Valyrakis et al. [59]. This procedure was conducted to increase the number of impulses above the critical impulse and to provide more data for investigating the type of flow structure. In other words, the goal of this step was to investigate the events that are less energetic than the events that have been investigated previously, i.e., using the critical impulse value of 0.0455 m^2/s, and assess their flow structures.

It is clear from Figure 16 that the occurrences of Q_4 during the occurrences of impulses above the critical impulse are more dominant than Q_1 in the case of a critical impulse of 0063 m^2/s as is clear from Figure 15. This is a similar finding to the case of the critical impulse of 0.0445 m^2/s.

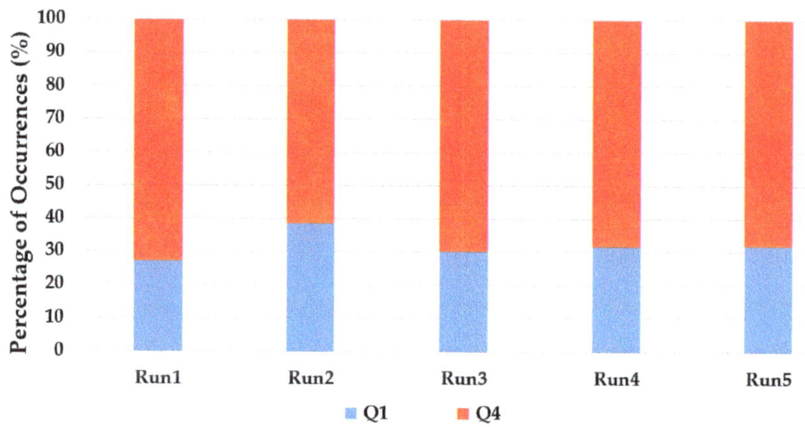

Figure 16. The percentages of the quadrant readings that have occurred during the events with an impulse above the critical impulse (I_{cr} = 0.0063 m^2/s).

It is clear from Figure 17 that in the case of the critical impulse of 0.0063 m^2/s, most of the transitions between one quadrant reading that has occurred before the occurrence of an impulse above the critical impulse and the first quadrant reading that has occurred during an impulse above the critical impulse are transitions between Q_4 and Q_4, Q_1 and Q_4 and Q_1 and Q_1. As for the case of the last quadrant reading that has occurred during the occurrence of an impulse above the critical impulse and one quadrant reading that has occurred after that, most are transitions between Q_4 and Q_4, Q_1 and Q_4 and Q_1 and Q_1 as is clear from Figure 18. This is the similar observation to the case of transitions between one quadrant before and one quadrant after the first and the last quadrants that have occurred during the occurrence of impulses above the critical impulse. Another observation that could be made is that there is a slight increase in the transitions between Q_2 and Q_4 and Q_3 and Q_4 that have occurred one quadrant before and after the first and the last quadrant readings that have occurred during the occurrence of an impulse above the critical impulse. Similar to the case of a critical impulse of 0.0445 m^2/s, it could be observed that in the case of transitions between two quadrant readings before and after the first and the last quadrant reading, they have occurred during the impulse above the critical impulse. In other words, the transitions are between Q_4 and Q_4, Q_1 and Q_4 and Q_1 and Q_1 as shown in Figures 19 and 20.

Finally, according to Figures 21 and 22, it is clear that most of the transitions between the first and last quadrant reading occurred during an impulse above the critical impulse and three, four and five quadrants before and after that are transitions between Q_4 and Q_4, Q_1 and Q_4 and Q_1 and Q_1.

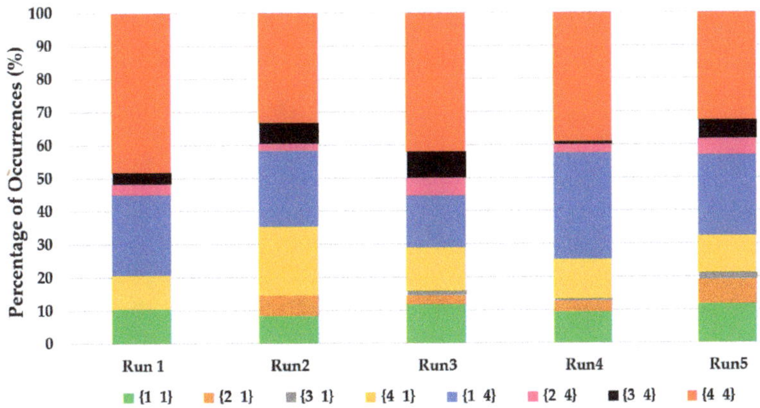

Figure 17. The percentages of the transitions between 1 quadrant reading before the impulse above the critical impulse and the first quadrant reading during the impulse above the critical impulse (for I_{cr} = 0.0063 m^2/s).

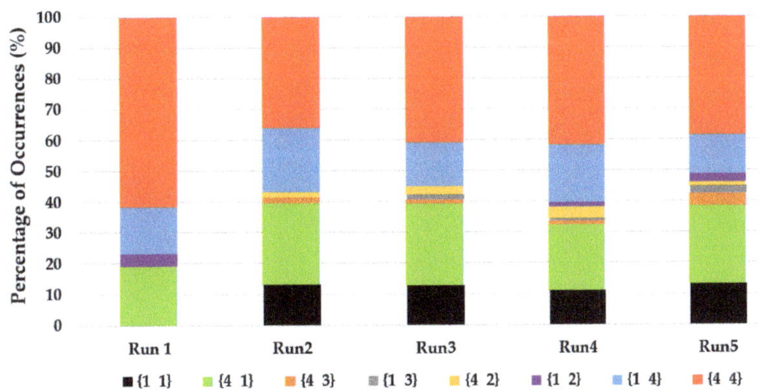

Figure 18. The percentages of the transitions between the last quadrant reading during the impulse above the critical impulse and 1 quadrant reading after that (for I_{cr} = 0.0063 m^2/s).

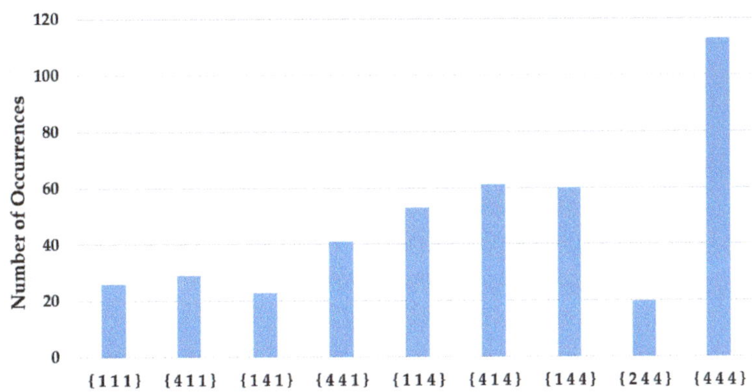

Figure 19. The percentages of the occurrences of the most dominant transitions between two quadrant readings before the impulse above the critical impulse and the first quadrant reading during the impulse above the critical impulse (for I_{cr} = 0.0063 m^2/s).

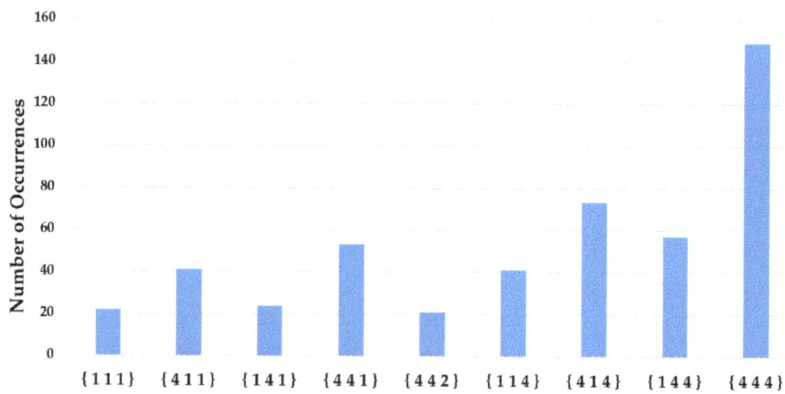

Figure 20. The percentages of the occurrence of the most dominant transitions between the last quadrant reading during the impulse above the critical impulse and 2 quadrant readings after that (for I_{cr} = 0.0063 m^2/s).

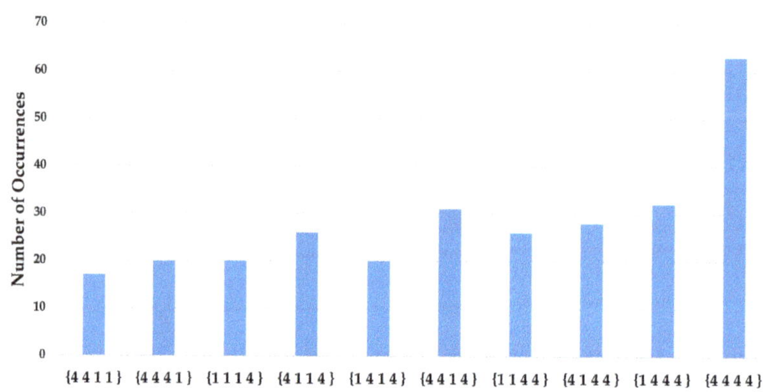

Figure 21. The percentages of the occurrences of the most dominant transitions between three quadrant readings before the impulse above the critical impulse and the first quadrant reading during the impulse above the critical impulse (for I_{cr} = 0.0063 m^2/s).

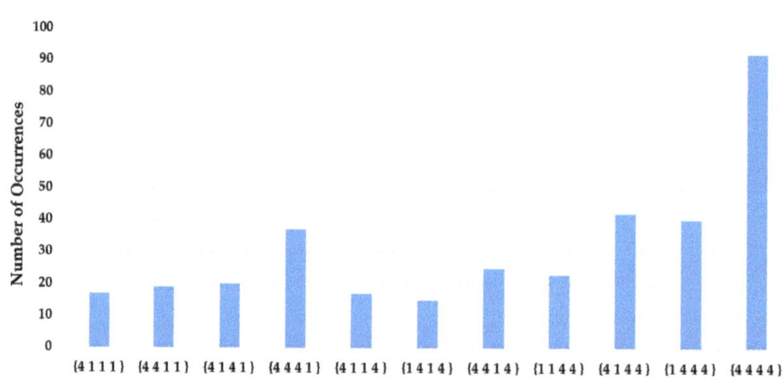

Figure 22. The percentages of the occurrence of the most dominant transitions between the last quadrant reading during the impulse above the critical impulse and 3 quadrant readings after that (for I_{cr} = 0.0063 m^2/s).

4. Discussion

In this work, the frequency of entrainment of the instrumented particle, 7 cm in diameter, that was estimated using its logged readings, is linked to the frequency of flow impulses above the critical impulse at the positions in the middle of the top half of the particle's diameter, which is the position of investigation. Different threshold values were assessed to determine the critical impulse that results in a good agreement between the frequencies of entrainment of the instrumented particle, estimated using its readings, and the flow impulses above the critical impulse as shown in Figure 2. The main aim of this study was to assess whether there was any trend in terms of the type of flow structures that occur in sequence before and after the occurrences of the flow impulses that have resulted in the coarse particle's entrainment.

The ADV logging frequency that was used in the experiment of this work for flow velocity measurements was 25 Hz. The ADV was able to detect events with a timescale of 0.04 s and higher. Based on the conclusion of the experimental work of Valyrakis et al. (2013), more than 50% of the turbulent flow events that have resulted in the dislodgement of the particle that was used in their research in the stream-wise direction have length scales of the particle's diameter. Additionally, only a very small fraction of the turbulent flow events that have resulted in the dislodgement of the coarse particle that was used in the stream-wise direction have length scales four times the particle's diameter. Hofland et al. [48] reported a similar observation, specifically that the length scale of the events that have resulted in sediment entrainment are sweeps of two to four grain diameters in the stream-wise direction.

In the experiment in this work, the instrumented particle that was used was 7 cm in diameter. The critical velocity used to overcome the resistance and to entrain the instrumented particle that was placed on the micro-topography used in this experiment, based on the moment balance equations [21], was 0.623 m/s. Therefore, assuming the length scale of a turbulent flow event of the coarse particle's diameter, the mean time scale of such an event could be estimated using Equation (3):

$$t_{tfe} = \frac{0.07\ \text{m}}{0.623\ \text{m/s}} = 0.1123s \tag{3}$$

where t_{tfe} is the time-scale of the turbulent flow event.

The ADV that was used in the experiment of this work for flow velocity measurements has a logging frequency of 25 Hz, therefore it was able to detect turbulent flow events of length scales of 35% of the particle's diameter and longer, assuming their velocities are equal to the critical velocity. Therefore, the logging frequency of the ADV used in this work is sufficient for the study of coherent flow structures that can result in the entrainment of the instrumented particle placed on the micro-topography used in this work based on the conclusions of the experimental work presented in [22,48]. Further investigation was performed by assessing the histogram of the durations of the impulses above the critical impulse (I_{cr} = 0.0445 m^2/s), as shown in Figure 23 below.

Close investigation of the histogram of the durations of impulses above 0.0445 m^2/s indicates that about 75% of the impulses have durations of less than 0.2 s. This duration, 0.2 s, is equivalent to five ADV readings. Therefore, it has been decided to investigate the quadrant readings for a window of 1–5 readings in the case of I_{cr} = 0.0445 m^2/s.

Another investigation that was performed was lowering the critical impulse to a value of 0.0063 m^2/s. This is similar to the critical impulse used by Valyrakis et al. [59]. This procedure increases the number of impulses above the critical impulse and provides more data for the quadrant analysis. To define the window of investigation of the quadrant readings for the case of I_{cr} = 0.0063 m^2/s, the histogram of the durations of the impulses above the critical impulse, Figure 24, was assessed.

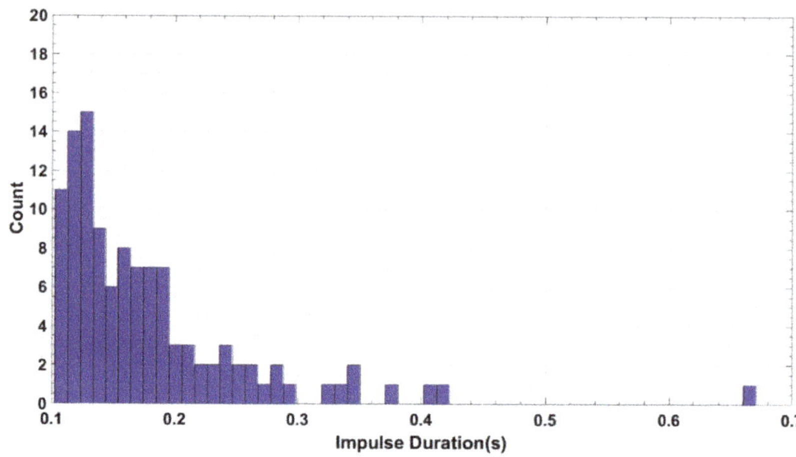

Figure 23. Histogram of the durations of the flow impulses for all five runs of the experiment (for I_{cr} = 0.0445 m^2/s).

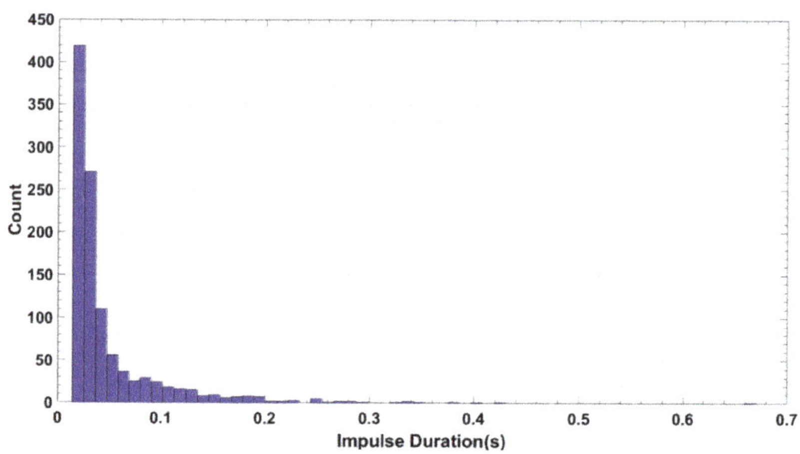

Figure 24. Histogram of the durations of the flow impulses for all five runs of the experiment (for I_{cr} = 0.0063 m^2/s).

Close investigation of the histogram of the durations of impulses above 0.0063 m^2/s indicates that about 90% of the impulses have durations of less than 0.12 s. This duration, 0.12 s, is equivalent to up to three ADV readings. Therefore, it has been decided to investigate the quadrant readings for a window of the first few readings, in the case of I_{cr} = 0.0063 m^2/s. The mean time scale of the flow structure that is of comparable size to the instrumented particle, and thought to be relevant to its entrainment [21], has been identified in Equation (3). The authors of [21] have shown that at near threshold flow conditions, the impulses that may lead to coarse particle entrainment have timescales that correspond to about 1–3 times the particle's representative length-scale (diameter). Thus, in our analysis it is considered that the near threshold impulse events are mostly short-lived and as a consequence better represented by a sequence of one or two readings (given the ADV recording frequency and the entrainment inducing impulses). We could also be assessing sequences of three or more readings, but these would primarily correspond to impulses above critical, much less than threshold and near threshold impulses that are relatively more long-lived (e.g., only a small percentage according to [21,22]).

As for the quadrant readings in the flow, it is clear from Figure 4 that there is a slight dominance in the occurrences of Q_4 and Q_2 quadrant readings in comparison to the occurrences of Q_1 and Q_3 quadrant readings in the flow. This is similar to some of the observations that have been reported in the literature for a similar type of flow (Cantwell, 1981; Robinson, 1991b). As for the quadrant readings within the impulses, sweeps have occurred more than outward interactions during the events with impulses above the critical impulse (for both cases I_{cr} = 0.0445 m^2/s and I_{cr} = 0.0063 m^2/s) as can be noticed from Figures 5 and 16. The difference between the percentages of occurrence of Q_4 readings and Q_1 readings in the flow (shown in Figure 4) does not justify the difference between the percentages of the occurrences of Q_4 and Q_1 during the events with impulses above the critical impulse. This conclusion is similar to the findings of Nelson et al. [43], i.e., Q_4 moves the majority of the sediment followed by Q_1 and to a much lesser extent, Q_2 and Q_3. However, according to Nelson et al. [43], the main reason that Q_4 moves the majority of the sediment compared to Q_1 is due to its common occurrence. This is not the case in this study. In other words, and as is clear from Figure 4, for the five runs of the experiment, the percentages of quadrant readings that have occurred in the flow during the periods of ADV recordings at the position of investigation are almost the same with the slight dominance of Q_2 and Q_4 (i.e., sweep and ejection events). Dwivedi et al. [47] suggested that a higher probability of occurrence of a high-magnitude force is induced by sweep events.

Of importance is to observe that even though the Q_2 and Q_4 events are somewhat more commonly observed than the Q_1 and Q_3 events (Figure 4), the majority of the high magnitude impulses are due to Q_4 events (according to e.g., Figure 5). A possible explanation for this is due to the much higher magnitude of the Q_2 and Q_4 events compared to the Q_1 and Q_3 events, eventually resulting at higher stress values [60,61]. This can be clearly seen for example from the joint probability distributions of the velocity component fluctuations (u' and v'), according to Wallace [60,62]. The joint probability distribution has an elliptical shape which is significantly more elongated in the Q_2 and Q_4 quadrants. As a consequence, the Q_2 and Q_4 events are having more high magnitude values while the Q_1 and Q_3 events will be limited to relatively smaller magnitude values. This observation aligns with the fact that impulses above a critical level, linked to entrainments, should be of a higher value, thus the contribution of the higher magnitude Q_4 events will be more pronounced compared to the Q_1 events.

Additionally, Dey et al. [63] suggested that sweeps are the more dominant mechanism in sediment entrainment while Q_4 and Q_2 cancel each other out on an immobile bed, giving rise to the Q_1. Other researchers [12,64] have also reported the predominance of the role of Q_4 and to a lesser extent Q_1 in sediment entrainment. According to the rest of the figures in the results section, considering all five runs, for both cases of critical impulse, i.e., I_{cr} = 0.0445 m^2/s and I_{cr} = 0.0063 m^2/s, for both investigation windows of readings, i.e., 1–5 readings and 1–3 readings and for before and after the occurrences of impulses above the critical impulse, it is clear that most of the transitions between the quadrant readings are Q_4 to Q_4, Q_1 to Q_4 and Q_1 to Q_1. In other words, it is clear from the results of this study that the impulses above the critical impulse, i.e., the impulses that have caused the instrumented particle's entrainment, are relatively short-lived sweep events and to a lesser extent, relatively short-lived outward interactions. This conclusion is supported by the observation that the majority of transitions between quadrant readings occur before and after the first and last quadrant readings of the impulses above the critical impulse being Q_4 to Q_4 and Q_1 to Q_1. This is in addition to the transitions between sweeps and outward interactions, i.e., Q_4 and Q_1.

The presence of even a weak but non-negligible secondary flow in a rectangular flume can affect the structure and intensity of the shear flow and subsequently influence the probability of sediment entrainment rates. Secondary flow is significantly affected via the flume's aspect ratio. In this study, the aspect ratio of the rectangular channel for the flows tried, is about 8.2 and it is almost fixed, given that the flow depth is changing less than 5%. Jing et al. [65], assessed the impact of secondary flow in an open channel for ratios ranging from 3 to 7.5 and found the presence of more secondary flow cells for the higher aspect ratios, which is also closer to the aspect ratio in our experiment. These can affect the flow velocity and flow energy distribution, resulting for example in possible deviations from the mean logarithmic flow velocity profiles, which we have also observed. These are expected to change as the aspect ratio of the flume is altered but given its infinitesimal change for the flow rates trialled herein, it is expected that any contribution of secondary flows to the entrainment rate of the target coarse particle, will remain the same.

Finally, do the impulses correspond to coherent flow structures and vice versa? The answer to that is yes and no. An impulse includes the magnitude and duration of the turbulent flow events. However, the impulse criterion is a resistance dependent criterion. If the resistance is small, small fluctuations can result in particle entrainment but if the resistance is big, the small fluctuations cannot result in the entraining of a particle. In other words, for the same flow field, having two particles of different diameters results in a different distribution of impulses, even though the same distribution of coherent flow structures exists in the flow field. Thus, for a non-uniform bed surface, having a range of entrainment resistance thresholds, the amount of overlap between the instantaneous hydrodynamic forcing (represented by the distribution of impulses above critical), assessed for the different critical impulses (representing the distinct resistance levels), will be indicative of the probability of entrainment. However, closer to the bed surface the contribution of Q_2 and Q_4 events may increase, which leaves more space for future research to explore how the contribution of these quadrants may change for different particle-size-to-flow-depth ratios (i.e., grain bed surface exposure ratios).

5. Conclusions

In this work, an instrumented particle of 7 cm has been used to investigate the incipient entrainment from a hydraulically rough bed surface. Five runs of a well-controlled laboratory flume experiment were performed for a flowrate range near the critical threshold of motion. ADV was used to record the local flow hydrodynamics from which flow impulses above critical can be estimated, as well as define the quadrant that each reading corresponds to. The critical impulse level herein was defined by stochastically linking the frequency of a coarse instrumented particle entrainment estimated using the sensor readings, to the frequency of the flow impulses above the critical impulse for each corresponding critical impulse level. The main objective of this study was to identify trends in the sequence of flow structures before and after the occurrence of impulses in excess of the critical level, linked to coarse particle entrainment, using quadrant analysis.

Our results indicate that there is a slight dominance in the Q_4 and Q_2 quadrant events compared to the Q_1 and Q_3 quadrant readings, in the flow. Additionally, sweep quadrant readings have occurred more often compared to outward interaction quadrant events during the occurrence of impulses above the critical level. However, given the greater prevalence of Q_2 and Q_4 events near the bed surface, it is expected that for cases of smaller particle-size-to-flow-depth ratios, these events will contribute even more towards coarse particle entrainment.

Finally, the results, considering all five runs, indicate the primary contribution of sweep events (Q_4 to Q_4, Q_1 to Q_4) and secondarily the ejection events (Q_1 to Q_1, for both cases of critical impulse, i.e., I_{cr} = 0.0445 m^2/s and I_{cr} = 0.0063 m^2/s). In essence, the findings of this study clearly demonstrate that the impulses above the critical impulse, which are responsible for the entrainment of the coarse instrumented particle, predominantly consist of relatively short-lived sweep events and to a lesser extent, relatively short-lived outward interactions. This conclusion is confirmed by the finding that the bulk of quadrant reading transitions take place prior to and following the first and last quadrant readings of the impulses. This is in addition to the transitions between sweeps and outward interactions, i.e., Q_4 and Q_1.

Author Contributions: Conceptualization, M.V.; methodology, M.V. and K.A.; software, K.A. and M.V.; validation, M.V. and K.A.; formal analysis, K.A.; investigation, K.A.; resources, K.A.; data curation, M.V. and K.A.; writing—original draft preparation, K.A.; writing—review and editing, M.V. and K.A.; visualization, M.V. and K.A.; supervision, M.V.; project administration, M.V.; funding acquisition, M.V. All authors have read and agreed to the published version of the manuscript.

Funding: This work has been supported in part by the Royal Society (Research Grant RG2015 R1 68793/1), the Royal Society of Edinburg (Crucible Award) and the Carnegie Trust for the Universities of Scotland (project 7066215).

Data Availability Statement: All data are available in the main text body and can be made available upon submitting a request to the corresponding author.

Acknowledgments: We are grateful to the University of Glasgow lab technician, Timothy Montgomery, and undergraduate research student, Athanasios (Thanos) Alexakis, for their help in preparing the experimental setup, as well as the editor and the anonymous reviewers for their recommendations.

Conflicts of Interest: The authors declare no conflict of interest.

References

1. Gomez, B.; Church, M. An assessment of bed load sediment transport formulae for gravel bed rivers. *Water Resour. Res.* **1989**, *25*, 1161–1186. [CrossRef]
2. Leopold, L.B.; Emmett, W.W. Bedload measurements, East Fork River, Wyoming. *Proc. Natl. Acad. Sci. USA* **1976**, *73*, 1000–1004. [CrossRef]
3. Talukdar, S.; Kumar, B.; Dutta, S. Predictive Capability of Bedload Equations Using Flume Data. *J. Hydrol. Hydromech.* **2012**, *60*, 45–56. [CrossRef]
4. Shields, A. *Application of Similarity Principles and Turbulence Research to Bed-Load Movement*; California Institute of Technology: Pasadena, CA, USA, 1936.
5. Bagnold, R.A. *An Approach to the Sediment Transport Problem from General Physics*; Physiographic and Hydraulic Studies of Rivers, Geological Survey Professional Paper 422-I; United States Government Printing Office: Washington, DC, USA, 1966; pp. 1–37.
6. Bagnold, R.A. Transport of solids by natural water flow: Evidence for a worldwide correlation. *Proc. R. Soc. Lond. A Math. Phys. Sci.* **1986**, *405*, 369–374.
7. Reynolds, O. On the dynamical theory of incompressible viscous fluids and the determination of the criterion. *Philos. Trans. R. Soc. Lond. A* **1895**, *186*, 123–164.
8. Einstein, H.A.; El-Samni, E.A. Hydrodynamic forces on a rough wall. *Pap. Present. Rev. Mod. Phys.* **1949**, *21*, 520–524. [CrossRef]
9. Einstein, H.A. The Bed-Load Function for Sediment Transportation in Open Channel Flows. (Technical Report No. 1026). United States Department of Agriculture, Soil Conservation Service. 1950. Available online: https://naldc.nal.usda.gov/download/CAT86201017/PDF (accessed on 1 January 2020).
10. Grass, A.J. Initial Instability of Fine Bed Sand. *J. Hydraul. Div.* **1970**, *96*, 619–632. [CrossRef]
11. Grass, A.J. Structural features of turbulent flow over smooth and rough boundaries. *J. Fluid Mech.* **1971**, *50*, 233–255. [CrossRef]
12. Heathershaw, A.D.; Thorne, D.P. Seabed noises reveal role of turbulent bursting phenomenon in sediment transport by tidal currents. *Nature* **1985**, *316*, 339–342. [CrossRef]
13. Leighly, J.B. Turbulence and the transportation of rock debris by streams. *Geogr. Rev.* **1934**, *24*, 453–464. [CrossRef]
14. Sutherland, A.J. Proposed mechanism for sediment entrainment by turbulent flows. *J. Geophys. Res.* **1967**, *72*, 6183–6194. [CrossRef]
15. Schmeeckle, M.W.; Nelson, J.M. Direct numerical simulation of bedload transport using a local, dynamic boundary condition. *Sedimentology* **2003**, *50*, 279–301. [CrossRef]
16. Sumer, B.M.; Lloyd HC, C.; Cheng, N.S.; Fredsøe, J. Influence of Turbulence on Bed Load Sediment Transport. *J. Hydraul. Eng.* **2003**, *129*, 585–596. [CrossRef]
17. Xu, Y.; Valyrakis, M.; Gilja, G.; Michalis, P.; Yagci, O.; Przyborowski, L. Assessing riverbed surface destabilization risk downstream isolated vegetation elements. *Water* **2022**, *14*, 2880. [CrossRef]
18. Williams, J.J.; Thorne, P.D.; Heathershaw, A.D. Measurements of turbulence in the benthic boundary layer over a gravel bed. *Sedimentology* **1989**, *36*, 959–971. [CrossRef]
19. Diplas, P.; Dancey, C.L.; Celik, A.O.; Valyrakis, M.; Greer, K.; Akar, T. The Role of Impulse on the Initiation of Particle Movement Under Turbulent Flow Conditions. *Science* **2008**, *322*, 717–720. [CrossRef]
20. Valyrakis, M.; Diplas, P.; Dancey, C.L.; Greer, K.; Celik, A.O. Role of instantaneous force magnitude and duration on particle entrainment. *J. Geophys. Res. Earth Surf.* **2010**, *115*, 1–18. [CrossRef]
21. Celik, A.O.; Diplas, P.; Dancey, C.L.; Valyrakis, M. Impulse and particle dislodgement under turbulent flow conditions. *Phys. Fluids* **2010**, *22*, 046601. [CrossRef]
22. Valyrakis, M.; Diplas, P.; Dancey, C.L. Entrainment of coarse particles in turbulent flows: An energy approach. *J. Geophys. Res. Earth Surf.* **2013**, *118*, 53. [CrossRef]

23. Kline, S.J.; Reynolds, W.C.; Schraub, F.A.; Runstadler, P.W. The structure of turbulent boundary layers. *J. Fluid Mech.* **1967**, *30*, 741–773. [CrossRef]
24. Adrian, R.J. Hairpin vortex organization in wall turbulence. *Phys. Fluids* **2007**, *19*, 041301. [CrossRef]
25. Robinson, S.K. Coherent Motions in the Turbulent Boundary Layer. *Annu. Rev. Fluid Mech.* **1991**, *23*, 601–639. [CrossRef]
26. Smith, C. Coherent flow structures in smooth-wall turbulent boundary layers: Facts, mechanisms and speculation. In *Coherent Flow Structures in Open Channels*; Ashworth, P.J., Ed.; J. Wiley: Chichester, UK, 1996; pp. 1–39.
27. AlHusban, Z.; Valyrakis, M. Assessing sediment transport dynamics from energy perspective by using the smart sphere. *Int. J. Sediment Res.* **2022**, *37*, 833–846. [CrossRef]
28. Wallace, J.M.; Eckelmann, H.; Brodkey, R.S. The wall region in turbulent shear flow. *J. Fluid Mech.* **1972**, *54*, 39–48. [CrossRef]
29. Corino, E.R.; Brodkey, R.S. A visual investigation of the wall region in turbulent flow. *J. Fluid Mech.* **1969**, *37*, 1–30. [CrossRef]
30. Best, J. On the entrainment of sediment and initiation of bed defects: Insights from recent developments within turbulent boundary layer research. *Sedimentology* **1992**, *39*, 797–811. [CrossRef]
31. Cleaver, J.W.; Yates, B. The effect of re-entrainment on particle deposition. *Chem. Eng. Sci.* **1976**, *31*, 147–151. [CrossRef]
32. Dyer, K.R.; Soulsby, R.L. Sand Transport on the Continental Shelf. *Annu. Rev. Fluid Mech.* **1988**, *20*, 295–324. [CrossRef]
33. Grass, A.J. The influence of boundary layer turbulence on the mechanics of sediment transport. In *Mechanics of Sediment Transport*; Müller, A., Mutlu Sumer, B., Euromech Colloquium, Eds.; Balkema: Rotterdam, The Netherlands, 1983.
34. Kaftori, D.; Hetsroni, G.; Banerjee, S. The effect of particles on wall turbulence. *Int. J. Multiph. Flow* **1998**, *24*, 359–386. [CrossRef]
35. Lapointe, M. Burst-like sediment suspension events in a sand bed river. *Earth Surf. Process. Landf.* **1992**, *17*, 253–270. [CrossRef]
36. Nakagawa, H.; Nezu, I. Prediction of the contribution to Reynolds stress from bursting events in open-channel flow. *J. Fluid Mech.* **1977**, *80*, 99–128. [CrossRef]
37. Ninto, Y.; Garcia, M.H. Experiments on particle—Turbulence interactions in the near–wall region of an open channel flow: Implications for sediment transport. *J. Fluid Mech.* **1996**, *326*, 285–319. [CrossRef]
38. Rashidi, M.; Hetsroni, G.; Banerjee, S. Particle-turbulence interaction in a boundary layer. *Int. J. Multiph. Flow* **1990**, *16*, 935–949. [CrossRef]
39. Sechet, P.; Le Guennec, B. Bursting phenomenon and incipient motion of solid particles in bed-load transport. *J. Hydraul. Res.* **1999**, *37*, 683–696. [CrossRef]
40. Sumer, B.M.; Deigaard, R. Particle motions near the bottom in turbulent flow in an open channel. Part 2. *J. Fluid Mech.* **1981**, *109*, 311–337. [CrossRef]
41. Cantwell, B.J. Organized Motion in Turbulent Flow. *Annu. Rev. Fluid Mech.* **1981**, *13*, 457–515. [CrossRef]
42. Jiménez, J. Turbulent flows over rough walls. *Annu. Rev. Fluid Mech.* **2004**, *36*, 173–196. [CrossRef]
43. Nelson, J.M.; Shreve, R.L.; McLean, S.R.; Drake, T.G. Role of Near-Bed Turbulence Structure in Bed Load Transport and Bed Form Mechanics. *Water Resour. Res.* **1995**, *31*, 2071–2086. [CrossRef]
44. Celik, A.O.; Diplas, P.; Dancey, C.L. Instantaneous turbulent forces and impulse on a rough bed: Implications for initiation of bed material movement. *Water Resour. Res.* **2013**, *49*, 2213–2227. [CrossRef]
45. Detert, M.; Nikora, V.; Jirka, G.H. Synoptic velocity and pressure fields at the water–sediment interface of streambeds. *J. Fluid Mech.* **2010**, *660*, 55–86. [CrossRef]
46. Dwivedi, A.; Melville, B.; Shamseldin, A.Y. Hydrodynamic Forces Generated on a Spherical Sediment Particle during Entrainment. *J. Hydraul. Eng.* **2010**, *136*, 756–769. [CrossRef]
47. Dwivedi, A.; Melville, B.W.; Shamseldin, A.Y.; Guha, T.K. Flow structures and hydrodynamic force during sediment entrainment. *Water Resour. Res.* **2011**, *47*, 1–18. [CrossRef]
48. Hofland, B.; Battjes, J.; Booij, R. Measurement of Fluctuating Pressures on Coarse Bed Material. *J. Hydraul. Eng.* **2005**, *131*, 770–781. [CrossRef]
49. Schmeeckle, M.W.; Nelson, J.M.; Shreve, R.L. Forces on stationary particles in near-bed turbulent flows. *J. Geophys. Res. Earth Surf.* **2007**, *112*, 1–21. [CrossRef]
50. Smart, G.M.; Habersack, H.M. Pressure fluctuations and gravel entrainment in rivers. *J. Hydraul. Res.* **2007**, *45*, 661–673. [CrossRef]
51. Paiement-Paradis, G.; Buffin-Bélanger, T.; Roy, A.G. Scalings for large turbulent flow structures in gravel-bed rivers. *Geophys. Res. Lett.* **2003**, *30*, 1773. [CrossRef]
52. AlObaidi, K.; Xu, Y.; Valyrakis, M. The Design and Calibration of Instrumented Particles for Assessing Water Infrastructure Hazards. *J. Sens. Actuator Netw.* **2020**, *9*, 36. [CrossRef]
53. AlObaidi, K.; Valyrakis, M. A sensory instrumented particle for environmental monitoring applications: Development and calibration. *IEEE Sens.* **2021**, *21*, 10153–10166. [CrossRef]
54. AlObaidi, K.; Valyrakis, M. Linking the explicit probability of entrainment of instrumented particles to flow hydrodynamic. *Earth Surf. Process. Landf.* **2021**, *46*, 2448–2465. [CrossRef]
55. Liu, D.; AlObaidi, K.; Valyrakis, M. The assessment of an Acoustic Doppler Velocimetry profiler from a user's perspective. *Acta Geophys.* **2022**, *70*, 2297–2310. [CrossRef]
56. Goring, D.G.; Nikora, V.I. Despiking Acoustic Doppler Velocimeter Data. *J. Hydraul. Eng.* **2002**, *128*, 117–126. [CrossRef]
57. Wahl, T. Discussion of "Despiking Acoustic Doppler Velocimeter Data" by Derek G. Goring and Vladimir I. Nikora. *J. Hydraul. Eng.* **2003**, *129*, 484–487. [CrossRef]

58. Mori, N.; Suzuki, T.; Kakuno, S. Noise of Acoustic Doppler Velocimeter Data in Bubbly Flows. *J. Eng. Mech.* **2007**, *133*, 122–125. [CrossRef]
59. Valyrakis, M.; Diplas, P.; Dancey, C.L. Entrainment of coarse grains in turbulent flows: An extreme value theory approach. *Water Resour. Res.* **2011**, *47*, 1–17. [CrossRef]
60. Wallace, J.M. Quadrant Analysis in Turbulence Research: History and Evolution. *Annu. Rev. Fluid Mech.* **2016**, *48*, 131–158. [CrossRef]
61. Raupach, M.R. Conditional statistics of Reynolds stress in rough-wall and smooth-wall turbulent boundary layers. *J. Fluid Mech.* **1981**, *108*, 363–382. [CrossRef]
62. Wallace, J.M.; Brodkey, R.S. Reynolds stress and joint probability density distributions in the u-v plane of a turbulent channel flow. *Phys. Fluids* **1977**, *20*, 351–355. [CrossRef]
63. Dey, S.; Sarkar, S.; Solari, L. Near-Bed Turbulence Characteristics at the Entrainment Threshold of Sediment Beds. *J. Hydraul. Eng.* **2011**, *137*, 945–958. [CrossRef]
64. Cecchetto, M.; Tregnaghi, M.; Bottacin-Busolin, A.; Tait, S.; Marion, A. Statistical Description on the Role of Turbulence and Grain Interference on Particle Entrainment from Gravel Beds. *J. Hydraul. Eng.* **2017**, *143*, 06016021. [CrossRef]
65. Jing, S.; Yang, W.; Chen, Y. Smooth Open Channel with Increasing Aspect Ratio: Influence on Secondary Flow. *Water* **2019**, *11*, 1872. [CrossRef]

applied sciences

MDPI

Article

Migration and Diffusion of Surface Sediments in Bohai Bay: Evidence from Grain Size and Elements

Bo Zhao [1,2,3], Lianjie Zhang [1,2,*], Jishun Yan [1,2], Xia Lin [1,2], Peng Wang [4], Pan Zhang [1,2], Yonghai Yu [1,2] and Shuai Yu [5]

1 State Environmental Protection Key Laboratory of Marine Ecosystem Restoration, Dalian 116023, China
2 National Marine Environmental Monitoring Center, Dalian 116023, China
3 College of Harbour, Coastal and Offshore Engineering, Hohai University, Nanjing 210098, China
4 Marine Academy of Zhejiang Province, Hangzhou 310012, China
5 Dalian Compass Ocean Engineering Consulting Corporation Ltd., Dalian 116023, China
* Correspondence: ljzhang@nmemc.org.cn; Tel.: +86-0411-84783422

Abstract: Grain size and element content of surface sediment from Bohai Bay were analyzed to study the sediment migration and diffusion based on grain size trend analysis (GSTA) and discriminant function (DF). The sediment in the southern, central and western part of Bohai Bay mainly originates from the Yellow River, while that in the northern part of Bohai Bay mainly originates from the Luanhe River. The influence boundary between the Yellow River and the Luanhe River is estimated to be at 38°50′ N. In both the southern and northern parts of Bohai Bay, sediment is transported into the bay under the influence of prevailing waves, strongest waves and tidal remnants, resulting in sediment accumulation in western Bohai Bay. The coastal sediment of Bohai Bay is generally in a state of offshore movement, which is consistent with the large-scale coastline retreat in past decades found by previous studies.

Keywords: Bohai Bay; sediment migration; diffusion; grain size; element

Citation: Zhao, B.; Zhang, L.; Yan, J.; Lin, X.; Wang, P.; Zhang, P.; Yu, Y.; Yu, S. Migration and Diffusion of Surface Sediments in Bohai Bay: Evidence from Grain Size and Elements. *Appl. Sci.* **2022**, *12*, 10738. https://doi.org/10.3390/app122110738

Academic Editors: Gordon Gilja, Manousos Valyrakis, Panagiotis Michalis, Thomas Pahtz and Oral Yagci

Received: 9 September 2022
Accepted: 20 October 2022
Published: 23 October 2022

1. Introduction

Source, transport and accumulation of sediment in the ocean are components in the study of the "source-sink" process, and are also an important part of land-sea interaction research. Transportation of sediment in the ocean is the result of complex dynamic processes. Their flux and fate directly affect the evolution characteristics of seabed topography and are closely related to the stability of coast and service life of marine engineering, especially around Bohai Bay, where it is usually accompanied by intense marine development and human activities. Sediment transportation has always been one of the widely concerning oceanographic issues.

Bohai Bay is one of the three major bays in the Bohai Sea (Figure 1). Due to the discharges of three large rivers (Yellow River, Luanhe River and Haihe River), Bohai Bay has a rich source of sediment, and large deltas have developed along much of the bay's shoreline. In recent years, the runoff and sediment input from the Yellow River have gradually decreased, due to the massive construction of dams, power stations, increased water consumption, climate change and other factors. Subsequently, the evolutionary state of the Yellow River underwater delta has changed from constructive to destructive [1]. Sediment eroding from the subaqueous Yellow River Delta (YRD) is causing the retreat of the adjacent coast [2] as well as shoaling offshore in southern and western Bohai Bay (Figure 1). These conditions reveal that sediment transport in Bohai Bay has changed significantly. Bohai Bay is famous for its well-developed harbor transportation and oil exploitation industries. There are several large ports and oil exploitation areas along the coast. Therefore, the sediment transport and the resulting siltation and erosion in Bohai Bay hold great scientific significance, not only to enhance research on the source and sink process of bay sediment, but also to protect transportation and oil exploitation in Bohai Bay.

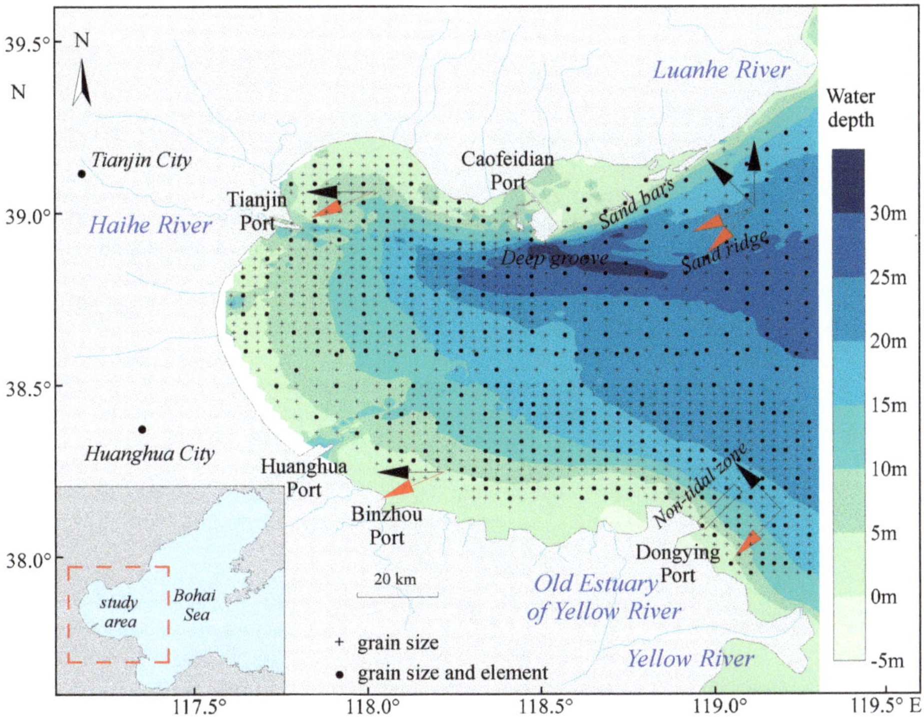

Figure 1. Location, circulation structure and wave direction of Bohai Bay and distribution of sampling sites. Black arrows indicate prevailing wave direction, and red arrows indicate strongest wave direction.

Earlier work by Qin et al. [3], Han et al. [4] and Zhang et al. [5] from 1985 to 2015 provided a basis for the work described here, and an indication that sediment transport in Bohai Bay has changed significantly over the past 30 to 40 years. Detrital sediments from the Yellow River and Luanhe River can affect the central and eastern parts of the Bohai Sea [6]. About 88% of the Yellow River sediment is deposited on the front edge of the underwater delta [7], and there are three directions of diffusion: northeast, northwest and southeast [8]. Due to the existence of the shear front, the estuary sediment is confined to the shore side of the shear front and transported to the north and south with the current [9]. Sediment discharge from the Luan River is much smaller than that of the Yellow River, and most of it is deposited in the coastal area from the mouth of the Luanhe River west to Caofeidian (Figure 1). The sediment outside the Caofeidian Sandbar is transported from sea to land, which is believed to be related to the coastal sediment transport caused by winds in the NW, SW and SE directions [10]. The sediment distribution of the Bohai Sea is consistent with the circulation trend [11]. However, the study areas of earlier studies are all local parts of Bohai Bay, such as the Yellow River estuary waters and Caofeidian waters. Therefore, a clear understanding of sediment transport in Bohai Bay overall is still lacking.

This study clarifies the sediment transport trend and its dynamic mechanism in Bohai Bay through a large-scale investigation on the grain size and element content of surface sediment. The results can provide guidance and reference for coastal disaster prevention and safety.

2. Materials and Methods

2.1. Study Area

Bohai Bay is located in the west of the Bohai Sea (Figure 1), and is one of three bays of the Bohai Sea. The submarine slope is smaller in the west of Bohai Bay, but larger in the northeast of Bohai Bay and the Yellow River Estuary. The maximum water depth of the Caofeidian Deep Trough can reach 40 m. The three large rivers—the Yellow River, Luanhe River, and Hai River—flow into Bohai Bay. The Yellow River is world famous for its large sediment discharge, with an average annual sediment discharge of about 7.45×10^8 t. This sediment originates mainly in the Loess Plateau. The average annual sediment discharge of the Luanhe River is about 0.21×10^8 t, mainly stemming from Yansha granite and metamorphic rock [12]. The average annual sediment discharge of the Haihe River is only about 0.06×10^8 t, much smaller than that of the Yellow River and Luanhe River. Some tributaries in the upper reaches of the Haihe River traverse or originate from the Taihang Mountains [13]. With the Yellow River being the primary contributor to the sediment of Bohai Bay, multiple deltas have accumulated along the coast of Bohai Bay.

The sediment types in central Bohai Bay are silt and muddy silt. The sediment types from coast to sea are silty sand and sandy silt in south Bohai Bay, and sand and sandy silt in north Bohai Bay. The sediment type of the underwater sand ridge is silty sand. The suspended sediment concentration (SSC) is high in south Bohai Bay (along the delta coast) and low in north Bohai Bay [2]. The average SSC along the delta exceeded 20 mg L^{-1} and was 200–3000 mg L^{-1} higher in winter than in summer [14].

2.2. Sample Collection and Testing

1154 surface sediment samples (Figure 1) were collected in Bohai Bay using a grab (Shuguang HNM12 0.05 m^2, CHN) in July 2010. Sediment was sampled at 3 km intervals in almost the entire bay, but at 5 km intervals in the northeastern part of the bay. The upper 2 cm of the sediment was collected and sealed for storage at room temperature. Sediment samples were tested in the Key Laboratory of Submarine Geosciences and Prospecting Techniques, Ministry of Education, Ocean University of China. Grain size tests were performed on samples from all stations, and geochemical tests were performed on samples from every other station. Data from several stations were excluded due to missing samples or poor data quality.

The grain size of 1154 samples was measured by a laser diffraction grain size analyzer (Malvern Mastersizer 2000, Malvern, UK). After sample pre-treatment and preparation [15], the particle size was tested with a Malvern Mastersizer 2000. The relative error of repeated measurement was less than 3%. The grain size parameters were calculated using a moment method [16].

The geochemical composition of 330 samples was measured by a SPECTRO XEPOS energy-dispersive X-ray fluorescence spectrometer (XRF, SPECTRO Analytical Instruments, Kleve, Germany). After sample pre-treatment and preparation [15], the content of the element was tested with the XRF. The relative error of repeated measurement was less than 2%.

2.3. Grain Size Trend Analysis (GSTA)

Sediment trend study based on grain size parameters was carried out using a GSTA Fortran program [17] based on a two-dimensional model proposed by [18]. The model is based on the comparison of the three statistical grainsize parameters (mean size, sorting and skewness) of sediment samples [19]. The combination of three parameters will result in eight possible grain size trends, but only two trends are reasonable [20]. That is, the sediment should become: (i) finer, better sorted, more negatively skewed (case: FB-); (ii) coarser grain size, better sorted, and more positively skewed (case: CB+) in the transport direction. The calculation process of GSTA is as follows. First, a critical distance (Dcr) is defined, which is generally the length of the maximum sampling interval or the diagonal length of the sampling grid. A station whose distance from the computing station is less than Dcr is identified as a neighbor. Judge one by one whether the two trends (FB- and

CB+) exist between the computing site and its neighbors. If the result is yes, a length 1 trend vector from the computing site to the neighboring site is generated. The trend vector of any site is the sum of all trend vectors of the site with its neighbors:

$$\vec{R}(x,y) = b\sum_{1}^{n}\vec{r}(x,y) \tag{1}$$

In order to eliminate "noise", GSTA smooths the trend vector by calculating the average vector of the site and its neighbors. The mathematical transformation formula for smoothing is as follows:

$$\vec{R_m}(x,y) = \frac{1}{1+k}\left[\vec{R}(x,y) + \sum_{1}^{k}\vec{R_j}(x,y)\right] \tag{2}$$

Finally, a map of the remaining vectors of all stations is drawn. The GSTA model is considered affected by two main sources of uncertainty: (i) different feature distances and (ii) edge effects of edge sampling points [18,21].

2.4. Discriminant Function (DF) Analysis

Sediment trend study based on the element diffusion index was carried out using the discriminant function (DF) method [22]. Significant differences were identified in the element contents of materials from the Yellow River and Luanhe River according to the earlier work of other researchers (Table 1). The main stream of the Yellow River flows through the Loess Plateau, and the sediment carried by the Yellow River is characterized by a high calcium (Ca) content [23]. The Ca content in the material from the Luanhe River is much lower than that of the Yellow River. Therefore, Ca can be used to analyze the sediment contribution of the rivers, which can also be used to study the transport and diffusion of river material. The element aluminum (Al) was used to eliminate the controlling effect of grain size. DF is used to judge the similarity between two samples, and the calculation formula is as follows:

$$DF = |(Ca/Al)_{sample}/(Ca/Al)_{river} - 1| \tag{3}$$

Table 1. Element Content of Sediment in the Yellow River and Luanhe River.

River	Al_2O_3	SiO_2	Fe_2O_3	MgO	MnO	CaO	Na_2O	References
YR	9.20	62.68	3.15	1.39	0.06	4.60	2.20	Zhao and Yan [24]
LR	9.05	73.93	2.10	0.80	0.04	1.53	3.19	Liu [25]
River	K_2O	TiO_2	Cu	Zn	Cr	Sr	Ba	References
YR	1.94	0.60	13.00	40.00	60.00	220.00	540.00	Zhao and Yan [24]
LR	2.35	0.22	10.67	35.07	36.53	375.60	816.90	Liu [25]

Note: The unit of Cu, Zn, Cr, Sr and Ba is ppm, and that of the rest is %. YR: Yellow River, LR: Luanhe River.

Therefore, the DF index indicated the proximity of a sample to some source. The closer the DF index was to 0, the closer the composition of sediment was to the source. As the annual sediment transport of the Haihe River is much smaller than that of the Yellow River and Luanhe River, the Yellow River and Luanhe River were selected to calculate the DF.

3. Results

3.1. Grain Size Trends

Since the sampling interval in this study is relatively regular, the critical distance is determined to be no shorter than the diagonal distance of the sampling grid and no longer than twice the distance of the sampling grid. Finally, we decided to use Dcr = 0.08 for grain size trend analysis. Before running GSTA, the data was checked through scatter plot in order to eliminate obvious outliers. The distribution of remaining vectors is shown in

Figure 2. In northeast Bohai Bay, sediment from the modern Luanhe River migrated in the southwest direction and gathered at a sandbank and underwater bank slope (Area A). In the Caofeidian Deep Trough, sediment migrated from east to west. In northwest Bohai Bay, sediment was moved from the coast and sunk with materials from the Deep Trough (Area C). In southwest Bohai Bay, sediment was removed from the coast and sunk in the outer sea area of the Ziyaxin Estuary (Area D). In southern Bohai Bay, offshore sediment transportation occurred on all coasts, but differences in the direction of sediment transport were evident when compared with the two sides of the Old Yellow River Estuary. In the sea area to the west of the Old Yellow River Estuary, the offshore sediment moved westward, crossed the north side of Huanghua Port, migrated toward the northwest, and converged in the middle of Bohai Bay (Area B). East of the Old Yellow River Estuary, the offshore sediment migrated to the southeast and converged northeast of Dongying Port (Area E).

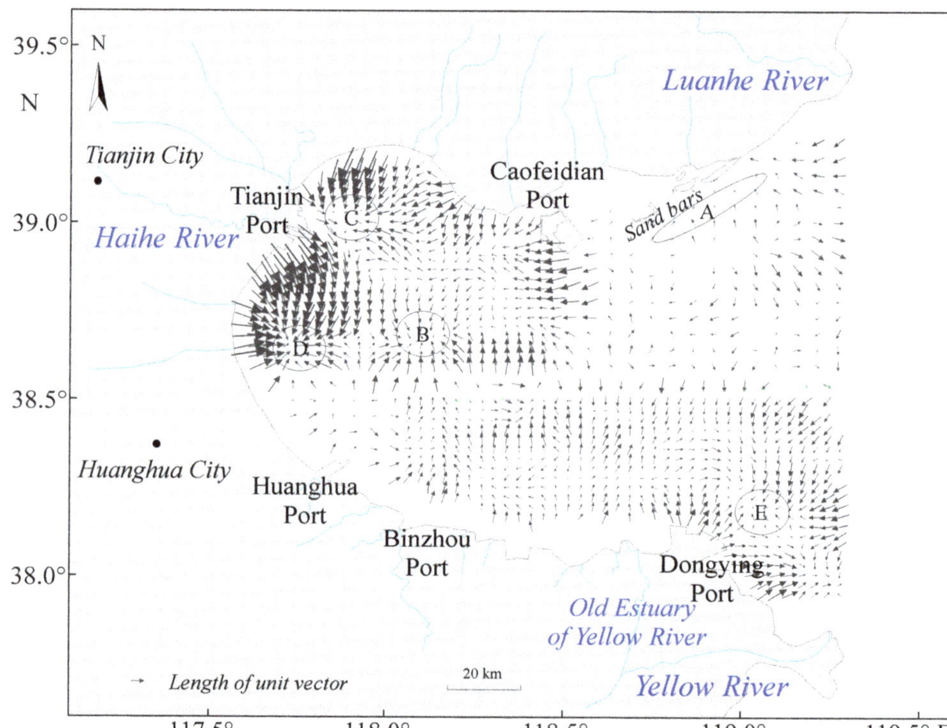

Figure 2. Distribution of the grain size trend vectors in Bohai Bay. Letters A, B, C, D and E indicate sediment accumulation areas.

3.2. Distribution of Discriminant Function Indicator

Figure 3A illustrates the distribution of the Ca/Al discriminant function between sediment samples and the Yellow River materials (DFYR). The DFYR of most sea areas in southwest Bohai Bay was less than 0.5, which was close to the provenance of the Yellow River. The DFYR in the northeast and northern coastal waters of Bohai Bay were greater than 0.5. The distribution of the Ca/Al discriminant function between sediment samples and Luanhe River materials (DFLR) is shown in Figure 3B. The DFLR in the northeast and north coast of Bohai Bay was less than 0.5 and was at its smallest in the lagoons and the modern Luanhe Estuary. The data indicated that the sample was close to the material from Luanhe River. In most areas of southwest Bohai Bay, the DFLR was larger than 0.5.

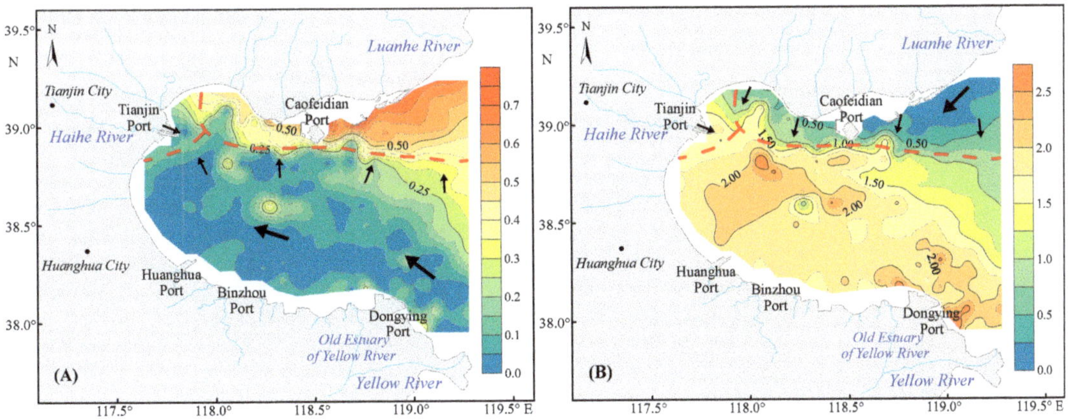

Figure 3. Contour map of DF in Bohai Bay based on Ca/Al. (A) DFYR: DF of surface sediments with Yellow River source. (B) DFLR: DF of surface sediments with Luanhe River source. Red dotted lines indicate the approximate range of influence of different river sources. Black arrows indicate the direction of sediment diffusion.

4. Discussion

4.1. River Contribution and Sediment Diffusion

The DFYR was less than 0.5 in southern, middle and western Bohai Bay, which indicates that materials from the Yellow River dominated almost the entire bay. The DFLR in northeast Bohai Bay was less than 0.5, which indicates that material from the Luanhe River dominated only in northeast Bohai Bay. The DFLR in the tidal sand ridge outside Caofeidian was also small, serving as the evidence that its material came from the Luanhe River. The sediment in south and west Bohai Bay had a small DFYR with Yellow River provenance, which may have been related to the accumulation history of the Yellow River Delta. In the past, the Yellow River frequently changed its course to the sea. The area from the north bank of Bohai Bay to Tianjin, Huanghua, Binzhou, Kenli and Lijin was once an estuary of the Yellow River [26,27]. Currently, the west bank and south bank of Bohai Bay are deltas formed by materials from the Yellow River.

The DFYR distribution showed that the influence of the Yellow River materials gradually weakened from southwest to northeast, an indicator for the dispersion process of the material from the Yellow River as it was transported from the Yellow River Delta to its surrounding areas. The DFLR distribution showed that the influence of Luanhe River material gradually weakened from northeast to southwest, an indicator for the dispersion process of Luanhe River material as it was transported from the Luanhe River Delta to the southwest. Thus, the results of DF and GSTA are unified. Areas C and D in the western part of Bohai Bay (Figure 2) are influenced by mixed material sources from different rivers.

The sediment transport trends derived from GSTA and DF analyses in this study are consistent with the results obtained from geochemical analyses [5], grain size [28] and remote sensing inversion of suspended sediment [29,30].

Approximate influence range of different river sources is shown in Figure 3 and is generally consistent with the previous research [3–5]. Zhang et al. [5] gave the location of the influence boundary between the Yellow River and Luanhe River (38°42′ N). However, this study indicates that the influence of the Yellow River source is still underestimated, and the influence boundary between the Yellow River and Luanhe River is estimated to be at 38°50′ N.

The sediment diffusion details of the local sea area were finely described in this study. For example, although we neglected the source of the Haihe River in the discriminant function, the effect boundary between the Haihe River and the Yellow River can still be found from the map. Figure 3 shows that the influence of the Haihe River source is limited to the west of 118° E.

4.2. Coastal Erosion and Shoreline Retreat

The distribution of grain size trend vectors (Figure 2) indicates that sediment is transported from the shore to the sea in most coastal areas of Bohai Bay, implying that coastal erosion is currently an important source of sediment in Bohai Bay. Earlier studies showed that the submerged delta of the Yellow River [1] and the upper part of the submerged shore slope of the delta [31] have become destructive, and the shoreline of the Diaokou River (old Yellow River channel) has retreated severely towards land [32]. The sediment in the southern part of Bohai Bay has a very high content of sand components, while the clay and chalk components, which are easily carried away, have been almost eroded away [33]. Overall, this study shows that the coastal sediments of Bohai Bay are generally in a state of offshore movement, which is consistent with the large-scale coastline retreat [34] in past decades.

The Yellow River is known for its large sediment discharge, carries a large amount of sediment to the sea every year, causing the delta plain to gradually expand [27]. However, in recent decades, the Yellow River has discharged significantly less runoff and sediment [35,36]. The main reason for the retreat of the shoreline is that the Yellow River supplies less sediment than is being carried away [28]. This also leads to the fact that material eroded from the shore is another important source of sediment in Bohai Bay, in addition to the rivers.

4.3. Relationship between Sediment Transport and Ocean Dynamics

Waves are the main driving force for sediment transport in coastal waters. In the process of wave propagation to the coast, waves will break and form a coastal flow when the wave orthogonal and the coastline have a certain intersection angle, resulting in coastal sediment transport [34]. In the northern part of Bohai Bay, waves usually propagate from S–SE and the strongest waves mainly propagate from ENE–NE (Figure 1), resulting in westward transport of coastal materials from the modern Luanhe River and the ancient Luanhe River delta. In the southern part of Bohai Bay, waves usually propagate from NE–SE and the strongest waves mainly propagate from the NE (Figure 1), resulting in westward transport of coastal material from the Yellow River estuary and delta. Since the wave direction is almost perpendicular to the shoreline (Figure 1), waves disperse in the western part of Bohai Bay, leading to insignificant sediment transport along the coast (Figure 2).

Another important role of waves is to re-suspend sediment from the sea floor, as sediment can move with seawater only once it enters the water column. Sediment transport pattern in the Bohai Sea is seasonally variable and has been described as "stored in summer and transported in winter" [37]. Early studies have shown that the amount of suspended sediment along the Yellow River Delta is tens or even hundreds of times higher in winter than in summer, as strong winds and waves occur much more often in winter than in summer [14,38]. The sediment is re-suspended under wave disturbance and is constantly carried away by tidal currents into Bohai Bay when seawater flows to the northwest and into Laizhou Bay when seawater flows to the southeast.

Tidal currents also play an important role in sediment transport. Both in the northern [39] and southern [40] parts of Bohai Bay, the residual flow is westward, which leads to the transport of discharged material from the Yellow River estuary and the Luanhe River estuary to the western part of Bohai Bay. Figure 3 clearly indicates that the sediment in the wide area from the Yellow River estuary westward to the west coast of the bay has great similarity with the Yellow River source, and the sediment in the wide area from the Luanhe River estuary westward to the west coast of the bay has great similarity with the Luanhe River source. In addition, the circulation in Bohai Bay is in approximately the same direction as the residual flow (Figure 1), resulting in an environment conducive to sediment sinking formed by the circulation centers in the middle of the northwest of Bohai Bay and the middle of the southwest of Bohai Bay.

4.4. Impact of Sediment Transport on Offshore Oil Extraction

The oil exploration and exploitation industry is well developed in Bohai Bay, and numerous offshore platforms have been built. This study shows that the sediment in the south coast of Bohai Bay are strongly moved away from the shoreline (Figure 2), leading to changes in the seafloor, which poses a threat to the offshore platforms in the nearby Chengdao Oilfield and Chengbei Oilfield. The high-intensity erosion may adversely affect the stability of the offshore platform pile foundations, so it is necessary to monitor the dynamics of seafloor erosion and siltation.

5. Conclusions

Based on the analysis of the grain size and geochemical characteristics of the surface sediment in the Bohai Bay, this paper studied the provenance contribution of rivers, sediment migration and diffusion, and discussed their relationship with ocean dynamics. The main conclusions are as follows:

1 The results of grain size trend analysis showed that coastal sediment in Bohai Bay is generally removed from the coast, which is consistent with the widely observed shore retreat.
2 Discriminant function (DF) index based on Ca/Al can indicate the sediment contribution of the Yellow River and Luanhe River to Bohai Bay. Materials from the Yellow River dominated southern, western and central Bohai Bay, while materials from the Luanhe River dominated northeast Bohai Bay. The DF analysis indicated that the dominance of the Yellow River source was greater than previously thought, and the influence boundary between the Yellow River and Luanhe River is estimated to be at 38°50′ N. The influence of the Haihe River source is limited to the west of 118° E.
3 With the diversion of the Yellow River channel and the reduction of material feed into the sea, sediment transport in Bohai Bay is now controlled mainly by marine dynamic processes. In both south and north Bohai Bay, sediment was transported into the Bay under the influence of constant wave directions, strongest wave directions and tidal residual currents. The central sea area and the underwater bank slope along the west coast of Bohai Bay have become the main sink areas of sediment. The coast has become an important source of sediment in Bohai Bay, and the source is higher up to include the coast itself.

Author Contributions: Conceptualization, B.Z.; methodology, L.Z. and P.W.; formal analysis, B.Z.; investigation, B.Z., P.Z. and J.Y.; writing—original draft preparation, B.Z. and L.Z.; writing—review and editing, L.Z., P.W. and Y.Y.; visualization, J.Y.; supervision, Y.Y.; data curation, X.L. and S.Y.; and project administration, X.L. and S.Y. All authors have read and agreed to the published version of the manuscript.

Funding: This research received funding from the Foundation of State Environmental Protection Key Laboratory of Marine Ecosystem Restoration (No. 202106) and the Doctoral Scientific Research Foundation of Liaoning Province (No. 2021-BS-300).

Institutional Review Board Statement: Not applicable.

Informed Consent Statement: Not applicable.

Data Availability Statement: Data sharing not applicable.

Acknowledgments: The authors would like to thank the editors and reviewers for constructive suggestions and comments that helped improve the quality of the article.

Conflicts of Interest: The authors declare no conflict of interest.

References

1. Wu, X.; Bi, N.; Xu, J.; Nittrouer, J.A.; Yang, Z.; Saito, Y.; Wang, H. Stepwise morphological evolution of the active Yellow River (Huanghe) delta lobe (1976–2013): Dominant roles of riverine discharge and sediment grain size. *Geomorphology* **2017**, *292*, 115–127. [CrossRef]
2. Li, J.; Kang, H.; Wang, H.; Pei, Y. Modern geological action and discussion of influence facters on the west coast of Bohai Bay, China. *Geol. Surv. Res.* **2007**, *30*, 295–301.
3. Qin, Y.; Zhao, Y.; Zhao, S. *Geology of the Bohai Sea*; Science Press: Beijing, China, 1985.
4. Han, Z.; Zhang, J.; Zou, H.; Yi, W.; Li, M. Characteristics and provenance of clay mineral assemblage of sediments from the northern part of the Bohai Bay. *Period. Ocean Univ. China* **2011**, *41*, 95–102.
5. Zhang, A.; Liu, M.; Liao, Y.; Zhang, Y.; Fan, D. The sedimentary geochemical trace of the Yellow River sediments diffusion in the Bohai Bay. *Adv. Mar. Sci.* **2015**, *33*, 246–256.
6. Qin, Y.; Mei, X.; Jiang, X.; Luan, X.; Zhou, L.; Zhu, X. Sediment provenance and tidal current-driven recycling of Yellow River detritus in the Bohai Sea, China. *Mar. Geol.* **2021**, *436*, 106473. [CrossRef]
7. Li, G.; Yue, S.; Zhao, D.; Sun, Y. Rapid deposition and dynamic processes in the modern Yellow River Mouth. *Mar. Geol. Quat. Geol.* **2004**, *24*, 29–36.
8. Li, G.; Tang, Z.; Yue, S.; Zhuang, K.; Wei, H. Sedimentation in the shear front off the Yellow River mouth. *Cont. Shelf Res.* **2001**, *21*, 607–625. [CrossRef]
9. Wang, H.; Yang, Z.; Bi, N. 3-Dsimulation of the suspended sediment transport in the Yellow River mouth I: Shear front off the Yellow River mouth. *J. Sediment Res.* **2006**, *2*, 1–9.
10. Jia, Y.; Ke, X.; Xu, Y.; Wang, Y. Sedimentary transport trends of within a sand bar/lagoon system in the Bohai Sea. *Mar. Sci.* **1999**, *3*, 56–59.
11. Zhao, B.; Zhuang, G.; Cao, D.; Lei, F. Circulation, tidal residual currents and their effects on the sedimentations in the Bohai Sea. *Oceanol. Limnol. Sin.* **1995**, *26*, 466–473.
12. Zhang, Y.; Li, F. The characteristics of material component and the material resources in Huanghe (yellow) River, Luanhe River. *Mar. Sci.* **1983**, *7*, 15–18.
13. Wang, Y. Analysis of Sediment Characteristics in the Tail Section of Haihe River and Compared with the Tail Section of Yellow River. Master's Thesis, Ludong University, Yantai, China, 2017.
14. Yang, Z.; Ji, Y.; Bi, N.; Lei, K.; Wang, H. Sediment transport off the Huanghe (Yellow River) delta and in the adjacent Bohai Sea in winter and seasonal comparison. *Estuar. Coast. Shelf Sci.* **2011**, *93*, 173–181. [CrossRef]
15. Wang, P.; Zhang, L.; Lin, X.; Yan, J.; Zhang, P.; Zhao, B.; Zhang, C., Yu, Y. Spatial distribution, control factors and sources of heavy metal in the surface sediments of Fudu Estuary waters, East Liaodong Bay, China. *Mar. Pollut. Bull.* **2020**, *156*, 111279. [CrossRef] [PubMed]
16. Mcmanus, J. Grain size determination and interpretation. In *Techniques in Sedimentology*; Tucker, M.E., Ed.; Blackwell Scientific Publications: Hoboken, NJ, USA, 1988.
17. Gao, S. A Fortran program for grain-size trend analysis to define net sediment transport pathways. *Comput. Geosci.* **1996**, *22*, 449–452. [CrossRef]
18. Gao, S.; Collins, M.B. Analysis of grain size trends, for defining sediment transport pathways in marine environments. *J. Coast. Res.* **1994**, *10*, 70–78.
19. McLaren, P. An interpretation of trends in grain size measures. *J. Sediment. Petrol.* **1981**, *51*, 611–624.
20. McLaren, P.; Bowles, D. The effects of sediment transport on grain-size distributions. *J. Sediment. Petrol.* **1985**, *55*, 457–470.
21. Poizot, E.; Méar, Y.; Biscara, L. Sediment Trend Analysis through the variation of granulometric parameters: A review of theories and applications. *Earth-Sci. Rev.* **2008**, *86*, 15–41. [CrossRef]
22. Lan, X.; Zhang, X.; Zhao, G.; Zhang, Z. REE distributions and provenance discrimination of sediments from South Yellow Sea. *Geochimica* **2009**, *38*, 123–132.
23. Yang, Z.; Wang, Z.; Ju, J.; Zheng, A. Study on carbonates from the coastal zone of the Yellow River Delta and adjacent Bohai Gulf. *J. Ocean Univ. Qingdao* **1989**, *19*, 91–99.
24. Zhao, Y.; Yan, M. *Geochemistry of China's Shallow Sea Sediments*; Science Press: Beijing, China, 1994.
25. Liu, J. Material Compositions of Sediments in the Bohai Sea mud Area during the Holocene and Their Environmental Implications. Ph.D. Thesis, Chinese Academy of Sciences, Qingdao, China, 2007.
26. Qiao, S.; Shi, X.; Saito, Y.; Li, X.; Yu, Y.; Bai, Y.; Liu, Y.; Wang, K.; Yang, G. Sedimentary records of natural and artificial Huanghe (Yellow River) channel shifts during the Holocene in the southern Bohai Sea. *Cont. Shelf Res.* **2011**, *31*, 1336–1342. [CrossRef]
27. Xue, C. Historical changes in the Yellow River delta, China. *Mar. Geol.* **1993**, *113*, 321–330. [CrossRef]
28. Yuan, P.; Bi, N.; Wu, X.; Zhang, Y.; Wang, H. Surface sediments at the subaqueous Yellow River Delta: Classification and distribution. *Mar. Geol. Quat. Geol.* **2016**, *36*, 49–57.
29. Sun, S.; Xu, J.; Li, G.; Liu, X.; Qiao, L.; Zhou, S. Analysis of suspended sediments concentration in the Bohai Bay in winter based on remote sensing. *Oceanol. Limnol. Sin.* **2020**, *51*, 258–264.
30. Liu, X.; Qiao, L.; Zhong, Y.; Wan, X.; Xue, W.; Liu, P. Pathways of suspended sediments transported from the Yellow River mouth to the Bohai Sea and Yellow Sea. *Estuar. Coast. Shelf Sci.* **2020**, *236*, 106639. [CrossRef]

31. Wu, S.; Shen, X.; Zang, Q.; Zhang, Q.; Li, P.; Zhong, D.; Huang, Y. Preliminary study on the changes of sediment erosion and deposition in the Wuhaozhuang area of the Yellow River Delta. *Ocean Dev. Manag.* **1991**, *8*, 57–63.
32. Wu, X.; Fan, Y.; Wang, H.; Bi, N.; Xu, C.; Zhang, Y.; Liu, J.; Lu, T. Evolution of abandoned deltaic river channel—A case from the Diaokou channel of the Yellow River. *Mar. Geol. Quat. Geol.* **2021**, *41*, 22–29.
33. Zhang, L.; Hu, R.; Zhu, L.; Zhao, G.; Zhong, W. Characteristics of detrital minerals in Bohai Bay and their implications for provenance and sedimentary dynamical environment. *Period. Ocean Univ. China* **2019**, *49*, 60–70.
34. Mi, B.; Yan, J.; Zhuang, L.; Luan, Z. The topographical and morphological features of modern Yellow River mouth area and their relation to erosional/depositional processes. *Mar. Geol. Quat. Geol.* **2010**, *30*, 31–38. [CrossRef]
35. Wu, X.; Wang, H.; Bi, N.; Saito, Y.; Xu, J.; Zhang, Y.; Lu, T.; Cong, S.; Yang, Z. Climate and human battle for dominance over the Yellow River's sediment discharge: From the Mid-Holocene to the Anthropocene. *Mar. Geol.* **2020**, *425*, 106188. [CrossRef]
36. Wu, X.; Wang, H.; Bi, N.; Xu, J.; Nittrouer, J.A.; Yang, Z.; Lu, T.; Li, P. Impact of Artificial Floods on the Quantity and Grain Size of River-Borne Sediment: A Case Study of a Dam Regulation Scheme in the Yellow River Catchment. *Water Res.* **2021**, *57*, e2021W–e29581W. [CrossRef]
37. Ren, M.; Shi, Y. Sediment discharge of the Yellow River (China) and its effect on the sedimentation of the Bohai and the Yellow Sea. *Cont. Shelf Res.* **1986**, *6*, 785–810. [CrossRef]
38. Wang, H.; Wang, A.; Bi, N.; Zeng, X.; Xiao, H. Seasonal distribution of suspended sediment in the Bohai Sea, China. *Cont. Shelf Res.* **2014**, *90*, 17–32. [CrossRef]
39. Ji, R.; Lu, Y.; Zuo, L. Formation Mechanism and Stability of Caofeidian Channel in the Bohai Bay. *Acta Geogr. Sin.* **2011**, *66*, 348–355.
40. Shi, M.; Zhao, J. The analysis of hydrographical characteristics in the nontidal region M2 near the delta of the Huanghe River. *J. Shandong Coll. Oceanol.* **1985**, *15*, 127–136.

applied sciences

MDPI

Article

Nature-Based Solutions for Flood Mitigation and Soil Conservation in a Steep-Slope Olive-Orchard Catchment (Arquillos, SE Spain)

Patricio Bohorquez [1,2,*], Francisco José Pérez-Latorre [2,3], Inmaculada González-Planet [1], Raquel Jiménez-Melero [2,4] and Gema Parra [2,4]

1 Department of Mechanical and Mining Engineering, University of Jaén, 23071 Jaén, Spain
2 Centre of Advanced Studies in Earth Science, University of Jaén, 23071 Jaén, Spain
3 Department of Mechanical and Mining Engineering, University of Jaén, 23700 Linares, Spain
4 Department of Animal Biology, Plant Biology and Ecology, University of Jaén, 23071 Jaén, Spain
* Correspondence: patricio.bohorquez@ujaen.es

Abstract: The frequency and magnitude of flash floods in the olive orchards of southern Spain have increased because of climate change and unsustainable olive-growing techniques. Affected surfaces occupy > 85% of the rural regions of the Upper Guadalquivir Basin. Dangerous geomorphic processes record the increase of runoff, soil loss and streamflow through time. We report on ripple/dune growth over a plane bed on overland flows, deep incision of ephemeral gullies in olive groves and rock-bed erosion in streams, showing an extraordinary sediment transport capacity of sub-daily pluvial floods. We develop a novel method to design optimal solutions for natural flood management and erosion risk mitigation. We adopt physical-based equations and build a whole-system model that accurately reproduces the named processes. The approach yields the optimal targeted locations of nature-based solutions (NbSs) for active flow-control by choosing the physical-model parameters that minimise the peak discharge and the erosion-prone area, maximising the soil infiltration capacity. The sub-metric spatial resolution used to resolve microtopographic features of terrains/NbS yields a computational mesh with millions of cells, requiring a Graphics Processing Unit (GPU) to run massive numerical simulations. Our study could contribute to developing principles and standards for agricultural-management initiatives using NbSs in Mediterranean olive and vineyard orchards.

Keywords: nature-based solutions; distributed hydrological model; IBER+; Guadalquivir Basin; European Union Directive 2007/60; flash flood; gully erosion; olive and vineyard orchards

Citation: Bohorquez, P.; Pérez-Latorre, F.J.; González-Planet, I.; Jiménez-Melero, R.; Parra; G. Nature-Based Solutions for Flood Mitigation and Soil Conservation in a Steep-Slope Olive-Orchard Catchment (Arquillos, SE Spain). *Appl. Sci.* **2023**, *13*, 2882. https://doi.org/10.3390/app13052882

Academic Editors: Gordon Gilja, Manousos Valyrakis, Panagiotis Michalis, Thomas Pahtz and Oral Yagci

Received: 19 January 2023
Revised: 13 February 2023
Accepted: 20 February 2023
Published: 23 February 2023

1. Introduction

The frequency and intensity of sub-daily floods have increased in the Upper Guadalquivir Basin (southern Spain) during the last twenty years due to the actual scenario of climate change [1]. This kind of extreme hydrological event has become more and more relevant in many Mediterranean areas of Europe [2], corroborating the predictions of the Intergovernmental Panel on Climate Change (IPCC) [3] and the European Environmental Agency (EEA) [4]. High precipitation rates combined with anthropogenic effects of soil use and management have provoked unsustainable soil-loss rates and large-scale gully erosion in some European agricultural catchments [5]. Simultaneously, floods have developed at the catchment scale because of deep runoff [6]. Both problems are linked because they depend, in leading order, on the dynamics of overland flows [7]. The Mediterranean Experts on Climate and Environmental Change (MedECC) [8] recommend the implementation of NbS and adaptive practices of sustainable land management because climate change is expected to increase soil erosion in a complex, non-linear way. Mitigating flood risk and preventing soil erosion at large scales cannot be achieved with isolated civil engineering actions; the economic costs and environmental impacts would be prohibitive. Alternatively, we adopt Nature-Based Solutions (NbSs) [9–11] that seek to delay the flooding, reduce the peak water discharge and erosion capacity of overland flows, and promote infiltration at the basin scale.

We considered a pilot study site of the Upper Guadalquivir Basin where the current land management practice and the increase in extreme rainfall rates (up to approximately 30 mm·h^{-1}) have caused soil loss rates of 20–50 t ha^{-1} yr^{-1} (see Figure 2 in [12]). Such values are nearly a maximum in Europe, so the Guadalquivir watershed has attracted the scientific community's interest in gully erosion processes [13,14]. More than 80% of the catchment is dedicated to traditional olive groves (i.e., the vegetation surrounding the crops was removed), a practice extended across the entire Mediterranean basin [15]. In such environments, we have shown the potential of the two-dimensional Saint-Venant Equations (2D-SVEs) to map the extent of the inundations and describe fluvial geomorphological processes at the river-stretch scale [16,17]. High-spatial resolution numerical simulations using 2D-SVE-distributed hydrological models yield further details of the rainfall-runoff-inundation processes at the watershed [6]. The output products for a given rainfall event are the flow depth and velocity not only in the main river but also in smaller spatial-scale elements of the whole drainage network. If the intensity of the precipitation is high enough to provoke water discharges above the drainage capacity of the streams and rivers, the 2D-SVE-distributed numerical simulation gives all the flooded areas across the whole watershed. Furthermore, the interpretation of the flow-velocity map serves to predict the locations where fine sediments transported as wash load settle and provoke slackwater deposits [1].

The main objective of this article is to find the optimal NbS (including its location) which provides the most benefits in terms of hydraulic effectiveness to prevent flooding and soil erosion in steep-slope olive groves. We address both problems at the same time instead of independently. Subsidiary—but no less important—objectives are the increase in the retention capacity of overland flows and the soil infiltration capacity to avoid the undesirable phenomenon of soil-water repellence and higher surface runoff volumes in olive groves [18]. All of them contribute to mitigating street-scale urban floods in the site study. So, this work extends the capabilities and applicability of our former model, initially calibrated and planned only for flood-risk mapping in ungauged olive-orchard basins [1,6].

The developed methodology has two stages: (i) first, the diagnosis of the causes leading to inundation and soil erosion using the highly resolved Computational Fluid Dynamics (CFD) simulation of the current situation; (ii) second, the quantification of the hydraulic effects of the possible measures for natural flood management (NFM) and soil loss mitigation by simulating parameter variation in future scenarios. The output products identify the optimal model parameters across the catchment (i.e., controlled variables) that minimise the objective functions, namely: the peak water discharge at the basin outlet and the shear stresses responsible for soil erosion. The approach is standard in Active Flow-Control (AFC), a "multidisciplinary science and technology thrust aimed at altering a natural flow state into a more desired state" [19]. The last step is the selection of a realistic NbS that produces the optimal values of the controlled physical variables at the targeted placement. We consider NbSs compatible with sustainable production techniques in olive growing [20] and complementary innovations developed worldwide, as described below. Isolated NbSs cannot fix all the problems alone, and the ideal future scenario should exploit several actions' synergetic benefits.

To this end, we apply standard techniques in AFC [19] and CFD [21], which consist of parameter variation and geometric modelling with sub-metric spatial resolution, uncommon yet in evaluating the effectiveness of an NbS [10]. There is presently no standard approach to determining how many NbSs to use and where to locate them in NFM [22]. A limited number of studies applied the typical engineering protocol with spatially-resolved, physically-based 2D-SVE simulations [23]. A frequent limitation of the existing design routines is the coarse resolution of the computational grid because of two reasons: first, the coarse mesh impedes the detailed geometrical modelling and hydraulic simulation of an individual NbS at a given location; second, the grid size affects the simulation results of rainfall/runoff events in a magnitude as large as physical factors [24]. We overcome such drawbacks using the latest advances in distributed hydrological modelling that allow simulating catchment-scale problems at extremely high spatial resolution in the Graphics Processing Unit (GPU) [25]. We adopt the software IBER+ [26], among alternatives as SERGHEI-SWE [27], TRITON [28], and LISFLOOD-FP [29]. Here we show that IBER+ allows the development of a whole system model incorporating the agricultural areas, the drainage network, urban areas, and civil engineering infrastructures by a multiple-scale grid resolution and proper mapping of physical parameters. Hence, the following issues are addressed at once: simulating rainfall-runoff processes [30]; urban flood mapping [31,32]; soil-erosion risk [14]; and the effectiveness of nature-based mitigation measures [33–35]. However, the current limitation of single GPU computations in IBER+ restricts the catchment area to 20 km^2 for extremely high spatial resolutions of 1 m^2. Scaling up the analysis to larger draining basins (<1000 km^2) is now possible with SERGHEI-SWE through exascale High-Performance Computing (HPC) technology and multi-GPU [27]. The proposed approach permits fully exploiting the available Digital Elevation Model (DEM), resolving many microtopographic features controlling surface flow, soil erosion and sediment transport connectivity.

To the best of the authors' knowledge, the design of nature-based interventions and the quantification of their hydraulic benefits in reducing flood and soil loss hazards in catchments with soil uses dominated by olive growing have not been undertaken beforehand. Such objectives lie in the scope of NbSs, which are "actions to protect, sustainably manage, and restore natural or modified ecosystems, that address societal challenges effectively and adaptively, simultaneously providing human well-being and biodiversity benefits" [36]. Plausible NbSs for our targeted physical goals are the creation of an online wetland [37–39] or a naturalised irrigation pond [40], gullies infilling with dry bush or small boulders [20,41], gravels seeding in streambeds [23,42], flexible submerged vegetation planting in streams [43], and the restoration of the river floodplains [44]. Additionally, we suggest adopting cultivation systems in olive orchards based on plant residue mulches [45] and live plant covers along the middle of the orchard lanes [46]. Such tools are suitable for water conservation and erosion control in Mediterranean olive tree cropping [47] and are also compatible with water collection pits [20]. The outputs of our study contribute to developing principles and standards for agricultural-management initiatives using NbSs [48,49] and could be applied in other Mediterranean areas of Europe with similar problems under the current climate change scenario [8].

2. Materials and Methods

2.1. Study Site

The selected study area is located in the southwest of Europe within the upper Guadalquivir Basin (Spain). The basin has a small size of $A_{\mathrm{drain}} = 6.9$ km^2, a mean slope of 19% and a maximum (minimum) height greater (lower) than 750 m MSL (360); see Figure 1. The catchment mimics an ideal laboratory with a homogeneous surface but at a much larger, real scale. Olive groves dominate the soil use, occupying 93.4% of the drainage area. There are approximately 46,500 olive trees (green dots in Figure 1a), with a density of 70 trees·hectare^{-1}. The combination of steep slopes (Figure 1b) and land use of olive groves under conventional tillage prevent infiltration [18]. Additionally, the lack of natural cover exposes sediment to erosion and causes a high development of laminar and gully processes.

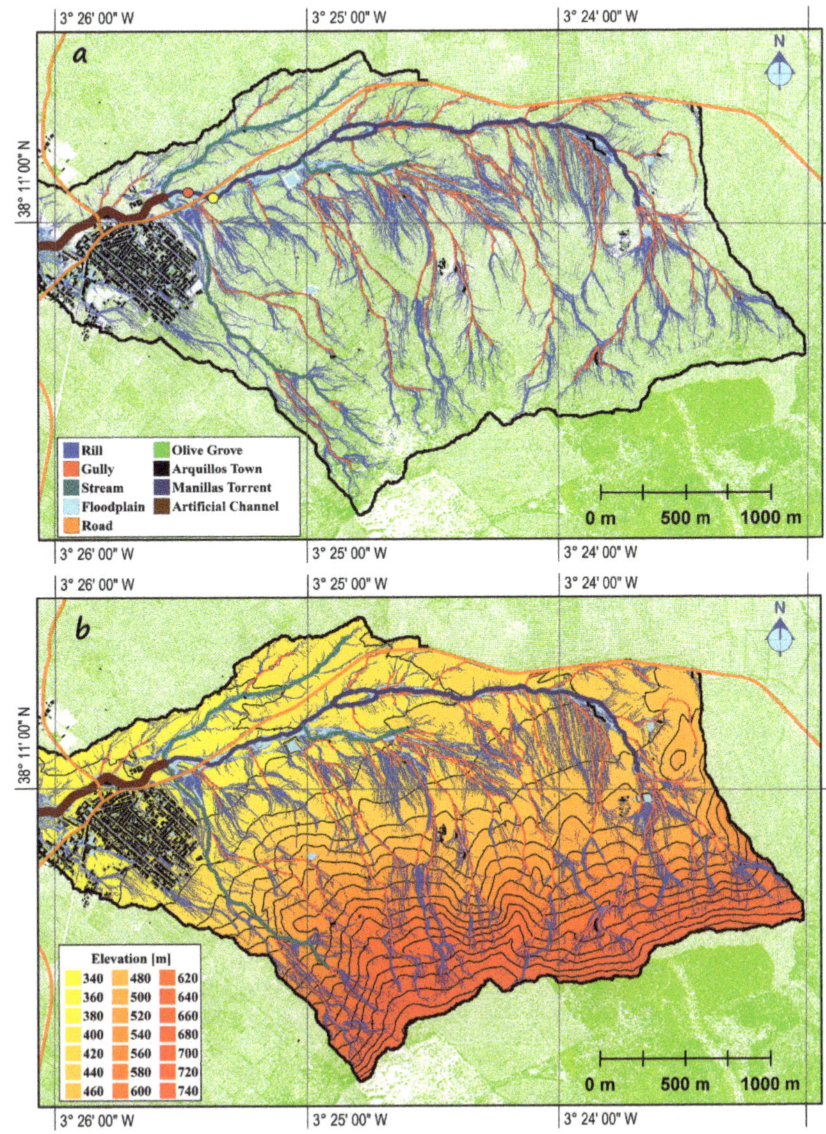

Figure 1. General characteristics of the Manillas basin in Arquillos (Jaén, Southeast Spain). (**a**) Catchment overview (perimeter highlighted with the black solid line) with three regions of interest: the urban areas (black), the man-made channel that should protect the town from inundation (thick line in brown), and olive trees (green dots). The drainage network is composed of rills, gullies, and streams. (**b**) Hypsometric map and drainage network. For completeness, main traffic roads are coloured in orange. Maps created from scratch with Matlab and GlobalMapper based on a Light Detection and Ranging (LiDAR) dataset acquired on May 2021 with 1.5 points per square meter (https://www.ign.es/, accessed on 1 July 2022). Rills, gullies, and streams were obtained from the distributed hydrological simulations presented in this paper.

We identified the following sediment sources according to the classification by Charlton [50], see Figure 1: rill (light blue), gully (red), and stream (dark blue and greenish blue). Rills are small micro-channels where overland flow exerts a shear stress on the soil close to the threshold value for erosion with flow depths of up to 30 mm. Gullies are relatively permanent ephemeral channels in semi-arid environments without vegetation cover, eroded during highly seasonal precipitation. The gully channel is narrow, with a steep bed slope and a characteristic flow depth larger than 0.5 m. Smaller features, intermediate in size between gullies and rills, are referred to as ephemeral gullies in Soil Science Society of America [51]. Stream channels collect and transfer the sediment budget produced in rills and gullies [52]. In the Manillas basin, streams develop in regions with much smaller bottom slopes than for ephemeral gullies. All these elements of the drainage network are connected to the primary river system, referred to as the Manillas stream. In general, the channel form of the streams can adapt to altered environmental conditions [53].

The absence of civil infrastructure in most of the watershed ensures that runoff responds naturally to torrential rainfall. Hence, overland flows erodes many gullies and shallow streams that drain transverse to the Manillas stream. The Manillas stream catches overland flows from the gullies and shallow streams along 4.6 km with a mean slope of 4.2% and two nick points. The stepper reach is located in the headwater (i.e., first 500 m with a 6.7% slope). Later, the bed slope decreases to 3.3% for 1000 m. Further downstream, it reaches a constant value of 1.8%, which remains for 3 km. During its fluvial course, the main channel exhibits non-uniform bankfull dimensions. On occasion, the water has sculpted a much wider and deeper channel than other river stretches. Therefore, the channel-reach morphology may correspond to a single confined flow or a floodplain channel, depending on the river stretch [52]. The wash load is the dominant transport process due to the small sediment size and straight planform [54]. The fine granulometry of available sediments favours transport processes dominated by wash load that settle slackwater deposits with low flow velocity [55].

The Mediterranean climate, characterised by seasonal and irregular rainfall distribution, dominates the Upper Guadalquivir Basin, including the Manillas sub-basin. The complex topography of the landscapes and the geographical position favour two climate influences coming from the Mediterranean Sea and the Atlantic Ocean. So, two kinds of precipitation events usually develop. First, low intensity and continuous rainfalls longer than one week, which are not dangerous in small basins [16,17]. Second, sub-daily rainfalls, which are produced by reduced area storms and cause local short-rain floods at small basins [1,6]. The increase in frequency and magnitude of short-pluvial floods reported in the present study case further corroborates the predictions of the IPCC [3] and the EEA [4]. Nowadays, short rains are becoming more frequent and more severe in the Manillas basin and other basins of the Guadalquivir headwaters.

Because of orographic constraints, since 1970, Arquillos town has grown, occupying the meanders floodplain that exists at the basin outlet. A regular train of two meanders surrounds the urban area, with a linear wavelength and amplitude of 500 m and 100 m, respectively. In the last decades, storms with an accumulated precipitation depth of 90 mm provoked flash floods for a few hours nearly once a year. To protect the town from inundations, an artificial channel with a uniform rectangular section of 1.4 m depth and 3 m width was built in the 1980s (Figure 2). Presently, the reinforced-concrete base of the channel is eroded, which illustrates the high-speed flows that develop along the flume during extreme rains. This point implies that the flow velocity is above the threshold of erosion for a concrete pavement (i.e., 6 $m \cdot s^{-1}$), which would be corroborated with the numerical simulations conducted in this study. Furthermore, there is considerable evidence of channel overtopping because several inundations occurred in the paved road crossing the town shown in Figure 2b.

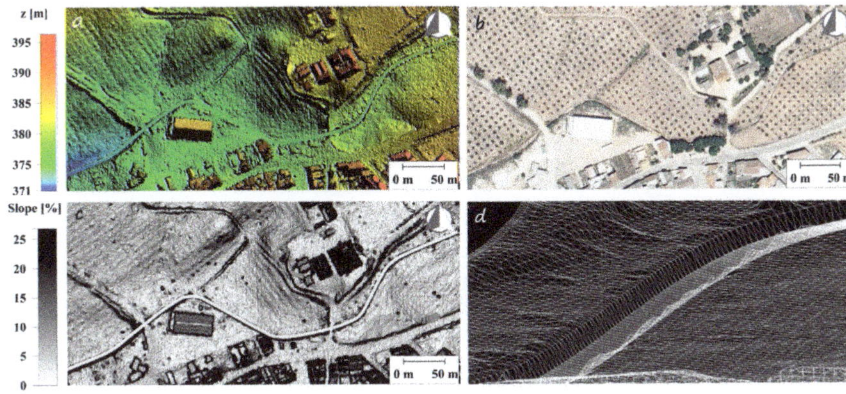

Figure 2. (**a**) DEM, (**b**) orthophoto, and (**c**) DEM slope in a small area surrounding the Arquillos town. We built the DEM from scratch using filtered LiDAR data and in-situ measurements with Leica Zeno 20 Global Positioning System (GPS). (**d**) Zoom of the computational mesh for the artificial channel and the floodplains. Note the hybrid topology of the cells with rectangular elements in the channel and triangles otherwise.

2.2. Distributed Hydrological-Hydraulic Modelling

The two-dimensional shallow-water equations, better known as the Saint-Venant equations, were adopted to describe the motion of surface waters. To this end, we configured the hydrological module of IBER+ [26]. It solves numerically the depth-averaged inviscid conservation laws of mass and momentum for clear water in a two-dimensional, Cartesian system of coordinates, given in compact form by

$$\frac{\partial h}{\partial t} + \nabla \cdot (h\,\boldsymbol{u}) = P - I\,, \tag{1}$$

$$\frac{\partial h\,\boldsymbol{u}}{\partial t} + \nabla \cdot (h\,\boldsymbol{u}\,\boldsymbol{u}) + \nabla\left(\frac{g\,h^2}{2}\right) = -g\,h\,\nabla z - \frac{\boldsymbol{\tau}_b}{\rho}\,, \tag{2}$$

where t is the time, h is the depth of the water measured along the vertical coordinate, $\boldsymbol{u}$ is the depth-averaged velocity vector, z is the bed altitude, and g is the acceleration due to gravity. The source terms in the continuity Equation (1) represent the rainfall intensity, P, and the soil infiltration rate, I. In the momentum balance Equation (2), the source terms are the bed slope ∇z and the bottom shear stress $\boldsymbol{\tau}_b/\rho$, in which ρ is the water density. Manning's friction law was used for evaluating the hydraulic resistance in (2) by setting

$$\frac{\boldsymbol{\tau}_b}{\rho} = \frac{g\,n^2}{h^{1/3}}\boldsymbol{u}|\boldsymbol{u}|\,, \tag{3}$$

where n represents Manning's roughness coefficient.

The assumption of small bottom slopes is the basic hypothesis of the two-dimensional shallow-water Equations (1) and (2), which imply that the pressure distribution on the vertical is hydrostatic. Although such a constraint is not formally satisfied in all the situations, as in mountain areas, the most sophisticated GPU numerical codes for distributed hydrological simulations solve standard Equations (1) and (2) [26–29]. Nowadays, there is no consensus on the theoretical formulations for steep slopes, and several theories have been proposed (see the review by Maranzoni and Tomirotti [56]). So, on the numerical side, the existing numerical codes are in-house and still based on CPU instead of GPU [57,58], preventing their use for the current purpose.

The two-dimensional shallow water model (1)–(3), together with homogeneous initial conditions for h and $\boldsymbol{u}$, were discretised using a finite volume method based on the DHD solver [59] implemented in the free-software IBER+ [26]. The Reynolds stresses were neglected as suggested by Cea and Bladé [59] for overland flow applications. We ran all the simulations using single-precision calculations in a Quadro RTX5000 GPU. In so doing, we achieved a reasonable computational time of three days per simulation. We set the Courant-Friedrich-Lewy number of 0.45 to satisfy the stability constraint of the DHD scheme.

In the following subsections, we describe the configuration of the physical parameters of the hydrological model. We give details on the computational grid, and the DEM in Section 2.2.1. Then, in Section 2.2.2, we used historical data of extreme hydrological events to characterise the design storm parametrised with P in (1). Lastly, we configured the Manning roughness n (2) and the infiltration rate I (1) in Section 2.2.3 and Section 2.2.4, respectively, borrowing existing results from well-established studies. The analysis of the possible values of I and n for the present-day situation and the future with NbSs served to vary them systematically and configure a set of fifteen numerical simulations.

2.2.1. Computational Mesh

We built a computational mesh across the whole drainage area of 7 km^2 subdividing it into three regions. The whole system model requires a high spatial resolution to capture with accuracy the topographic flow control exerted across the catchment pathways and urban areas. First, we discretised the man-made channel of 1 km length and 3 m width (brown line in Figure 1) with 60,000 structured elements of 0.1 m $\times$ 0.5 m size. Elevations of the channel thalweg and lateral walls were measured in situ with a Leica Zeno 20 Global Positioning System (GPS) combined with laser Leica DISTO^{TM} and AS11 multi-frequency Global Navigation Satellite System (GNSS) antenna. Subsequently, a digital elevation model of the channel area was constructed, and the elevations were assigned to the computational grid. Second, the urban area (black squares in Figure 1) was discretized with triangular cells with a characteristic size of 1 m. The nodes elevation were obtained from a home-made DEM based on Light Detection and Ranging elevations acquired on May 2021 by the Spanish Geographical Institute (https://www.ign.es/, accessed on 1 July 2022). The average distance of a point from its neighbours in the LiDAR dataset is 0.82 m (corresponding to the point density of 1.5 points per square meter) with a root mean squared error between the ground reference and estimated position of 0.3 m and 0.15 m in the horizontal and vertical coordinates, respectively. Third, the computational grid is coarser in the greatest rural area of the Manillas catchment (green region in Figure 1) because the unstructured grid has a characteristic edge length of 1.5 m. Elevations were also obtained from the LiDAR data that had to be filtered to delete the vegetation and olive trees, keeping only the ground points. The elevations of the gullies and streams are accurately represented in the LiDAR data thanks to the seasonal behaviour of the precipitation. The computational mesh amounts to 7 million cells and replicates real topography in the simulated region.

Figure 2 shows details of the geometrical model in a zoomed area that includes the channel, buildings and olive groves. The elevations (Figure 2a) accurately represent the civil infrastructures visible in the orthophoto (Figure 2b). The man-made channel that surrounds the town can be better observed in the shadowgraph (Figure 2c). Lastly, Figure 2d shows a zoom of the computation grid near the channel. We ensured that the surface in the olive-grove floodplains was smooth by removing the olive trees when constructing the DEM from LiDAR, as described above.

2.2.2. Design Storm

To represent the distribution of rainfall intensity over time, parametrised by $P(t)$ in (1), we analysed the extreme values of available rainfall data since 1950. The main objective was to calibrate the rainfall rate $\bar{P}$ and duration D of the step hyetograph:

$$P(t) = \begin{cases} \bar{P} & \text{if } 0 \leq t \leq D \\ 0 & \text{if } t > D \end{cases} \tag{4}$$

Figure 3a shows the accumulated precipitation for 24 h (referred to as daily rainfall depth from now on) measured by a rain gauge located in the Guadalén Reservoir at a distance of 4 km downstream of the town of Arquillos. The maximum value was 95 mm on 15 August 1996 due to a sudden storm in the night. It caused the most catastrophic flood reported in the town of Arquillos. Real-time hydrologic monitoring has provided hourly average values since 2010. Figure 3b depicts the accumulated precipitation for 1 h (referred to as hourly rainfall depth), with a peak value of 30 mm·h^{-1} on 8 June 2011 at 7:00 a.m. There are other extreme events with hourly rainfall depths close to the peak value, for instance, 27.9 mm·h^{-1} on 21 October 2012 at 4:00 a.m. and 28.9 mm·h^{-1} on 7 September 2016 at 8:00 a.m. Usually, the storm lasts for three hours and leads to daily rainfall depths of 80–95 mm (Figure 3a). Subsequently, we set the design storm parameters given by $\bar{P} = 30$ mm·h^{-1} and $D = 3$ h that lead to daily/hourly rainfall depths similar to the highest values measured since 1950.

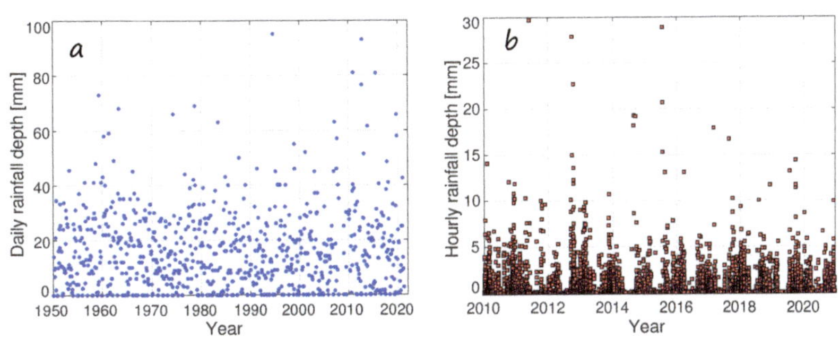

Figure 3. Maximum values of daily (**a**) and hourly (**b**) precipitation depth recorded 1950–2022 and 2010–2022, respectively. Data source: http://www.chguadalquivir.es/saih/, accessed on 1 July 2022.

The dangerous nature of short-rain floods in small-area catchments is related to the short period of concentration T required to achieve the peak streamflow Q_{ss}, given in the absence of infiltration by Brutsaert [60]:

$$Q_{ss} = A_{\text{drain}} \bar{P}\,, \tag{5}$$

$$T = \frac{L}{U_{\bar{P}}}\,, \tag{6}$$

with the runoff velocity $U_{\bar{P}}$:

$$U_{\bar{P}} = (\bar{P}\,L)^{\frac{2}{5}} \cdot S_0^{\frac{3}{10}}\, n^{-\frac{3}{5}}\,. \tag{7}$$

In (6) and (7), L is the characteristic length of the catchment (e.g., $L = A_{\text{drain}}^{1/2}$ in a nearly square basin), S_0 is the mean basin slope and n corresponds to Manning's roughness coefficient. Setting $n = 0.02$ s·m$^{-1/3}$ as for similar basins [1,6], we get $U_{\bar{P}} = 1.3$ m·s^{-1}, $T = 33$ min and $Q_{ss} = 57.4$ m^3·s^{-1}. The theoretical value of T can be interpreted as the minimum duration of the rain event T needed to provoke Q_{ss}.

Such an extreme value of the peak water discharge was not observed before 1996. Conversely, it has become usual in the last decade because of sudden storms with a duration shorter than one day and longer than the concentration time T, referred to as a short-rain flood [61]. Short-rain floods have acquired relevancy in small catchments under the actual climate change scenario not only in the Upper Guadalquivir River [1,6] but also in the Mediterranean [2,62]. Furthermore, the reported increase in the frequency and the magnitude of short-rain inundations corroborates the predictions of the IPCC [3] in the current study site.

2.2.3. Surface Roughness

Manning's equation has been used intensively in riverine hydraulics since its formulation in 1889. Extensive tables of the roughness coefficient n are available for different channel characteristics [63,64], natural channels, and flood plains [65]. More recently, distributed hydrological simulations have shown that Manning's resistance law is also plausible to predict the runoff dynamics in roads, urban floods and agriculture areas [32,66–68].

We classified the catchment area into six types of soils, as shown in Figure 4a, and assigned the Manning coefficient described below. Urban areas (dark green), roads (orange), and existing irrigation ponds (brown) were delimited from the most recent orthophoto. They occupy about 7% of the drainage area. The analysis of the water distribution for paved roads [69] and the urban regions [70] showed that the coefficient $n = 0.015\ \text{s}\cdot\text{m}^{-1/3}$ is appropriate in both of them. For irrigation ponds finished in plastic materials, the accepted values is $n = 0.012\ \text{s}\cdot\text{m}^{-1/3}$ [64].

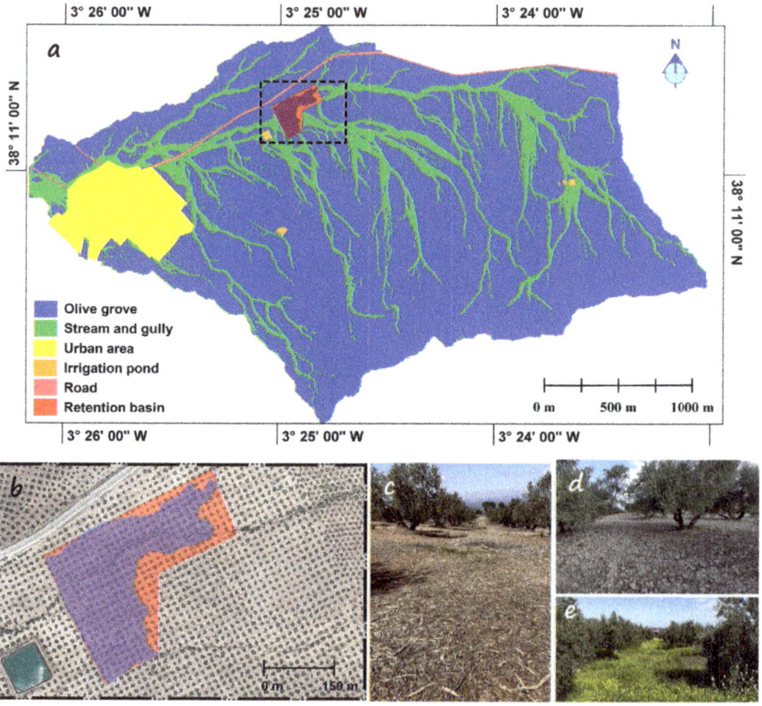

Figure 4. (**a**) Regions defined in Section 2.2.3 to assign the Manning roughness. (**b**) Orthophoto in the dashed rectangular area of panel a showing sandy clay loam on the surface outside the olive tree canopy projection and inside streams because of the herbicide uses. The blue and red polygons are possible retention basins based on an artificial wetland or a large irrigation pond. (**c**–**e**) Possible NbS using ground branches, gravels and natural full-cover crop.

The largest region occupied by olive grove (81%), see the blue area in Figure 4a, has a density of 70 trees·hectare^{-1} and the soil management is nearly homogeneous. The traditional use implies that the vegetation surrounding the crops was removed, as observed in the orthophoto of Figure 4b. To simulate hydrological processes in the present day, we set the value $n = 0.02$ s·m$^{-1/3}$ corresponding to fine sand (colloidal) as calibrated in our previous work [6]. We will see that the ensuing runoff velocities are high, so we would consider a possible NbS for soil erosion mitigation that seeks rising resistance to the flow. Some examples are: integrating the ground branches back into the soil after the pruning process (Figure 4c), redistributing gravels to form coarse surface structures (Figure 4d), and temporary spontaneous cover crops (Figure 4e). Following Crompton and Thompson [71], we set $n \approx 0.2$ s·m$^{-1/3}$ to imitate an NbS that provokes macro-roughness effects in possible future scenarios.

The absence of natural cover has exposed sediment to erosion and caused a high incision of gullies and streams (green area in Figure 4a). Such regions were unknown a priori. To delimit them, we run a preliminary simulation with a constant Manning value of $n = 0.02$ s·m$^{-1/3}$. Subsequently, we selected the regions with flood depths of $h > 0.01$ m to extract the water mask depicted in green in Figure 4a. It occupies 12% of the basin area and lacks riparian vegetation; note the sandy colour in the streams of Figure 4b. To simulate hydrological processes in the present day, we set $n = 0.025$ s·m$^{-1/3}$ as calibrated in a catchment with similar geographical characteristics and precipitation regime [1]. Further, we would consider possible NbSs for increasing the floodplains retention area, delaying the concentration time and attenuating the peak water discharge for short rainfalls [37]. Plausible NbSs for increasing the in-channel roughness are riparian buffer installation, rock armour, willow in-ditch barriers, and timbers in a cross formation, among others. So, we simulated scenarios with the large (and realistic as well) value of $n = 0.2$ s·m$^{-1/3}$ [65].

Lastly, the red region in Figure 4a represents a possible flood retention basin that (does not exist in present days but) would store overland waters, preventing the inundation of downstream urban areas. The present day simulation helps find the optimal siting and sizing of the flood barrier [38]. Figure 4b shows a zoom of the location and shapes of two possible solutions: the light blue corresponds to an artificial wetland that uses a 406 m height flood barrier to store the water; the light red represents a larger 6 m depth irrigation pond. The Manning coefficient $n = 0.012$ s·m$^{-1/3}$ [64] was used for the impermeable retention area when the simulated scenario included it.

2.2.4. Infiltration

We adopted the Soil Conservation Service (SCS) Curve Number (CN) approach for evaluating the infiltration parameter I (1) in the distributed hydrological simulation of the Manillas agricultural watershed. In particular, we used a continuous version of the curve number methodology [72]. The theoretical foundations of the SCS-CN are well described in the book by Hawkins et al. [73]. Details on the numerical implementation in the distributed-hydrological model IBER+ can be found in the software manual [74]. For brevity, we describe the parametrisation of the model for our simulations concisely.

The model relies on a single parameter—called a Curve Number (CN)—that depends on the hydrologic soil group and the soil use and management. In the urban area, irrigation ponds and roads of Figure 4, we set the impermeable condition $I = 0$ corresponding with $CN = 100$. Based on the surface texture of the olive grove region, gullies, and streams, the hydrologic soil group is sandy clay loam (group C [73]). Such soils have low infiltration rates when thoroughly wetted. Indeed, Burguet et al. [18] reported soil water repellence, i.e., $I = 0$, under conventional tillage in olive-orchard catchments. Here we borrow values for the CN parameter in olive orchards from Romero et al. [75], who calibrated it under different soil management in the Upper Guadalquivir watershed. Currently, the soil use and management correspond with no-tillage (Figure 4c), i.e., the soil is weed-free using herbicide. Hence, we set $CN = 89$ (see CNII in Table 6 in [75]). Such a scenario is denoted by CN89 from now on.

The risk of high runoff depths (associated with values of CN close to 100) could be mitigated using an NbS that rises the soil surface roughness for a possible future scenario (Figure 4e). The surface outside the olive tree canopy projection could be covered by a well-implanted cover crop providing 30–70% ground cover. Hence, the CN would decrease from 89 to 64 [75]. We denote the future scenario as CN64.

In addition, we consider a third scenario (CN100) for impermeable soils, i.e., $I = 0$ or $CN = 100$, to check the assumption that simplified the numerical modelling in our previous work [1,6].

To evaluate the infiltration capacity of overland flows, we accounted for the local basin slope S_0. We corrected the CN values for scenarios CN89 and CN64 with the algebraical formula per Ajmal et al. [76]:

$$CN \leftarrow CN\left[1 + \frac{50 - 0.5\,CN}{CN + 75.43}\left(1 - e^{-7.125(S_0 - 0.05)}\right)\right]. \quad (8)$$

Figure 5 shows the CN map in the Manillas stream's watershed. For completeness, we also included the computed histograms. In the present day (CN89), CN varies in the narrow range of 88–92, and in the future (CN64), CN could change to a broader interval of 60–72. In both scenarios, the most probable value tends to be the highest boundary, i.e., 92 for CN89 and 72 for CN64, highlighting the effect of the steep slopes. Indeed, the comparison between the CN maps (Figure 5a,c) with the hypsometric map (Figure 1b) indicates a good correlation between the highest values of CN and the steepest slopes.

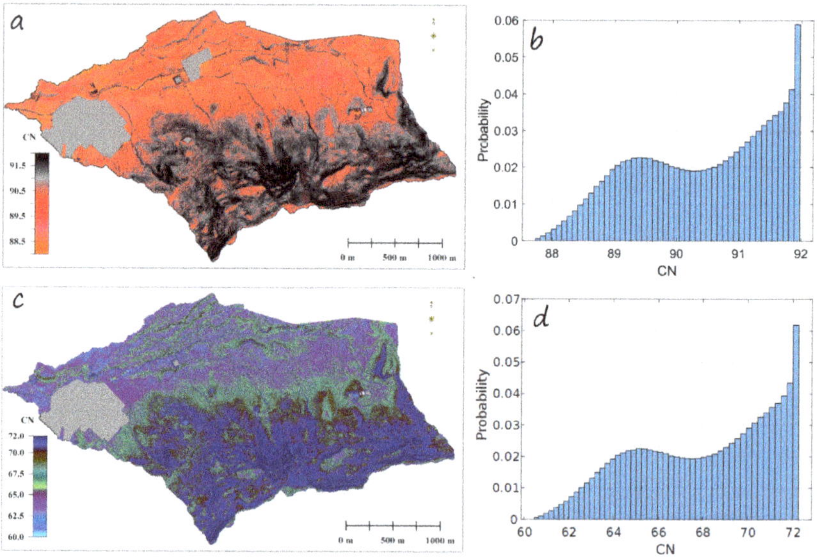

Figure 5. Map and histogram of the local values of the Curve Number (CN) for the simulation setups CN64 (**a**,**b**) and CN89 (**c**,**d**) in the Arquillos Basin, computed and drawn by the authors using the Methods detailed in Section 2.2.4.

3. Results

We simulated the hydraulic response of overland flows for six hours for the design storm given in Section 2.2.2, i.e., for a spatially-uniform rainfall $\bar{P} = 30$ mm·h^{-1} of three-hours duration. We ran fifteen simulations corresponding to the combinations of the three CNs $\{64, 89, 100\}$ in the olive-grove area and the A-to-E scenarios that varied the Manning roughness or altered the digital elevation model in the retention basin as follows:

(A) Without NbS, i.e., as in the present day with the Manning roughness mapping described in Section 2.2.3;
(B) Macro-roughness effects in the olive-grove region and the drainage network, i.e., $n = 0.2$ in the blue and green areas of Figure 4a;
(C) Macro-roughness effects only in the olive-grove region (i.e., $n = 0.2$), keeping the drainage network as in the present day (i.e., $n = 0.025$);
(D) The artificial wetland, see the blue area in Figure 4b, by constructing a barrier at the level of 406 m with a maximum local height of 6 m, which yields an inundated area of 0.044 km^2 with a storage capacity of $V_w = 0.097$ hm^3. The optimal location (dashed square in Figure 4a) catches three streams that drain 3.9 km^2 (56% of the total basin area);
(E) A reservoir pond with a uniform depth of 6 m to store storm runoff. The trapezoidal shape of the reservoir (red area in Figure 4a,b), surrounding the ground line at 406 m, has a perimeter of 1.5 km, a surface of 0.065 km^2, the bottom carved at 397 m, and a pond capacity of $V_p = 0.39$ hm^3.

The following subsections summarise the results for the fifteen scenarios simulated numerically. In Section 3.1, we start describing the dynamics of overland flows for the actual situation of the draining basin. Regarding the flood risk analysis, we shall verify the outputs of the computational model with available paleohydrological information on the simulated flood event. Further, we discuss the effects of surface waters on soil erosion risk at the catchment scale. Subsequently, in Section 3.2, we identify the most beneficial physical parameters concerning the mitigation of urban floods. Additionally, in Section 3.3, we analyse the simulated shear stresses to select the optimal model parameters for mitigating soil erosion.

Lastly, in Section 3.4, we site and select the optimal nature-based solutions that provide the most benefits for minimising the peak streamflow, the runoff depth, and the erosion-prone area, which is the desired state. The fifteen scenarios simulated previously serve to delimit the specific locations where a change of the model's physical parameters is required to achieve the optimal values. Next, the particular NbS for the targeted placement is selected according to the simulation results.

3.1. Distributed Hydrological Simulation of the Catchment in the Present Day

Figure 6 shows the maps of the flow depth h (Figure 6a) and the magnitude of the velocity vector $|\mathbf{u}|$ (Figure 6b) simulated at the time of the peak flow $t = 3$ h for the scenario CN89-A. The flow depth in the tributary streams that drain to the main Manillas stream (green are in Figure 4) reaches values close to 1 m in agreement with the bankfull depth of the actual channels. There, the flow velocity is higher than 1 m·s^{-1}, see the orange region in Figure 6b, which is fast. Deeper and even faster flows develop in the downstream reach of the Manillas stream with flow depths and velocities of approximately $h_{\max} = 2.5$ m and $|\mathbf{u}_{\max}| = 5$ m·s^{-1}. The flow regime becomes critical with the Froude number $Fr = |\mathbf{u}_{\max}|/\sqrt{g\,h_{\max}} = 1.01$. Regarding the runoff depth, characteristic values are around 2.5 mm (purple colour in Figure 6a) in the southern region draining to the Manillas stream, with high velocities in the range of 0.1–0.4 m·s^{-1} (Figure 6b). Conversely, in the northern area, where the basin slope decreases and the terrain is flatter than in the southern region, the runoff depth is much shallower, and the velocity is slower. Hence, we observe a clear topographic control between the runoff dynamics correlated well with the terrain elevation and slope (Figure 1b). Then, we evaluated the shear-stress vector $\boldsymbol{\tau}_b$ from (3) using the simulated hydraulic variables h and $\mathbf{u}$, and drew the map of its magnitude in Figure 6c. The shear stress varies in the wide range of the order of magnitudes from 10^{-4} to 10^4 N·m^{-2}.

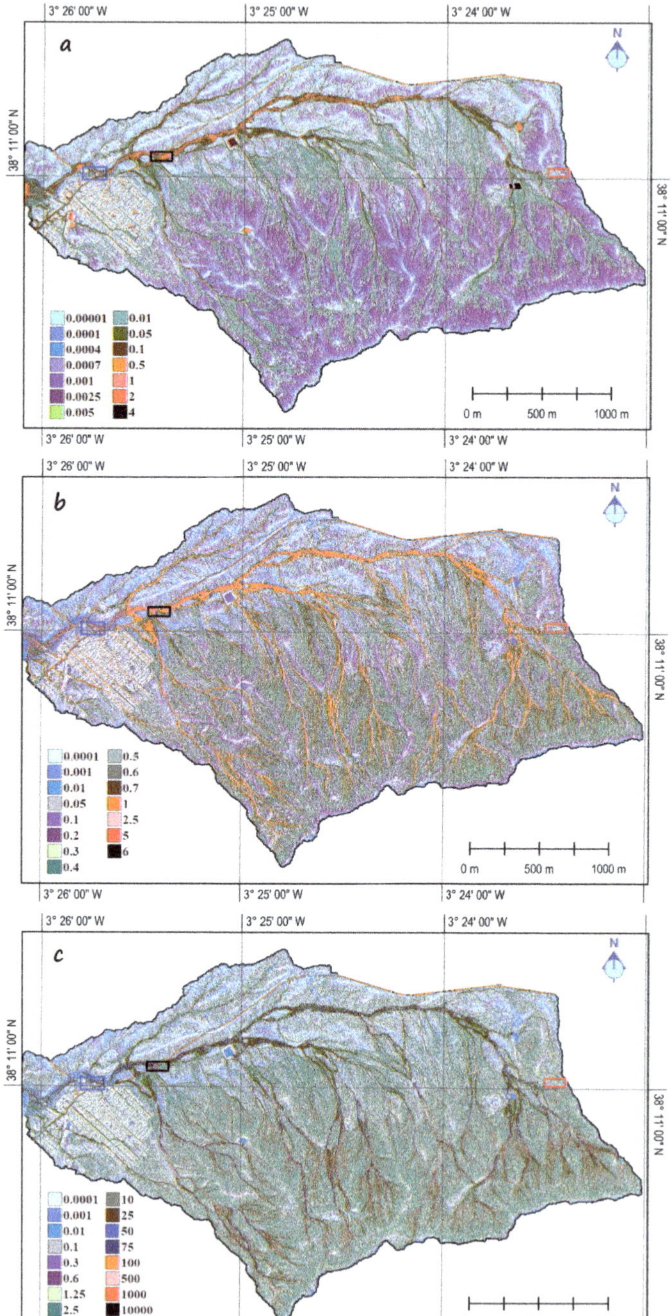

Figure 6. Map of the maximum values of the (**a**) flow depth h [m], (**b**) magnitude of the velocity vector $|\mathbf{u}|$ [m·s^{-1}], and (**c**) magnitude of the shear stress vector $|\boldsymbol{\tau}_b|$ [N·m^{-2}] at peak flow (i.e., $t = 3$ h) for the present-day soil use and management (simulation CN89-A).

In agreement with the simulated flow velocities, the runoff exerts high stresses in the southern region to the Manillas stream because of the steep slopes. The values are more significant than the critical threshold $\tau_c = 0.083\ \mathrm{N{\cdot}m^{-2}}$ [77] for silt motion in most watersheds. The highest values occur in the headwater streams, $|\boldsymbol{\tau}_b| \sim O(10^2)$, decreasing progressively downstream and keeping high values of $O(10)$ until the tributary streams reach the main Manillas stream. In the olive-grove region of the southern watershed, erosional processes are also dominant because the shear stress is above the critical value for erosion. The high shear stress and the fine granulometry of available sediments favour transport processes dominated by wash load. Slackwater sediments may settle in the inundated floodplains with a low flow velocity of the Manillas stream where $|\boldsymbol{\tau}_b| < \tau_c$ [55]. The high spatial resolution of the numerical simulation allows the identification of multiple-scale geomorphological processes, see Figure 7, illustrating such sediment transport regimes.

Figure 7. Examples of fluvial geomorphological features in the Manillas watershed that serve to verify the accuracy of the simulation CN89-A. (**a**) Gully network in the eastern headwater (red rectangle in Figure 6c): the left column shows a zoom of the shear-stress map, an orthophoto and the slope direction map for the DEM; the photo in the right-hand-side depicts an upstream view of the existing gullies. (**b**) Confined single-thread channel eroded in a rocky bed due to high shear stresses. (**c**) Paleostage indicators of flood levels and bedforms developing in unconfined flow on a floodplain. Photographs b and c were taken in the locations marked with the red and yellow dots in Figure 1a few meters downstream and upstream of the cross between the Manillas stream and the road, respectively.

For instance, extreme erosion or gullies can be located from the simulated map of shear stress (Figure 6c). In the small rectangular area in red of Figure 6c, there is a train of gullies. They are visible in the orthophoto of this region (see the panel at the top of Figure 7a). The five gullies are in-parallel outside the olive tree canopy projection. The simulated shear stress inside the gullies is approximately 5 N·m^{-2}, as shown in the shear-stress map in the middle of Figure 7a. Thus, the contour level $|\boldsymbol{\tau}_b| = 5$ N·m^{-2} yields the extension and the path of the gully. The predicted trajectory of the gully is indeed well-correlated with the breaks in the slope of the terrain depicted in the panel at the bottom of Figure 7a. After locating the gullies on the map, we visited the place. We found the deeply incised gullies visible in the in situ photo of Figure 7a, proving the accuracy of the numerical simulation.

Another example illustrating the erosional power of the high-speed flow is the rock-bed incision depicted in Figure 7b. The precise location corresponds with the red square in Figure 1a (also, with the left-upper corner of the black rectangle in Figure 6). A few meters downstream of the first cross between the Manillas stream and the road, where the shear stress was as high as 100 N·m^{-2} (Figure 6c), the water flow sculpted a single confined channel. The flow eroded a first layer (i.e., 0.7 m) of poorly consolidated sediments (mostly silt and clay) and a second rocky bed (i.e., 1.7 m), leading to the rectangular cross-section of depth $H = 2.4$ m and with $B = 6$ m. The authors of [78] showed that the bankfull dimensions B and H of straight rivers adapt to the regime-based bankfull capacity Q_{1D} and satisfies the classical Manning equation for a uniform flow in a rectangular channel [63]:

$$Q_{1D} = \frac{B\,H}{n_{1D}}\,R_h^{2/3}\sqrt{S_0} \quad \text{with} \quad R_h = \frac{B\,H}{2\,H + B}\,. \tag{9}$$

For dimensionally-similar channel planforms, Bohorquez [16] calibrated the one-dimensional roughness parameter $n_{1D} = 0.055$ s·m$^{-1/3}$, which accounts not only for the hydraulic resistance of the walls and the bed but also for the presence of sparse riparian vegetation and secondary head losses. We measured the thalweg elevation using Leica Zeno 20 GPS and obtained the local slope value $S_0 = 0.0196$. Substituting these values into (9), yields the peak water discharge $Q_{1D} = 44.4$ m^3·s^{-1}. Taking into account that the draining area to this cross-section is $A_{\text{drain}} = 4.76$ km^2, the theoretical value of the streamflow given by (5) is $Q_{ss} = 39.7$ m^3·s^{-1} for the rainfall intensity $\bar{P} = 30$ mm·h^{-1}, which differs only by 11%.

Upstream of the single channel described above and before the cross between the Manillas stream and the road, see the specific location marked with a yellow dot in Figure 1a, a singular sedimentary process was identified during fieldworks on 25 August 2021. We found ripples over parabolic dunes and ripples above the plane bed, see Figure 7c, provoked by the previous day's flood. The growth of ripples over dunes, see the review on dunes by Bohorquez et al. [79], is usual in deep river flow and reflects a severe inundation in the floodplains of the Manillas stream. Indeed, the high-water marks on the road ditch and the top level of the vegetation debris on the olive-tree trunk yield flow depths of 0.4–0.5 m. Such values are in agreement with the simulated flow depth in Figure 6b, which provides further validation of the model input data and the simulation results. Additionally, the formation of a plane-bed occurs only at high shear stresses [77], which implies a notable sediment transport capacity of overland flows in olive groves. The transport-stage parameter $\Gamma \equiv |\boldsymbol{\tau}_b|/\tau_c - 1$ achieved the value of 29 in the simulation, which lies in the interval of the upper-regime plane bed $20 < \Gamma < 40$ for silt [54,77]. So, dunes ($2 < \Gamma < 10$) and ripples ($\Gamma < 4$) grew during the recession stage of the flood.

3.2. Mitigation of Urban Flooding through Basin-Scale NbSs

The distributed hydrological simulation also provides the inundation map in the urban area near the catchment outlet. First, we use the simulated values of flow depth to verify the accuracy of the numerical simulation regarding historical information on flood levels. Later, we describe the benefits of NbSs for flood risk mitigation by analysing the outputs of the simulated future scenarios.

Figure 8a depicts a zoom of the flow depth map for the present-day hydrological simulation introduced in Section 3.1. The flooded urban area corresponds with the blue rectangle in Figure 6a. The flow depth was more profound than the 1.4 m channel height in most of the reach shown on the map. Overtopping of the right bank provoked the inundation of the olive-grove floodplains. In the 45-degree channel bend upstream of the area of Figure 8a, the water flow abandoned the channel over the left bank and inundated the road and main street of the town of Arquillos. The numerical simulation predicts the flow accumulation on a corner (highlighted with a black rectangle in Figure 8a), leading to a maximum depth of about 0.5 m. The pictures taken in 1996 and 2013 in that place corroborates the numerical simulation result. Note the brown marks left by the slurry flow on the walls (Figure 8b) or the deep water flow reaching the windows (Figure 8c). Lastly, the flood left the town through the level ground surrounding the sports hall and returned to the channel.

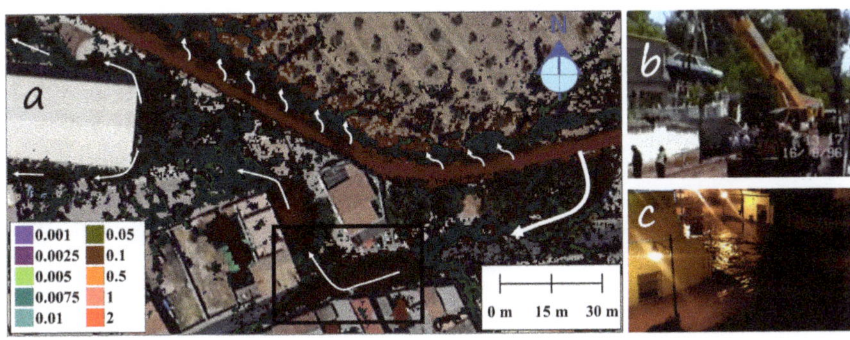

Figure 8. (**a**) Map of the simulated flow depth (h in meter), for the present-day scenario CN89-A at the time of peak flow (i.e., $t = 3$ h), in the urban area inundated downstream of the overtopped artificial channel. Flow from right to left. The region in panel a occupies the blue rectangle in the flow-depth map shown in Figure 6a. The pictures on the right-hand side correspond to the inundations on (**b**) 15 August 1996 and (**c**) 8 March 2013 in the boxed area of panel a when the artificial channel was unable to protect the town from flooding.

To draw an overview of the benefits of possible future scenarios that implement an NbS for flood management, we compared the simulated hydrographs $Q(t)$ in a cross-section of the Manillas stream at the town inlet. The hydrograph was computed from the hydraulic variables $h(\mathbf{x},t)$ and $\mathbf{u}(\mathbf{x},t)$ during the transient numerical simulation:

$$Q(t) = \int_{\mathbf{x}_i}^{\mathbf{x}_e} h(\mathbf{x},t)\mathbf{u}(\mathbf{x},t) \cdot \mathbf{n}\, dl\,, \tag{10}$$

where $\mathbf{x}_i$ and $\mathbf{x}_e$ denotes the initial and end points of the cross-section, respectively, and $\mathbf{n}$ is the unit vector pointing in the direction perpendicular to the cross-section.

Figure 9 shows the hydrographs for the fifteen simulations. The water discharge grew monotonously during the rising limb and attained a steady state at the end of the rainfall, i.e., $t = 3$ h, when the water discharge was at a maximum. The key variables used to evaluate the benefits of the NbS are the values of the peak water discharge and the time to peak flow, also known as the time of concentration or the duration of the rising limb. In the simulation, we defined the time of concentration T_{sim} as the instant of time when the simulated discharge $Q(t)$ reached the 90% of the maximum value (i.e. $Q = 0.9 \times Q_{\text{max}}$ at $t = T_{\text{sim}}$) [60]. Table 1 gives both the peak water discharge Q_{max} and the concentration-time T_{sim} for all the simulated scenarios.

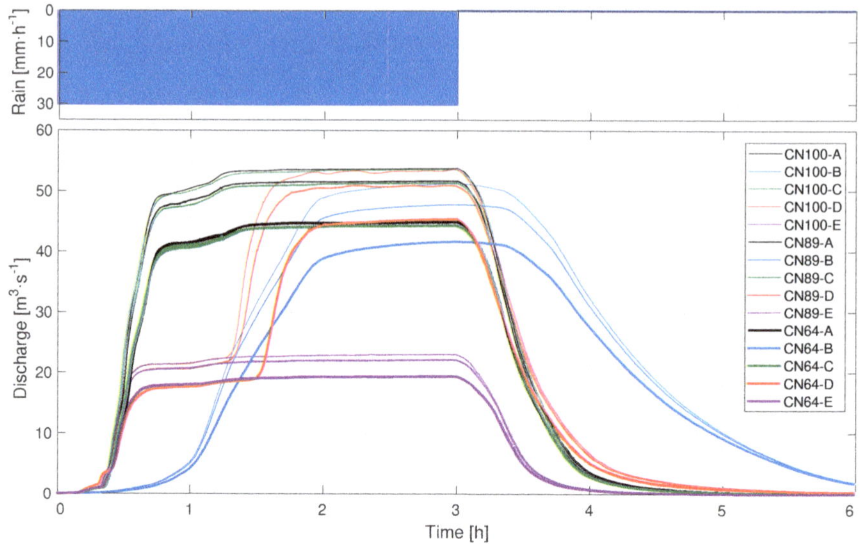

Figure 9. Hydrograph at the basin outlet for the fifteen physical scenarios.

Table 1. Peak water discharge $Q_{\max}$ and time of concentration T_{sim} for the simulated hydrographs in Figure 9. The attenuation of the peak discharge regarding the maximum value of scenario CN100-A and the delay factor of the time to peak flow are also given.

	A	B	C	D	E	A	B	C	D	E
	Q_{max} [m^3·s^{-1}]					T_{sim} [min]				
CN100	53.8	51.3	53.7	53.7	23.1	44	120	45	94	37
CN89	51.7	47.8	51.5	51.0	22.2	45	120	46	96	38
CN64	44.9	41.7	44.3	45.4	19.5	46	120	47	108	41
	Attenuation (%)					Delay (×)				
CN100		4.6	0.3	0.2	57.1		2.7	1.0	2.1	4.1
CN89	3.9	11.2	4.3	5.1	58.7	1.0	2.7	1.0	2.2	4.1
CN64	16.6	22.5	17.6	15.5	63.8	1.1	2.7	1.1	2.5	4.1

The peak streamflow in the absence of infiltration (i.e., 53.8 m^3·s^{-1} for CN100-A) lies very close to the theoretical value predicted using the rational method for the whole basin, i.e., $Q_{ss} = 57.4$ m^3·s^{-1} (5), due to the proximity of the town of Arquillos to the basin outlet. The time of concentration $T_{\text{sim}} = 44$ min is also close to the theoretical prediction $T = 33$ min (6). Interestingly, in the current scenario with the actual soil uses and management (i.e., simulation CN89-A), the infiltration was negligible and attenuated the peak discharge only by 3.9%. Even for the lowest curve number (simulation CN64-A), we found that $Q_{\max}$ decreased slightly from 53.8 to 44.9 m^3·s^{-1}, which represents a small relative factor of 16.6%. Furthermore, hydrographs A (black lines in Figure 9) nearly overlap those of the simulations type C (green lines in Figure 9). Subsequently, acting on the olive-grove area to improve the infiltration (i.e., decreasing the curve number) and raise the surface roughness (i.e., increasing the Manning parameter), as shown in Figure 4c–e, does not help to attenuate the peak discharge substantially nor delay the inundation in the urban area. In the best case, the streamflow was attenuated by 17.6% (CN64-C) and reached $Q_{\max} = 44.3$ m^3·s^{-1} with the same concentration time as for scenario A. The increase in the Manning roughness in the olive-tree area (coloured in blue in Figure 4) did not delay the inundation because the water drops had to travel a short distance to reach elements of the drainage network. Such a distance was much shorter than the characteristic lengths of the gullies and streams (green area in Figure 4) that remained unaltered in scenario C concerning A.

Promising results were obtained for NbS type B that alters the roughness of the drainage network through riparian buffer installation, boulders, and willow in-ditch barriers, among others. The most noticeable result was the delay of the rising limb, which can be appreciated in the hydrographs of Figure 9 coloured in blue. The concentration-time increased from about 45 min to 120 min, i.e., by a factor of 2.7. On the one hand, the benefits of NbS type B are great during the first hour of the rainfall independently of the CN value because $4.2 \leq Q \leq 5.3$ m^3·s^{-1} at $t = 1$ h for $64 \leq CN \leq 100$. Such streamflow values are much lower than for scenarios A and C at the same instant of time, which ranges in the interval $40.9 \leq Q \leq 50.5$ m^3·s^{-1}. On the other hand, for rainfall durations of 2–3 h, the peak discharge is about $38.9 \leq Q \leq 51.3$ m^3·s^{-1}, which is closer to the peak values for scenarios A and C. Hence, NbS type B is practical for mitigating the risk of flooding in the town for extreme rains shorter than one hour, but longer events require other NbSs.

The installation of a flood barrier to create an artificial wetland (NbS D) and the construction of a large reservoir pond (NbS E), recall Figure 4a,b, stopped the growth of the streamflow at $t \approx 30$ min as shown by the hydrographs in red and purple (Figure 9), respectively. The discharge reached a constant value of $Q = 17.7$ m^3·s^{-1} (21.5) for $CN = 64$ (100) up to $t = 1$ h. The reservoir pond was able to store stormwater for the three-hour rainfalls, keeping the peak discharge in the town as low as $Q_{\max} = 19.5$ m^3·s^{-1} (CN64-E). The wetland was filled in 90 min (77 min) in scenario CN64-D (CN100-D), and the hydrograph shows a second stage of the rising limb achieving the same peak discharges as for present scenario A (Table 1). Hence, the wetland could delay the flood and protect the town from inundation for about 1 h and 30 min, whilst the reservoir pond would be a solution for rainfalls of 3 h.

3.3. Benefits of NbS to Mitigate Soil Erosion Risk

We seek to find the NbS that decreases the component of the shear stress responsible for soil erosion τ_{bs} below the critical threshold τ_c. The main objective is to impede that $\tau_{bs} > \tau_c$ as occurs actually in most of the olive-grove area (simulation CN89-A described in Section 3.1 and Figure 6c). Otherwise, the significant shear stresses provoke unsustainable soil losses and the formation of deep gullies in most of the catchment (e.g., Figure 7a). If we achieve the goal $\tau_{bs} < \tau_c$, we stop the actual erosion processes and reduce the catchment sediment budget. Furthermore, we prevent the silting of the Guadalén Reservoir, where the Manillas stream drains at a distance of 4 km from the basin outlet. Sediments exposed to erosion over the Manillas basin are mostly coarse silt, with a uniform grain size in the range of $[0.031, 0.0625]$ mm. The critical shear stress for the onset of sediment motion was determined from Table 7.1 in Julien [77], which yields $\tau_c = 0.083$ N·m^{-2} in water at 20 °C. For evaluating the local shear stress responsible for erosion τ_{bs}, we used the classical shear-stress partitioning that decomposes the effective boundary shear stress $|\boldsymbol{\tau}_b| = \tau_{bf} + \tau_{bs}$ into the form drag τ_{bf} and the skin friction τ_{bs} (see details in § 2.8.1 by García [54]).

A simple and compact way to visualise the benefits of NbS in the whole watershed is to analyse the changes in the probability of the flow depth (Figure 10a), velocity magnitude (Figure 10b), and shear stress (Figure 10c) regarding the current situation. To compute the probabilities, we accounted only for the regions of interest: the olive groves and the drainage network (blue and red areas in Figure 4a, respectively). Hence, we excluded the non-erodible urban areas and civil infrastructure from the analysis.

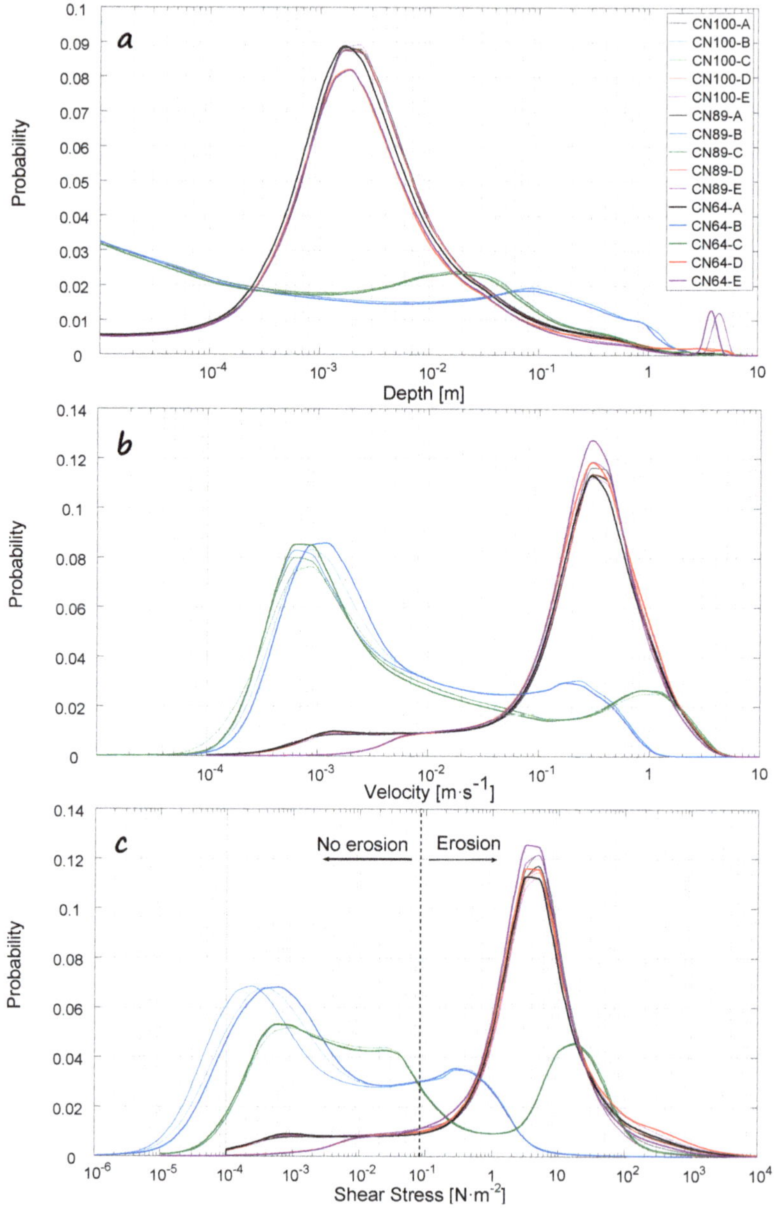

Figure 10. Histograms of the simulated (**a**) flow depth h, (**b**) magnitude of the velocity vector $|\boldsymbol{u}|$, and (**c**) the skin friction τ_{bs} for the fifteen scenarios explained in Section 3.3.

In the absence of an NbS (solid black lines in Figure 10), the most probable values of the flow depth, velocity, and shear stress vary in the narrow intervals $[1.6, 2.4]$ mm, $[0.3, 0.4]$ m·s^{-1}, and $[3, 5]$ N·m^{-2} for all the infiltration rates (simulations CN100-A, CN89-A, and CN64-A). Indeed, the three curves in black nearly overlap. The wetland (NbS D in purple) and the reservoir pond (NbS E in red) do not provoke a substantial modification of the probability curves that are similar to the actual ones (scenario A in black). A significant part of the shear-stress curves for the simulations type A, D, and E lie out of the target region $\tau_{bs} < \tau_c$. Table 2 shows the cumulative probability for the catchment area where erosion occurs due to $\tau_{bs} > \tau_c$. Independently of the infiltration rate, erosion develops in more than 86.4% in scenarios A, D, and E. The values in Table 2 show that the infiltration does not play a role in controlling erosion because of the same reasons that did not substantially attenuate the streamflows (Figure 9a). The steep slopes of the watershed and the high precipitation intensity favour runoff formation to the detriment of infiltration processes.

Table 2. Relative area (%) where the dangerous threshold for erosion $\tau_{bs} > \tau_c$ holds.

	A	B	C	D	E
CN100	87.8	25.4	39.9	86.9	86.7
CN89	87.2	25.1	40.1	86.9	87.3
CN64	86.4	25.3	39.4	91.7	93.1

The interventions on the agricultural area (NbS C, green lines in Figure 10c), including the channels of the drainage network (NbS B, blue lines in Figure 10c), changed the shear-stress distribution by a great extent. The distribution switched to a multimodal distribution. The major mode decreased from $[3,5]$ N·m^{-2} (present-day) to $[2.4, 6.5] \times 10^{-4}$ (NbS B) and $[5.1, 9.2] \times 10^{-4}$ (NbS C) when the mean curve number varied in the range of $[64, 100]$. The most probable values of the shear stress developed in the agricultural area of the watershed. This fact highlights the soil-loss mitigating effect of NbS B and NbS B C because the major modes are lower than the threshold for erosion. The minor mode for NbS B, about $[0.27, 0.44]$ N·m^{-2}, is also one order of magnitude lower than presently. Consequently, the relative area where the dangerous threshold for erosion $\tau_{bs} > \tau_c$ occurred decreased from about 86% to 25% in NbS B. In the case of NbS C, the potential area for erosion also falls to 39%. The minor peak arising in the probability of the shear stress at $\tau_{bs} \sim 15$ N·m^{-2} for NbS C corresponds to the high velocities, of the order of 1 m·s^{-1} (Figure 10b), in the unaltered channels of the drainage network (recall the velocity map in Figure 6b for the present day). Note the positive aspects of increasing the resistance to flow in the channels (NbS B), which reduces the minor mode of the velocity to $[0.16, 0.24]$ m·s^{-1} (Figure 10b) and could favour sediment trap processes to restore deep incised channels.

3.4. Synergetic Benefits of the Wetland, In-Channel Macro-Roughness and Cultivation Systems in Olive Orchards

Considering the independent benefits of each nature-based solution, we combine three to maximise their effectiveness in terms of flood mitigation and soil conservation. Furthermore, we choose the optimal targeted placement of NbS because funding is always limited. We seek to delay the flood, attenuate the peak discharge in the urban area and favour infiltration processes across the watershed. At the same time, we want to prevent soil erosion in olive orchards, promote mechanisms for gully aggradation rather than gully erosion and reduce the catchment sediment budget.

Figure 11 shows a map of the optimal location of an NbS to provide the most benefits. The results of the distributed hydrological simulations described in the previous sections helped site and classify the actions (which were unknown a priori) as follows:

- First, the black rectangle in Figure 11 depicts the installation of a flood barrier with the crest of the earthen embankment at 406 m, the maximum height of 6 m and the dam crest length of 245 m. It should create an artificial wetland (purple area in Figure 11) as in previous simulations of type D.
- Second, from simulation type B, we delimited the extension of the main channel and the inundated floodplains along the Manillas stream (yellow area). Further, the tributary streams in green correspond to simulated flow depths larger than 0.5 m

outside of the Manillas stream. Then, ephemeral gullies in red were located using criterion $0.03 < h < 0.5$ m from Soil Science Society of America [51].
- Third, the river floodplains in blue, located in the headwater and near the town, are an extinct, ephemeral wetland and a stream confluence requiring restoration.
- Fourth, in the olive groves, we propose different cultivation systems for bed slopes lower (brown area) and larger (greenish blue area) than 10 degrees.

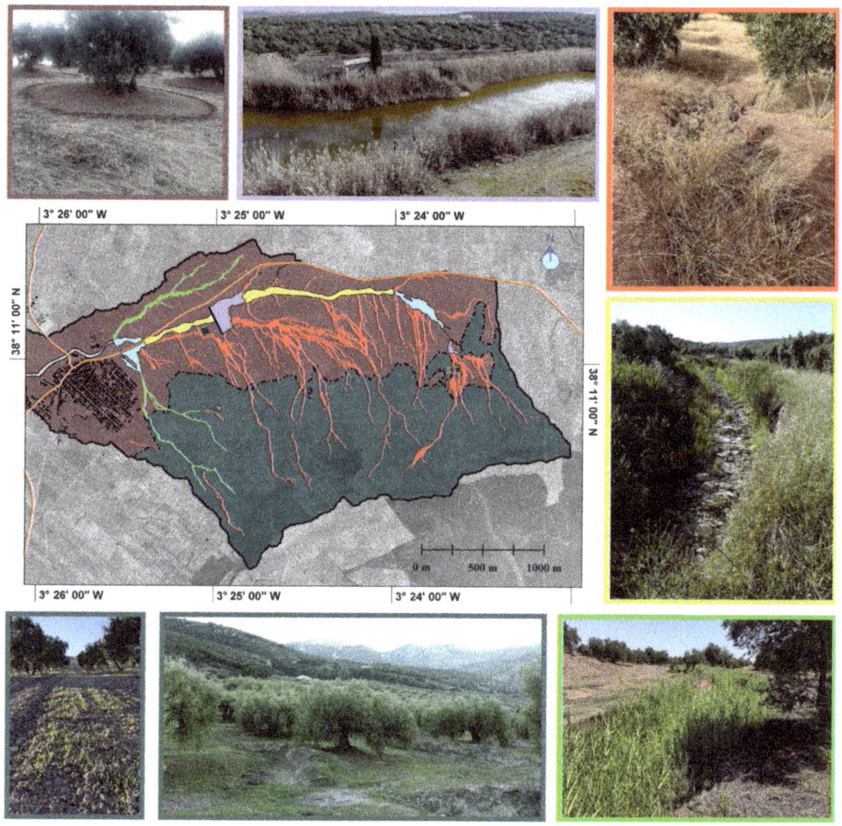

Figure 11. Optimal targeted placements of specific NbS to provide the most benefits and maximise their effectiveness for flood management and soil conservation, as described in Section 3.4. The photos illustrate the actions at each location, as indicated by the frame colour.

The following step in AFC [19] is choosing the model physical parameters at the targeted placements to achieve the optimal values of the objective functions. According to the minimum values of the peak water discharge and the erosion-prone area given, respectively, in Tables 1 and 2, we set the Manning roughness parameter $n = 0.2\ \mathrm{s{\cdot}m^{-1/3}}$ in the green, yellow, and red areas, and $n = 0.025\ \mathrm{s{\cdot}m^{-1/3}}$ in the olive groves for the CN64 setup (recall Section 2.2.4). The NbS that can reproduce the desired model parameters can only be selected and sized after the highly resolved CFD simulation of the flow depth h (1) and the velocity vector $\boldsymbol{u}$ (2).

The simulated hydrograph in Figure 12a (blue line) highlights the dynamic and transient nature of the surface waters across the draining basin. Its shape changes substantially regarding the current situation (black line).

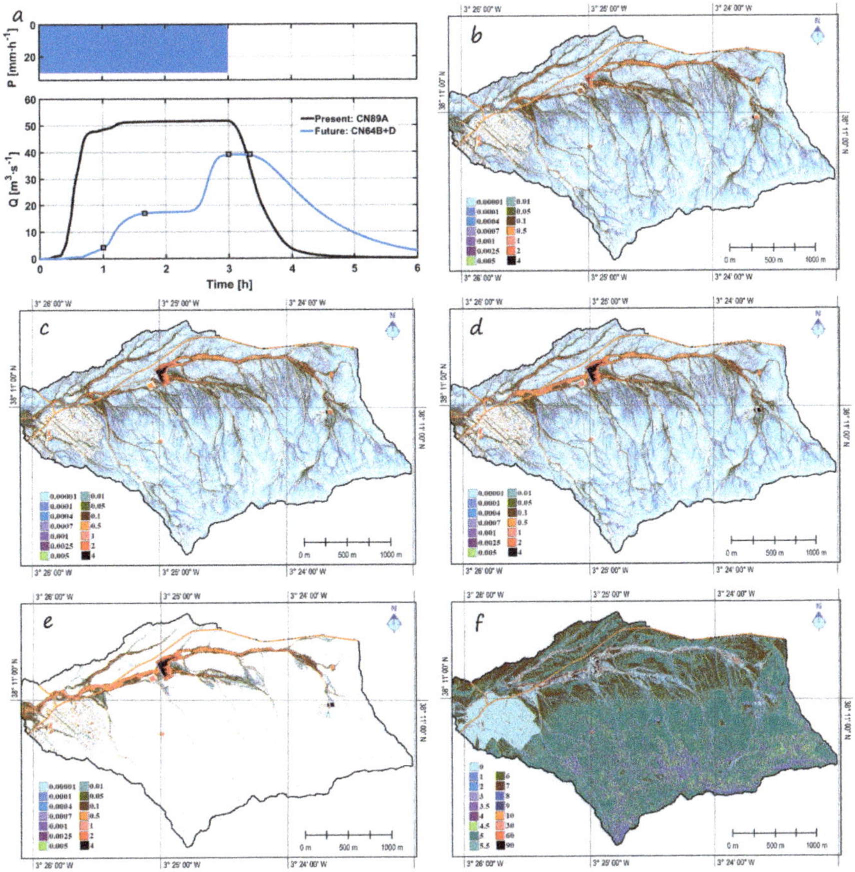

Figure 12. (**a**) Simulated hydrographs in the Manillas stream, near the town, for the current state (black) and the possible future (blue) after implementing the combined NbS B and D. (**b**–**e**) Snapshots of the simulated flow depth (in meters) with NbS B and D at t = 60, 100, 180, and 200 min corresponding, respectively, with the streamflows Q = 4.2, 17.6, 39.2, and 39.1 $m^3 \cdot s^{-1}$ (squares in panel a). (**f**) Maximum infiltration rate (in $mm \cdot h^{-1}$) achieved at the end of the rain (i.e., t = 180 min) for the future scenario with NbS B and D.

Overall, we observe the mixed benefits of increasing the in-channel roughness in the drainage network and the installation of a flood barrier to create the artificial wetland. They protect against flooding by slowing down the flow and storing surface runoff. Hence, the rising limb exhibits two stages separated by a steady plateau associated with the wetland filling. After one hour of rainfall, the peak discharge in the urban area is only $Q = 4.2\ m^3 \cdot s^{-1}$, much lower than for the actual scenario CN89-A, which is $Q = 48.5\ m^3 \cdot s^{-1}$. The map of water depth at t = 1 h (Figure 12b) shows that the Manillas stream did not inundate the town. The early infilling of the wetland and the absence of flow downstream can also be appreciated in such a map. Forty minutes later, the water discharge reaches a steady state with $Q = 17.6\ m^3 \cdot s^{-1}$ until $t \approx 2.5$ h. For this low value of the water discharge, the town remains again protected against flooding; see the map of flow depth in Figure 12c. At $t \approx 2.5$ h, the wetland achieves its maximum capacity, and the water starts to flow downstream of the wetland barrier. Then, the second stage of the rising limb occurs up to reaching the peak water discharge $Q_{max} = 39.2\ m^3 \cdot s^{-1}$ at the end of the rainfall (i.e., t = 3 h). The map of flow depth at that time is shown in Figure 12d. Later, the water discharge remains constant near the urban area for about 30 min (i.e., up to $t = 200$ min). However, at that time, the rain had already ceased, runoff vanished in the watershed (see the flow depth map in Figure 12e), and the discharge monotonously decreased, vanishing approximately at $t = 6$ h.

The peak discharge for the combined NbSs B and D, $Q_{\max} = 39.2\ \mathrm{m}^3{\cdot}\mathrm{s}^{-1}$, decreased with respect to the present value of $51.6\ \mathrm{m}^3{\cdot}\mathrm{s}^{-1}$ as much as using uniquely NbS B, which yields $41.7\ \mathrm{m}^3{\cdot}\mathrm{s}^{-1}$ (CN64-B in Table 1). The infiltration was responsible for the attenuation of the peak streamflow. The infiltration rate was at a maximum at that time, reaching the values depicted in the map of Figure 12f. In the agricultural area, the highest infiltration rates were around $7\ \mathrm{mm{\cdot}h^{-1}}$ and occurred in the northern part. In the headwaters, the infiltration decreased to 2–5 $\mathrm{mm{\cdot}h^{-1}}$ because of the effects of the slope in the curve number (recall Figure 5c). Though such values are low regarding the larger precipitation rate of $30\ \mathrm{mm{\cdot}h^{-1}}$, the cumulative volume of water infiltrated during the whole hydrological event of 6 h is relevant.

To quantify the benefits of NbSs concerning infiltration, we computed the percentage of the groundwater volume relative to the total rainfall depth as

$$\eta = \frac{100}{V_r}\left[V_r - \int Q(t)\,dt - V_w\right]. \tag{11}$$

In (11), $V_r \equiv Q_{ss}\,D$ represents the volume of water of the design storm (4) and (5), $\int Q(t)\,dt$ is the volume of surface water that leaves the drainage basin, and V_w is the capacity of the wetland reservoir in the existing case. Evaluating (11), we concluded that the groundwater volume could be increased from $\eta = 4\%$ in the present day to $\eta = 20.5\%$ in the possible future.

To close the cycle, we need to choose the specific NbS that provoke the desired modification of the model's physical parameters. The flood barrier can be built with a vegetated crib wall using inert material and plants, see Figure 1 in Rey et al. [42]. The desired increase in hydraulic resistance can be achieved along the drainage network by adopting well-known actions for NFM. For instance, see Figure 11: (i) placing dry bush, small boulders, or another material across the ephemeral gullies (red frame photo), e.g., Quinn et al. [41]; (ii) flexible submerged vegetation in deeper streams with a confined flow (green frame photo), see Darby [43]; (iii) combined actions of gravel seeding in the streambed, see Hassan and Church [80], and riparian buffer installation in the Manillas stream (yellow frame photo) [44]. The proper sizing of plants and boulders requires physically-based equations, e.g., Wilson [81] and Nitsche et al. [82], respectively, using as input data the simulated values of flow depth and velocity for the optimal macro-roughness parameter $n = 0.2\ \mathrm{s{\cdot}m^{-1/3}}$. This goal belongs to soil and water bioengineering, see Zaimes et al. [23], and lies out of the scope of this work. We propose adopting sustainable production techniques in the olive grove; see Tombesi et al. [20] and recent innovations explicitly designed for erosion management. For small bed slopes, we suggest the application of plant residue mulches, leaving the soil untilled and covering it with prunings or other plant residues outside of the tree canopies (brown frame photo). For medium-gradient slopes, water collection pits collect large amounts of water during intense rainfall. The live plant covers, e.g., *Brachypodium Hybridum* BHJHIN [46], along the middle of the orchard lanes also prevent soil erosion and improve infiltration (greenish blue frame photo).

4. Conclusions

The set of fluvial geomorphological features presented in the sections above proves the development of extreme soil erosion and sedimentary processes in a complex, non-linear way over an agricultural watershed dedicated to traditional olive growing. The formation of ripple/dune over a plane bed in the floodplains of the main stream is uncommon in overland flows, as these bedforms are characteristic of deep rivers rather than shallow flows. It reflects the high sediment load and transport capacity of sub-daily pluvial floods. Overland flows have sculpted extreme gullies in the olive groves because of cultivation systems incompatible with the growth in the magnitude of extraordinary rainfall events under the current climate change scenario.

Active flow control through nature-based solutions can drive olive groves to a more desired state. Highly resolved CFD simulations of hydrological and hydraulic processes using the 2D Saint-Venant equations yield the optimal location of an NbS to provide the most benefits for minimising the water discharges and the erosion-prone area, maximising the soil infiltration capacity. Sub-metric spatial resolutions are required to resolve many microtopographic features of the terrain and the NbS, controlling overland flow and sediment transport connectivity. Affordable computational times can be achieved only by using GPU-accelerated numerical codes. The software IBER+ adopted in this study is suitable for catchment areas up to 20 km^2 with spatial resolutions of 1 m^2 now common in LiDAR-DEM.

Future works can overcome several limitations that impede scaling up the proposed method over larger draining basins (<1000 km^2) by means of exascale multi-GPU HPC technology as the recent SERGHEI-SWE [27]. However, as commented throughout this paper, the available software packages adopt the assumption of small bottom slopes intrinsic to the Saint-Venant equations. Further research is required for overland flows on steep slopes, which need a unified theory and an efficient numerical scheme and its implementation in an open-source code. We could thus improve the effectiveness and acceptance of NbS at larger scales for the natural management of floods and soil erosion.

We have proven, theoretically, the effectiveness of adaptive practices for sustainable olive growing and NbS actions in gullies and streams for mitigating the risk of flood and erosion in steep-slope basins (mean value of 20%) with high rainfall rates (around 30 $mm \cdot h^{-1}$). As funding is always limited, the optimal targeted placements of NbS have to be delimited from the simulated flow depth, velocity and shear stress maps. To ensure NFM in the long term, private landowners and communities must adopt sustainable cultivation systems in olive orchards, mitigating soil erosion and preventing the silted-up state of NbSs. These actions help protect from flood, favour soil conservation and are aligned with sustainably managing and restoring modified ecosystems. Furthermore, they address societal challenges effectively and adapt current practices to climate change, simultaneously providing human well-being and biodiversity benefits. Future works on the real verification of the proposed NbS would require regional policies to break some barriers that limit effective implementation at local scales.

Author Contributions: Conceptualization, P.B., F.J.P.-L., R.J.-M. and G.P.; methodology, P.B.; software, P.B. and I.G.-P.; validation, I.G.-P. and P.B.; formal analysis, P.B.; writing—original draft preparation, P.B.; writing—review and editing, All Authors; project administration, P.B.; funding acquisition, P.B., F.J.P.-L., R.J.-M. and G.P. All authors have read and agreed to the published version of the manuscript.

Funding: This work was funded by: "Programa Operativo FEDER 2014-2020" and "Consejería de Economía y Conocimiento de la Junta de Andalucía" grant number 138096; the European Union NextGenerationEU/PRTR and MCIN/AEI/10.13039/501100011033 grant number TED2021-129910B-I00.

Institutional Review Board Statement: Not applicable.

Informed Consent Statement: Not applicable.

Data Availability Statement: Not applicable.

Acknowledgments: P.B and F.J.P.L acknowledge Miguel Ángel Manrique Peinado (the mayor of Arquillos) for the support and fruitful discussions on the topics described in this paper.

Conflicts of Interest: The authors declare no conflict of interest.

Abbreviations

The following abbreviations are used in this manuscript:

AFC	Active Flow-Control
CN	Curve Number
CFD	Computational Fluid Dynamics
DEM	Digital Elevation Model
EEA	European Environmental Agency
GNSS	Global Navigation Satellite System
GPS	Global Positioning System
GPU	Graphics Processing Unit
HPC	High-Performance Computing
IPCC	Intergovernmental Panel on Climate Change
LiDAR	Light Detection and Ranging
MDPI	Multidisciplinary Digital Publishing Institute
MedECC	Mediterranean Experts on Climate and Environmental Change
MSL	Mean Sea Level
NbSs	Nature-based Solutions
SCS	Soil Conservation Service
2D-SVE	Two-Dimensional Saint-Venant Equations

References

1. Moral-Erencia, J.; Bohorquez, P.; Jimenez-Ruiz, P.; Pérez-Latorre, F. Slackwater sediments record the increase in sub-daily rain flood due to climate change in a European Mediterranean catchment. *Water Resour. Manag.* **2020**, *34*, 4431–4447. [CrossRef]
2. Alfieri, L.; Thielen, J. A European precipitation index for extreme rain-storm and flash flood early warning. *Meteorol. Appl.* **2015**, *22*, 3–13. [CrossRef]
3. Portner, H.O.; Roberts, D.C.; Adams, H.; Adler, C.; Aldunce, P.; Ali, E.; Begum, R.A.; Betts, R.; Kerr, R.B.; Biesbroek, R.; et al. *Climate Change 2022: Impacts, Adaptation and Vulnerability. Contribution of Working Group II to the Sixth Assessment Report of the Intergovernmental Panel on Climate Change*; Cambridge University Press: Cambridge, UK; New York, NY, USA, 2022. [CrossRef]
4. EEA. *Climate Change, Impacts and Vulnerability in Europe 2016: An Indicator-Based Report*; 1/2017, Publications Office of the European Union: Luxembourg, 2017. [CrossRef]
5. Vanmaercke, M.; Panagos, P.; Vanwalleghem, T.; Hayas, A.; Foerster, S.; Borrelli, P.; Rossi, M.; Torri, D.; Casali, J.; Borselli, L.; et al. Measuring, modelling and managing gully erosion at large scales: A state of the art. *Earth-Sci. Rev.* **2021**, *218*, 103637. [CrossRef]
6. Moral-Erencia, J.; Bohorquez, P.; Jimenez-Ruiz, P.; Pérez-Latorre, F. Flood hazard mapping with distributed hydrological simulations and remote-sensed slackwater sediments in ungauged basins. *Water* **2021**, *13*, 3434. [CrossRef]
7. Huang, R.; Ni, Y.; Cao, Z. Coupled modeling of rainfall-induced floods and sediment transport at the catchment scale. *Int. J. Sediment Res.* **2022**, *37*, 715–728. [CrossRef]
8. MedECC. *Climate and Environmental Change in the Mediterranean Basin—Current Situation and Risks for the Future. First Mediterranean Assessment Report*; Cramer, W., Guiot, J., Marini, K., Eds.; Union for the Mediterranean, Plan Bleu, UNEP/MAP: Marseille, France, 2020; p. 632. [CrossRef]
9. Nesshöver, C.; Assmuth, T.; Irvine, K.; Rusch, G.; Waylen, K.; Delbaere, B.; Haase, D.; Jones-Walters, L.; Keune, H.; Kovacs, E.; et al. The science, policy and practice of nature-based solutions: An interdisciplinary perspective. *Sci. Total Environ.* **2017**, *579*, 1215–1227. [CrossRef]
10. Palomo, I.; Locatelli, B.; Otero, I.; Colloff, M.; Crouzat, E.; Cuni-Sanchez, A.; Gómez-Baggethun, E.; González-García, A.; Grêt-Regamey, A.; Jiménez-Aceituno, A.; et al. Assessing nature-based solutions for transformative change. *ONE Earth* **2021**, *4*, 730–741. [CrossRef]
11. Simelton, E.; Coulier, M.; Damen, B.; Howell, J.; Tran, H. NBS framework for agricultural landscapes. *Front. Environ. Sci.* **2021**, *9*, 678367. [CrossRef]
12. Panagos, P.; Borrelli, P.; Poesen, J.; Ballabio, C.; Lugato, E.; Meusburger, K.; Montanarella, L.; Alewell, C. The new assessment of soil loss by water erosion in Europe. *Environ. Sci. Policy* **2015**, *54*, 438–447. [CrossRef]
13. Castillo, C.; Gómez, J. A century of gully erosion research: Urgency, complexity and study approaches. *Earth-Sci. Rev.* **2016**, *160*, 300–319. [CrossRef]
14. Borrelli, P.; Alewell, C.; Alvarez, P.; Anache, J.A.A.; Baartman, J.; Ballabio, C.; Bezak, N.; Biddoccu, M.; Cerdà, A.; Chalise, D.; et al. Soil erosion modelling: A global review and statistical analysis. *Sci. Total Environ.* **2021**, *780*, 146494. [CrossRef]
15. Loumou, A.; Giourga, C. Olive groves: "The life and identity of the Mediterranean". *Agric. Hum. Values* **2003**, *20*, 87–95. [CrossRef]
16. Bohorquez, P. Paleohydraulic reconstruction of modern large floods at subcritical speed in a confined valley: Proof of concept. *Water* **2016**, *8*, 567. [CrossRef]

17. Bohorquez, P.; Moral-Erencia, J. 100 years of competition between reduction in channel capacity and streamflow during floods in the Guadalquivir River (Southern Spain). *Remote Sens.* **2017**, *9*, 727. [CrossRef]
18. Burguet, M.; Taguas, E.; Cerdà, A.; Gómez, J. Soil water repellency assessment in olive groves in Southern and Eastern Spain. *Catena* **2016**, *147*, 187–195. [CrossRef]
19. Collis, S.; Joslin, R.; Seifert, A.; Theofilis, V. Issues in active flow control: Theory, control, simulation, and experiment. *Prog. Aerosp. Sci.* **2004**, *40*, 237–289. [CrossRef]
20. Tombesi, A.; Tombesi, S.; d'Andria, R.; Lavini, A.; Saavedra, M.; Jardak, T. *Production Techniques in Olive Growing*; International Olive Council: Madrid, Spain, 2007; ISBN 978-84-931663-6-6.
21. Blocken, B.; Gualtieri, C. Ten iterative steps for model development and evaluation applied to Computational Fluid Dynamics for Environmental Fluid Mechanics. *Environ. Model. Softw.* **2012**, *33*, 1–22. [CrossRef]
22. Hamers, E.; Maier, H.; Zecchin, A.; van Delden, H. Effectiveness of Nature-Based Solutions for mitigating the impact of pluvial flooding in urban areas at the regional scale. *Water* **2023**, *15*, 642. [CrossRef]
23. Zaimes, G.; Tardio, G.; Iakovoglou, V.; Gimenez, M.; Garcia-Rodriguez, J.; Sangalli, P. New tools and approaches to promote soil and water bioengineering in the Mediterranean. *Sci. Total Environ.* **2019**, *693*, 133677. [CrossRef]
24. Caviedes-Voullième, D.; García-Navarro, P.; Murillo, J. Influence of mesh structure on 2D full shallow water equations and SCS Curve Number simulation of rainfall/runoff events. *J. Hydrol.* **2012**, *448–449*, 39–59. [CrossRef]
25. Nones, M.; Caviedes-Voullième, D. Computational advances and innovations in flood risk mapping. *J. Flood Risk Manag.* **2020**, *13*, e12666. [CrossRef]
26. García-Feal, O.; González-Cao, J.; Gómez-Gesteira, M.; Cea, L.; Domínguez, J.; Formella, A. An accelerated tool for flood modelling based on Iber. *Water* **2018**, *10*, 1459. [CrossRef]
27. Caviedes-Voullième, D.; Morales-Hernández, M.; Norman, M.R.; Özgen-Xian, I. SERGHEI (SERGHEI-SWE) v1.0: A performance-portable high-performance parallel-computing shallow-water solver for hydrology and environmental hydraulics. *Geosci. Model Dev.* **2023**, *16*, 977–1008. [CrossRef]
28. Morales-Hernández, M.; Sharif, M.B.; Kalyanapu, A.; Ghafoor, S.; Dullo, T.; Gangrade, S.; Kao, S.C.; Norman, M.; Evans, K. TRITON: A Multi-GPU open source 2D hydrodynamic flood model. *Environ. Model Softw.* **2021**, *141*, 105034. [CrossRef]
29. Shaw, J.; Kesserwani, G.; Neal, J.; Bates, P.; Sharifian, M.K. LISFLOOD-FP 8.0: The new discontinuous Galerkin shallow-water solver for multi-core CPUs and GPUs. *Geosci. Model Dev.* **2021**, *14*, 3577–3602. [CrossRef]
30. Caviedes-Voullième, D.; Fernández-Pato, J.; Hinz, C. Performance assessment of 2D Zero-Inertia and Shallow Water models for simulating rainfall-runoff processes. *J. Hydrol.* **2020**, *584*, 124663. [CrossRef]
31. Olcina-Cantos, J.; Díez-Herrero, A. Technical evolution of flood maps through Spanish experience in the European framework. *Cartogr. J.* **2021**, *59*, 55–68. [CrossRef]
32. Guo, K.; Guan, M.; Yu, D. Urban surface water flood modelling – a comprehensive review of current models and future challenges. *Hydrol. Earth Syst. Sci.* **2021**, *25*, 2843–2860. [CrossRef]
33. Ferreira, C.S.; Mourato, S.; Kasanin-Grubin, M.; Ferreira, A.J.D.; Destouni, G.; Kalantari, Z. Effectiveness of Nature-Based Solutions in mitigating flood hazard in a Mediterranean peri-urban catchment. *Water* **2020**, *12*, 2893. [CrossRef]
34. Ferguson, C.; Fenner, R. Evaluating the effectiveness of catchment-scale approaches in mitigating urban surface water flooding. *Philos. Trans. R. Soc. A* **2020**, *378*, 20190203. [CrossRef]
35. Hankin, B.; Page, T.; McShane, G.; Chappell, N.; Spray, C.; Black, A.; Comins, L. How can we plan resilient systems of nature-based mitigation measures in larger catchments for flood risk reduction now and in the future? *Water Secur.* **2021**, *13*, 100091. [CrossRef]
36. Cohen-Shacham, E.; Walters, G.; Janzen, C.; Maginnis, S. Nature-based solutions to address global societal challenges. *IUCN* **2016**, *97*, 2016–2036.
37. Forbes, H.; Ball, K.; McLay, F. *Natural Flood Management Handbook*; Scottish Environment Protection Agency: Stirling, UK, 2015; ISBN 978-0-85759-024-4.
38. Palmeri, L.; Trepel, M. A GIS-Based score system for siting and sizing of created or restored wetlands: Two case studies. *Water Resour. Manag.* **2002**, *16*, 307–328. [CrossRef]
39. Stefanakis, A. The role of constructed wetlands as green infrastructure for sustainable urban water management. *Sustainability* **2019**, *11*, 6981. [CrossRef]
40. Casas, J.; Toja, J.; Bonachela, S.; Fuentes, F.; Gallego, I.; Juan, M.; León, D.; Peñalver, P.; Pérez, C.; Sánchez, P. Artificial ponds in a Mediterranean region (Andalusia, southern Spain): Agricultural and environmental issues. *Water Environ. J.* **2011**, *25*, 308–317. [CrossRef]
41. Quinn, P.; Hewett, C.; Wilkinson, M.; Adams, R. The Role of Runoff Attenuation Features (RAFs) in Natural Flood Management. *Water* **2022**, *14*, 3807. [CrossRef]
42. Rey, F.; Bifulco, C.; Bischetti, G.; Bourrier, F.; De Cesare, G.; Florineth, F.; Graf, F.; Marden, M.; Mickovski, S.; Phillips, C.; et al. Soil and water bioengineering: Practice and research needs for reconciling natural hazard control and ecological restoration. *Sci. Total Environ.* **2019**, *648*, 1210–1218. [CrossRef]
43. Darby, S. Effect of riparian vegetation on flow resistance and flood potential. *J. Hydraul. Eng.* **1999**, *125*, 443–454. [CrossRef]
44. Delangue, J.; Teillac-Deschamps, P. *Nature-Based Solutions for Climate Change Adaptation and Disaster Risk Reduction*; IUCN French Committee: Paris, France, 2019; ISBN 978-2-918105-77-0.

45. Bombino, G.; Denisi, P.; Gómez, J.; Zema, D. Mulching as best management practice to reduce surface runoff and erosion in steep clayey olive groves. *Int. Soil Water Conserv. Res.* **2021**, *9*, 26–36. [CrossRef]
46. Manzaneda-Avila, A.; Rey-Zamor, P. Brachypodium Hybridum BHJHIN. Patent 20195515, 17 January 2022.
47. Novara, A.; Cerda, A.; Barone, E.; Gristina, L. Cover crop management and water conservation in vineyard and olive orchards. *Soil Tillage Res.* **2021**, *208*, 104896. [CrossRef]
48. Cohen-Shacham, E.; Andrade, A.; Dalton, J.; Dudley, N.; Jones, M.; Kumar, C.; Maginnis, S.; Maynard, S.; Nelson, C.R.; Renaud, F.G.; et al. Core principles for successfully implementing and upscaling Nature-based Solutions. *Environ. Sci. Policy* **2019**, *98*, 20–29. [CrossRef]
49. Balzan, M.; Geneletti, D.; Grace, M.; De Santis, L.; Tomaskinova, J.; Reddington, H.; Sapundzhieva, A.; Dicks, L.; Collier, M. Assessing nature-based solutions uptake in a Mediterranean climate: Insights from the case-study of Malta. *Nat.-Based Solut.* **2022**, *2*, 100029. [CrossRef]
50. Charlton, R. *Fundamentals of Fluvial Geomorphology*; Routledge: London, UK; New York, NY, USA, 2008. [CrossRef]
51. Soil Science Society of America. *Glossary of Soil Science Terms*; Soil Science Society of America: Madison, WI, USA, 1996. [CrossRef]
52. Buffington, J.M.; Montgomery, D.R. Geomorphic classification of rivers: An updated review. In Proceedings of the Treatise on Geomorphology (2nd); Shroder, J., Wohl, E., Eds. Academic Press: San Diego, CA, USA, 2021; Volume 9, pp. 730–767. [CrossRef]
53. Jacobson, R.B.; O'Connor, J.E.; Oguchi, T. Surficial geological tools in fluvial geomorphology. In *Tools in Fluvial Geomorphology*; John Wiley & Sons, Ltd: Hoboken, NJ, USA, 2016; pp. 11–39. [CrossRef]
54. García, M. (Ed.) Sediment Transport and Morphodynamics. In *Sedimentation Engineering: Processes, Measurements, Modeling, and Practice*; Number 110 in ASCE Manuals and Reports on Engineering Practice; American Society of Civil Engineers: Reston, VA, USA, 2008; pp. 11–191. [CrossRef]
55. Baker, V.R.; Benito, G.; Brown, A.G.; Carling, P.A.; Enzel, Y.; Greenbaum, N.; Herget, J.; Kale, V.S.; Latrubesse, E.M.; Macklin, M.G.; et al. Fluvial palaeohydrology in the 21st century and beyond. *Earth Surf. Process. Landf.* **2022**, *47*, 58–81. [CrossRef]
56. Maranzoni, A.; Tomirotti, M. New formulation of the two-dimensional steep-slope shallow water equations. Part I: Theory and analysis. *Adv. Water Resour.* **2022**, *166*, 104255. [CrossRef]
57. Denlinger, R.P.; O'Connell, D.R. Computing nonhydrostatic shallow-water flow over steep terrain. *J. Hydraul. Eng.* **2008**, *134*, 1590–1602. [CrossRef]
58. Maranzoni, A.; Tomirotti, M. New formulation of the two-dimensional steep-slope shallow water equations. Part II: Numerical modeling, validation, and application. *Adv. Water Resour.* **2023**, *in press*. [CrossRef]
59. Cea, L.; Bladé, E. A simple and efficient unstructured finite volume scheme for solving the shallow water equations in overland flow applications. *Water Resour. Res.* **2015**, *51*, 5464–5486. [CrossRef]
60. Brutsaert, W. *Hydrology: An Introduction*; Cambridge University Pres: Cambridge, UK, 2005.
61. Merz, R.; Blöschl, G. A process typology of regional floods. *Water Resour. Res.* **2003**, *39*, SWC51–SWC520. [CrossRef]
62. Dottori, F.; Szewczyk, W.; Ciscar, J.C.; Zhao, F.; Alfieri, L.; Hirabayashi, Y.; Bianchi, A.; Mongelli, I.; Frieler, K.; Betts, R.; et al. Increased human and economic losses from river flooding with anthropogenic warming. *Nat. Clim. Change* **2018**, *8*, 781–786. [CrossRef]
63. Chow, V. *Open Channel Hydraulics*; McGraw-Hill: New York, NY, USA, 1959.
64. Yen, B. Hydraulic Resistance in Open Channels. In *Channel Flow Resistance: Centennial of Manning's Formula*; Water Resource Publications: Highlands Ranch, CO, USA, 1991; pp. 1–135.
65. Arcement, G.; Schneider, V. Guide for selecting Manning's roughness coefficients for natural channels and flood plains. *United States Geol. Surv. Water-Supply Pap.* **1989**, *2339*, 1–38. [CrossRef]
66. Cea, L.; Legout, C.; Darboux, F.; Esteves, M.; Nord, G. Experimental validation of a 2D overland flow model using high resolution water depth and velocity data. *J. Hydrol.* **2014**, *513*, 142–153. [CrossRef]
67. Naves, J.; Anta, J.; Puertas, J.; Regueiro-Picallo, M.; Suárez, J. Using a 2D shallow water model to assess Large-Scale Particle Image Velocimetry (LSPIV) and Structure from Motion (SfM) techniques in a street-scale urban drainage physical model. *J. Hydrol.* **2019**, *575*, 54–65. [CrossRef]
68. Sanz-Ramos, M.; Bladé, E.; González-Escalona, F.; Olivares, G.; Aragón-Hernández, J. Interpreting the Manning roughness coefficient in overland flow simulations with coupled hydrological-hydraulic distributed models. *Water* **2021**, *13*, 3433. [CrossRef]
69. Ma, Y.; Geng, Y.; Chen, X. Water distribution influenced by pavement alignment design. *J. Transp. Eng. Part B* **2020**, *146*, 4020058. [CrossRef]
70. Guan, M.; Sillanpää, N.; Koivusalo, H. Modelling and assessment of hydrological changes in a developing urban catchment. *Hydrol. Process.* **2015**, *29*, 2880–2894. [CrossRef]
71. Crompton, O.; Thompson, S. Sensitivity of dryland vegetation patterns to storm characteristics. *Ecohydrology* **2021**, *14*, e2269. [CrossRef]
72. Kannan, N.; Santhi, C.; Williams, J.; Arnold, J. Development of a continuous soil moisture accounting procedure for curve number methodology and its behaviour with different evapotranspiration methods. *Hydrol. Process.* **2008**, *22*, 2114–2121. [CrossRef]
73. Hawkins, R.; Ward, T.; Woodward, D.; Van Mullem, J. *Curve Number Hydrology. State of the Practice*; American Society of Civil Engineers: Reston, VA, USA, 2009. [CrossRef]

74. Sanz-Ramos, M.; Cea, L.; Bladé, E.; López-Gómez, D.; Sañudo, E.; Corestein, G.; García-Alén, G.; Aragón-Hernández, J. *Iber v3. Manual de Referencia e Interfaz de Usuario de las Nuevas Implementaciones*; Centre Internacional de Mètodes Numèrics a l'Enginyeria (CIMNE): Barcelona, Spain, 2022.
75. Romero, P.; Castro, G.; Gómez, J.; Fereres, E. Curve number values for olive orchards under different soil management. *Soil Sci. Soc. Am. J.* **2007**, *71*, 1758–1769. [CrossRef]
76. Ajmal, M.; Waseem, M.; Kim, D.; Kim, T.W. A pragmatic slope-adjusted curve number model to reduce uncertainty in predicting flood runoff from steep watersheds. *Water* **2020**, *12*, 1469. [CrossRef]
77. Julien, P. *Erosion and Sedimentation*, 2nd ed.; Cambridge University Press: Cambridge, UK, 2010. [CrossRef]
78. Williams, G. Paleohydrologic Equations for Rivers. In *Developments and Applications of Geomorphology*; Costa, J., Fleisher, P., Eds. Springer: Berlin/Heidelberg, Germany, 1984; pp. 343–367. [CrossRef]
79. Bohorquez, P.; Cañada Pereira, P.; Jimenez-Ruiz, P.J.; del Moral-Erencia, J.D. The fascination of a shallow-water model for the formation of megaflood-scale dunes and antidunes. *Earth Sci. Rev.* **2019**, *193*, 91–108. [CrossRef]
80. Hassan, M.A.; Church, M. Experiments on surface structure and partial sediment transport on a gravel bed. *Water Resour. Res.* **2000**, *36*, 1885–1895. [CrossRef]
81. Wilson, C. Flow resistance models for flexible submerged vegetation. *J. Hydrol.* **2007**, *342*, 213–222. [CrossRef]
82. Nitsche, M.; Rickenmann, D.; Kirchner, J.; Turowski, J.; Badoux, A. Macroroughness and variations in reach-averaged flow resistance in steep mountain streams. *Water Resour. Res.* **2012**, *48*, W12518. [CrossRef]

applied sciences

MDPI

Article

Torrential Hazards' Mitigation Measures in a Typical Alpine Catchment in Slovenia

Jošt Sodnik [1,2], Matjaž Mikoš [2] and Nejc Bezak [2,*]

[1] Tempos, Environmental Civil Engineering Ltd., 1000 Ljubljana, Slovenia; jost.sodnik@tempos.si
[2] Faculty of Civil and Geodetic Engineering, UNESCO Chair on Water-Related Disaster Risk Reduction, University of Ljubljana, 1000 Ljubljana, Slovenia; matjaz.mikos@fgg.uni-lj.si
* Correspondence: nejc.bezak@fgg.uni-lj.si; Tel.: +386-1-4768-685

Featured Application: This study presents an overview of the several mitigation measures, including several check dams and flexible net barriers constructed in a typical Alpine torrential catchment in northern Slovenia (Europe) in order to protect infrastructure (i.e., ski resort cabin lift station, access road to the station, downstream settlements) against sediment-related disasters due to torrential hazards, such as flash floods and debris floods, debris flows, and shallow landslides. This newly introduced combination of technical countermeasures in Slovenia withstood the extreme floods in August 2023.

Abstract: Different sediment-related disasters due to torrential hazards, such as flash floods, debris flows, and landslides, can occur in an Alpine torrential catchment. When protecting infrastructure and human lives, different structural and non-structural protection measures can be used to mitigate permanent and future risks. An overview of the mitigation measures constructed near the Krvavec ski resort in northwest Slovenia (Central Europe) is presented. In May 2018, an extreme debris flood occurred in this area, causing significant economic damage. After the May 2018 event, different field investigations (i.e., geological and topographic surveys) and modeling applications (e.g., hydrological modeling, debris flow) have been conducted with the purpose of preparing the required input data for the design of protection measures against such disasters in future—due to climate change, more disasters are expected to happen in this torrential watershed. The mitigation includes the restoration of local streams, the construction of a large slit check dam for sediment retention, the construction of several smaller check dams and the construction of 16 flexible net barriers with an estimated ~8000 m^3 retention volume for controlling in-channel erosion in steep torrential streams. Additionally, in order to observe and monitor potential future extreme events, an extensive monitoring system has been established in the investigated area. This monitoring system will cover measurements of flexible net corrosion, the estimation of concrete abrasion at check dams, periodical geodetic surveys using small drones (UAV), hydro-meteorological measurements using rainfall gauges and water level sensors. The recent extreme floods of August 2023 also hit this part of Slovenia, and this combination of technical countermeasures withstood the event and prevented large amounts of coarse debris from being transported to the downstream section and devastating infrastructure, as was the case in May 2018 during a less extreme event. Therefore, such mitigation measures can also be used in other torrential catchments in the Alpine environment.

Keywords: torrential hazards; debris flows; flash floods; extreme events; rainfall; mitigation; flexible net barriers; check dams; monitoring system

Citation: Sodnik, J.; Mikoš, M.; Bezak, N. Torrential Hazards' Mitigation Measures in a Typical Alpine Catchment in Slovenia. *Appl. Sci.* **2023**, *13*, 11136. https://doi.org/10.3390/app132011136

Academic Editors: Gordon Gilja, Manousos Valyrakis, Panagiotis Michalis, Thomas Pahtz, Oral Yagci and Athanasios Sfetsos

Received: 1 August 2023
Revised: 2 October 2023
Accepted: 4 October 2023
Published: 10 October 2023

1. Introduction

Alpine areas around the world, including the Alps in Europe, are susceptible to different gravitational and rainfall-triggered natural hazards, such as shallow and deep-seated landslides [1,2], dry and wet debris flows [3–5], flash floods [6–9] or a combination of these hazards that can be initiated during regular but extreme catastrophic rainfall events [10–12]. The Alps in Europe cover more than 300,000 km^2 and are located in several European countries. Slovenia is one of the countries located in Central Europe, and its north and northwestern parts are within the Alps. It is expected that climate change will increase both the magnitude and frequency of extreme torrential events [13,14] that will cover larger areas during single disaster events. Therefore, different structural and non-structural mitigation measures should be applied in order to protect existing infrastructure and human lives. Different types of measures can be used, ranging from so-called grey measures [1,15] to green-blue [16] solutions and a combination of them (i.e., hybrid solutions [17]). Green or blue solutions are often promoted in different disciplines [18,19] since they can be related to multiple benefits, such as increased biodiversity. Moreover, even in torrential risk management in the Alpine environment, wooden dams were often used in the past for the stabilization of unstable slopes and torrential channels and limiting sediment transport from upland parts of torrential watersheds to urban areas. However, classical engineering grey measures, such as concrete or reinforced concrete check dams [1], are still very frequently built to mitigate torrential risks. When enhancing knowledge about torrential processes, detailed observations of natural phenomena are needed, asking for the collection of relevant field data, such as information about triggering mechanisms, magnitudes of debris flow and peak discharges of flash floods, or debris flood characteristics, such as velocities and pressures on infrastructure [20,21]. The pressure and dynamic impact forces are the most important parameters used for the design of protection measures, such as flexible net barriers or check dams against debris floods and debris flows [22,23]. The same applies for some other infrastructure, such as coastal structures used for tsunami protection [24,25]. Therefore, the main objectives of this study is to: (i) present different counter measures proposed to mitigate torrential hazards in the Krvavec ski resort area in northern Slovenia, (ii) to describe the monitoring system used to observe and detect potential future torrential events, and (iii) present some preliminary results with respect to functioning of the mitigation countermeasures during the recent extreme August 2023 flood that occurred in Slovenia [26,27]. In Krvavec area, flexible net barriers have been constructed in torrential channel beds for sediment retention and channel stabilization for the first time in Slovenia, and, therefore, this cases study shows that this type of torrential structures can be used as technical countermeasures in steep torrential streams in the Alpine environment in Slovenia.

The original contribution of the presented case study lies in the thoughtfully selected combination of structural measures of different types (flexible ring net barriers in steep channels, bank stabilization and low check dams in less steep channels, and a reinforced concrete slit check dam) that can be used as an open-air monitoring site for the effectiveness of such measures with respect to flexible barriers behavior using: (i) innovative monitoring devices (i.e., Guards), and (ii) combining estimated sediment transport quantities from retained debris behind barriers and LiDAR measurements using drones, with concrete abrasion experiments of diverse concrete specimens subjected to real field conditions. The successful functioning of the system during the extreme August 2023 event (if compared to the 2018 debris flood) confirms the proper design of the system and the possibility of using torrential countermeasures for field research on interactions between natural processes and torrential structures.

2. Materials and Methods

2.1. Case Study Description: Krvavec Ski Resort Area

The Krvavec ski resort area is part of the Kamnik-Savinja Alps, a mountain range located in the northern part of Slovenia (Figure 1). The total area of Kamnik-Savinja Alps in Slovenia is around 1000 km^2, and there are several mountains with elevations above 2000 m a.s.l., with the highest one being Grintovec Mountain with an elevation of 2558 m a.s.l. The Krvavec ski resort stretches from around 1500 m a.s.l. to around 2000 m a.s.l. The area is characterized by steep slopes and narrow valleys (Figure 2). The cabin lift is the main means of transport for skiers in winter during the ski season. As can be seen from Figure 1, there are two steep torrential ravines (i.e., the Brezovški graben and the Lukenjski graben) in the upstream slopes from the station of the cabin car. The average longitudinal channel slope of these two torrents is ~25–30%, and the total catchment area is close to 2 km^2 [22]. Around three-quarters of both catchment areas are forested, while the rest are agricultural areas [22].

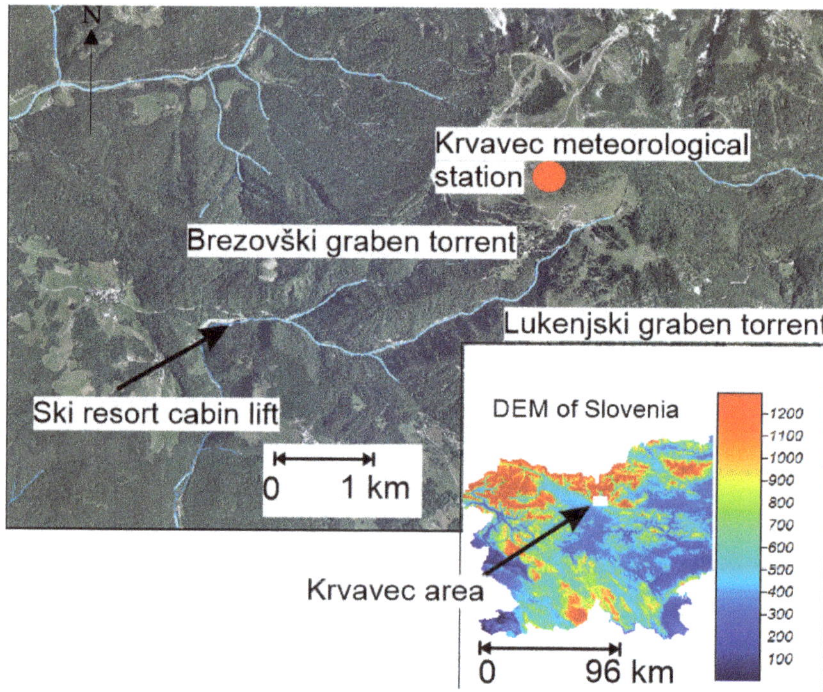

Figure 1. Location of the Krvavec ski resort area on the elevation map of Slovenia (legend is in meters above sea level) and an overview of the situation in the torrents upstream of the valley station of the Krvavec ski resort cabin lift. Slovenia is positioned between 45.50526° to 46.83694° (latitude) and 13.52639° to 16.45091° (longitude). The Krvavec meteorological station is positioned at 14.5333° longitude and 46.2973° latitude and at 1742 m a.s.l.

Figure 2. Two photos of the situation in the Brezovški graben and Lukenjski graben torrents, respectively. The channel width is between 1–2 m to several meters.

The mean annual precipitation at the top of the Krvavec ski resort area (approx. 1700 m a.s.l.) is around 1620 mm (in the period 2010–2020), with no clear seasonal pattern (Figure 3). Moreover, there is a clearer pattern in the monthly rainfall erosivity distribution with a clear peak in the summer period (Figure 3). The rainfall erosivity was calculated based on the 30-m rainfall data for the 2010–2020 period based on the RUSLE methodology [28] and using the RIST software [29]. The average annual rainfall erosivity in the 2010–2020 period was around 3531 MJ $\times$ mm $\times$ ha^{-1} $\times$ h^{-1} $\times$ year^{-1}, which can be regarded as relatively large [30]. Therefore, it is evident that hydro-meteorological events can occasionally be relatively extreme in this alpine part of Slovenia. In combination with steep slopes and other torrential characteristics, this can lead to rainfall events, such as the one that occurred in May 2018 (Section 2.2). This local thunderstorm triggered a local debris flood that partially devastated the valley station of the Krvavec ski resort cable car and interrupted the local access road to the station [22]—for protection against such torrential events, several structural protection measures were built in the past [22,23], but most of these torrential structures were already in relatively bad condition before, and especially after the May 2018 event (Section 2.3).

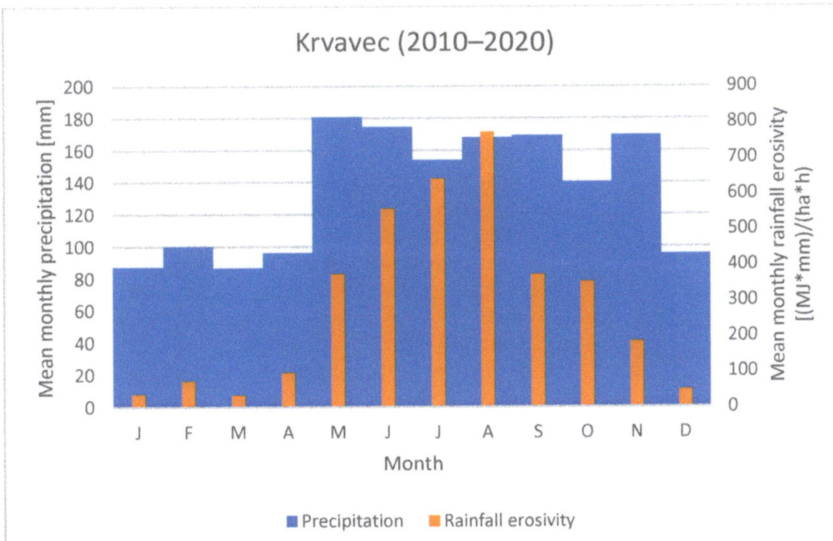

Figure 3. Mean monthly precipitation and rainfall erosivity at the Krvavec meteorological station in the period 2010–2020.

2.2. *Extreme Debris Flood in 2018*

On 30 May 2018, a relatively extreme rainfall event occurred in this area [22]. Around 50 mm of rainfall fell in around 30 min, corresponding to the statistically estimated return period of over 50 years. The total event duration was very short (less than 3 h), and the majority of rainfall was concentrated in around 15–30 min (Figure 4). The rainfall erosivity for this specific event based on the 30-m rainfall data can be estimated to be over 1420 MJ $\times$ mm $\times$ ha^{-1} $\times$ h^{-1} $\times$ year^{-1}, which is almost half of the total annual rainfall erosivity and almost twice as large as the average monthly erosivity at the Krvavec meteorological station (Figure 3). Furthermore, in the 2010–2020 period, this was the event with the second-highest rainfall erosivity. More specifically, the average erosivity per event at the Krvavec meteorological station was around 90 MJ $\times$ mm $\times$ ha^{-1} $\times$ h^{-1} $\times$ year^{-1}.

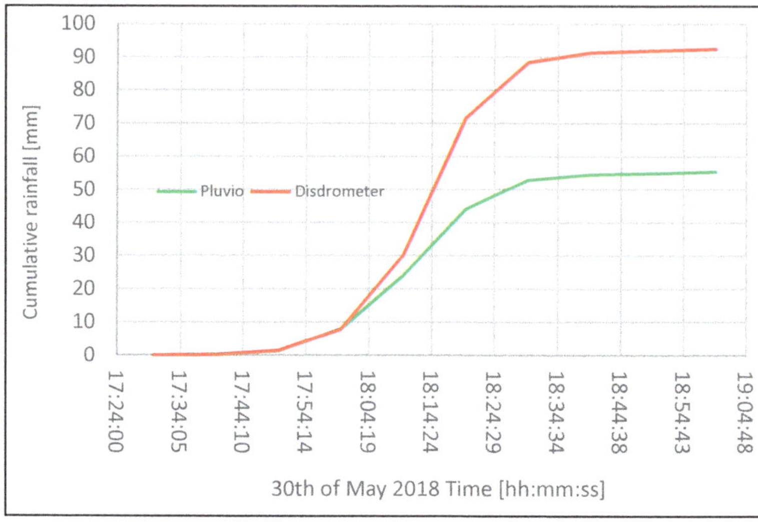

Figure 4. Measured cumulative rainfall using an optical disdrometer and a pluviograph at the Krvavec meteorological station at 1742 m a.s.l. during the extreme event that occurred on 30 May 2018.

As indicated by Bezak et al. [22], the extreme rainfall in May 2018 triggered the local debris flood and caused intense erosion and sedimentation processes in both Brezovški graben and Lukenjski graben torrents. At the valley station of the Krvavec cable car, around 10,000 m^3 of relatively coarse debris was deposited, and it was assumed that the main source of the debris material was the Brezovški graben since the debris material from the Lukenjski graben was deposited upstream (before the confluence of the Brezovški graben and Lukenjski graben torrents) from the valley station of the Krvavec cable car (i.e., in the debris flow fan area) [22]. The debris flood caused extensive damage to the existing torrential infrastructure in both torrents (Figure 5). Hence, in order to protect the valley station of the cable car and the downstream area along the access road to the station, new countermeasures needed to be implemented (Section 3.3).

Figure 5. One example of the consequences of the May 2018 event on the existing infrastructure in both torrents: a partially destroyed already filled flexible barrier.

2.3. Overview of the Mitigation Measures

In both torrents and prior to the May 2018 debris flood, there were several countermeasures, such as smaller check dams, gabion walls, wooden dams and low weirs already built to mitigate torrential hazards. However, these objects were destroyed to a large extent during the May 2018 event (Figure 5). Therefore, in order to mitigate the torrential risk in the future, new structural measures were planned for both torrents [22,23].

After the May 2018 event, a geological survey was carried out to determine geological conditions for planning and executing new mitigation measures (Figure 6). The main objective of the field survey was to estimate the possibility and magnitude of future mass movement and debris flood events. Areas with deposited debris material and areas with high erosion potential (Figure S1) were identified to secure input data for debris flood modeling and countermeasures design process.

Figure 6. Engineering geological map of the area with the identification of the main sediment sources.

Mitigation measures were roughly divided into three sections, respectively, phases (Figure 7). The first phase was a restoration of existing concrete structures (channel regulation works) on the Reka torrent, i.e., downstream of the Brezovški and Lukenjski torrents confluence. The second phase was planning and building a large slit check dam just downstream at the confluence of both torrents to temporarily retain future mass flows and slowly dose debris to the downstream section. In this location, there was already a small check dam, but the capacity of the sediment trap was only about 500 m^3, which is a relatively small capacity compared to possible future events magnitude (volumes). A new slit check dam is designed with the largest possible sediment capacity in this location of close to 14,000 m^3. The third phase includes measures to reduce erosion potential in the Brezovški and Lukenjski graben catchments. Both upstream sections have a large erosion potential, and a large amount of mixed coarse and fine debris material can be mobilized during future rainfall events. Since the area is hardly assessable even on foot, non-conventional countermeasures were sought. As an appropriate structure, a flexible ring-net debris flow barrier was selected, which can be transported to a construction site with helicopters or forestry cableways. No heavy machinery (such as excavators) is needed for installation—as it could not be applied—and light machinery was applied (such as so-called spider excavators). The main objective of the third phase was to reduce in-channel erosion intensity via stabilization of the torrent channels (side-slopes reduction at the same time). Geological information shown in Figure 6 was used for the selection of the optimal locations of the flexible net barriers. All three phases together present a holistic technical solution for this endangered torrential area and will reduce mass-flow-related hazards for downstream areas (cabin lift valley station, access road, Grad village, etc.).

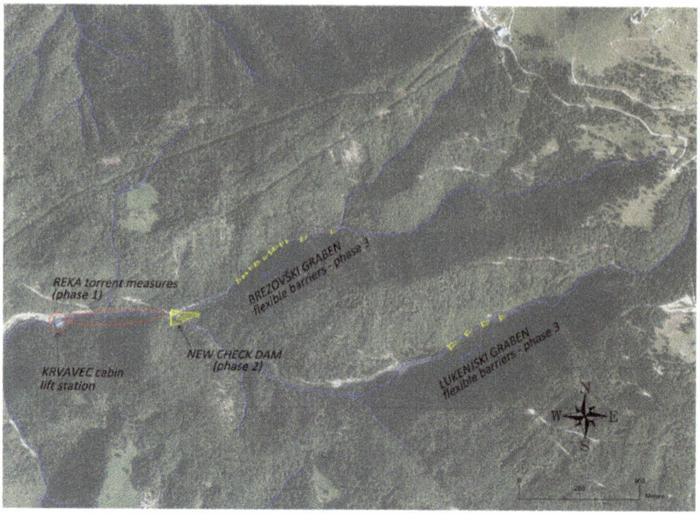

Figure 7. Location of proposed measures and an indication of the three main phases used to mitigate the future risk of torrential hazards.

3. Results and Discussion

3.1. Reka Torrent Mitigation Measures (Phase 1)

The Reka torrent was already regulated in the past (in the last few decades), but mainly concrete and wooden structures were in relatively poor condition, and the May 2018 event severely damaged or destroyed most of them. This section of the Reka torrent is steep, and the structures such as small check dams, low weirs and bank protections are a necessity for channel stabilization. During construction phase 1, executed in 2019 and 2020 (pre-COVID-19), some of the structures were renewed, but most of them were completely reconstructed (Figure 8). Banks with erosion were protected with stone rip-rap structures.

Figure 8. An example of the Reka torrent measures (carried out in 2019, before COVID-19).

3.2. New Slit Check Dam (Phase 2)

Due to the high risk of similar or even more severe mass flow events in the future, a large 10.5 m high check dam with approximately 14,000 m^3 sediment trap capacity was designed on the confluence of Brezovški and Lukenjski torrent. The check dam was designed as a slit dam to have minimal impact on geomorphological processes in this section and with the aim of ensuring the sediment connectivity and not to completely stop the sediment transport process. Therefore, the sediment transport process will not be completely interrupted, except for extreme events during which large masses of coarse sediments will be mobilized and could present a significant hazard for downstream areas. In the dam design phase, geotechnical research (boreholes, laboratory) was carried out to obtain reliable input data for the dam structural design. Check dam with a construction height of 10.5 m is designed for the dynamic impact of mass flows. It will be built as a reinforced concrete structure, covered with a mixture of concrete and rocks, to ensure a more nature-friendly look.

Together with a new slit check dam, some additional structures will be built to ensure the stability of the upstream section of the confluence (Figure 9). These sections of both torrents will be modified to provide larger sediment trap capacity and erosion stability. The check dam is designed in a location with an access road so that the sediments can be removed regularly, and the whole sediment trap volume is available.

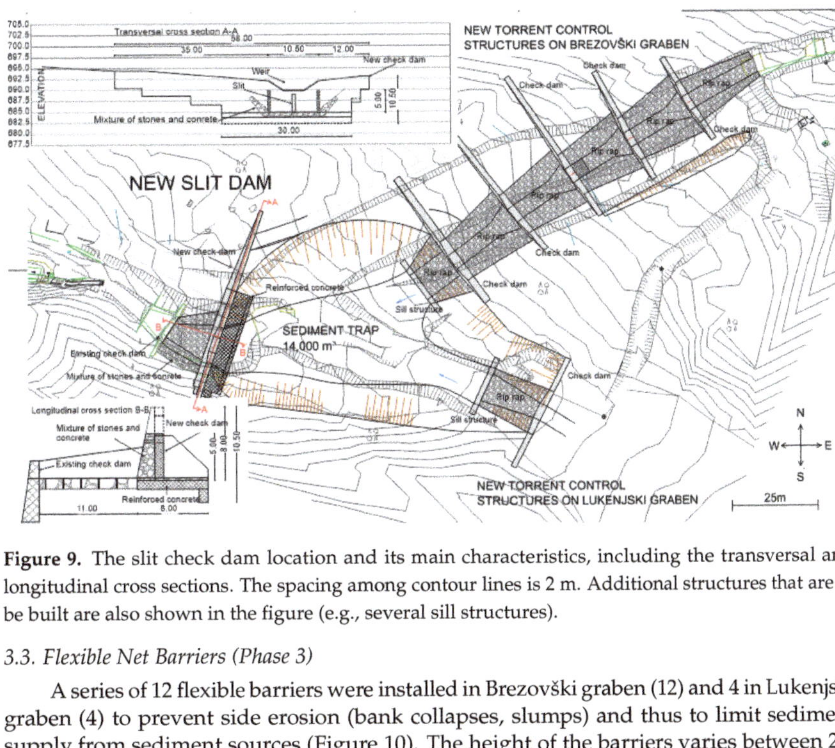

Figure 9. The slit check dam location and its main characteristics, including the transversal and longitudinal cross sections. The spacing among contour lines is 2 m. Additional structures that are to be built are also shown in the figure (e.g., several sill structures).

3.3. Flexible Net Barriers (Phase 3)

A series of 12 flexible barriers were installed in Brezovški graben (12) and 4 in Lukenjski graben (4) to prevent side erosion (bank collapses, slumps) and thus to limit sediment supply from sediment sources (Figure 10). The height of the barriers varies between 2.5 and 6 m, and the sediment trap capacity is between 100 and 800 m^3. The top width of the barriers varies between 9 and 25 m. The total retention volume was estimated for 4 barriers in the Lukanjski graben to be over 3000 m^3 and for 12 barriers in the Brezovški graben to be close to 5000 m^3.

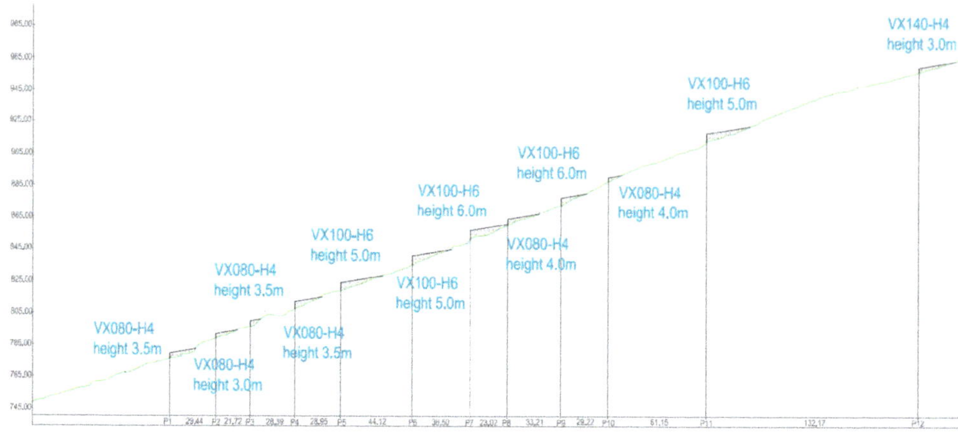

Figure 10. A longitudinal profile of the Brezovški graben with the locations of the 12 flexible net barriers.

All the barriers were designed and dimensioned using the DEBFLOW online tool developed by Geobrugg [31]. The design of the barriers was modified because the deposits are not intended to be removed after the events, but the main purpose of the barriers is to reduce channel slope, reduce velocities and support unstable channel banks. Because of this specific design, not all barriers are designed for debris flow impact, but all of them are designed for overflowing forces. The upstream barriers and the barriers with larger distances to the next upstream barriers are designed for debris flow impact, which is one of the possible scenarios in the phase where barriers will be filling up with sediments. The final state of the barriers will be, when all of them are filled up, torrential bed slope reduced and bank erosion mostly mitigated. Based on the topographic characteristics, VX-type (Geobrugg) barriers were selected [31] (Figure 11). Some of them were customized with additional ropes and anchors since there was no need to use stronger, more resistant nets, but only to add some ropes to fulfill all the design requirements (stability, resistance).

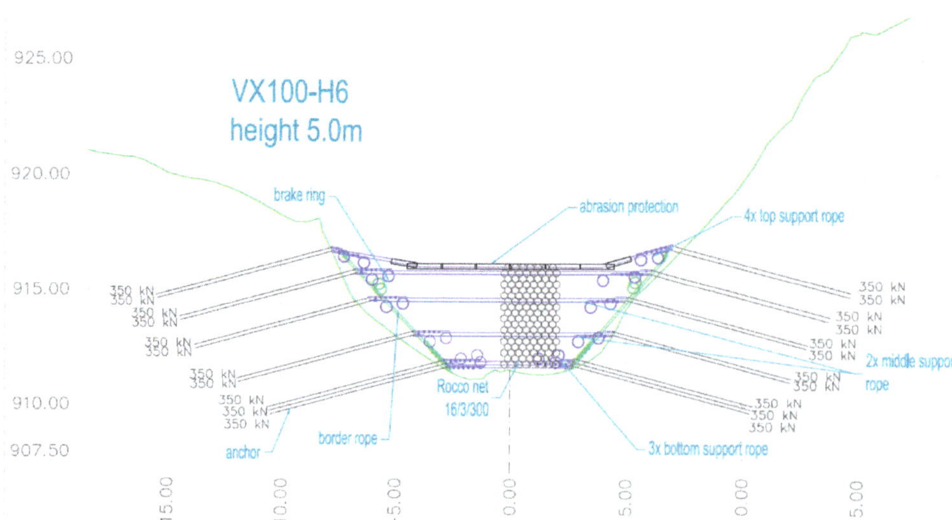

Figure 11. Example of a designed VX barrier.

Special equipment was used for the installation of the barriers. Logging lifts were installed to transport the construction material to the construction sites. Most of the work was performed manually (drilling, installation) (Figures S2–S5).

3.4. Ongoing Monitoring

The monitoring system of torrential processes in the Krvavec area can be divided into the following parts:

- Measuring local rainfall using one tipping bucket as a rainfall gauge (resolution 0.1 mm) (will be installed at the top of Brezovški graben torrent);
- The Brezovški graben is regularly surveyed using Unmanned Aerial Vehicles (UAV drones), and surface changes (eroded slopes and new deposits upstream of the flexible barriers) between the two field measurements are determined from a difference between two consecutive LiDAR maps of the torrent. These measurements will explicitly tell us about erosion and sedimentation dynamics in the torrential channel and reveal the efficacy of the built flexible net barriers to retain coarse torrential debris. After they are filled up with coarse debris, they will support unstable torrential flanks, and sediment transport to downstream sections closer to the cable car station will increase again;
- Six devices, called Guards®, were installed on four flexible barriers in the Brezovški graben (barriers 1, 3, 7, 8, 9, and 12) and on two flexible barriers in the Lukanjski graben (barriers 2 and 6). Introduced in late 2020, a Guard® is a waterproof, small, battery-powered autonomous device (2.6 kg, battery life estimated to be between 7 and 10 years), easily installed on the suspension rope of a selected protection barrier for permanent monitoring (24/7) of its environmental and physical status under the toughest conditions [32]. It monitors corrosion (due to air pollution) and movements (due to debris or rock impacts) of the flexible nets, as well as reporting events, such as a rockfall or a debris flow via a mobile network (i.e., GSM, UMTS, LTE) to a dashboard on your computer—without a need to go to the field site. It is, in essence, an IoT (Internet-of-Things) smart data logger developed to estimate corrosion damage to a single rope within flexible net barriers [33], but it is additionally equipped with an impact sensor unit, measuring accelerations of up to 200 g and forces to up of 294 kN. The Guards® report events when a barrier is moved due to inflowing debris or sediments (even single surges in the case of a wet debris flow), or a single falling/rolling large rock hits it, but also due to strong winds or even wild animals. Such reports are then compared to local weather (wind, air temperature, and rainfall data) to differentiate among causes for barrier movements. The installed Guards® started operating in March 2023, and until the end of July, occasional activities were reported. In total, in this period, there were 42 entries in the cloud-based database that were associated with an increase in force on the flexible net barriers. The measured force ranged from around 1 g to around 17 g. In this period, there were 6 days when the reported number of events was larger than 1 (Table S1). It can be seen that on days with more than one activity detected by Guards, there was either strong wind or a storm in that area (Table S1). Furthermore, the maximum 30-m wind speed and rainfall were above average (Figure S6). However, the reported 30-m rainfall values shown in Table S1 all have a return period of less than 5 years according to the intensity-duration-frequency (IDF) curves for that location (i.e., the 30-m rainfall rate with a 5-year return period equals 25 mm). It can also be seen that there are some days where the activity detected by the Guards® was not driven by extreme wind or precipitation (Table S1). Some events were also detected on the days without rainfall (Table S1). During the period of Guard® operation, there were some days when wind speed was relatively high, or 30-m rainfall rates were relatively high, but there was no activity detected by the Guards® (Figure S6). The Guards® also continuously measure some other parameters, such as temperature, humidity or corrosion (Figure S7);

- A field test site for concrete abrasion measurements due to sediment transport in torrential streams; in the bottom of the Brezovški graben a concrete flat ramp appr. 2.5 × 4.5 m was built in with an array of the total eight concrete square 5 cm thick plates (4 × 2 smaller concrete plates of the dimensions 50 × 50 cm prepared from four different concrete mixtures—two plates of each mixture) (Figure 12). One concrete mixture was a typical one used in torrent control works in Slovenia, and the other ones were prepared with additives, such as microsilica and steel fibers, to enhance their abrasion resistance. The plates are leveled in height with the ramp, not cast into the ramp but put in place so that they can be easily removed after the end of the field test. The ramp and the plates are regularly measured by terrestrial laser scanning to estimate abrasion depth on the plates due to overflowing sediments. Since flexible barriers were put in place in late 2022, concrete abrasion is rather limited and cannot be reported in this article yet. We expect that in situ measurements of concrete plate abrasion will correspond well with the standardized laboratory tests for underwater abrasion of concrete [34]. Positive experiences regarding concrete abrasion resistance of different concrete mixtures typically used for hydraulic structures were gathered in the past on the Lower Sava River in Slovenia [35].

Figure 12. A field test site in the lower Brezovški graben for monitoring concrete abrasion—an upstream view.

3.5. *August 2023 Extreme Flood*

On 3–4 August 2023, extreme rainfall caused intense erosion processes and mass movements, such as debris flows, slumps and landslide slips, and especially large and devastating floods in many parts of Slovenia [26,27]. The Krvavec meteorological station measured 30 mm in 30 min, 165 mm in 12 h and 196 mm of rainfall in 24 h, and rainfall erosivity for this event was over 1500 MJ × mm × ha^{-1} × h^{-1} × $year^{-1}$—approximately 50% of the mean annual rainfall erosivity for this station and larger than during the previous extreme event (i.e., 2018). The return period of 12 h rainfall amount was estimated to be over 250 years. The 12 h rainfall intensity was also larger than the one associated with empirical rainfall threshold curves for the initiation of shallow landslides and debris flows—e.g., the one proposed by Caine in 1980 [36–38]. Hence, numerous smaller landslides, slumps and soil slips occurred in this area, and flexible net barriers were able to retain the eroded material (Figure 13) and stabilize the channel bed without excessive channel incision. Consequently, the cabin car ski lift was not affected, as was the case during the May 2018 event, and no significant damage was caused in this part of the catchment (Figure 13). Based on the field geodetic survey using drones, it was estimated that around 19,000 m^3 of material was eroded during the August 2023 event, and the majority of the debris material was trapped by the flexible net barriers.

Figure 13. A situation in both torrents after the extreme August 2023 event.

4. Conclusions

This study presents an overview of torrential hazard mitigation measures used to mitigate the torrential hazards in a typical Alpine catchment in Slovenia based on post-extreme-event surveys and modeling results. The implementation of countermeasures was divided into three phases: (i) channel regulation work in the downstream part of the watershed, (ii) the construction of a large slit check dam with a retention volume of approximately 14,000 m^3 for sediment retention, and (iii) the implementation of 16 flexible net barriers with an estimated retention volume of 8000 m^3 in both torrents in the upstream part of the watershed. It should be noted that phase (ii) is currently ongoing, while phases (i) and (iii) have already been completed. In the scope of the conducted measures, an extensive monitoring system has also been established that will enable us to explore characteristics of torrential hazards in the future. The monitoring system includes a rain gauge, a video recording system, periodical UAV surveys, Guards®, and a field test site to measure concrete abrasion. The preliminary results have indicated that activity detected by Guards® was caused either by strong wind or heavy rainfall. However, not all thunderstorms caused any torrential erosion activities. Hence, further research is needed to better evaluate the triggering factors of torrential hazards. Field abrasion experiments need more time to be effective since, so far, flexible net barriers have substantially reduced sediment transport over the concrete plates—this will be a rather long-term experiment.

The main scientific and applied contribution of this study is to clearly present and show examples of good practices in sediment management in a typical Alpine catchment that could also be used in other parts of the country. The recent August 2023 flood event has shown that the executed structural (grey) mitigation measures effectively limited excessive sediment erosion and transport from steep torrential channels to downstream areas; they withstood high dynamic loads and, as such, are able to cope with extreme hydro-meteorological events above design discharges determined for a 100-year return period [28]. After the August 2023 event, there was no significant damage to the Krvavec cabin car ski lift (Figure 13)—contrary to substantial damages during the less intense May 2018 event [23]. Also, the rainfall erosivity for the 2023 event was larger than the one for the 2018 event, when there was significant damage at the Krvavec cabin car ski lift ([23]). Therefore, such structural (grey) measures (e.g., flexible net barriers) could be applied in steep torrential catchments in similar geological settings and those that are difficult to access with heavy machinery and hence the construction of torrential concrete check-dams is more challenging.

Supplementary Materials: The following supporting information can be downloaded at: https://www.mdpi.com/article/10.3390/app132011136/s1, Figure S1: Torrential unstable conditions in the Brezovški torrent; Figure S2: Drilling process (left) and steel anchors (right); Figure S3: Construction site in the Brezovški graben torrent; Figure S4: Series of flexible barriers on the Brezovški torrent; Figure S5: Series of flexible barriers on the Brezovški torrent; Figure S6: Maximum 30-m wind speed and rainfall rate for the period of Guard®operation and identification of days when the Guards®detected some activity. Wind speed and rainfall data are obtained from the nearby Krvavec station; Figure S7: Some additional parameters measured by the Guards® during the one-month period; Table S1: Main characteristics of events detected by Guards®. The information about strong winds and storms was obtained from the climatological station at Ljubljana Brnik, while 30-m precipitation and wind speed data were obtained from the nearby meteorological station at Krvavec.

Author Contributions: J.S., M.M. and N.B.: conceptualization, methodology, writing—original draft preparation. All authors have read and agreed to the published version of the manuscript.

Funding: The check dam design and design of other mitigation measures were financed by the Slovenian Water Agency. Geobrugg AG, Romanshorn, Switzerland, supported the installation of the GUARD measuring equipment. The authors' research work was supported by the Slovenian Research and Innovation Agency (ARIS) through grants P2-0180, J1-3024, J1-2477, J6-4628, and N2-0313.

Institutional Review Board Statement: Not applicable.

Informed Consent Statement: Not applicable.

Data Availability Statement: Not applicable.

Acknowledgments: We would like to acknowledge the Slovenian Environment Agency's (ARSO) support for making available the hydro-meteorological data. The slit check dam design and the design of other mitigation measures were financed by the Slovenian Water Agency. Geobrugg AG supported the installation of measuring equipment (i.e., Guards).

Conflicts of Interest: The authors declare no conflict of interest.

References

1. Bezak, N.; Sodnik, J.; Maček, M.; Jurček, T.; Jež, J.; Peternel, T.; Mikoš, M. Investigation of potential debris flows above the Koroška Bela settlement, NW Slovenia, from hydro-technical and conceptual design perspectives. *Landslides* **2021**, *18*, 3891–3906. [CrossRef]
2. Hilker, N.; Badoux, A.; Hegg, C. The swiss flood and landslide damage database 1972–2007. *Nat. Hazards Earth Syst. Sci.* **2009**, *9*, 913–925. [CrossRef]
3. Bertoldi, G.; D'Agostino, V.; McArdell, B.W. An integrated method for debris flow hazard mapping using 2D runout models. *12th Congr. Interpraevent* **2012**, 435–446.
4. De Finis, E.; Gattinoni, P.; Marchi, L.; Scesi, L. Anomalous Alpine fans: From the genesis to the present hazard. *Landslides* **2018**, *15*, 683–694. [CrossRef]
5. Rickenmann, D.; Laigle, D.; McArdell, B.W.; Hübl, J. Comparison of 2D debris-flow simulation models with field events. *Comput. Geosci.* **2006**, *10*, 241–264. [CrossRef]
6. Sikorska, A.E.; Viviroli, D.; Seibert, J. Flood-type classification in mountainous catchments using crisp and fuzzy decision trees. *Water Resour. Res.* **2015**, *51*, 7959–7976. [CrossRef]
7. Andres, N.; Badoux, A. The Swiss flood and landslide damage database: Normalisation and trends. *J. Flood Risk Manag.* **2019**, *12*, e12510. [CrossRef]
8. Rusjan, S.; Kobold, M.; Mikoš, M. Characteristics of the extreme rainfall event and consequent flash floods in W Slovenia in September 2007. *Nat. Hazards Earth Syst. Sci.* **2009**, *9*, 947–956. [CrossRef]
9. Kvočka, D. One-dimensional sediment transport modelling with engelund-hansen and ackers-white transport equations for the lower danube river; [Enodimenzijsko Modeliranje Premeščanja Plavin s Transportnima Enačbama Engelund-Hansen in Ackers-white za Spodnjo Donavo]. *Acta Hydrotech.* **2021**, *34*, 103–117. [CrossRef]
10. Mikoš, M. After 2000 Stože landslide: Part I—Development in landslide research in Slovenia—Po zemeljskem plazu Stože leta 2000: Del I—Razvoj raziskovanja zemeljskih plazov v Sloveniji. *Acta Hydrotech.* **2020**, *33*, 129–153. [CrossRef]
11. Lala, J.M.; Rounce, D.R.; McKinney, D.C. Modeling the glacial lake outburst flood process chain in the Nepal Himalaya: Reassessing Imja Tsho's hazard. *Hydrol. Earth Syst. Sci.* **2018**, *22*, 3721–3737. [CrossRef]
12. Zscheischler, J.; Westra, S.; Van Den Hurk, B.J.J.M.; Seneviratne, S.I.; Ward, P.J.; Pitman, A.; Aghakouchak, A.; Bresch, D.N.; Leonard, M.; Wahl, T.; et al. Future climate risk from compound events. *Nat. Clim. Chang.* **2018**, *8*, 469–477. [CrossRef]
13. Saez, J.L.; Corona, C.; Stoffel, M.; Berger, F. Climate change increases frequency of shallow spring landslides in the French Alps. *Geology* **2013**, *41*, 619–622. [CrossRef]
14. Viani, C.; Chiarle, M.; Paranunzio, R.; Merlone, A.; Musacchio, C.; Coppa, G.; Nigrelli, G. An integrated approach to investigate climate-driven rockfall occurrence in high alpine slopes: The Bessanese glacial basin, Western Italian Alps. *J. Mt. Sci.* **2020**, *17*, 2591–2610. [CrossRef]
15. Kryžanowski, A.; Brilly, M.; Rusjan, S.; Schnabl, S. Review Article: Structural flood-protection measures referring to several European case studies. *Nat. Hazards Earth Syst. Sci.* **2014**, *14*, 135–142. [CrossRef]
16. Debele, S.E.; Kumar, P.; Sahani, J.; Marti-Cardona, B.; Mickovski, S.B.; Leo, L.S.; Porcù, F.; Bertini, F.; Montesi, D.; Vojinovic, Z.; et al. Nature-based solutions for hydro-meteorological hazards: Revised concepts, classification schemes and databases. *Environ. Res.* **2019**, *179*, 108799. [CrossRef] [PubMed]
17. Nakamura, F. Concept and Application of Green and Hybrid Infrastructure. In *Green Infrastructure and Climate Change Adaptation: Function, Implementation and Governance*; Nakamura, F., Ed.; Springer: Singapore, 2022; pp. 11–30. ISBN 978-981-16-6791-6.
18. Alves, A.; Vojinovic, Z.; Kapelan, Z.; Sanchez, A.; Gersonius, B. Exploring trade-offs among the multiple benefits of green-blue-grey infrastructure for urban flood mitigation. *Sci. Total Environ.* **2020**, *703*, 134980. [CrossRef]
19. Kabisch, N.; Frantzeskaki, N.; Pauleit, S.; Naumann, S.; Davis, M.; Artmann, M.; Haase, D.; Knapp, S.; Korn, H.; Stadler, J.; et al. Nature-based solutions to climate change mitigation and adaptation in urban areas: Perspectives on indicators, knowledge gaps, barriers, and opportunities for action. *Ecol. Soc.* **2016**, *21*, 39. [CrossRef]
20. Graf, C.; Christen, M.; McArdell, B.W.; Bartelt, P. An overview of a decade of applied debris-flow runout modeling in Switzerland: Challenges and recommendations. In Proceedings of the Debris-Flow Hazards Mitigation: Mechanics, Monitoring, Modeling, and Assessment—Proceedings of the 7th International Conference on Debris-Flow Hazards Mitigation, Golden, CO, USA, 10–13 June 2019; pp. 685–692.

21. Peternel, T.; Kumelj, Š.; Oštir, K.; Komac, M. Monitoring the Potoška planina landslide (NW Slovenia) using UAV photogrammetry and tachymetric measurements. *Landslides* **2017**, *14*, 395–406. [CrossRef]
22. Bezak, N.; Jež, J.; Sodnik, J.; Jemec Auflič, M.; Mikoš, M. An extreme May 2018 debris flood case study in northern Slovenia: Analysis, modelling, and mitigation. *Landslides* **2020**, *17*, 2373–2383. [CrossRef]
23. Sodnik, J.; Mikoš, M. The Krvavec bottom cabin lift station protection against torrential hazards by a new slit check dam and a series of flexible net barriers. In Proceedings of the 5th Regional Symposium on Landslides in the Adriatic–Balkan Region (ReSyLAB), Rijeka, Croatia, 23–26 March 2022; pp. 219–224.
24. Huang, Y.; Zhu, C. Numerical analysis of tsunami–structure interaction using a modified MPS method. *Nat. Hazards* **2015**, *75*, 2847–2862. [CrossRef]
25. Chongqiang, Z.; Zhiyi, C.; Yu, H. Coupled Moving Particle Simulation–Finite-Element Method Analysis of Fluid–Structure Interaction in Geodisasters. *Int. J. Geomech.* **2021**, *21*, 4021081. [CrossRef]
26. ARSO Report about the Extreme Rainfall between 3rd and 6th of August 2023. Available online: https://meteo.arso.gov.si/uploads/probase/www/climate/text/sl/weather_events/padavine_3-6avg2023.pdf (accessed on 25 August 2023).
27. Bezak, N.; Panagos, P.; Liakos, L.; Mikoš, M. Brief communication: A first hydrological investigation of extreme August 2023 floods in Slovenia, Europe. *EGUsphere* **2023**, *2023*, 1–13. [CrossRef]
28. Renard, K.G.; Foster, G.R.; Weesies, G.A.; McCool, D.K.; Yoder, D.C. *Predicting Soil Erosion by Water: A Guide to Conservation Planning with the Revised Universal Soil Loss Equation (RUSLE)*; U.S. Department of Agriculture: Washington, DC, USA, 1997.
29. USDA Rainfall Intensity Summarization Tool (RIST). Available online: https://www.ars.usda.gov/southeast-area/oxford-ms/national-sedimentation-laboratory/watershed-physical-processes-research/research/rist/rist-rainfall-intensity-summarization-tool/ (accessed on 25 August 2023).
30. Panagos, P.; Ballabio, C.; Borrelli, P.; Meusburger, K.; Klik, A.; Rousseva, S.; Tadić, M.P.; Michaelides, S.; Hrabalíková, M.; Olsen, P.; et al. Rainfall erosivity in Europe. *Sci. Total Environ.* **2015**, *511*, 801–814. [CrossRef]
31. Geobrugg. *DEBFLOW*; Geobrugg: Romanshorn, Switzerland, 2021.
32. Geobrugg Geobrugg GUARD. Available online: https://www.geobrugg.com/en/Geobrugg-GUARD-157047.html (accessed on 25 August 2023).
33. Geobrugg Überwachungsvorrichtung und Verfahren zur Überwachung einer Korrosion eines Drahtnetzes. DE Patent 102017123810 A1, 18 April 2019.
34. *ASTM C 1138M-05*; Standard Test Method for Abrasion Resistance of Concrete (Underwater Method). American Society for Testing and Materials: West Conshohocken, PA, USA, 2006.
35. Kryžanowski, A.; Mikoš, M.; Šustersic, J.; Planinc, I. Abrasion Resistance of Concrete in Hydraulic Structures. *ACI Mater. J.* **2009**, *106*, 349–356. [CrossRef]
36. Caine, N. The rainfall intensity-duration control of shallow landslides and debris flows. *Geogr. Ann. Ser. A* **1980**, *62*, 23–27. [CrossRef]
37. Guzzetti, F.; Peruccacci, S.; Rossi, M.; Stark, C.P. The rainfall intensity-duration control of shallow landslides and debris flows: An update. *Landslides* **2008**, *5*, 3–17. [CrossRef]
38. Rosi, A.; Peternel, T.; Jemec-Auflič, M.; Komac, M.; Segoni, S.; Casagli, N. Rainfall thresholds for rainfall-induced landslides in Slovenia. *Landslides* **2016**, *13*, 1571–1577. [CrossRef]

applied sciences

MDPI

Article

Sedimentological Analysis of the Turbidite Sequence in the Northern Part of the West Crocker Formation, Northwest Sabah

Nurul Afifah Mohd Radzir *,†,‡, Che Aziz Ali ‡ and Kamal Roslan Mohamed ‡

Basin Research Group, National University of Malaysia, Bangi 43600, Selangor, Malaysia
* Correspondence: nurulafifah@ukm.edu.my
† Current address: Geology Program, Department of Earth Science and Environment, Faculty of Sciences and Technology, National University of Malaysia, Bangi 43600, Selangor, Malaysia.
‡ These authors contributed equally to this work.

Abstract: Gravity-flow deposits form the northern part of the Crocker Formation (Oligocene–Early Miocene), with the most significant interpretation as a sand-rich system in the proximal and a mud-rich system in the distal area of the deep-water turbidite depositional setting. Seven outcrop localities in the northern-part area were selected for mapping and sampling, starting from Kota Kinabalu up to the Telipok area to evaluate the sedimentary sequence. This study used mapping, field observation, and log sketches in the field, as well as extensive analysis and interpretation of sedimentological methods to investigate the sequence of sediment outcrops in the Crocker Formation area of northwest Sabah. During the fieldwork, five main facies were found, namely, massive sandstone facies (f1), graded sandstone facies (f2), laminated sandstone facies (f3), interbedded sandstone and mudstone facies (f4), and mudstone facies (f5). These northern-part outcrops are interpreted as being deposited from the highest to the lowest turbidity currents and the actuality of pelagic mudstone deposition, based on their fining-coarsening-upward pattern. The five geometrical bodies were proposed as laterally contiguous depositional environments, namely, (1) inner fan channel, (2) inner fan channel–levee complex, (3) mid-fan channelized lobes, (4) non-channelized lobes/distal lobes, and (5) basin plains. The facies interpretation shows that the study area consists of lobes, channel–levee complexes, and levees formed in a fan of a deep-water basin setting, with the basinal plain enveloped by thick mudstone deposits. This northern part of the Crocker Formation is interpreted as a multiple-sourced sediment, shelf-fed, Type II, low-efficiency, and sand-rich turbidite depositional system.

Keywords: Crocker Formation; turbidite sequence; deep-water fan system; Sabah

Citation: Mohd Radzir, N.A.; Ali, C.A.; Mohamed, K.R. Sedimentological Analysis of the Turbidite Sequence in the Northern Part of the West Crocker Formation, Northwest Sabah. *Appl. Sci.* **2022**, *12*, 12149. https://doi.org/10.3390/app122312149

Academic Editors: Gordon Gilja, Manousos Valyrakis, Panagiotis Michalis, Thomas Pahtz and Oral Yagci

Received: 27 October 2022
Accepted: 23 November 2022
Published: 28 November 2022

1. Introduction

Sediment gravity flow or mass transport process [1] is the movement of sediment that occurs due to the pull of gravity from the top to the bottom of a slope. It is also defined as a sediment flow or sediment fluid mixture influenced by sediment grains moving due to the influence of gravity [2]. The gravity flow and sediment distribution can be affected by the basinal geometries, sediment supply, and also tectonic activity which becomes part of the triggering mechanisms [3,4]. There are two types of gravity-flow sediment which are (i) sediment flow, and (ii) mass transport flow [5–9]. The end product is classified into (i) complex mass transport, (ii) debris flow, (iii) turbidite and debris co-genetics, and (iv) turbidite current.

Turbidite is a dominant sequence of sediments deposited where this process can transport sediments from the shallower to the deep-water basin [10] and can travel further distances and be deposited in large areas if related to a submarine landslide [11]. Turbidite sediments have various characteristics [4], are usually classified based on several dominant components, have different facies characteristics, and can be identified through field observations and also borehole recordings. The three main components are (i) sand and sediment quantity [12], (ii) flow efficiency [13], and (iii) characteristics of sediment contributors and sediment supply [14], as summarized in Table 1.

Table 1. Summary of characteristics and depositional system [12–14].

Component	Type of System	Characteristics of System
Sand component and quantity of deposits [12]	Type 1: Sand dominant Type 2: Sand dominant and lack of mud Type 3: Mud dominant	Formed as a result of sandstone deposition in non-channel or outer lobe areas. Produces sandstone bodies and lobes. Formed as a result of the platform or littoral zone contribution. Produces channel–lobes complex and fan–supra lobes. Formed from overflows of mud sediment from platforms and slopes. Produces a channel–levee complex.
Flow efficiency [13]	Type 1: Low efficiency Type 2: Medium efficiency Type 3: High efficiency	Rich in sand contributed by the shelf. Forming lobes that are dominated by sand sediments. A mixture of sand and silt contributed by fluvial and deltas. Forms aggradation features with mixed lithology. Rich in fluvial and delta mud. Forming a dominant progradation of mud and sand in the distal area.
Fed-system characteristics and sediment supply [14]	Type 1: Hinterland-fed system Type 2: Shelf-fed system Type 3: Delta-fed system	Sediment from toe-slope erosion. Does not have a shelf or a very narrow shelf. Sediments resulting from river transport, coastal currents, and tidal currents. Dominant with sand. Experience the original transport by waves, tides, and ocean floor currents. Sediment is contributed by rivers. Rich in mud grains.

Understanding gravity-flow sediments is very important in understanding their natural phenomena [15–17] and in the application of science, especially in petroleum exploration and production, determining potential hazards during petroleum production and also determining the risk of geohazards in the future [18]. In northwest Borneo, specifically the Crocker Formation (Figure 1), a large expanse of Paleogene clastic sedimentary rocks of mainly deep-water origin has been outlined by [19–30] and many more. The turbidite succession represents deposition in a tectonically active margin during the subduction of the Proto-South China Sea with the Sunda Shelf opening from Eocene to Holocene [31–33]. In this paper, the sedimentological characteristics of the northern part of the formation are described, based on field observation in the new cutting outcrops of the Pan Borneo Highway, which extends from Kota Kinabalu to the Telipok areas, and includes seven (7) sedimentary rock outcrop localities. The seven localities are as follows: (1) Hap Seng Quarry, Telipok, (2) KFC, UMS, (3) Prime Kondo University, (4) Lorong Suria Inanam, (5) Kg. Divatto, (6) Taman Okk Lojungah, and (7) Pekan Kinarut (Figure 1). Detailed sedimentological work in conjunction with the occurrence of sediment deposits with associated turbidites gives further insight into the sedimentary processes, the depositional setting, and the type of turbidite depositional systems in this Paleogene basin.

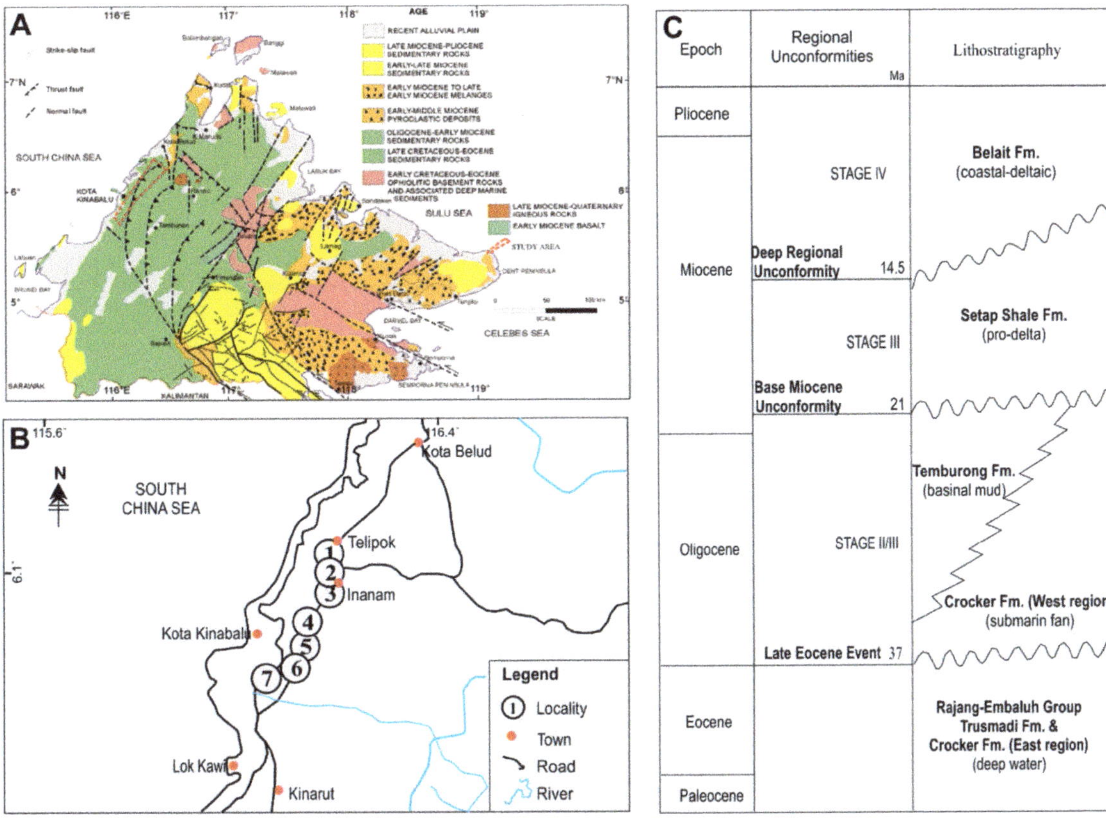

Figure 1. (**A**) Location of the study area within the geological distribution of Sabah. (**B**) Location of the outcrops, (1) Hap Seng Quarry, (2) KFC UMS, (3) Prime Kondo University, (4) Lorong Suria Inanam, (5) Kg. Divatto, (6) Taman Okk Lojungah, and (7) Pekan Kinarut. (**C**) Chronostratigraphic chart of the Sabah area.

2. Geological Setting

Turbidite and deep-water sedimentary deposits were formed in the Sabah deep-water basin from the late Cretaceous to early Eocene age because of continuous subduction events [34,35]. Late Cretaceous to early Eocene sediments are classified as Trusmadi and the Crocker Formation. This formation consists of sandstone, shale, limestone, and tuff. These sediments were deposited in the central part of the deep-water basin and were then covered by the Rijang-Split Formation of the ophiolite sequence, consisting of ultramafic and mafic rocks [34,36–40]. The sediments were contributed to by drainage from Indo-China [29,41]. During this period, the late Cretaceous Lupar Line, located in the eastern part of Borneo, served as a marker for the shallow-sea platform [42]. The deposited sediments were then deformed and metamorphosed, resulting in phyllites and slates [36,39].

During the Sarawak Orogeny formation event [43], the Rajang Group was thrust and uplifted from the middle Eocene to the late Eocene (Figure 1). During this age, northwest Sabah was dominated by deep-sea marine sediments [29], with the shallow-sea platform advancing northward but still controlled by the Lupar Line. The formation and deposition of sediments in the basin continued to form the Temburong, Setap Shale, and Western Crocker Formations until the Oligocene age. During the early Oligocene age, the shallow-sea platform moved east until it reached the Mersing Line, which is located between the Lupar and the Baram Delta Lines, causing the deposition of delta sediments forming the Meligan Formation, and limestone shallow marine sediments forming the Melinau Formation around the edge of the Bruneian basin and Sabah [29]. A thick sequence of ocean sediments and deep seabed sediments was deposited in deep troughs in the west and northwest of Sabah. The Crocker Formation is a sequence of deep-sea sediments that have deformed and thrust in this area [34,40,44].

The Crocker Formation in the west consists of sandstone with mudstone/shale, whereas the Crocker Formation in the east (part of the Rajang Group) consists of argillite. The Crocker Formation in the west is younger than the Crocker Formation in the east. During the Sarawak Orogeny (late Eocene), the uplift of the tectonic plates caused and triggered the formation of deep-water sediments, including Crocker, Temburong, and Setap Shale, which are younger than the turbidite deposits of the Rajang Group [28,45] (Figure 1).

3. Materials and Methods

This study involves two stages which are, (i) extensive fieldwork, and (ii) sedimentological work.

3.1. Fieldwork

Extensive fieldwork was performed to get the input of mapping, sedimentological logging, and sampling of the potential exposed outcrops. The data collected included the location of the localities, the determination of the sedimentary structures and fossils, and the measurement of the strike and dip of outcrops. In addition, the thickness of each bed was measured and the changes in the beds were recorded, with photographs being taken to document the outcrops present.

3.2. Sedimentalogical Work

This stage was performed using facies analysis, facies association analysis, and depositional environment interpretation. This analysis focused on several facies based on the parameters of texture, lithology, sediment structure, and the presence of fossils and trace fossils in the study outcrop sequence. Facies association analysis was performed by classifying the published facies into specific groups and interpreting the depositional environment of the study outcrop sequence.

4. Results

4.1. Facies Analysis

The study area was classified into five main facies, namely, massive sandstone facies (f1), graded sandstone facies (f2), parallel laminated sandstone facies (f3), interbedded sandstone and mudstone facies (f4l), and mudstone facies (f5).

4.1.1. Massive Sandstone Facies (f1)

Description: This is a dominant rock layer with coarse to fine sand grains, silt, or mud-sized matrix (<20 percent) and is occasionally found with coarse to very coarse sand grains (<5 percent). These facies are present in single layers and amalgamations with a thickness starting from 10 cm to more than 1 m (Figure 2). The bottom border is sharp and flat, with an eroded and irregular border. Sorting is moderate to good with mud matrix support clasts. The presence of sedimentary structures, such as parallel laminations, fluted casts, load casts, flame structures, tool marks, and rip-up mud clasts was observed. *Cruziana* sp., *Nereites* sp., and *Chondrites* sp. are examples of bioturbation structures (Figures 3 and 4).

Figure 2. Characteristic examples of lithofacies encountered in the study area. (**A**) Outcrop section showing the sandstone and mudstone with incline bedded at L1. (**B**) Outcrop section showing the thickness of the interbedded sandstone and mudstone at L1. (**C**) Interbedded sandstone and mudstone at L1. (**D**) Syncline structure in the L1. (**E**) Outcrops of interbedded sandstone and mudstone in L2. (**F**) Massive sandstone in L2.

Figure 3. Examples of sedimentary structures in the Crocker Formation. (**A**) Flame structure found in L1. (**B**) Complete Bouma sequence from Ta to Tf found in L3. (**C**) Rip-up mud clasts in L2. (**D**) *Cruziana* sp. of trace fossils in L2.

Figure 4. Examples of sedimentary structures in the Crocker Formation. (**A**) Trace fossils were established in the sandstone bed at L1. (**B**) *Ophiomorpha* sp. originates in the sandstone bed at L3. (**C**) Water escape structures and convolute structures in the sandstone bed at L3. (**D**) Flute casts form at the base of a thick sandstone bed at L5. (**E**) Tool marks trace in the sandstone bed at L7. (**F**) Coarse sand grain size floated at the base of the graded sandstone bed in L7.

Interpretation: The massive sandstone facies can be formed through two depositional methods, namely, high-density turbidity currents or debris flows [26,43].

4.1.2. Graded Sandstone Facies (f2)

Description: These facies have a fining-upward grain size trend and form a complete to incomplete Bouma sequence (Figures 3–5). The bed at the bottom is sharp, eroded, or irregular, with coarser-sized grains. These facies show poor sorting at the bottom which improves vertically upwards. This facies layer contains parallel, laminated, cross-laminated, and ripple structures.

Interpretation: These facies were deposited because of high to low-concentration turbidite current flow processes, where the movement was not controlled by the matrix concentration [5].

4.1.3. Parallel Laminated Sandstone Facies (f3)

Description: These facies are dominated by fine sand- to silt-sized grains, with a small amount of silt or mud-sized matrix (<20 percent). Parallel lamination is observed in the outcrop sequence layers, but no trace fossils can be found in these facies (Figure 6).

Interpretation: The presence of a powerful current over ripples or dune structures can re-transport fine sand sediment, causing the sediment grains to rearrange in a new layer with a horizontal plane.

4.1.4. Interbedded Sandstone and Mudstone Facies (f4)

Description: The sand-to-mud ratio of these facies is 1:1 or more (>1) according to the layer thicknesses. The sedimentary structures found are wavy and parallel laminations (Figures 6 and 7).

Interpretation: Fine-grained sand sediments were deposited by low-density turbidity currents at the bottom of the water, whereas mud grains are synonymous with suspended transportation in calm energetic currents. Each interbedded deposit formed in conjunction with sand and mudstone showing that sediment deposition occurred at the same time and underwent a transition process [46] or a change in current energy and velocity [47,48].

4.1.5. Mudstone Facies (f5)

Description: These facies contains silt to mud fine sediments. There are parallel lamination-like structures, and no bioturbation traces were found (Figures 6 and 7).

Interpretation: Usually, pelagic or hemipelagic mud sediments are transported through suspension processes [49] before being deposited in open and quiet waters.

Figure 5. Characteristics of outcrop lithofacies in the study area. (**A**) Interbedded sandstone and mudstone and also thick sandstone outcrops at the L3. (**B**) Interbedded sandstone and mudstone, mudstone and massive sandstone with fault structure (red dashed line) at the upper area of L3 outcrop. (**C**) Sandstone with fault structure (red dashed line) at L3 area. (**D**) Massive sandstone was found in the L7 area. (**E**) Interbedded sandstone and mudstone are found in L6. (**F**) Thick sandstone outcrops show the amalgamation beds in L6.

4.2. Facies Associations

4.2.1. Channels

These facies associations show a fining-upward sequence with sharp (Figures 6 and 7), and well-defined grain size transition sites between each sand-to-mud layer. The upper bed (mudstone facies) occasionally contained mud clasts in the beds (Figure 3). Parallel laminated structures are also found within the sandstone beds. The lower part of these facies associations is dominated by coarse-sized grains (sandstone) transported by high-density turbidity currents. It can also be identified by the presence of current effects such as ripples, flutes, and erosion at the bottom of the sediment bed. The presence of a basal erosive surface represents the multi-reactivation and lateral migration of deposits [50] which is a characteristic usually found in channels.

4.2.2. Levees

These facies associations are characterized by a thin to medium thickness followed by fine-grain-sized sedimentary deposits interspersed with mud sediments in the uppermost part (Figures 6 and 7). Generally, the transportation of abundant and concentrated sediments in turbidity currents can slow and overflow, forming levee deposits [51], which have a fining-upward sequence with varying amalgamation and sediment thickness.

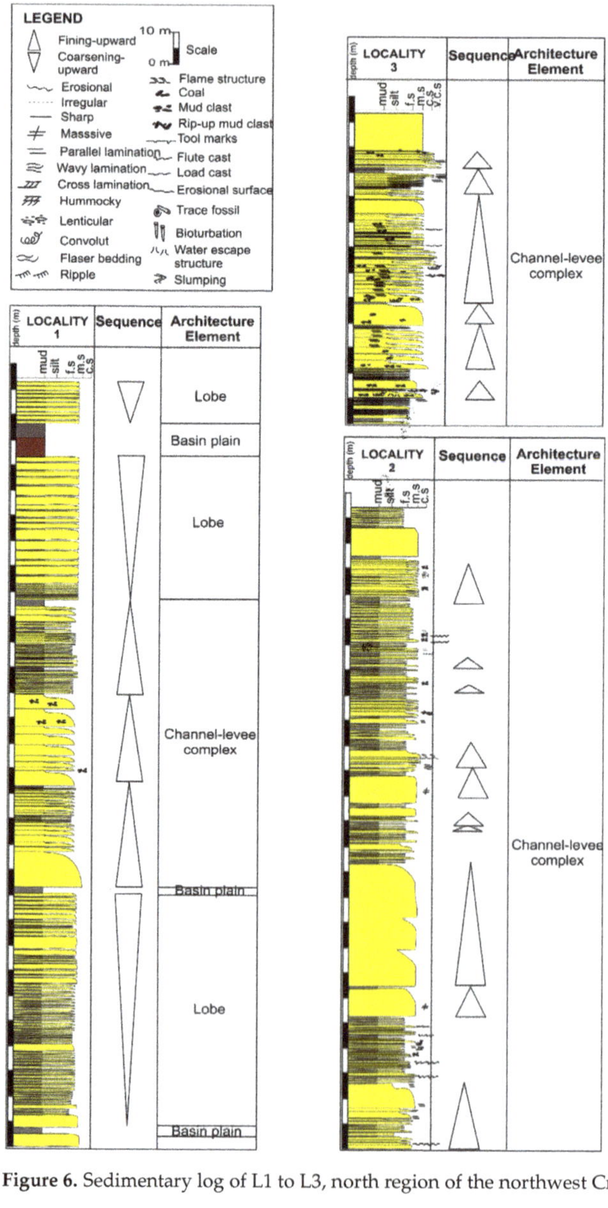

Figure 6. Sedimentary log of L1 to L3, north region of the northwest Crocker Formation.

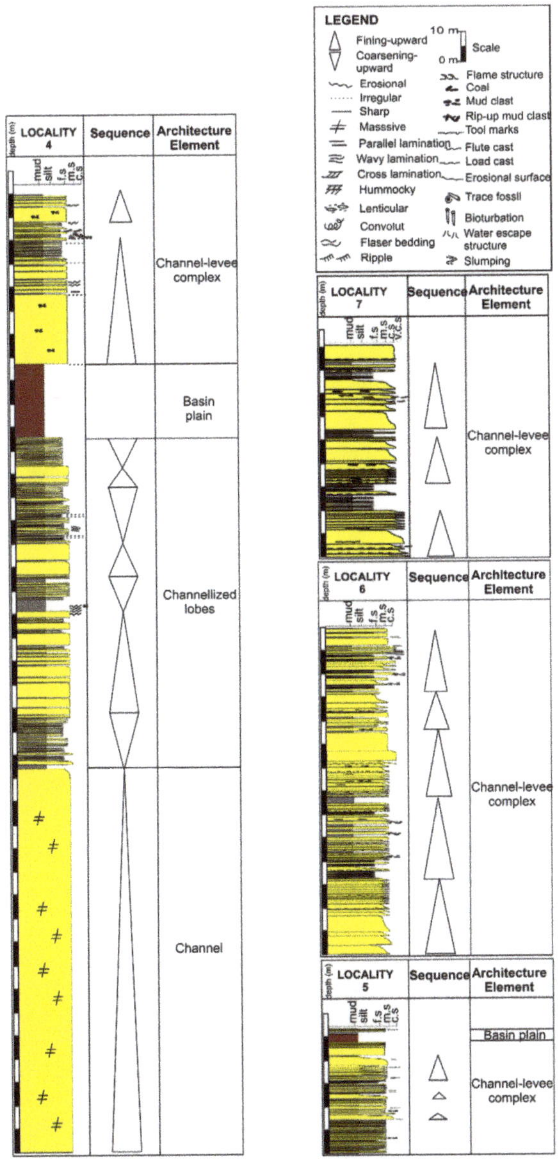

Figure 7. Sedimentary log of L4 to L7, north region of the northwest Crocker Formation.

4.2.3. Lobes

The entire facies association shows an upward thickening and coarsening sequence (Figures 6 and 7). The base structures, such as load casts and flutes are not found at the bottom of the layer, but consist of sand grains with a thin thickness. A vertical sequence that thickens and coarsens upwards indicates that these deposits formed in the lobes.

4.2.4. Basin Plains

These facies associations are pelagic sediments from a very fine sediment fall that are transported in suspension. The deposited pelagic sediments originated from various directions and form a wide and thick spread.

5. Discussion

5.1. Depositional Setting

The study area consists of a deep-water fan system starting from the area of the channels, levees, lobes, and basin plains. All facies associations collect to form a depositional environment based on zones 1, 2, 3, and 4 of the deep-water depositional environment (Figure 8).

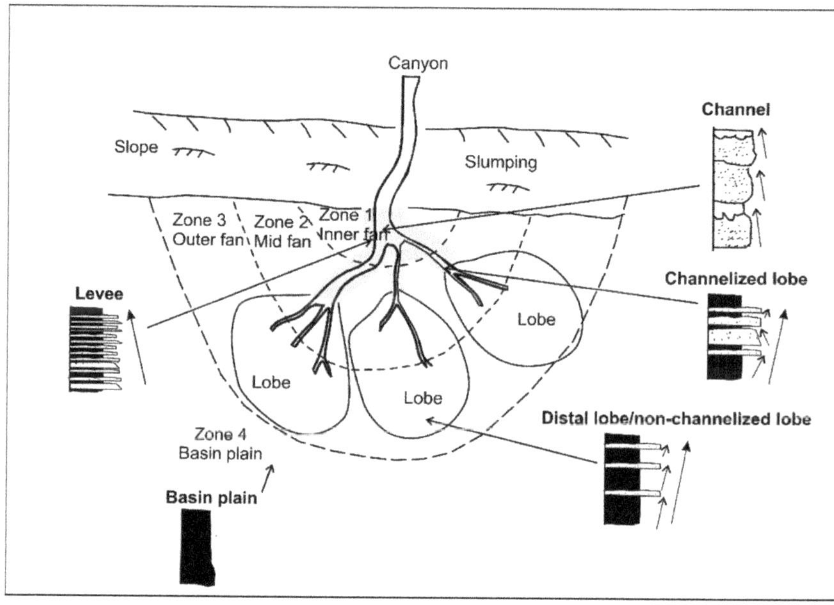

Figure 8. The depositional setting for each component of the facies association interpreted in the study area. Channel shows a fine-upward sequence located in the inner fan (zone 1) part; the levee shows a fine-upward sequence located in the outer channel in the inner and mid fan (zone 1 and zone 2); the channelized lobe shows a coarsening-upward sequence located in the proximal lobe with the presence of a distributary channel, which can be found in the mid fan (zone 2); the distal lobe/non-channelized lobe shows the coarsening-upward sequence located in the distal lobe and no distributary channel are present, which can be found in the outer fan (zone 3); the basin plain is located in the outer of the turbidite fan system.

Zone 1 is the inner fan, consisting of the channel–levees complexes (Figure 8). This proximal area receives its sediments from the canyon and experiences high-density turbidite currents that can transport and deposit sediments ranging in size from coarse sand to medium sand, forming a fining-upward sequence pattern (Figure 8). The proximal area faces a hypoconcentrated density flow where the flow is associated with the matrix strength and grain friction [15]. The density of the flow causes the presence of erosion and records the sharpness and eroded bed (Figure 9). This zone can be a good hydrocarbon accumulation area because it has a high sand content, thick sediment thickness, and good sorting properties with a wider diameter distribution [52].

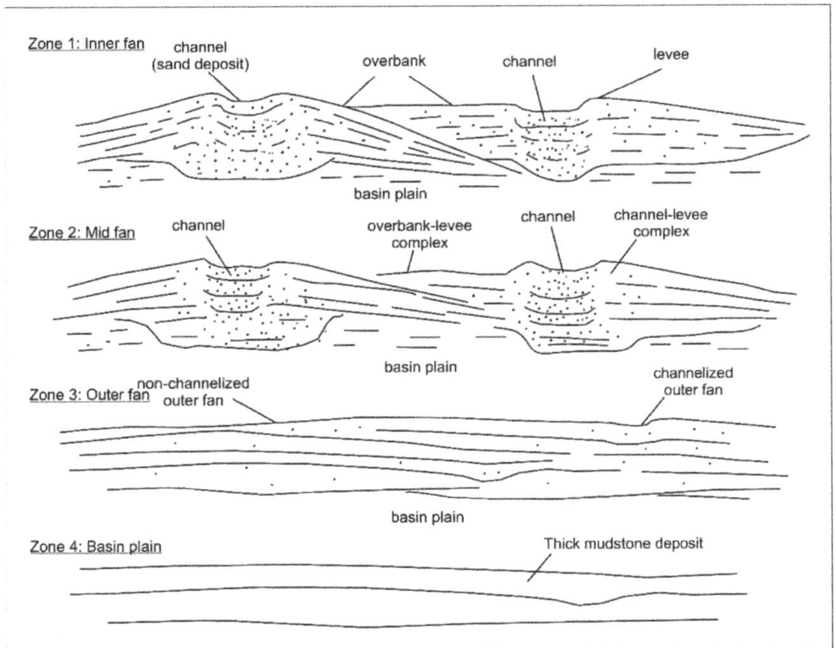

Figure 9. Summary of a depositional model for the northern part of the Crocker Formation. The schematic drawing shows a complete proximal turbidite fan to the basin plain of the deep-water depositional environment in a lateral view.

Zone 2 is the middle fan section consisting of lobes (Figure 8). The lobes in zone 2 consist of channelized lobes and non-channelized lobes. Channelized lobes are distinguished by the presence of channels/distributary channels in the vertical sequence upward of the sedimentary log, as shown in L4 (Figure 7), whereas non-channelized lobes are perceived by the absence of channel sequences in the interpreted vertical sequence. These channelized lobes have a sandstone lithology (f1 and f2) with a medium to thin thickness and a fining-upward sequence pattern and are usually located in the upper lobe (lobe axis) (Figure 8). However, non-channelized lobes are usually located around the proximal lobe to the end lobe, as shown in L1 (Figure 6). These lobes do not show any combination of fining-upward sequence patterns in the entire vertical sequence. This pattern is usually formed by the spreading of the tips of the distributary channels in the upper part of the lobe. However, if there are any edges or slopes in that area, zone 2 is reachable by the tips of distributary channels.

Zone 2 is located in front of zone 1, and their sediment transportation depends on medium to high-density turbidity currents, as well as debris currents (Figure 9). The basis of their sediments support mechanism is equivalent to the concentrated density flows which possess both hypoconcentrated and turbidity current characteristics [15,53]. The medium to high-density turbidite currents was proficient in depositing medium-sized sand sediments with a medium thickness (thickness less than zone 1), good to moderate sorting, and low-matrix content (Figure 10). The transportation of this medium-density turbidity current results in sediment characteristics that have a low-matrix content, high authigenic content, high porosity, and permeability [54]. The entire zone exhibits coarsening-upward sequence characteristics, whereas the presence of distributary channels in the lobes indicates a fine-upward sequence pattern within a vertical coarsening-upward sequence.

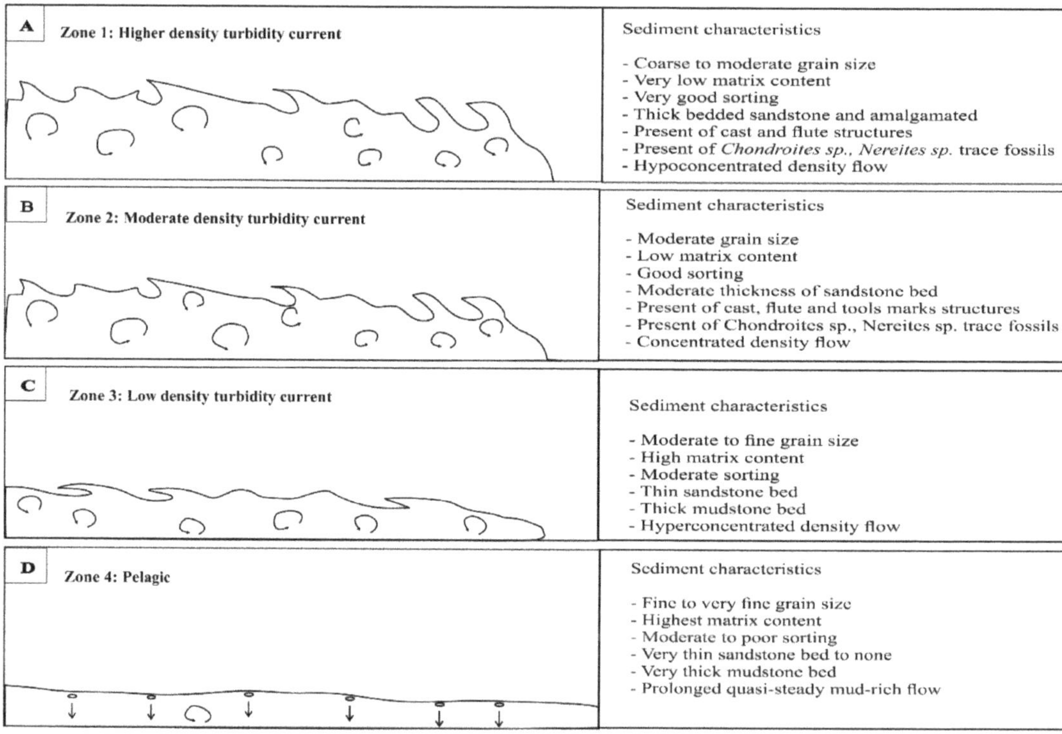

Figure 10. Summary of sediment transport behaviors and sediment characteristics of each zone established in this research. (**A**) Highest turbidity current related to the proximal zone 1 in the inner fan area. (**B**) Moderate turbidity current related to the mid fan of the zone 2 area. (**C**) Low turbidity current illustrated by zone 3 in the outer fan area. (**D**) The pelagic deposited in the basin plain area.

Zone 3 is the area of the distal lobe located in the outer fan of the deep-water depositional environment (Figure 8). This zone is dominated by fine- to very fine-grained layers of thin thickness, with repeated layers, coarsening upwards with the overall sequence. Low-density turbidite currents influence sediment deposition in this zone, which can transport fine-sized sediments to very fine-sized sediments (Figures 9 and 10). Sandstone and mudstone contents are almost equal in the proximal part of the middle lobe, and the mud content is greater than the sandstone in the distal lobe area. The outer lobe is the farthest part of the lobe (distal) and is dominated by fine sediment layers, such as fine sand, silt, and clay [55].

Zone 4 consists of basin plain areas that are characterized by hemipelagic mudstone sediments (with less than a 0.125 or 0.063 mm grain size). Mudstone deposits range in thickness from 2 to 10 m, and coarse grains, such as sand and coarse silt grains, are rarely found (Figure 10). This section is the farthest part of the deep-water fan, and faces a prolonged quasi-steady mud-rich turbidity flow [53,56], forming a broad and widespread expanse of thick mudstone deposits.

5.2. Depositional System

The facies sequence in the fan system in the northern part is dominated by coarse grain sizes such as sandstone with a thick to very thick thickness that forms a channel–levee complex, as well as channel and lobe geometries. The analysis and interpretation of outcrop localities in the northern part of the Crocker Formation show a sand-rich (Type I), low- efficiency (Type I), and shelf-fed (Type II) deep-water turbidite system [12–14] (refer

Table 1). This system is a fan that formed due to the sediment contribution from the shelf, which is rich in the sand but has a small amount of mud sediment. This is in line with the two previous studies which stated that the sediment supply to Crocker Formation is caused by the Sunda Continental Shelf that experienced sediment bypass during the Cenozoic era (middle Eocene to late Oligocene) and Mount Schwaner to the east of Sarawak [57,58]. The supply sediments would have been temporarily deposited in the sediment depocenter on the shelf before being transported again by energy currents such as waves, tides, and turbidite currents, causing numerous sand sediments to be deposited in front of the shelf [59]. With the conjunction of density and turbidity current, the basin provides enough space to accommodate the accumulation and deposition of sediments resulting in the formation of various geometrical bodies such as channels, channels–levee complexes, lobes, and channel–lobes complexes. The distal part of the basinal area is the place of accumulation for the thick mudstone.

6. Conclusions

Based on the sedimentary facies and their evaluation, most of the outcrops in the northern part of the northwest Crocker Formation mostly represent a complete deep-water basin environment. Occasionally, as observed in the outcrops, the intervals of massive sandstone facies (f1), graded sandstone facies (f2), parallel laminated sandstone facies (f3), cross-laminated sandstone facies (f4), interbedded sandstone and mudstone facies (f4) and mudstone facies (f5) form a package of turbidity current influences which are representative of a large-scale turbidite deep-water basin. The turbidite basin, interpreted to consist of geometrical bodies such as channels, levees, lobes, and basin plains, is categorized as a sand-rich, low- efficiency, and shelf-fed system. This system represents the bypass sediment supply during the active subduction of the Proto-South China Sea, with the Sunda Shelf opening from the Eocene to the Holocene age. Such results are in line with previous theories and may be useful as depositional indicators in deep-water depositional settings where the outcrops are sparse, and when integrated with other field data evidence, may provide additional clues to the deposition setting and depositional systems of the Crocker Formation.

Author Contributions: Conceptualization, N.A.M.R.; methodology, N.A.M.R., C.A.A. and K.R.M.; formal analysis, N.A.M.R.; investigation, N.A.M.R., C.A.A. and K.R.M.; resources, N.A.M.R.; writing—original draft preparation, N.A.M.R.; writing—review and editing, N.A.M.R.; visualization, N.A.M.R.; supervision, C.A.A. and K.R.M. All authors have read and agreed to the published version of the manuscript.

Funding: This research received no external funding.

Informed Consent Statement: Not applicable.

Data Availability Statement: Not applicable.

Acknowledgments: This research data are part of the author's research data during the Doctor of Philosophy degree supervised by Che Aziz Bin Ali and Kamal Roslan Bin Mohamed who has retired from academia. The author would like to thank the two supervisors for their guidance during their service in the academic field.

Conflicts of Interest: The authors declare that there are no known conflicts of interest associated with this publication and there has been no significant financial support for this work that could have influenced its outcome.

References

1. Cook, H.E.; Mullins, H.T. Basin margin environments. In *Carbonate Depositional Environments: American Association of Petroleum Geologist Memoir*; Schole, P.A., Bebout, D.G., Moore, C.H., Eds.; American Association of Petroleum Geologist: Tulsa, OK, USA, 1983; Volume 33, pp. 539–618.
2. Dasgupta, P. Sediment gravity flow—the conceptual problems. *Earth Sci. Rev.* **2003**, *62*, 265–281. [CrossRef]
3. Yang, T.; Cao, Y.; Li, Y.; Zhang, S. Status and trends in research on deep-water gravity flow deposits. *Acta Geol. Sin.* **2015**, *89*, 610–631.

4. Walker, R.G. Deep-water sandstone facies and ancient submarine fans: Models for exploration for stratigraphic traps. *AAPG Bull.* **1978**, *62*, 932–966.
5. Lowe, D.R. Sediment gravity flows: II. Depositional models with special reference to the deposits of high-density turbidity currents. *J. Sedimentol. Soc. Econ. Paleontol. Mineral.* **1982**, *52*, 279–297.
6. Middleton, G.V.; Hampton, M.A. Sediment gravity flows: Mechanics of flow and deposition. In *Turbidites and Deep-Water Sedimentation*; Society of Economy Paleontologist Mineralogist; Pacific Section SEPM: Los Angelas, CA, USA, 1973; pp. 1–38.
7. Sanders, J.E. Primary sedimentary structures formed by turbidity currents and related re-sedimentation mechanism. In *Primary Sedimentary Structures and Their Hydrodynamic Interpretation: SEPM*; Middleton, G.V., Ed.; Special Publication: Los Angelas, CA, USA, 1965; Volume 12, pp. 192–219.
8. Shanmugam, G.; Lehtonen, L.R.; Straume, T.; Syversten, S.E.; Hodgkinson, R.J.; Skibeli, M. Slump and debris flow dominated upper slope facies in the Cretaceous of the Norwegian and Northern North Seas (61–67° N): Implication for sand distribution. *AAPG Bull.* **1994**, *78*, 910–937.
9. Shanmugam, G.; Miola, R.J. Reinterpretation of depositional processes in a classic flysch sequence (Pennsylvanian Jackfork Group) Ouauchita Mountains, Arkansas and Oklahoma. *Am. Assoc. Pet. Geol. Bull.* **1995**, *79*, 672–695.
10. Keunen, P.H.; Migliorini, C.I. Turbidity currents as a cause of graded bedding. *J. Geol.* **1950**, *58*, 91–127. [CrossRef]
11. Shan, Z.; Wu, H.; Ni, W.; Sun, M.; Wang, K.; Zhao, L.; Lou, Y.; Liu, A.; Xie, W.; Zheng, X. Recent technological and methodological advances for the investigation of submarine landslides. *J. Mar. Sci. Eng.* **2022**, *10*, 1728. [CrossRef]
12. Mutti, E. Turbidit systems and their relations to depositional sequences. In *Provenance of Arenites*; NATO-ASI Series; Zuffa, G.G., Ed.; Springer: Houston, TX, USA, 1985; pp. 65–93.
13. Richards, M.; Bowman, M.; Reading, H. Submarine-fan systems I: Characterization and stratigraphic prediction. *Mar. Pet. Geol.* **1998**, *15*, 689–717. [CrossRef]
14. Piper, D.J.W.; Normark, W.R. Sandy fans: From Amazon to Hueneme and beyond. *Am. Assoc. Pet. Geol. Bull.* **2001**, *85*, 1407–1438.
15. Yang, Y.; Peng, J.; Chen, Z.; Zhou, X.; Zeng, Y.; Wang, Y.; Wang, X. Depositional models of deep-water gravity-flow in lacustrine basin and its petroleum geological significance—A case study of Chang 6 Oil Group in Heshui Area, Ordos Basin, China. *Front. Earth Sci.* **2022**, *9*, 786403. [CrossRef]
16. Shanmugam, G. 50 years of the turbidite paradigm (1950s-1990s): Deep-water processes and facies models: A critical perspective. *Mar. Pet. Geol.* **2000**, *17*, 285–342. [CrossRef]
17. Shanmugam, G. New perspectives on deep-water sandstones: Implications. *Pet. Explor. Dev.* **2013**, *40*, 294–301. [CrossRef]
18. Madon, M. Submarine mass-transport deposits in the Semantan Formation (Middle—Upper Triassic), central Peninsular Malaysia. *Bull. Geol. Soc. Malays.* **2010**, *56*, 15–26. [CrossRef]
19. Akhmal, S.; Umar, H.; Rahim, A.S.; Hariri, M.A.; Radzuan, J. Deep crustal profile across NW Sabah Basin: Integrated potential field data and seismic reflection. *ARPN J. Eng. Appl. Sci.* **2016**, *11*, 1401–1411.
20. Azfar, M.; Hadi, A.A.R.; Suhaili, M.I. Sedimentary facies of the West Crocker Formation north Kota Kinabalu-Tuaran area, Sabah, Malaysia. *J. Phys. Conf. Ser. 660* **2015**, *10*, 142–149.
21. Ingram, G.M.; Chisholm, T.J.; Grant, C.J.; Hedlund, C.A.; Stuart-Smith, P.; Teasdale, J. Deepwater North West Borneo: Hydrocarbon accumulation in an active fold and thrust belt. *Mar. Pet. Geol.* **2004**, *21*, 879–887. [CrossRef]
22. Jackson, C.A.-L.; Zakaria, A.A.; Johnson, H.D.; Tongkul, F.; Cravello, P.D. Sedimentology, stratigraphy occurrence and origin of linked debrites in the West Crocker Formation (Oligo-Miocene), Sabah, NW Borneo. *Mar. Pet. Geol.* **2009**, *26*, 1957–1973. [CrossRef]
23. Lambiase, J.J.; Cullen, A.B. Sediment supply systems of the Champion "Delta" of NW Borneo: Implication for deep water reservoir sandstones. *J. Asian Earth Sci.* **2013**, *76*, 356–371. [CrossRef]
24. Morley, C. Geometry of an oblique thrust fault zone in a deepwater fold belt from 3D seismic data. *J. Struct. Geol.* **2009**, *31*, 1540–1555. [CrossRef]
25. Tan, D.N.K.; Lamy, J.M. Tectonic evolution of the NW Sabah continental margin since the Late Eocene. *Geol. Soc. Malays. Bull.* **1990**, *27*, 241–260. [CrossRef]
26. Tongkul, F. The Sedimentology and Structure of the Crocker Formation in the Kota Kinabalu Area, Sabah. Ph.D. Thesis, Univesity of London, London, UK, 1987.
27. Tongkul, F. Mass transport deposit (MTD) in the Paleogene submarine fan of the Crocker Formation in Sabah, Malaysia. In Proceedings of the Geoscience Technology Workshop, Asia Pacific Region, Kota Kinabalu, Sabah, Malaysia, 26–27 May 2015; AAPG: Tulsa, OK, USA, 2015.
28. William, A.G.; Lambiase, J.J.; Back, S.; Jamiran, M.K. Sedimentology of the Jalan Salaiman and Bukit Melinsung outcrops, western Sabah: Is the West Crocker Formation an analogue for Neogene turbidites offshore? *Bull. Geol. Soc. Malays.* **2003**, *47*, 63–75. [CrossRef]
29. Zakaria, A.A.; Johnson, H.D.; Jackson, C.A.-L.; Tongkul, F. Sedimentary facies analysis and depositional model of the Paleogene West Crocker submarine fan system, NW Borneo. *J. Asian Earth Sci.* **2013**, *76*, 283–300. [CrossRef]
30. Jamil, M.; Siddiqui, N.A.; Rahman, A.H.A.; Ibrahim, N.A.; Ismail, M.S.; Ahmed, N.; Usman, M.; Gul, Z.; Imran, Q.S. Facies heterogeneity and lobe facies multiscale analysis of deep-marine sand-shale complexity in the West Crocker Formation of Sabah Basin, NW Borneo. *Appl. Sci.* **2021**, *11*, 5531. [CrossRef]

31. Hall, R. Cenozoic geological and plate tectonic evolution of SE Asia and the SW Pacific: Computer-based reconstructions, model and animations. *J. Asian Earth Sci.* **2002**, *20*, 353–434. [CrossRef]
32. Hutchison, C.S. Marginal basin evolution: The southern South China Sea. *Mar. Pet. Geol.* **2004**, *21*, 1129–1148. [CrossRef]
33. Hutchison, C.S.; Hall, R.; Blundell, D.J. The "Rajang Accretionary Prism" and "Lupar Line" problem of Borneo. In *Tectonic Evolution of SE Asia*; Geological Society London Special Publication: London, UK, 1996; Volume 106, pp. 247–261.
34. Liechti, P.; Roe, F.W.; Haile, N.S. The geology of Sarawak, Brunei and the western part of Borneo. Bulletin of British Borneo. *Geol. Surv.* **1960**, *3*, 360.
35. Wilson, M.E.J.; Moss, S.J. Cenozoic paleogeographic evolution of Sulawesi and Borneo. *Palaeogeogr. Palaeoclim. Palaeoecol.* **1999**, *145*, 303–337. [CrossRef]
36. Collenete, P. The geology and mineral resources of the Jesselton-Kinabalu area, north Borneo. *Br. Borneo Geol. Surv. Dept. Mem.* **1958**, *6*, 194.
37. Stauffer, P.H. Studies in the Crocker Formation, Sabah. *Malays. Geol. Surv. Bull.* **2008**, *8*, 1–13.
38. Tjia, H.D. Sense of tectonic transport in intensely deformed Trusmadi Crocker sediments, Ranau—Tenom area, Sabah. *Sains Malays.* **1974**, *3*, 129–166.
39. Tongkul, F. Regional geological correlation of Paleogene sedimentary rocks between Sabah and Sarawak, Malaysia. *GEOSEA '98 Proc. Geol. Soc. Malays. Bull.* **1999**, *45*, 31–39. [CrossRef]
40. Wilford, G.E. *Geological Map of Sabah, Scale 1:500,000*; Geological Survey of Borneo Region: Borneo Region, Malaysia, 1967.
41. Hall, R. Reconstructing Cenozoic SE Asia. In *Tectonic Evolution of SE Asia*; Geological Society London Special Publication: London, UK, 1996; Volume 106, pp. 153–184.
42. Honza, E.; John, J.; Banda, R.M. An imbrication model for the Rajang Accretionary Complex in Sarawak, Borneo. *J. Asian Earth Sci.* **2000**, *18*, 751–759. [CrossRef]
43. Baas, J.H.; Kesteren, W.V.; Postma, G. Deposits of depletive high-density turbidity currents: A flume analogue of bed geometry, structure and texture. *Sedimentology* **2004**, *51*, 1053–1088. [CrossRef]
44. Tongkul, F.; Benedick, H.; Chang, F.K. Geology of slope in the Crocker Range, Sabah. Malaysia. *J. Nepal Geol. Soc.* **2006**, *34*, 73–80. [CrossRef]
45. Tracy, G.L.; Sanudin, T.; Junaidi, A. The stratigraphy of southwest Sabah. *Bull. Geol. Soc. Malays.* **2018**, *66*, 65–74.
46. Basilici, G.; Vidal, A.C. Alternating coarse-and fine grained sedimentation in Precambrian deep-water ramp (Apiùna Formation, SE of Brazil): Tectonic and climate control or sea level variations? *Precambrian Res.* **2018**, *311*, 211–227. [CrossRef]
47. Leeder, M. Sedimentology and Sedimentary Basins. In *From Turbulence to Tectonis*; Blackwell Science: Oxford, UK, 1999; pp. 1–592.
48. Prothero, D.R.; Schwab, F.L. *Sedimentary Geology: An Introduction to Sedimentary Rocks and Stratigraphy*, 2nd ed.; W. H. Freeman: London, UK, 2004; pp. 1–557.
49. Boggs, S.J. *Principles of Sedimentology and Stratigraphy*, 3rd ed.; Pearson College Div., University of Oregon: Eugene, OR, USA, 2001; pp. 1–726.
50. Rohais, S.; Bailleul, J.; Brocheray, S.; Schmitz, J.; Paron, P.; Kezirian, F.; Barrier, P. Depositional model for turbidite lobes in complex slope settings along transform margins: The Motta San Giovanni Formation (Miocene—Calabria, Italy). *Front. Earth Sci.* **2021**, *9*, 766946. [CrossRef]
51. Bowen, A.J.; Normark, W.R.; Piper, A.J.W. Modelling of turbidity currents on navy Submarine Fan, California Continental Borderland. *Sedimentology* **1984**, *31*, 169–185. [CrossRef]
52. Browne, G.H.; King, P.R.; Higgs, K.E.; Slatt, R.M. Grain-size characteristics for distinguishing basin floor fan and slope fan depositional settings: Outcrop and subsurface examples from the late Miocene Mount Messenger Formation, New Zealand. *N. Z. J. Geol. Geophys.* **2005**, *48*, 213–227. [CrossRef]
53. Azpiroz-Zabala, M.; Cartigny, M.J.B.; Talling, P.J.; Parsons, D.R.; Sumner, E.J.; Clare, M.A.; Simmons, S.M.; Cooper, C.; Pope, E.L. Newly recognized turbidity current structure can explain prolonged flushing of submarine canyons. *Sci. Adv.* **2017**, *3*, e1700200. [CrossRef] [PubMed]
54. Bell, D.; Kane, I.A.; Ponten, A.S.; Flint, S.S.; Hodgson, D.M.; Barret, B.J. Spatial variability in depositional reservoir quality of deep-water channel-fill and lobe deposits. *Mar. Pet. Geol.* **2018**, *98*, 97–115. [CrossRef]
55. Garcìa, M.; Ercilla, G.; Alonso, B.; Estrada, F.; Jané, G.; Mena, A.; Alvés, T.; Juan, C. Deep-water turbidite systems: A review of their elements, sedimentary processes and depositional models. Their characteristics on the Iberian margins. *Bol. Geolgico Y Min.* **2015**, *126*, 189–218.
56. Lambiase, J.J.; Tan, Y.T.; William, A.G.; Bidgood, M.-D.; Brenac, P.; Cullen, A.B. The West Crocker Formation of northwest Borneo: A Paleogene accretionary prism. *Geol. Soc. Am. Spec. Pap.* **2008**, *436*, 171–184.
57. Van Hattum, M.W.A.; Hall, R.; Pickard, A.L.; Nichols, G.J. Southeast Asian sediments not from Asia: Provenance and geochronology of north Borneo sandstones. *Geol. Soc. Am.* **2006**, *34*, 589–592. [CrossRef]
58. Van Hattum, M.W.A.; Hall, R.; Pickard, A.L.; Nichols, G.J. Provenance and geochronology of Cenozoic sandstones of northern Borneo. *J. Asian Earth Sci.* **2013**, *76*, 266–282. [CrossRef]
59. Stow, D.; Smillie, Z. Distinguishing between Deep-Water Sediment Facies: Turbidites, Contourites and Hemipelagites. *Geosciences* **2020**, *10*, 68. [CrossRef]

MDPI
St. Alban-Anlage 66
4052 Basel
Switzerland
www.mdpi.com

Applied Sciences Editorial Office
E-mail: applsci@mdpi.com
www.mdpi.com/journal/applsci

www.ingramcontent.com/pod-product-compliance
Lightning Source LLC
Chambersburg PA
CBHW061817210926
51850CB00006B/727
9783039286027